Thorium—Energy for the Future

A. K. Nayak · Bal Raj Sehgal
Editors

Thorium—Energy for the Future

Select Papers from ThEC15

Springer

Editors
A. K. Nayak
Thermal Hydraulics Section, Reactor Engineering Division
Bhabha Atomic Research Centre
Mumbai, Maharashtra, India

Bal Raj Sehgal
Nuclear Power Safety Division, Alba Nova
Royal Institute of Technology
Stockholm, Sweden

ISBN 978-981-13-2657-8 ISBN 978-981-13-2658-5 (eBook)
https://doi.org/10.1007/978-981-13-2658-5

Library of Congress Control Number: 2018955175

This Springer imprint is published by the registered company Springer Nature Singapore Pte Ltd.
The registered company address is: 152 Beach Road, #21-01/04 Gateway East, Singapore 189721, Singapore

Preface

Thorium is one of the most precious nuclear fuels available in nature, more abundant than uranium which is employed currently for electricity production in nuclear reactors. Calculations have revealed that thorium fuel has the potential to provide energy to the world for centuries if used in a closed fuel cycle. Considering this, the technologies for thorium utilization for power generation in nuclear reactors are being developed worldwide.

The book on *Thorium—Energy for the Future* contains selected articles from the deliberations of distinguished scientists, engineers, and academicians who made significant contribution to the field of thorium utilization in the recently concluded International Thorium Energy Conference (ThEC15) held in Mumbai during October 12–15, 2015. ThEC15 was the fifth conference in the series of conferences hosted by the International Thorium Energy Organization (IThEO). Previously, IThEO organized ThEC10 at the Royal Institution of Great Britain, London, in the year 2010. ThEC11 was organized in collaboration with the Brookhaven National Laboratory, USA, at the City College of New York, New York, in the year 2011. ThEC12 was organized in collaboration with the Shanghai Institute of Applied Physics (SINAP), Shanghai, and Chinese Academy of Sciences in the year 2012. ThEC13 was organized in collaboration with CERN, Geneva, in the year 2013. ThEC15 was jointly hosted by BARC, the leader in the Indian thorium energy program; NPCIL which is involved in commercial nuclear power generation in India; and HBNI, a deemed university dedicated to research and development in the frontier areas of nuclear science and technology in association with IThEO.

This book discusses research results on various facets of thorium energy right from exploration and mining, thermo-physical and chemical properties of fuels, reactor physics, challenges in fuel fabrication, thorium fuel cycle, thermal hydraulics and safety, material challenges, irradiation experiences, to design of advanced thorium-fueled reactors. It also describes new developments, innovations, and philosophies on thorium technology as fuel for tomorrow's nuclear power reactors.

Since there is a strong thrust globally to design nuclear reactors with thorium-based fuel, the book shall be of particular interest to nuclear scientists, reactor designers, regulators, academicians, and students worldwide.

Mumbai, India A. K. Nayak
Stockholm, Sweden Bal Raj Sehgal

Contents

About the Editors

A. K. Nayak a graduate in mechanical engineering from NIT Rourkela, India, has been working at Reactor Design & Development Group in BARC since 1990 after completing a one-year orientation course in nuclear engineering from BARC Training School, Mumbai, India. He obtained his Ph.D. in nuclear engineering from Tokyo Institute of Technology, Japan, in 2000. He worked at Interfaculty Reactor Institute, Technical University Delft from 2002 to 2003 and in the Nuclear Power Safety Division, Royal Institute of Technology, Stockholm, from 2003 to 2004 as Postdoctoral Scientist in the field of design of natural circulation-based reactor systems and severe accident. He worked at the Institute of Nuclear Energy (IKE), University of Stuttgart, Germany during 2009–2010 in the DST-DAAD project on "Debris Coolability in Severe Accident." He has received several awards which include the DAE Scientific and Technical Excellence Award in 2006, DAE SRC Outstanding Research Investigator Award in 2010, and the Homi Bhabha Science and Technology Award in 2014. He has published more than 250 papers in academic journals and at conferences. His research contributions include advanced nuclear reactor design, thermal hydraulics, and safety, particularly in the context of severe accident. He served as Guest Editor of *Nuclear Engineering and Design journal, Sadhana journal*, and *ASME Journal of Nuclear Engineering and Radiation Science*. He is Editor-in-Chief of the journal "Multiphase Science and Technology" and Associate Editor of the journal "Frontiers in Energy." Currently, he is Head, Thermal Hydraulics Section, Reactor Engineering Division, BARC, and Professor, Homi Bhabha National Institute, India.

Bal Raj Sehgal is currently Professor Emeritus in the Nuclear Power Safety Division, Royal Institute of Technology, Stockholm, Sweden. Previously, he was Chair Professor at the same institution from December 1991 to December 2004. He came to Sweden from the University of California, Berkeley, where he was Professor in Residence, and concurrently worked at EPRI, where he was Senior Technical Advisor and Senior Program Manager for 18 years. He spent a year as Visiting Professor at Purdue University and MIT in 1973 and 1980, respectively. Prior to his employment in California, he worked at Argonne National Laboratory and Brookhaven National Laboratory. Since the TMI-2 accident, his main area of interest has been LWR severe accidents. He is Foreign Member of the Royal Swedish Academy of Engineering Sciences and the Lithuanian Academy of Sciences. In 2013, he was elected as Member of the US National Academy of Engineering. He is Fellow of the American Nuclear Society (ANS) and the American Society of Mechanical Engineers (ASME). He has received the Laurence Pioneering Award, the Seaborg Medal, the Tommy Thompson Reactor Safety Award, and the Thermal Hydraulics Technical Achievement Award from the American Nuclear Society. He also received the Thermal Hydraulics Fellow Award from the Thermal Hydraulics Division of the American Nuclear Society in 2013 and the Swedish Risk Academy Award in 2015. He has published more than 380 papers during his career.

Part I
General Articles

Harnessing Thorium for Clean Energy Future: Challenges Ahead

Anil Kakodkar

Abstract Resolution of conflict between growing energy needs and threat of climate change necessitates a rapid transition to non-fossil energy sources of which nuclear energy is an essential key component. For a country like India which is endowed with abundant Thorium resources and only modest Uranium resources, harnessing Thorium for her energy needs has been a key development objective. Even for other countries, Thorium presents an attractive nuclear energy alternative that has the advantage of greater safety and larger resource base as well as higher proliferation resistance. While the optimised solution for harnessing Thorium over the long term could take time, there are opportunities to get started with Thorium almost immediately using established nuclear power reactor technologies and address some of the barriers to growth of nuclear energy utilising the advantages that Thorium can offer. This would also enable scaling up our experience with Thorium and facilitate transition from Uranium to Thorium at the contemporary scale of operation whenever such a change is contemplated.

1 Introduction

"International Energy Outlook 2013" has projected the growth of world energy consumption from 524 quadrillion BTUs to 820 quadrillion BTUs between the years 2010 and 2040. This represents a rise of around 54%. Most of this growth is expected to take place in non-OECD countries. During the same period, the relative share of non-fossil (renewable + nuclear) energy was assessed to be increasing from 16.5 to

+ Even after the historic agreement at Paris in December 2015 and INDCs announced by all countries, the target of limiting global temperature rise to within 2 °C above preindustrial level is still eluding us.

Presentation at International Thorium Energy Conference—2015 (ThEC 2015), Mumbai, 12–15 October 2015.

A. Kakodkar (✉)
INAE Satish Dhawan Chair of Engineering Eminence, BARC, Mumbai, India
e-mail: kakodkaranil@gmail.com

A. K. Nayak and B. R. Sehgal (eds.), *Thorium—Energy for the Future*,
https://doi.org/10.1007/978-981-13-2658-5_1

21.4% only. This would correspond to an increase in CO_2 emission by around $40\%^{+}$. It is thus clear that deployment of nuclear energy would need to be accelerated, both in terms of non-fossil energy source in the grid as well as to have a reasonable share of base load generation using non-fossil energy. The reality is, however, different. Nuclear power generation capacity is growing but rather slowly. This is because of the barriers that exist in smoother deployment of new nuclear power. Thorium has a number of features that can soften some of these barriers and facilitate growth of nuclear power. We also need to recognise that globally, Thorium is 3 to 4 times more abundant as compared to Uranium.

As far as India is concerned, as against a minimum annual per capita energy use of around 2500 kgoe that would be necessary to support a high human development index similar to what obtains in advanced countries, the current level of energy use is just about a fourth. Thus, our current energy use of 0.82 Btoe would need to go up to a level of around 4.125 Btoe, taking into account likely increase in population. India's total primary (coal, oil and gas) energy resource is estimated at around 361.8 Btoe. In reality the actual mineable resource might be much less. There is thus a significant energy sustainability issue over and above the threat of climate change that is involved as far as the use of fossil energy available on Indian land mass is concerned. Burgeoning energy import bill particularly in the context of oil and gas is already a matter of serious worry that is expected to become worse in years to come. Clearly this demands serious efforts to shift India's energy basket to domestic non-fossil energy. Luckily, the only two energy resources that are available on Indian land mass and would be sufficient to sustainably meet India's long-term energy demands are Solar and Thorium. All other resources, though very important in the short run, would be far from being adequate in the long run. Use of both Solar and Thorium for energy production does not lead to any significant CO_2 emissions. Further Solar and Thorium are complimentary energy resources—intermittent versus base load generation, diffused versus concentrated energy source, etc., suggesting that both would be needed.

2 Going Forward

Thorium as it occurs in nature does not have any fissile component. To start energy production from Thorium therefore, one needs to use other available fissile materials or neutron irradiation platforms. Further, the scale at which Thorium utilisation can be pursued for energy production, pretty much depends on the prior inventory of fissile materials or capacity of irradiation platforms for conversion of Thorium. Thorium, while its utilisation can be designed to be in self-sustaining mode, would not support very significant growth through breeding by itself. This has been the basis of India's sequential three-stage nuclear power programme where large-scale utilisation of Thorium, even with modest availability of Uranium, has been the long-term core objective from day one. Development of technologies related to Thorium utilisation has thus been steadily pursued in India hand in hand with the development of the first two stages involving thermal and fast breeder reactors, respectively. For the world, at large however, the scarcity of Uranium is not yet in sight and significant nuclear reactor capacity exists already. Availability of Uranium or the capacity of platforms for irradiation of Thorium is thus not a constraint. There is thus an opportunity to start utilising Thorium immediately provided the advantages of doing so can be clearly seen.

According to WNA data, there are some 440 nuclear power reactors operating in 31 countries with a combined capacity of over 385 GWe. Over 60 power reactors corresponding to around 15% additional capacity are currently being constructed in 13 countries. Further, over 160 power reactors with a total net capacity of some 182 GWe (close to half the present capacity) are planned, and over 300 more are proposed.

Clearly, the projected growth of nuclear generation capacity is way short of the need for non-fossil capacity addition to contain global warming. At this stage, we need to recognise that growth of additional capacity is expected primarily in countries like China, India, Korea, Russia and USA. Also, there are a number of potentially emerging nuclear countries which need additional power generation capacity to support their economic growth. Option of nuclear power is under their active consideration. Experience so far suggests that the interest of emerging nuclear countries does not fully translate into actual deployment perhaps as a result of their perceived concerns. Their concerns primarily relate to (1) unmanageable liability that can arise out of large-scale displacement in the event of a major nuclear accident, (2) safety, security and waste disposal issues over short and long terms related to spent fuel management, (3) concerns over potential for diversion of fissile materials for malevolent purposes, (4) large capital investments necessary with large-size nuclear plants available on the market and (5) competitive unit energy cost.

Satisfactory resolution of these issues would go a long way in facilitating a faster growth of nuclear power worldwide. Thorium can indeed play an important role here in softening the barriers to nuclear power deployment [1]. Further, Thorium presents a much larger clean non-fossil energy potential available for large-scale power generation in conjunction with large accumulation of fissile inventory in several countries. This at the same time also alleviates concerns related to such an accumulation through its safe and secure elimination. We will discuss these aspects in subsequent paragraphs.

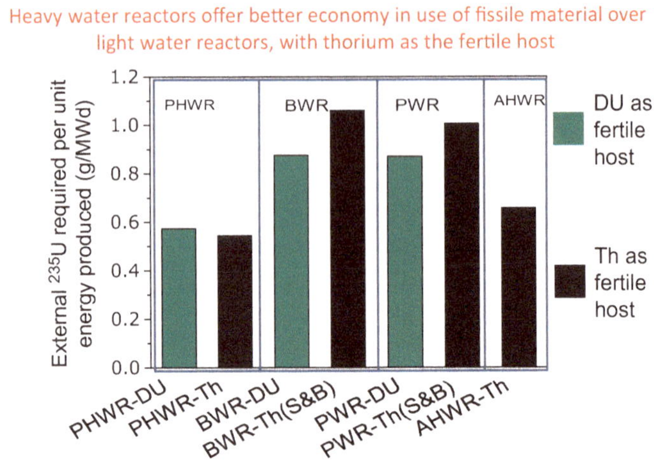

3 Thorium, an Efficient Fertile Host

In early stages of nuclear technology development, burn up realisable in water reactors was quite limited. Under such conditions, use of Thorium as fertile material in fuels meant the need for additional fissile material, since the benefit of in situ bred ^{233}U was not sufficient enough to compensate for the heavy absorption of neutrons in Thorium. Efficiency of fissile material use under such conditions was thus lower as compared to standard Uranium fuel, and there was no incentive to use Thorium as fertile host. With the improvements in nuclear fuel technology, things have changed. Today burn up in excess of 60,000 MWd/t is quite common in water reactors. At this level of burn up, a much larger share of energy can be obtained from fission of ^{233}U bred in situ, leveraging efficient conversion of fertile to fissile material in Thorium. Thus, in contrast to earlier times, today the fissile material use efficiency with ^{232}Th as fertile host is comparable with ^{238}U. While this is generally true with all reactor types, Heavy Water Reactors, known for their superior neutron economy, in fact deliver better fissile material use efficiency with Thorium as fertile host. There is thus incentive for use of Thorium in current-generation water reactors to derive benefits without any significant compromise on energy output from a given quantity of available fissile material (^{235}U).

4 Softening Barriers to Nuclear Power Deployment Leveraging Special Features of Thorium

Accidents like Chernobyl and Fukushima have underscored the need to minimise the impact of a severe accident in public domain. This is basically a matter of reactor design that ensures absence of core heat up even under conditions of multiple equipment failures. However, use of Thorium enables appropriate reactor core behaviour to facilitate such a performance. Indian Advanced Heavy Water Reactor is an example of how this can be achieved [2]. This reactor is an innovative configuration using present-day technologies and grossly reduces impact in public domain and offers far better resistance to insider/outsider malevolent acts. Design and development of this reactor is more or less complete and is ready for taking up construction. The front end as well as the back end of fuel cycle of AHWR 300 LEU version of this reactor fuelled by LEU + Thorium oxide fuel also offers a high degree of proliferation resistance. The design that has been presently worked out is for a 300 MWe capacity and is well suited to needs of smaller countries as also for countries where there are needs for marginal capacity addition.

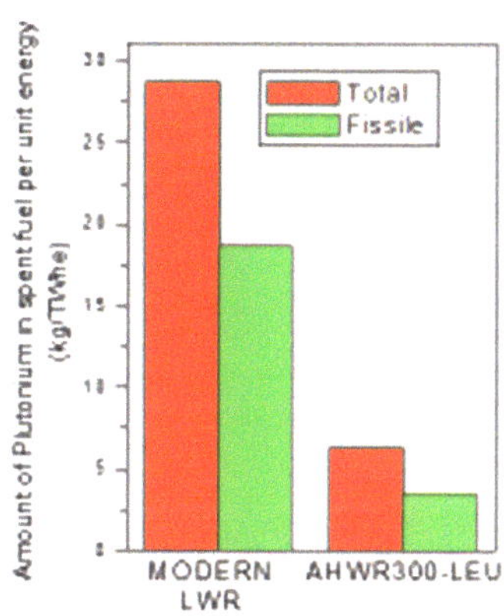

^{238}Pu	3.50	%	9.54	%
^{239}Pu	51.87	%	41.65	%
^{240}Pu	23.81	%	21.14	%
^{241}Pu	12.91	%	13.96	%
^{242}Pu	7.91	%	13.70	%
^{232}U	0.00	%	0.02	%
^{233}U	0.00	%	6.51	%
^{234}U	0.00	%	1.24	%
^{235}U	0.82	%	1.62	%
^{236}U	0.59	%	3.27	%
^{238}U	98.59	%	87.35	%

Significant reduction of proliferation risk can be realised with deployment of Thorium as fertile part of the fuel even in present-day designs of water reactors that are currently in use. This gain is maximum in PHWRs. It is interesting to note that Indian PHWRs are available in a range of capacities (220–700 MWe) to suit needs of individual countries with favourable economics as compared to other reactor designs in spite of the economy of scale. Adjustments in design would be necessary in fuel to cope with higher burn up and in reactivity control system in view of altered reactivity picture. This thus presents an interesting opportunity to smoothen one of the important barriers to large-scale deployment of nuclear energy in newly emerging nuclear countries.

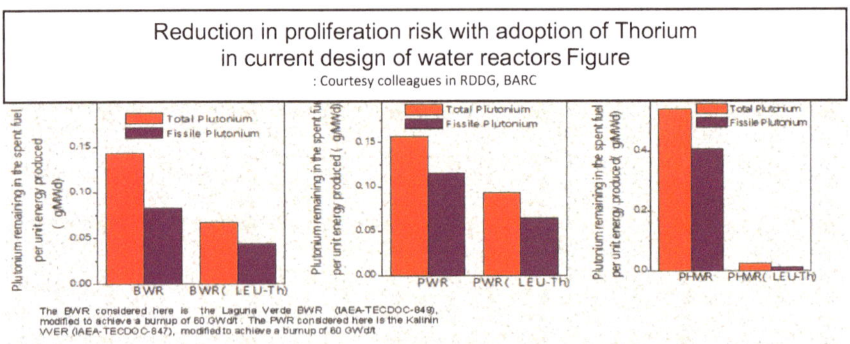

Closing of the fuel cycle is necessary for realisation of fuller energy potential energy available in Uranium and Thorium. It is this possibility that enables nuclear energy to offer a sustainable clean energy solution for the world at large. However, fear of diversion of separated fissile materials has resulted in many countries adopting open once through fuel cycle. Permanent disposal of used fuel coming out of nuclear reactors that have adopted once through fuel cycle has remained an unresolved issue, and it is likely to remain so. In contrast, the recycle option, while paving the way for consuming the fissile materials and produce additional energy, does leave some fear of possibility of diversion of fissile materials for malevolent purposes. Recycle option with Thorium, however, considerably reduces these fears because of inherent non-proliferation characteristics that get introduced in fissile materials in a Thorium-based fuel cycle.

Even if a country decides against recycle for the time being, Thorium-bearing spent fuel offers several advantages by way of greater stability and better thermo-physical and chemical properties besides comparable or better Uranium utilisation as brought out earlier. At a later date, the country could decide on differed recycle strategy. Either way the Thorium option enables a regime that is more secure from the risk of unanticipated diversion.

5 Way Ahead

The need to cut down on carbon dioxide emission is becoming more urgent every passing day. The historic Paris agreement which brought about all countries of the world together to declare their respective INDCs is still short of limiting global temperature rise to below the target of 2 °C above the preindustrial levels. There is thus a need to further accelerate deployment of non-fossil energy. That inevitably means that nuclear energy must go hand in hand with Solar energy to maintain a balance in terms of base load generation. While there is a realisation about the importance of urgent growth of nuclear power, apprehensions remain. As explained earlier, Thorium can play an important role here. Strategies for harnessing Thorium for our clean energy future can thus be seen evolving in phases. In the short term, Thorium can be used in most of present-day thermal reactor systems by incorporating it as the fertile part of fuel in place of ^{238}U with fissile part coming in the form of LEU with maximum permissible concentration (~20%). Fuel design would need to be rechecked and modified if necessary to sustain the highest possible burn up. Likewise, the reactivity control elements and refuelling pattern would need to be revised to remain within the permissible operational and safety limits. Clearly these changes are marginal but would allow considerable softening of barriers to deployment of nuclear energy particularly in the context of proliferation concerns. In the medium term, one could innovatively reconfigure reactors using Thorium based on present-day technologies to derive further advantage to minimise the impact in public domain in case of an unlikely accident as well as to enhance inherent security strength against external or internal malevolent acts. As mentioned earlier, AHWR 300—LEU is a good example of this strategy. In the longer term, one could see reactors designed specifically to exploit full potential of Thorium. These would need significant new technologies to be developed such as Molten Salt Reactors, High-Temperature Reactors and Accelerator-Driven Subcritical Reactor System. Such phased Thorium deployment strategy should enable enhanced deployment of nuclear energy in the nearer term averting the climate change threat and greater level of Thorium Utilisation in the longer term leading to greater sustainability with nuclear energy solution.

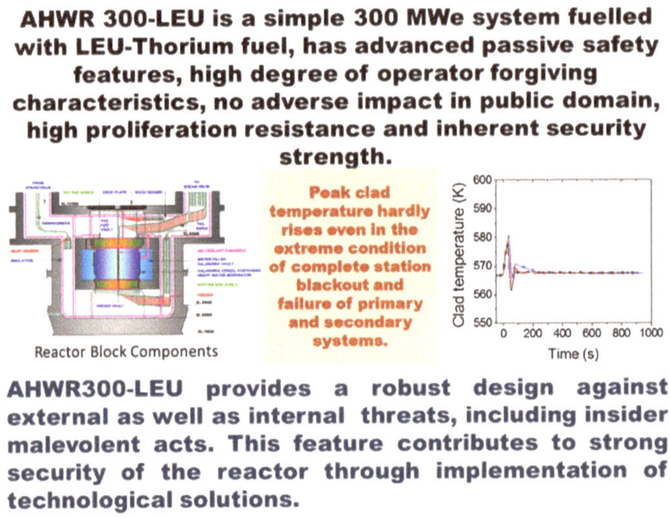

6 Closing Remarks

Despite awareness about impact of CO_2 emissions, imperatives of economy and development are leading to growth in fossil energy use and issues of sustainability. India is pursuing a three-stage nuclear power programme designed to achieve a large share of nuclear energy in the overall long-term energy mix leveraging her vast Thorium resources. While this sequential nuclear energy development programme is inevitable in Indian context and has its own time frame for deployment of Thorium on a large scale, there are also opportunities for deployment of Thorium with current thermal reactor systems in the shorter term with a view to soften barriers to deployment of nuclear power and address the global challenge posed by climate change threat. Indian experience with PHWRs and AHWR can be of advantage to world at large in this context.

References

1. A. Kakodkar, S.B. Degweker, *Towards Sustainable, Secure, and Safe Energy Future: Leveraging Opportunities with Thorium. Thorium Energy for the World* (Springer, 2016)
2. R.K. Sinha, A. Kakodkar, Design and development of AHWR—the Indian thorium fuelled innovative nuclear reactor. Nucl. Eng. Des. **236**, 683–700 (2006)

Nuclear Power from Thorium: Some Frequently Asked Questions

Srikumar Banerjee and Hari Prakash Gupta

Abstract From the inception of the nuclear power programme, the role thorium will play in sustaining supply of clean, safe and affordable nuclear energy for several centuries to come is being deliberated. The question is often raised why no commercial energy production unit utilising thorium has become operational till today. The present paper will essentially address this issue using a rather simplified language. The advantages offered by the Th-U^{233} fuel cycle in reducing the radioactive waste burden of the spent fuel and making it proliferation resistant are well known. In spite of these advantages, the absence of any fissile nuclide in the natural thorium makes it essential first to convert thorium into U^{233} by transmutation, and then, to utilise the latter as a fuel. It is for this reason thorium has all along remained as the nuclear fuel for the "future". A fairly steady value of η (above 2) of U^{233} in a wide range of neutron energy spectrum makes it usable in different types of nuclear reactors—thermal, epithermal and fast. It is argued here that utilisation of thorium for energy production can be initiated without much delay using essential technologies available today provided a sufficient inventory of fissile material as a driver fuel is made available. However, some challenging technological tasks such as remote fabrication of U^{233}-based fuel and industrial scale reprocessing of the corresponding spent fuel need to be addressed for operating a sustainable Th-U^{233} fuel cycle.

1 Introduction

During early nineteen fifties, when commercial exploitation of nuclear energy was at the inception stage there was a widespread expectation that it can provide clean and affordable energy for the whole world for many centuries to come. This expectation was based on commercially exploitable minable resources of uranium and thorium.

S. Banerjee (✉) · H. P. Gupta
Bhabha Atomic Research Centre, Mumbai 400085, India
e-mail: sbanerjee@barc.gov.in

S. Banerjee
Homi Bhabha National Institute, Mumbai 400085, India

A. K. Nayak and B. R. Sehgal (eds.), *Thorium—Energy for the Future*,
https://doi.org/10.1007/978-981-13-2658-5_2

Right at that time, it was clear that the total inventory of the fissile isotope U^{235} available in the uranium mines in the world is not adequate to sustain production of nuclear energy for more than a few centuries. The use of fertile isotopes for conversion into fissile was therefore contemplated. After nearly seventy years from that time, when about 11% of total electricity generation in the world come from nuclear, there is hardly any commercial energy production utilising thorium. The question that comes in the minds of common people is that why such a potential energy resource is not being exploited so far. In this article, authors would like to address the following questions to trace the real issues which came in the way of any significant utilisation of thorium for energy production.

- How large is the thorium reserve in the world?
- Is thorium a nuclear fuel?
- Are there any fundamental advantages in the thorium fuel cycle?
- Has power production from thorium reactors been demonstrated? What are previous experiences in generating energy from thorium?
- Are there any technological problems associated with thorium fuel cycle?
- Why the fuel cycle is considered proliferation resistant?
- Does thorium fuel cycle offer any solution to the long-lived radioactive waste?
- Is closed fuel cycle essential for thorium utilisation?
- What are different options for use of thorium?
- Does the option vary with different scenarios prevailing in different countries?
- What is the role of Accelerator-Driven Subcritical System (ADSS) in using thorium?
- Nuclear power from thorium is being talked about for over sixty years. Why is it taking so long to exploit it on a commercial scale?

An attempt is made here to address these questions in a simple language for a wider dissemination.

2 How Large Is the Thorium Reserve in the World?

Thorium was discovered in 1829 by a Norwegian mineralogist and identified by Swedish chemist who named it as THOR. Thorium has the atomic number 90. The natural abundance of Th^{232} (half-life ~ 1.4×10^{10} years) in natural thorium is 99.98%. The decay chain of thorium starts from Th^{232} and ends at the stable Pb^{208}. Uniform distribution of thorium and its abundance (3–4 times that of uranium) make it attractive for long-term energy supply. Thorium is found in rocks and soil. Monazite is the most important source of thorium. There are many estimates [1–3] of the reasonably assured resources (RAR) of thorium in the world varying between 1.17 [1] and 1.4 [1] Million Tonnes while the identified thorium resources are at least 6.35 Million Tonnes [2]. Countries such as USA, Canada, Brazil, India and South Africa have large deposits of thorium. However, according to a recent estimate (October

2016), deposits of about 1.07 Million Tonne [4] ThO_2 are established in 11.935 Million Tonne [4] of monazite (containing 9–10% ThO_2) [4] within India. This value is much larger than that reported in earlier estimates [1–3]. The monazite mineral is available in the sand of western and eastern coastline in India and is, therefore, relatively easy to mine.

3 Is Thorium a Nuclear Fuel?

Natural thorium (Th^{232}) does not have any fissile nuclide. It is a fertile material which undergoes fission only with high energy neutrons (> ~1 meV) and cannot have self-sustaining fission chain reaction on its own. To use thorium in a nuclear reactor, it is necessary to first convert Th^{232} into a fissile nuclide, U^{233} which can then be used as a fuel. The concept of converting fertile nuclides into fissile was conceived way back in the 1950s using neutrons from fission, fusion or spallation. Neutrons produced in nuclear reactors more than that required for sustaining chain reaction are utilised for conversion of Th^{232} to U^{233} which is fissionable with neutrons of a wide energy spectrum. Energy production from thorium in a nuclear reactor, therefore, requires the use of a driver fuel such as U^{235} or Pu^{239} till a certain quantity of U^{233} nuclides is accumulated through transmutation processes. Nuclear reactions in the radioactive chains that produce fissile materials such as U^{233} and Pu^{239} from Th^{232} and U^{238}, respectively, are shown in Fig. 1.

The Th-U^{233} fuel cycle once operationalised will not only provide energy for several centuries but also has the potential for annihilation of most of the long-lived

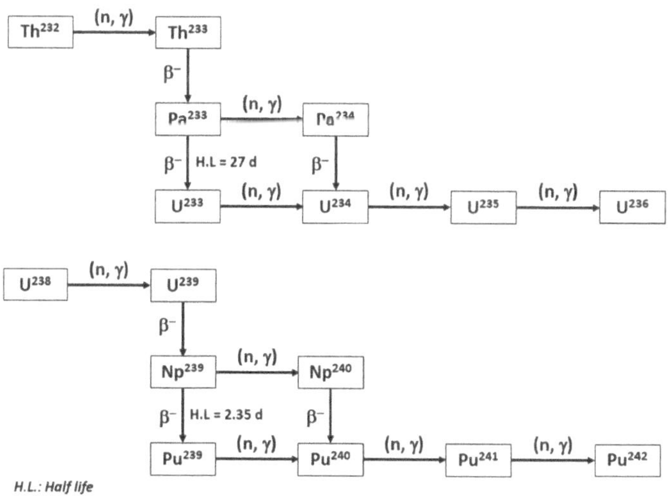

Fig. 1 Nuclear reactions for conversion from fertile to fissile material (Ref. [16], Curtsey Current Science Journal)

radioactive isotopes in the spent fuel. A long transition phase is required to produce a sufficient U^{233} inventory in existing or new design reactors to make the stand-alone Th-U^{233} fuel cycle operational.

4 Is There Any Fundamental Advantage in Thorium Fuel Cycle?

The thermal neutron (2200 m/s) capture cross sections [5–7] of U^{233}, U^{235} and Pu^{239}, respectively, are 46, 101 and 271 barns, while their fission cross sections [5–7] are 525, 577 and 742 barns. It can be observed from here that while the value of fission cross section of U^{233} is nearly same to that of U^{235}, and Pu^{239}, the capture cross section of U^{233} is much smaller than that of U^{235} and Pu^{239}. The lower probability of non-fissile capture makes multiple recycling of U^{233} more attractive compared to that of plutonium isotopes. The number of neutrons emitted per thermal neutron absorbed, denoted by η, averaged over thermal neutron spectrum are 2.29, 2.08 and 2.12, for U^{233}, U^{235} and Pu^{239}, respectively, while η values for these nuclides averaged over fast neutron spectrum are 2.31, 1.93 and 2.49, respectively. The variation of η with neutron energy is shown in Fig. 2. From the consideration of higher η value which leads to a higher breeding gain, Pu^{239} is undoubtedly a better fuel in the fast neutron spectrum. Whereas breeding with a smaller breeding gain is possible with U^{233} in a wide neutron energy spectrum. The breeding gain of U^{233} is highest among all fissile nuclides in the thermal and epithermal energies as can be seen from Fig. 2. The thermal capture cross sections of Th^{232} and U^{238} are 7.4 and 2.73 barns, respectively. Thermal reactors where Th^{232} is used as fertile material will require larger fissile inventory for achieving criticality compared to those in which U^{238} is the fertile material. However, the reactors using Th^{232} as fertile material will also produce larger amount of U^{233} compared to Pu^{239} produced in the thermal reactors using U^{238} as fertile material. The loss of neutrons in a nuclear reactor due to capture in structural material is low in relative terms due to larger thermal capture cross section of thorium.

Though U^{233} has the advantage of higher η in the thermal energy range, transmutation products originating from Th^{232} as shown in Fig. 1 include protactinium (Pa^{233}) which has a higher absorption cross section for thermal neutrons. Pa^{233} has a higher half-life (~27 days) compared to that of Np^{239} (~2.35 days) in the uranium cycle. It has also a high absorption cross section for thermal neutrons. Pa^{233} can lead to a reduction in the breeding ratio due to its conversion into U^{234} by absorbing a thermal neutron rather than decaying to U^{233}. To avoid this, in some reactors like molten salt reactors, Pa^{233} is continuously removed from fuel and kept in storage outside where it decays to U^{233}, which is subsequently fed back to the reactor. Saving of neutron helps in increasing the breeding gain in such a scheme.

In solid-fuelled reactors, thorium dioxide is the most acceptable fuel due to its higher melting point, better thermal conductivity, lower fission gas release rate and

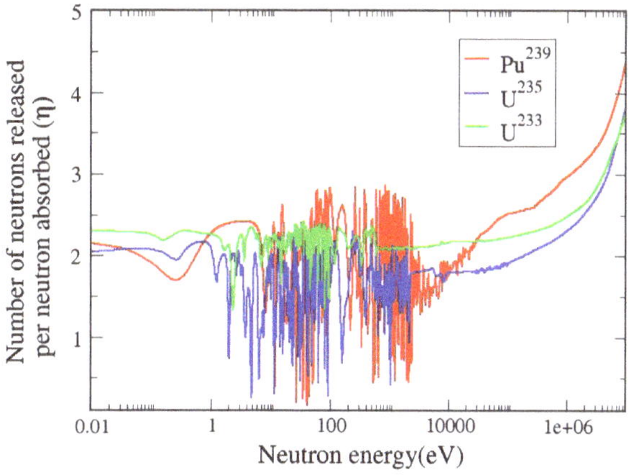

Fig. 2 Variation of η with neutron energy (Ref. [16], Curtsey Current Science Journal)

reduced fuel deterioration in the event of failure. Much reduced production of long-lived radioactive isotopes in the Th-U^{233} fuel cycle and higher stability of its compounds due to single valency of thorium make thorium fuel a very attractive option for generation of nuclear energy. Due to smaller atomic and mass number of $_{90}Th^{232}$ compared to that of $_{92}U^{238}$, the long-lived minor actinides resulting from the fuel burn-up in a nuclear reactor are in much lower quantity with the thorium fuel cycle. It has been shown [8] that the production of minor actinides like Np^{237}, Am and Cm is orders of magnitude smaller in U^{233}-Th^{232} fuel cycle compared to those in U^{235}-U^{238} cycle at 60 GWd/t burn-up for a pressurised water reactor (PWR).

5 Has Power Production from Thorium Reactors Been Demonstrated? What are Previous Experiences in Generating Energy from Thorium?

The high value of η of U^{233} in wider neutron energy spectrum (thermal and epithermal energy range) generated interest for thermal breeding the feasibility of which has been examined by several countries, e.g. high-temperature gas-cooled reactor (HTGR) in Germany and USA, BORAX-IV and Elk River boiling water reactors (BWR) in USA, shippingport reactor in USA and molten salt reactors (MSR) at Oak Ridge National Laboratory (ORNL) in USA. Table 1 summarises some parameters of these thorium reactors. Design features and operating experience of two of these reactors are briefly discussed as follows:

Table 1 Some examples of the thorium-based reactors in the world [9, 32, 33]

Reactor	Place and year	Fuel and moderator	Coolant	Power
Thorium high-temperature reactor	Germany (1985–89)	Th + Enriched uranium coated particles Graphite moderated	Helium cooled	300 MWe
High-temperature reactor	USA (1976–89)	Th + Enriched uranium Graphite moderated	Helium cooled	330 MWe
Shippingport	USA (1977–1982)	$Th + U^{233}$ H_2O moderated	H_2O cooled	60 MWe
KAMINI	India (1996–present)	U^{233} – Al alloy H_2O moderated	H_2O cooled	30 KW

5.1 *Use of Thorium in Shippingport Reactor in USA*

The pressurized light water reactor (PWR) named Shippingport reactor [9] in USA is the only reactor which has operated as thermal breeder reactor using thorium during 1977–1982, producing 2.5 billion kWh of electrical energy [9]. The core contained hexagonal-shaped modules composed of central movable seed surrounded by stationary blanket. The fuel pellets in seed and blanket composed of (U^{233}-Th) mixed oxide. The core was surrounded by reflector modules. The reflector and short sections at top and bottom of seed and blanket consisted of pellets made of thorium dioxide. The movable seed assemblies were used instead of placing absorber materials within the core for controlling reactivity. This was essential for minimising parasitic capture of neutrons and making breeding ratio greater than one. The destructive analysis of 17 fuel rods and non-destructive assay of 524 fuel rods demonstrated that 1.39% more fissile material [9] was present in the spent fuel compared to that in the beginning of cycle. This established that breeding had indeed occurred in the thermal energy spectrum [9].

5.2 *Use of Thorium in Molten Salt Reactors*

The molten salt reactor concept [10, 11] was introduced in the Aircraft Reactor Experiment (ARE) in 1954 with operating power of 2 MWt. ORNL developed an 8 MWt molten salt reactor under the leadership of Alvin Weinberg. The fuel was in the molten salt form (a solution of Li^7F, BeF_2, ThF_4 and UF_4). The nuclear heat generated was transferred by circulating the molten fuel through heat exchangers. The circulating fuel was also passed through online chemical reprocessing unit where the fission products were removed, and the required fissile material was replenished. Molten Salt Reactor Experiment (MSRE) was continued during the period of 1964–69. All the three fissile fuels namely U^{233}, U^{235} and Pu^{239} were used. ORNL later made conceptual design of two 1000 MWe Molten Salt Breeder Reactors (MSBR), one using single fluid [12] (with mixed fertile and fissile) MSBR (Breeding ratio 1.04) and the other using two fluids (separate fertile and fissile) (Breeding ratio 1.07).

6 Are There Any Technological Problems Associated with Thorium Fuel Cycle?

The technology required for use of U-Pu fuel cycles on commercial scale in thermal reactors is well proven and robust. This fuel cycle is delivering power for several decades. The same is not true with Th-U^{233} fuel cycle. Many technology issues are yet to be resolved before the deployment of Th-U^{233} fuel cycle in a commercial scale. In the thorium fuel cycle, U^{232} (half-life ~68.9 years) is generated by several nuclear reactions, two of which are indicated in the following:

$$\mathbf{U^{233} + n^1 \rightarrow U^{232} + 2\,n^1}$$

$$\mathbf{Pa^{231} + n^1 \rightarrow Pa^{232} \xrightarrow{\beta^-} U^{232}}$$

U^{232} itself does not emit gamma rays. However, two of its daughter products, namely, Bi^{212} (half-life ~60.5 min) and Tl^{208} (half-life ~3.05 min), emit intense high energy gamma rays which make fabrication and handling of U^{233} fuel technologically challenging.

Chemical reprocessing of spent thorium fuel has been demonstrated in small scales [13] and needs to be implemented in larger scale for establishing a robust technology which will involve either a three-stream reprocessing scheme for mixed U/Pu/Th-based fuel or an adoption of two different reprocessing streams one involving U/Pu and the other for pure thorium fuel pins.

7 Why the Thorium Fuel Cycle Is Considered Proliferation Resistant?

As mentioned in the preceding section, U^{233} separated from spent thorium fuel will invariably contain certain amount of U^{232}. The daughter products of U^{232} emit strong gamma rays. Such materials cannot be handled outside gamma tight hot cells. The possibility of diversion of this material for weapon applications and of pilferage by non-state actors is very remote. It is because of this reason thorium fuel cycle is considered proliferation resistant.

8 Does Thorium Fuel Cycle Offer Any Solution to the Long-Lived Radioactive Waste?

Nuclear fuel when discharged from the nuclear reactors is highly radioactive and contains radioactive fission products (FP), uranium (U), plutonium (Pu) and minor actinides (MA).

Typical values of these in the spent fuel discharged from light water reactors (LWR) are 4, 95 and 1% for FP, U and Pu and MAs combined, respectively. Typical values of these in the spent fuel discharged from light water reactors (LWR) are 4, 95 and 1% of FP, U and Pu and MAs combined, respectively. For the first 100 years, fission products dominate the radio-toxicity of spent fuel. Subsequently, plutonium and MAs are the major contributors towards the radio-toxicity of the spent fuel. Some of these MA isotopes take nearly two hundred thousand years to come to radio-toxicity level of natural uranium ore. Removal of these MAs during chemical reprocessing and their incineration enable bringing down the radio-toxicity to that of natural uranium ore in about 300 years. This will reduce the radioactive waste burden significantly. U^{238} is responsible to a good extent for production of plutonium and MAs isotopes which have very long radioactive life. In Th-U^{233} cycle, the production of long-lived MAs is reduced by orders of magnitude due to the lower mass number of Th^{232} compared to that of U^{238}. Switching over to the Th-U^{233} fuel cycle will, therefore, reduce the radio-toxicity of the spent fuel drastically [14].

The scheme of removal of MAs from spent fuel as planned in several countries where the closed fuel cycle is adopted will necessarily include nuclear incineration of MAs in fast reactors preferably in dedicated burners.

9 Is Closed Cycle Essential for Thorium Utilisation?

In case, U^{233} is used as the driver fuel in a reactor with thorium as the fertile material, there is a necessity of collecting U^{233} from the spent fuel by chemical reprocessing. In that scenario, the adoption of closed fuel cycle is essential. However, if one plans to burn fissile isotopes accumulated earlier, thorium can be used as a matrix for deep burning of U^{235} or Pu^{239} in a once-through fuel cycle. The biggest advantage of thorium matrix is that the spent fuel will contain isotopes with shorter radioactive life.

In the present generation light water reactors, using enriched fuel (4–5% U^{235}) in the once-through fuel cycle, the fuel utilisation [15] (generally defined as the ratio of mass of heavy metals of the fuel fissioned to the mass of natural uranium used in making the fuel) is typically about 0.6%. This value can be improved slightly to about 0.8% in a reactor with a better neutron economy as in a PHWR. Reprocessing of the spent fuel once and the use of Pu in the Mixed Oxide UO_2-PuO_2 (MOX) fuel (in place of enriched uranium) can enhance the fuel utilisation to about 1.5%. In breeder reactors, which operate on recycle mode, the fuel utilisation depends upon fuel burn-up and losses during chemical reprocessing of the fuel. It increases with increase in burn-up and reduction in losses during reprocessing. Deploying fast breeder reactors along with multiple reprocessing and recycling schemes, one can achieve nearly 60% fuel utilisation. The use of thorium as a fertile material from which fresh fissile material can be generated can further enhance fuel utilisation to achieve near sustainability of nuclear power.

In terms of the short-term gains, introduction of thorium in present generation reactors for in situ burning can bring about big savings in the consumption of freshly mined uranium. Some simple estimates of the savings of uranium consumption for two distinct scenarios have been presented in the next section. As mining of uranium is becoming increasingly difficult due to environmental issues, the saving in the mined uranium assumes a great importance.

10 What are Different Options for Use of Thorium?

In many occasions when one discusses the possibility of deriving energy from thorium, one says it is a source of energy for future. The same statement has been appearing for over six decades. While answering the above question, we will be examining whether energy from thorium can be tapped in near future using some of the well-established reactor technologies and what are the scenarios prevailing in different countries for realising the dream of energy from thorium.

One can visualise the utilisation of thorium for power production adopting one of the following technologies: (i) existing thermal reactors (PHWR, PWR, BWR) and in newly designed reactors (e.g. Advanced Heavy Water Reactors), (ii) fast reactors, (iii) molten salt reactors and (iv) accelerator-driven subcritical reactors.

10.1 In Existing PHWRs

The versatility of fuel management of PHWRs makes them suitable for introduction of thorium fuel in existing reactor systems. Of course, this will require use of enriched uranium or plutonium loaded MOX.

As has been discussed in detail earlier [16], substantial saving in natural uranium consumption can be achieved by introduction of thorium in PHWRs only when fissile content in the initial fuel exceeds about 4% and the homogeneous fuel burn-up is over 45,000 MWd/t.

For example, if 70 wt% of thorium dioxide is mixed with 30 wt% enriched Uranium dioxide (that is 10% enriched) in all the fuel pins, nearly 60,000 MWd/T discharge burn-up can be obtained in a once-through cycle [7, 16]. The annual requirement of natural uranium for equilibrium core in that case comes down from 169 Tonnes/GWe to 119 Tonnes/GWe. This corresponds to annual saving of nearly 50 Tons (30%) of natural uranium. In situ breeding of fissile material from thorium

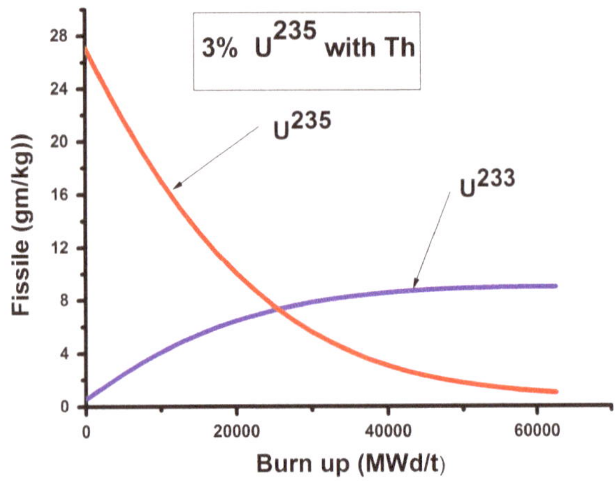

Fig. 3 Variation of fissile material (gm/kg of heavy metal) versus burn-up (3% enriched uranium case) (Ref. [16], Curtsey Current Science Journal)

is essentially responsible for this uranium saving. Figure 3 shows the depletion of U^{235} and production of U^{233} with increase in burn-up [16]. It has also been shown theoretically [7, 16–18] that though a self-sustaining Th-U^{233} cycle can be attained the corresponding burn-up is unacceptably low.

In case, a PHWR is used as a plutonium burner, one can consider a scheme of deploying plutonium dioxide as one of the driver fuel which can be mixed with thorium dioxide in the outer most ring of the fuel bundle while inner three rings can be constituted with oxides of Th^{232} and U^{233}. It can be shown that plutonium isotopes can be burnt nearly completely [16].

10.2 In Existing PWRs

Thorium can be used in existing PWRs in two ways: (i) thorium mixed homogeneously with enriched uranium and put in all fuel pins, (ii) fuel pins consisting of enriched uranium are kept in the central part of fuel assemblies which act as seed and surrounded by fuel pins consisting of thorium dioxide which act as blanket. A natural uranium saving of up to nearly 40% [19] can be obtained for the closed fuel cycle whereas it is marginal for the once-through cycle.

A Seed and blanket concept was proposed by Radkowsky, known as Radkowsky Thorium Reactor (RTR) for using thorium in PWRs in once-through cycle mode [20]. In this concept, metallic uranium–zirconium fuel is used in the seed and enriched uranium dioxide along with thorium dioxide is used in the blanket. The enriched uranium in the seed is to be replaced once in three years. The thorium in the outer blanket is to be replaced once in 10 years to increase in situ production [20] of U^{233}.

10.3 In Advanced Heavy Water Reactor (AHWR)

U^{233} fuelled reactors were operated in the past in USA and Germany. However, they were discontinued due to abondoning of the reprocessing of spent fuel of commercial reactors. The experience gained through these experiments not only provided the proof of the principle of thorium reactors but also generated meaningful data of nuclear cross sections, in-reactor behaviour of fuel, etc. India has designed an advanced heavy water reactor (AHWR) primarily as a technology demonstrator for thorium utilisation.

AHWR is designed as a vertical pressure tube light water cooled, heavy water moderated reactor with a power rating of 920 MWt (300 Mwe) [21, 22]. The boiling coolant removes the nuclear heat from the fuel by natural circulation. All the reactivity temperature coefficients including the coolant void reactivity coefficient are designed to be negative by keeping the neutron spectrum harder compared to that in PHWRs. In the design of AHWR, many innovative passive safety systems have been incorporated. Fuel cluster contains 54 fuel pins arranged in three rings. AHWR is designed such that it can accept different types of fuel including enriched U, (U-Pu) MOX, (Th-Pu) MOX and (U^{233}-Th) MOX in full core. Finally, when enough U^{233} is generated, the system can run with (Th, U^{233}) MOX. One of the recent designs of AHWR is based on mixed oxide (MOX) fuel of low enriched uranium (LEU) with thorium [23]. The design life of the reactor is 100 years.

10.4 In Fast Reactors

The breeding of fissile material can be obtained in fast reactors due to high value of η in the fast neutron spectrum. Therefore, a very high fuel utilisation can be obtained in fast breeder reactors compared to that in thermal converter reactors. The value of η of U^{233} is low compared to that of Pu^{239} in fast neutron spectrum as can be seen from Fig. 2. Therefore, the breeding ratio of the fast reactor is reduced if it is fuelled with U^{233} as against Pu^{239}. The breeding ratio will be significantly lowered if thorium is put in the core region. In view of this thorium is placed in both axial and radial blanket regions [24, 25] to produce U^{233}. This U^{233} when accumulated in sufficient quantity can be used in subsequent new reactors. The advantage of U^{233}-Th cycle in fast reactor arises from improvements in safety parameters: a reduction in sodium void coefficient and more negative Doppler coefficient [24].

10.5 In Molten Salt Reactors

The molten salt reactors which have the provision for online removal of fission products and feeding of required fissile material have some distinct advantages in thorium utilisation. The objective of fissile breeding in thermal reactors has indeed been realised in design of MSBRs. Very good neutron economy in these reactors makes it possible to have a breeding ratio exceeding unity. U^{233} having the highest η value in the thermal-epithermal neutron spectrum will be the best fuel as far as breeding is concerned. Online removal of protactinium (Pa^{233}) is also responsible in improving the neutron economy.

Despite its small breeding ratio (1.04–1.07), the doubling time of MSBRs can attain a value close to that of a fast reactor. This is because the fresh fuel inventory required to feed a thermal MSBR is much smaller (~1/4) than that required for a fast breeder reactor.

In the context of reactor safety, the MSBR technology is very attractive due to many inherent and passive safety features associated with it. In view of high fuel utilisation, possibility of thermal breeding using thorium fuel and inherently safe operation, these reactors are being considered very seriously in several countries [26–28].

11 Does the Option Vary with Different Scenarios Prevailing in Different Countries?

Issues related to the generation of nuclear power from thorium have been discussed in the preceding sections. Based on these general observations possible ways thorium utilisation has been discussed in detail in an earlier publication [16]. Table 2 summarises different options of thorium utilisation under different prevailing conditions.

Table 2 Different options of thorium utilisation under different prevailing conditions

• Large stockpile of fissile material and • No need for enhancement of nuclear capacity	• Large thorium and modest uranium reserves	• Reactor technology and reserve of fissile/fertile material not available
I. Deep burning of fissile material in thorium matrix with significantly reduced radio-toxicity burden in the spent fuel II. Substantial saving in use of freshly mined uranium	I. Operation of U-Pu fuel cycle for generating fissile inventory II. Rapid growth of nuclear power capacity by deployment of fast reactors and accumulation of U^{233} from Th blanket III. Sustaining nuclear power for a long time by deploying Th-U^{233} fuel cycle	I. Dependence on other countries on nuclear reactor technology and fuel supply II. Proliferation resistant nuclear fuel- no diversion of material from power programme to weapon programme
• Adoption of Th-U^{233} fuel cycle for long-term sustainability of nuclear power and significant reduction in radioactive waste burden of the spent fuel		

12 What is the Role of Accelerated-Driven Subcritical System (ADSS) in Using Thorium?

Accelerator-Driven Subcritical System (ADSS) [29, 30] consists of mainly three systems, namely, accelerator which produces high energy high current proton beam, the target where high energy neutron source is generated by spallation reaction initiated by high energy protons with high Z material and the subcritical reactor where these neutrons are multiplied.

Collision of high energy (1–2 GeV) proton beam with heavy nuclei such as lead or uranium causes spallation reaction leading to generation of twenty to forty neutrons per proton depending upon the mass number of target nucleus. The schematic of spallation reaction and ADSS is shown in Fig. 4a, b, respectively. The beam power is amplified by the neutron multiplication in the subcritical reactor. The energy amplification gain increases with effective multiplication factor (k_{eff}) of the subcritical reactor. Copious neutron production in a spallation process in ADSS helps not only in higher conversion of fertile to fissile but also in incineration of long-lived transuranic radioactive isotopes.

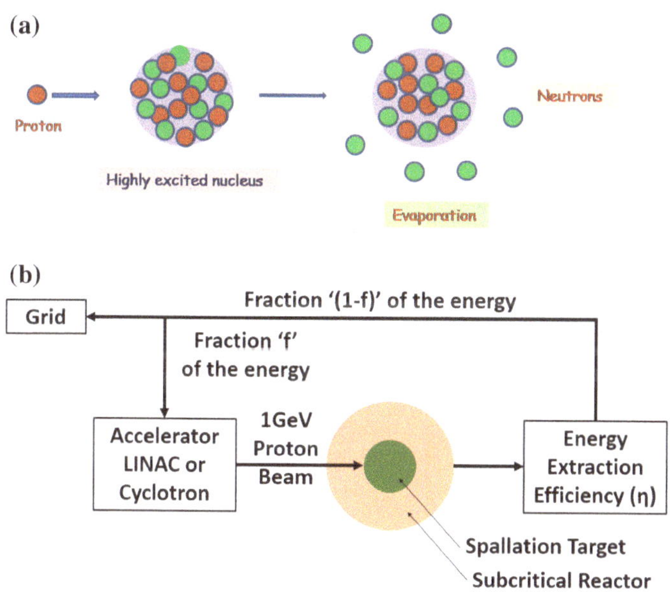

Fig. 4 **a** Schematic of Spallation reaction. **b** Schematic of accelerator-driven subcritical system (Ref. [16], Curtsey Current Science Journal)

13 Nuclear Power from Thorium is being Talked About for Over Sixty Years. Why Is It Taking So Long to Exploit It on a Commercial Scale?

Unlike natural uranium, natural thorium does not have any fissile isotope in it. Tapping energy from thorium will necessarily require conversion from fertile to fissile which needs neutrons. In a present generation power reactor thorium having a large capture cross section acts as an absorber material. Having been converted to U^{233} it becomes a fissile material and hence it is often called a "fissible poison".

Today, nuclear power is generated primarily from fission of U^{235}. The commercially exploitable resource of uranium in the world can sustain the currently installed nuclear power capacity for nearly 100 years. However, in the event of large-scale expansion of the nuclear programme this period will be reduced considerably. In the beginning of the nuclear era, there has been a great desire in attaining long-term sustainability of nuclear power and the use of thorium in achieving this has been strongly advocated. Nuclear power generation from thorium reactors has been demonstrated way back in nineteen sixties and the experience accumulated by irradiation of thorium-based fuel in present generation reactors is adequate for commercial exploitation of nuclear energy from thorium.

As has been discussed earlier, the real benefit of thorium utilisation can come when the closed nuclear fuel cycle involving chemical reprocessing of spent fuel is adopted. As there are serious restrictions on reprocessing in many countries, the development of associated technologies for separating U^{233} from irradiated thorium was halted. The accumulation of U^{233} inventory in the last few decades has therefore been so small that a major power programme cannot be initiated at this stage.

Considering the facts that Th-U^{233} fuel cycle has some advantages of inherent safety features, reduction of long-term radio-toxicity, proliferation resistance and possibility of thermal breeding, there is a recent renewal of interest in thorium-based reactors [31].

14 Concluding Remarks

The subject of nuclear power from thorium has been deliberated over a long time. The issues related are summarised in this paper in form of simplified answers to some frequently asked questions. The feasibility of generating nuclear power from thorium, possibility of fissile breeding in the thermal energy spectrum, relatively high stability of thoria fuel in nuclear reactors and reduced generation of long-lived radioactive nuclides in the spent fuel are proven beyond doubt. The use of thorium in present generation reactors and in molten salt reactors and reprocessing of thorium spent fuel in small scale have generated enough confidence in making a beginning of thorium fuel cycle. However, the technology of remotised fuel fabrication is still not matured. Considering the overall picture, the generation of nuclear power from thorium in once-through mode is implementable within a short time span provided enough inventory of fissile nuclides (U^{235} and/or Pu^{239}) is available. To derive all the advantages of Th-U^{233} fuel cycle it is imperative to accumulate enough of U^{233} fuel to be used as driver fuel in a thorium matrix and sustain power production by in situ fissile generation and the closed fuel cycle operation. As long as uranium resources are available for competitive commercial power generation, there is not enough incentive in many countries to start the use of thorium. This is not true for countries with meagre uranium resources and for them speedy accumulation of fissile nuclides through the operation of fast reactors is a priority. The reasons behind the delay of thorium introduction for power production can be explained on this basis.

Acknowledgements The authors wish to thank Board of Research in Nuclear Sciences (BRNS) of Department of Atomic Energy (DAE), India, for providing the financial assistance.

References

1. (a) Uranium 2007: Resources, production and demand. A Joint Report by Nuclear Energy Agency (NEA), and International Atomic Energy Agency (IAEA), 2008, p. 28, https://www.oecd-nea.org/ndd/pubs/2008/6345-uranium-2007.pdf (b) United States Geological Survey Mineral Commodity Summaries (2014), p. 167, https://minerals.usgs.gov/minerals/pubs/commodity/thorium/mcs-2014-thori.pdf. Accessed 27 April 2017
2. Uranium 2016: Resources, production and demand. A Joint Report by Nuclear Energy Agency (NEA), and International Atomic Energy Agency (IAEA) (2016), p. 40, https://www.oecd-nea.org/ndd/pubs/2016/7301-uranium-2016.pdf
3. Wikipedia, Occurrence of thorium. https://en.wikipedia.org/wiki/Occurrence_of_thorium. Accessed 27 April 2017
4. Reserves of Uranium and Thorium. http://www.dae.nic.in/writereaddata/parl/winter2016/lsus2377.pdf
5. P.G. Boczar, et al., *Thorium Fuel Cycle Studies for CANDU Reactors*, vol. 1319 (IAEA-TECDOC, 2002), p. 25
6. P.G. Boczar, et al., *Recent Advances in Thorium Fuel Cycles in CANDU Reactors*, vol. 1319 (IAEA-TECDOC, 2002), p. 104
7. H.P. Gupta, S.V.G. Menon, S. Banerjee, Advance fuel cycles for use in PHWRs. J. Nucl. Mater. **383**, 54–62 (2008)
8. A.R. Sundararajan, L.V. Krishnan, P. Rodriguez, Radiological and environmental aspects of Th-U fuel cycle facilities. Prog. Nucl. Energy **32**(3–4), 289–295 (1997)
9. G. Olson, et al., *Fuel Summary Report—Shippingport Light Water Breeder Reactor*, INEEL/EXT-98-00799, Rev. 2 (2002)
10. R.C. Briant, A.M. Weinberg, Molten fluorides as power reactor fuels. Nucl. Sci. Eng. **2**, 797–803 (1958)
11. P.R. Kasten, et al., Summary of molten salt breeder reactor design studies. ORNL-TM-1467 (1966)
12. R.C. Roberson, Conceptual design study of single—fluid molten—salt breeder reactor. ORNL-4541 (1971)
13. P.K. Wattal, Recycling challenges of thorium based fuel, in *International Thorium Energy Conference (IThEC-13) held in Cern, Geneva*, 27–31 Oct 2013
14. A. Rajora, et al., Generating one group cross-section for isotope depletion and generation calculation for thorium, in *A Paper Accepted for Presentation In International Thorium Energy Conference (ThEC-2015), to be held in Anushaktinagar, Mumbai, India*, 12–15 Oct 2015

15. Status and prospects of thermal breeders and their effect on fuel utilisation. Technical Reports Series No. 195, IAEA, 1979, p. 7
16. S. Banerjee, H.P. Gupta, S.A. Bhardwaj, Nuclear power from thorium: different options. Curr. Sci. **111**(10), 1607–1623 (2016)
17. H.P. Gupta, Use of thorium in PHWRs, in *National Conference on Power from Thorium: Present Status and Future Directions, Mumbai*, 22–24 Dec 2014
18. H.P. Gupta, R.D.S. Yadav, S.V.G. Menon, S. Banerjee, Utilisation of thorium in different reactors, in *German Annual Meeting of Nuclear Technology-2009 Dresden*, 12–14 May 2009
19. C. Grove, Comparison of thorium and uranium fuel cycles. NNL (11) **11593**(5), 9 (2012)
20. A. Radkowsky, A competitive thorium fuel cycle for pressurised water reactors of current technology. IAEA-TECDOC-1319 (2002), p. 25
21. R.K. Sinha, A. Kakodkar, Design and development of AHWR—The Indian thorium fuelled innovative nuclear reactor. Nucl. Eng. Des. **236**, 683–700 (2006)
22. K. Balakrishnan, A. Kakodkar, Preliminary physics design of advanced heavy water reactor (AHWR). IAEA-TECDOC-638, March 1990
23. A. Thakur, et al., Fuel cycle flexibility in advanced heavy water reactor (AHWR) with the use of Th-LEU fuel, in *International Conference on Future of HWRs, Ottawa, Canada*, 2–5 Oct 2011
24. R. Ramanna, S.M. Lee, The thorium cycle for fast breeder reactors. Pramana- J. Phys. **27**(1&2), 129–157 (1986)
25. D.K. Mohapatra et al., Physics aspects of metal fuelled fast reactors with thorium in blanket. Nucl. Eng. Des. **265**, 1232 (2013)
26. N. Hirakawa, et al., Molten salt reactor benchmark problem to constrain plutonium. IAEA-TECDOC-1319 (2002), p. 25
27. Trans atomic power, molten salt reactors. Technical White Paper, vol. 1.0.1, March 2014
28. H.P. Gupta, Preliminary design of molten salt converter reactor, in *Conference on Molten Salt Nuclear Technology (CMSNT)*, BARC, Mumbai, India, Jan 2013
29. C. Rubbia, et al., Conceptual design of a fast neutron operated high power amplifier. CERN/AT/95-44(ET) (1995)
30. S.S. Kapoor, Accelerator-driven sub-critical reactor system (ADS) for nuclear energy generation. Pramana- J. Phys. **59**(6), 941–950 (2002)
31. Introduction of thorium in nuclear fuel cycle, short to long term considerations. NEA No. 7224, Nuclear Science, OECD, 2015
32. J.B. Dee, J. Kupitz, International status of HTGRs. IAEA Bullet. **26**(4), 5 (1984)
33. B. Babu, V. Ramanathan, B. Rajendran, P.V. Ramalingam, B. Raj, Proposed modernisation and refurbishment of instrumentation and control system of the FBTR and Kamini reactors of India. IAEA—TECDOC-1625 (2009), p. 95

Thorium Technology Development in an Indian Perspective

A. K. Nayak, Arun Kumar, P. S. Dhami, C. K. Asnani and P. Singh

Abstract To meet the large energy demand in India, nuclear energy is vital which is not only a clean, green source of energy but also the high capacity factor and long life of plant make it a competitive source with other green sources of energy. India has one of the largest reserves of thorium in the world. It is well recognized that this fuel shall play a major role in India's nuclear power programme to produce a significant fraction of energy. However, thorium being a fertile material cannot be used directly unlike, for example, natural uranium which has U-235 fissile content. Thus, thorium needs an external feed of Pu-239 or U-233 for burning in the reactor. In view of the above, India has adopted a closed fuel cycle approach in its three-stage nuclear power programme. For burning thorium in a closed fuel cycle, several technologies need to be developed and demonstrated for commercial viability. India has worked a lot in this context right from fuel fabrication to reprocessing. This manuscript summarizes a brief review of the technology developed in India and past experiences for thorium utilization in Indian nuclear programme.

1 Introduction

India has a population close to 1.3 billion, at present. It per capita energy consumption is very low (1050 kWh per capita per year), and so its human development index (HDI) is 0.6 (130th in the world as seen in Fig. 1 [1]). The per capita energy consumption has a strong bearing on the socio-economic standard of a nation. India has an ambitious economic growth which is projected to be high according to the NITI Aayog of India [2] in the next 10 years. Energy shall play a key role in achieving the target.

A. K. Nayak · A. Kumar (✉) · P. S. Dhami · P. Singh
Bhabha Atomic Research Centre, Mumbai 400085, India
e-mail: arunths@barc.gov.in

C. K. Asnani
Indian Rare Earths Limited, Mumbai, India

A. K. Nayak and B. R. Sehgal (eds.), *Thorium—Energy for the Future*,
https://doi.org/10.1007/978-981-13-2658-5_3

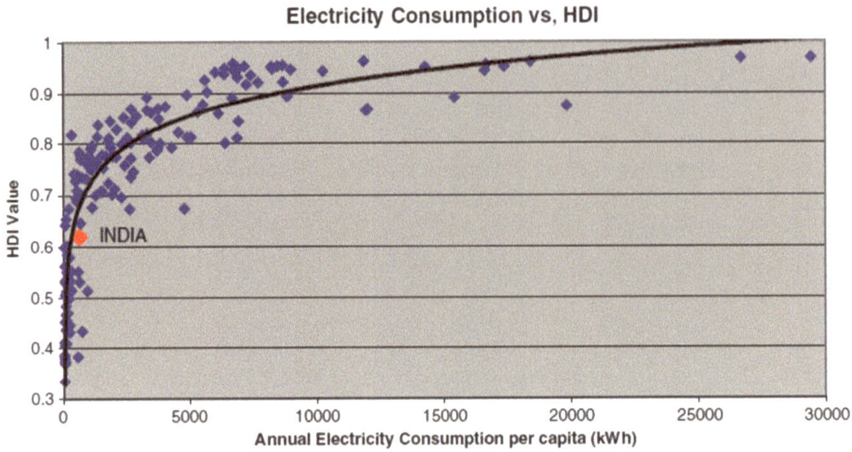

Fig. 1 Electricity consumption versus human development index

Currently, coal (~40%) and natural gas (~23%) dominate the energy share globally [3]. The share from nuclear energy is nearly 11.2%. In the Indian scenario, the carbon-based fuel is the main source of the energy production [3].

The global concern as discussed in the Paris summit is to reduce the carbon footprint and emission of the greenhouse gases to limit the environment temperature. It can be achieved by reducing the share of carbon-based fuel in energy production and increasing the participation of green energy. Renewable source of energy such as solar and wind has taken a gigantic step in India. By November 2017, India has already installed a total of 62 GW renewable power which includes solar and wind [4], while at the same time the share of nuclear, a clean and green source of energy, is relatively low.

The slow growth of nuclear energy production in India is primarily due to limited uranium reserves in India and high capital cost involved in building a nuclear power plant. However, it is well known that the energy cost is not only dependent on the capital cost and operating cost but also dependent on the capacity factor of the plant. Higher capacity factor reduces the cost of energy significantly. Operating experience of Indian nuclear plants shows that Indian PHWRs have very high capacity factor [5]. Even the capital cost is quite competitive with other sources of energy [5].

Considering the demand for a large amount of energy, India would require a significant fraction of energy from nuclear, primarily because not only the nuclear power plants run at higher capacity factors with almost no emission of greenhouse gases but also the fuel has a large power density to produce significant large power with relatively small amount of fuel. For example, a 500 MWe thermal power plant with coal as fuel requires 1.8 million ton of coal per year, while the fuel required for a 1300 MWe PWR is only 115 ton with 3.2% enrichment and for a BWR it is ~170 ton with 2% enrichment [6]. Thus, from an Indian perspective, this source of energy is considered extremely vital in near future.

2 Role of Thorium in Indian Energy Scenario

India has more than three times thorium than uranium. Utilization of thorium in India's three-stage nuclear power programme was conceptualized by the architect of our nuclear power programme, Homi Jehangir Bhabha, mainly because it has very large reserves of thorium and very limited reserves of uranium (Fig. 2). While natural uranium has a small percentage of U-235 (0.7%) which is a fissile isotope, thorium does not have any fissile content. Thus, it requires an external feed of fissile material to burn in a nuclear reactor. This was the main factor for India adopting a closed fuel cycle in its three-stage nuclear power programme. In the first stage, the PHWRs burn natural uranium fuel which results in the production of Pu-239. The spent fuel is reprocessed to separate plutonium. Pu-239 is a fissile material which shall be used as fuel in the second stage of Indian nuclear power programme in fast reactors. In fast reactors, the neutron energies are much higher than the thermal reactors. The advantage of fast breeder reactors is that it produces slightly more plutonium than it consumes. This is achieved by proper physics design of nuclear reactors with suitable choice of fuel and coolant in design. Over a period of time, with growing plutonium inventory, the number of fast breeder reactors (FBRs) in the country can be multiplied. Thus, the major share of electricity production from nuclear can be achieved through plutonium burning in FBRs. In addition, in plutonium-based FBRs, Th-232 is targeted to be introduced as a blanket material to be converted to U-233 in situ. In the third stage of nuclear power programme, Th-232 shall be burnt with U-233 to facilitate building the nuclear power capacity for a sustained power generation as long as thorium input is available (Fig. 3) [7].

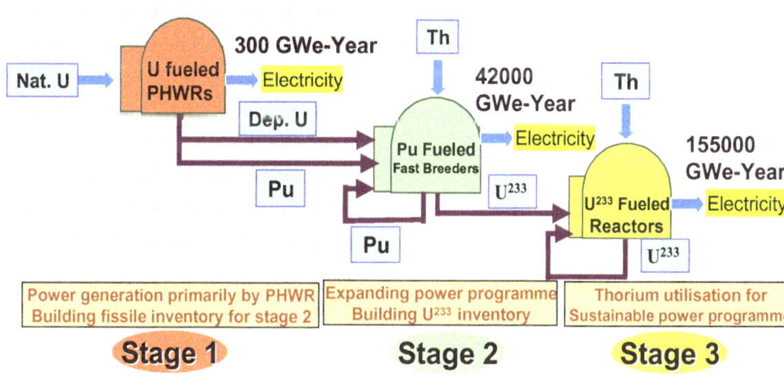

Fig. 2 Schematic of Indian three-stage nuclear power programme as originally envisaged on the basis of available uranium reserve to support a 10,000 MWe installed capacity in the first stage

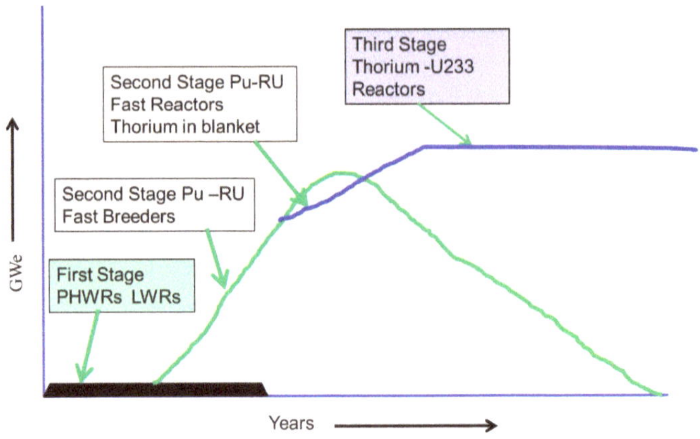

Fig. 3 Potential for nuclear power growth and energy sustainability in India using thorium

3 Advantages of ^{233}U-Th Cycle

^{233}U-Th fuel cycle has superior properties compared to Pu-U cycle for the following reasons:

- ^{233}U has η value (greater than 2) (Fig. 4) that remains fairly constant in thermal as well as epithermal regions. This facilitates
  - Th-232 as a better candidate compared to U-238 for neutron capture (Fig. 5) [8]
  - Higher fission to capture cross-sectional ratio (Fig. 6) [8].
- ^{232}Th–^{233}U fuel cycle can provide breeding characteristics with fast, epithermal or thermal spectra. Thermal reactors require less specific fissile inventory and could be a better option for this cycle.
- At high burnups beyond a threshold, thorium requires lower ^{235}U enrichment. Thus, it overtakes ^{238}U as fertile host. Hence, it is an ideal fuel for
  - Tolerating higher parasitic absorption

Fig. 4 Comparison of η value of different fuels

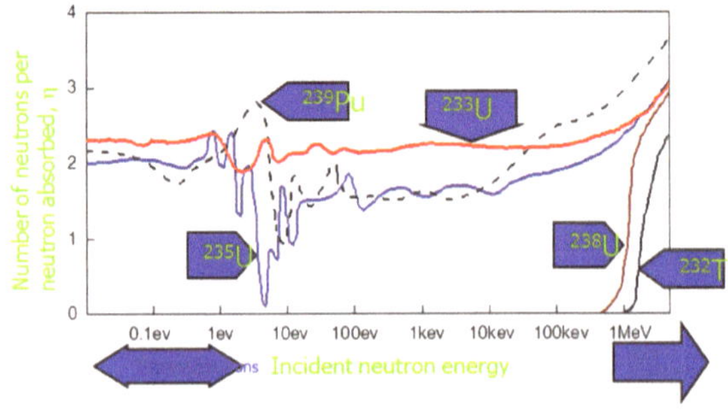

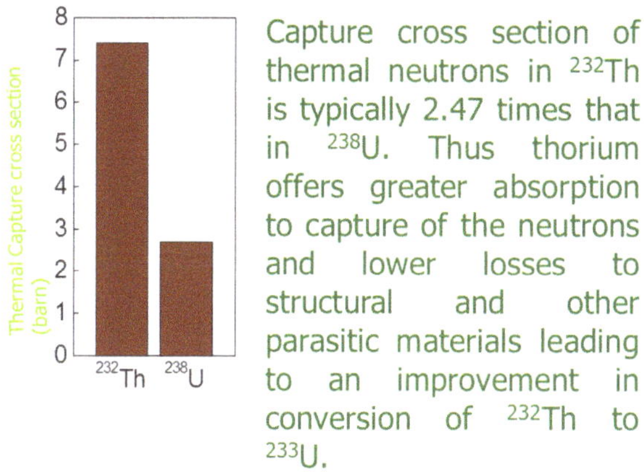

Fig. 5 Capture cross section of Th-232 and U-238 fuel

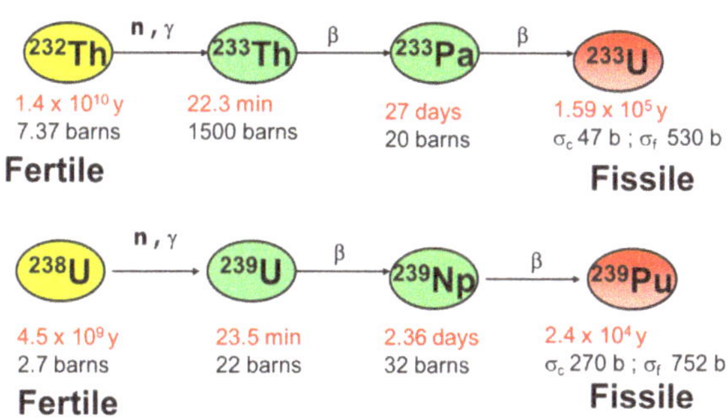

Fig. 6 Comparison of fertile species ^{238}U and ^{232}Th

- Better proliferation resistant due to the presence of U-232 (Fig. 7a, b) [9, 10].
- Lower minor actinide production (Fig. 8)
  ^{232}Th–^{233}U fuel cycle produces much lesser long-lived minor actinides (Np, Am and Cm) as compared to the ^{238}U–^{239}Pu fuel cycle. Hence, ^{232}Th–^{233}U fuel cycle creates reduced toxicity and issue of decay heat production. Thorium also does not add to long-lived waste.
- Thermo-physical characteristics of thorium are well suited for high burnup due to the following considerations:

(a) Better thermo-mechanical properties and slower fuel deterioration
(b) ThO_2 has a very high melting point of 3300 °C
(c) Lower fuel temperatures due to the better thermal conductivity of ThO_2
(d) Less fission gas release.
(e) Better dimensional stability due to high chemical stability and high FP retention in ThO_2 matrix, better characteristics for power distribution, lower reactivity loss

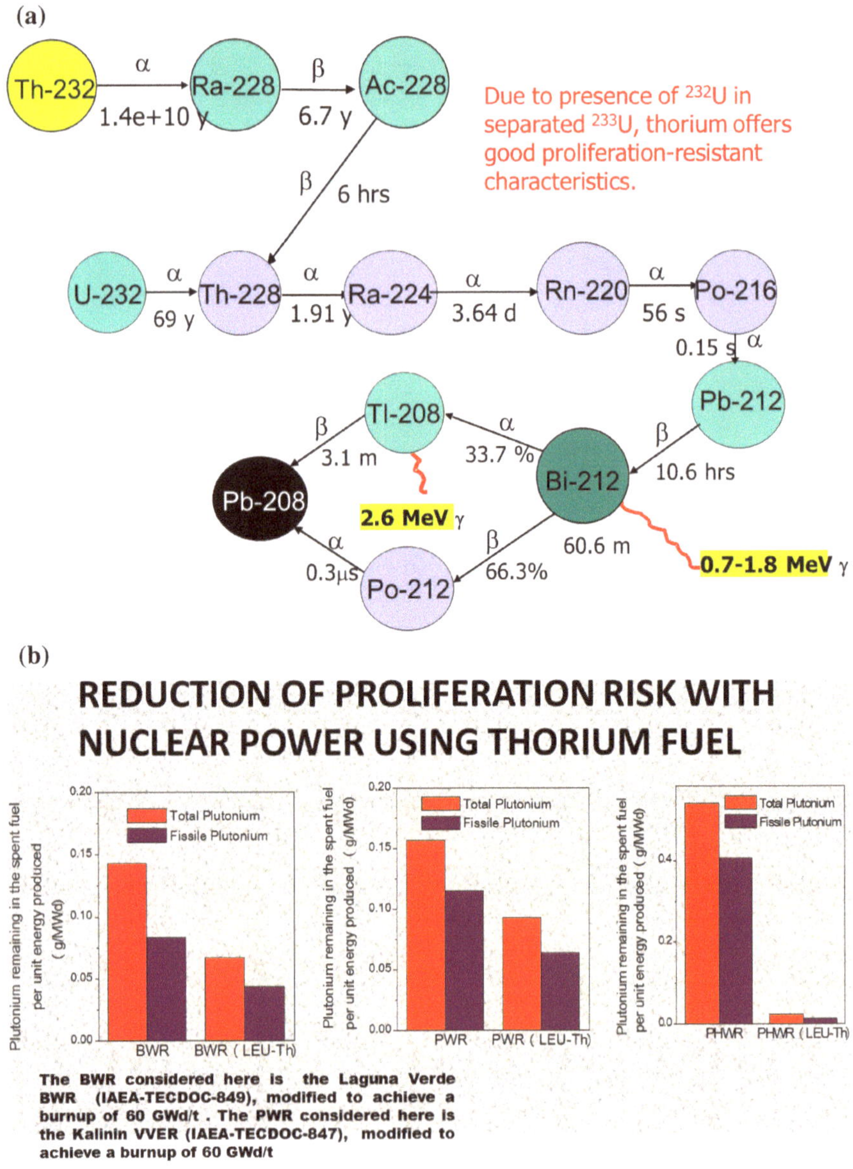

Fig. 7 **a** Thorium decay chain—the presence of U-232 makes high proliferation resistant; **b** comparison of proliferation risk of different fuels [11]

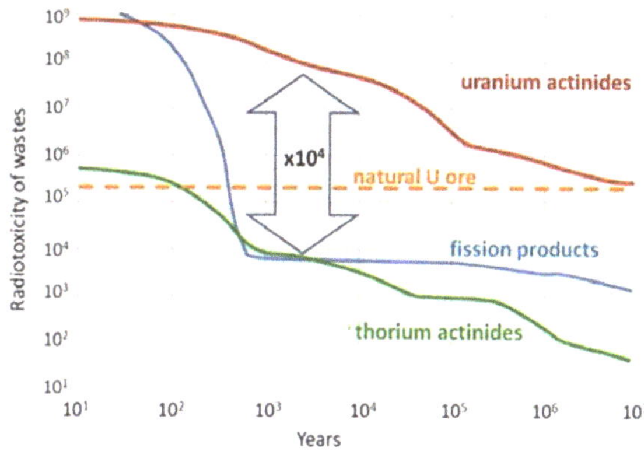

Fig. 8 Comparison of the production of MAs in thorium and uranium fuel cycle

Table 1 Resources of thorium in India

Minerals	Reserves (million tons)
Ilmenite	593
Sillimanite	226
Garnet	168
Zircon	34
Rutile	31
Monazite (thorium oxide)	12 (10 lakh tons)
Total	1065

with fuel depletion, better behaviour under irradiation which supports operating at higher burnup.

(f) More favourable safety characteristics are due to:

Overall temperature coefficient being negative and less chemical reactivity with water.

4 Thorium Reserves in India

In India, naturally occurring monazite mineral is the principal source of thorium (Fig. 9). It occurs in the sea beaches of coastal India (Fig. 10) and inland placer deposits, in association with six other heavy minerals such as ilmenite, rutile, leucoxene, zircon, sillimanite and garnet [11]. Concentrations of monazite in the suite of above-mentioned heavy minerals vary from 0.1 to 2.5%. Table 1 shows the resource availability of monazite along with other associated heavy minerals [11]. Monazite mineral is a phosphate of thorium and rare earth (RE), and in India it contains ~8.8% of ThO_2 [11]. Composition of monazite is shown in Table 2 [11].

Table 2 Typical composition of monazite

Composition	%
RE elements as RE_2O_3	59.37
P_2O_5	27.03
ThO_2	8.88
U_3O_8	0.35
CaO	1.24
SiO_2	1.0
MgO	0.63
Fe_2O_3	0.32
Al_2O_3	0.12
PbO	0.18
TiO_2	0.36
ZrO_2	0.49

4.1 Separation of Thorium from Monazite

Different technologies are being developed in India for separation of thorium from monazite. Separation is carried out by (1) the digestion with hot sulphuric acid and (2) the fusion with hot sodium hydroxide, i.e. caustic lye solution [11]. The separation by fusion with hot sodium hydroxide involves chemical attack of ground monazite sand with hot, concentrated caustic lye at a higher temperature to break the stable

Fig. 9 Monazite sand

Fig. 10 Thorium in beach sand (Kerala coast)

phosphate bond of the mineral by converting it to trisodium phosphate (TSP). The caustic fused mass is leached in water to leach out water-soluble TSP leaving behind uranium, thorium and rare earths as insoluble hydroxides. The insoluble mass is further reacted with dilute hydrochloric acid at pH-3 to extract most of the rare earth (RE) from the treated mass as rare earth chloride in the form of aqueous solution. The residual unreacted mass, called thorium cake, contains uranium, thorium and a small amount of RE as hydroxides [11].

The rare earth chloride solution is treated with barium chloride, sodium sulphate and sodium sulphide to remove the long half radioactive daughter products of thorium, i.e. lead as lead sulphide, which is co-precipitated along with barium sulphate. The treated RE chloride solution after settling and decantation is filtered to bring down its specific radioactive level below 1 Bq/g by removing its radioactivity as the precipitated lead sulphide along with barium sulphate (Pb–Ba cake).

The uranium and thorium elements contained in the thorium cake are completely extracted into solution along with small amount of remaining RE by reacting the thorium cake with dilute nitric acid in the presence of hydrofluoric acid. The acidic solution containing a few unreacted solid particles is filtered, and the small amount of insoluble mass (muck) is disposed off following low-level radioactive waste disposal guidelines.

The clear acidic solution after filtration is subjected to solvent extraction by using tributyl phosphate (TBP) as a solvent in two separate stages for extracting uranium and thorium in their purer forms and recovering them from TBP extractant as ammonium diuranate (ADU) nitrate and thorium nitrate, respectively [11]. Figure 11 shows a schematic of the process flow sheet for the extraction of rare earths, thorium, uranium and production of TSP from monazite. Figure 12 is a diagrammatic presentation of monazite attack by caustic lye to produce TSP and thorium cake containing all

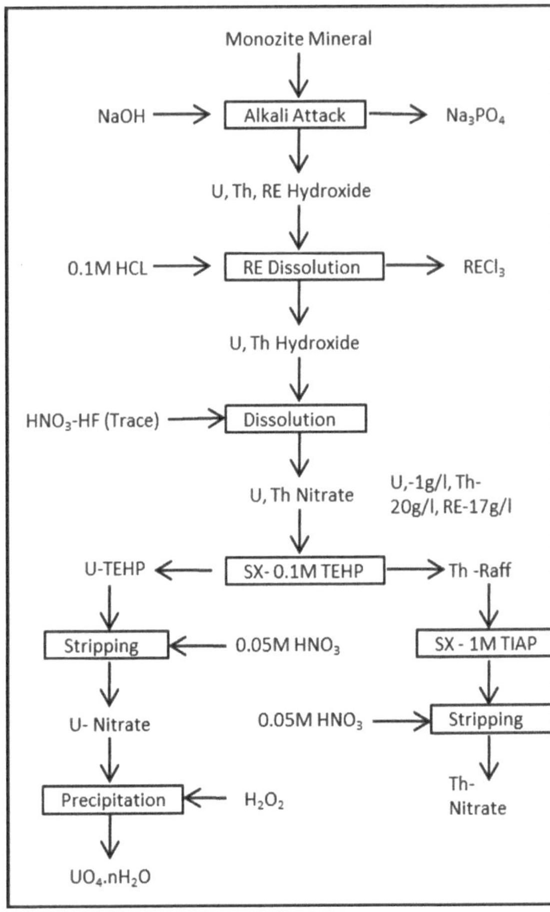

Fig. 11 Process flow sheet for extraction of thorium

the rare earth, uranium and thorium values of monazite. Figure 13 elucidates the scheme of processing thorium cake to produce thorium oxalate for long-term storage, thorium nitrate and nuclear grade thorium oxide. Thorium chloride is converted to thorium oxalate for its long-term storage in engineered trenches.

Post-solvent extraction of REs is left behind in the acidic raffinate solution, which is neutralized by caustic/alkali treatment in the effluent treatment plant (ETP). The precipitated mass (ETP cake), after caustic/alkali treatment of the acidic raffinate in ETP, is either disposed of as per radioactive waste disposal norms or recycled to the previous processing step to recover the RE as RE chloride. Neutralized liquid effluent from ETP is disposed as per statutory guidelines.

Based on the above-mentioned caustic attack process of monazite, Indian Rare Earths Limited (IREL), has set up a monazite processing plant [11]. The management of waste generated by processing monazite in this plant is given in Table 3. The

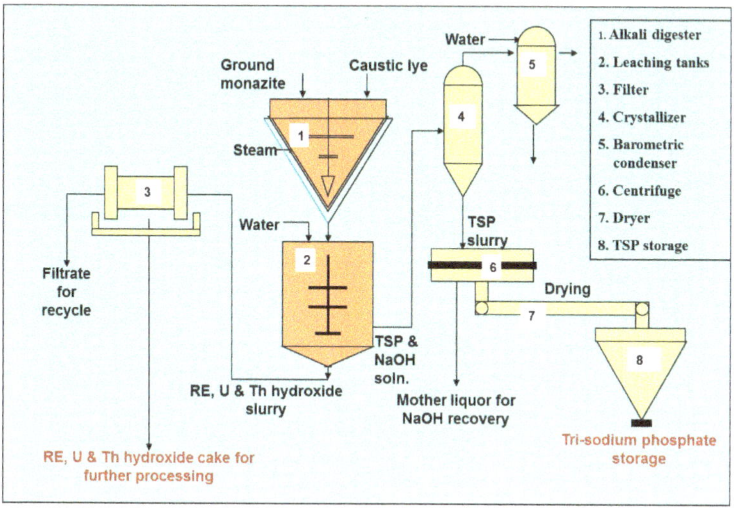

Fig. 12 Monazite processing—caustic attack plant

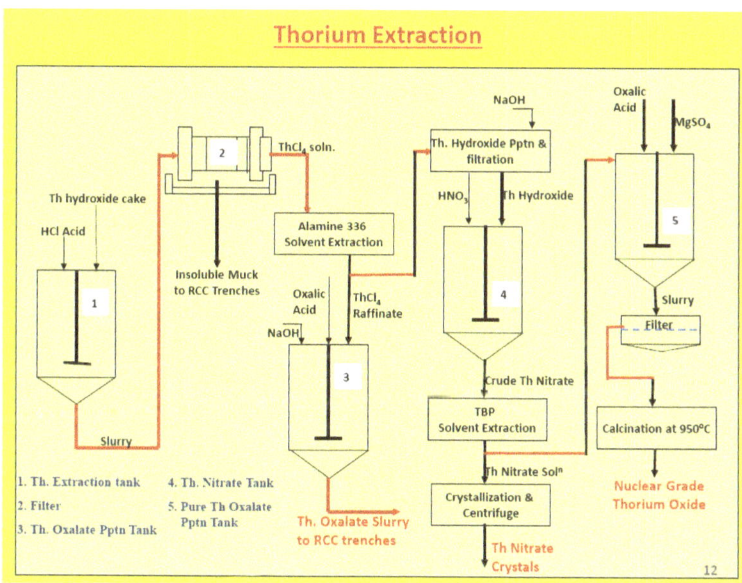

Fig. 13 Thorium extraction process

physical properties, chemical composition and the adopted methods of storing each of these radioactive materials are also presented in the same table.

Table 3 Management of waste generated during monazite processing

Solid waste	Physical properties	Chemical composition	Disposal methodology
Thorium oxalate (2000 tpy)	Bulk ρ: 1.27 g/cc pH: 7–7.5 Activity: 7400 Bq/g	Th oxalate: 60% Water: 40%	Underground RCC trenches
Muck (insoluble) (1200 tpy)	Bulk ρ: 1.2 g/cc Acidity: 1.1–1.5% Activity: 1800 Bq/g	Total oxide: 5–10% ThO_2: 0.5–3% Insoluble: 30–45% Moisture: 30–45%	Underground FRP lined RCC trenches
Pb–Ba cake (600 tpy)	Bulk ρ: 1.2 g/cc pH: 4–4.2 Activity: 8000 Bq/g	$BaSO_4$: 45–50% PbS: 5–10% Moisture: 45–50%	Underground FRP lined RCC trenches
ETP cake (50 tpy)	Bulk ρ: 1.25 g/cc pH: neutral Activity: 100 Bq/g	Moisture: 60–70% PO_4: 0.1–0.15% Fluorides: 0.05–0.1% Insoluble: 25–30%	Packed in HDPE bags and stored in earthen trenches

5 Extractive Metallurgy of Thorium

Being one of the most reactive elements, it is very difficult to obtain thorium in metallic form. Besides, its high melting point (1755 °C) adds to handling difficulties during extraction. Starting compound for extraction of thorium could be its oxide, chloride or fluoride.

5.1 *Oxide Route*

It is the most convenient route, since ThO_2 is easy to prepare and handle. However, this process requires calcium, which alone could be used as the reducing agent for the oxide. The resulting product is thorium powder, which contains about 1% oxygen. This process has been carried out in India up to 5 kg batch size [12]. Thorium is ductile despite high oxygen content and sheets of thorium metal have been made by powder metallurgy. Oxide route followed by powder metallurgy is cheaper, provided the purity of thorium is acceptable to the fuel designers and the desired sizes could be fabricated. Larger consolidated (i.e. without porosity) rods of pure thorium can be made by consumable arc melting; however, no experience exists in India. Some experience is gained using non-consumable arc melting for smaller size products. However, if alloying is to be done with refractory metals (e.g. Mo or Zr), the purity obtained in oxide route is definitely not sufficient. Although oxygen could be eliminated by electron beam melting, the loss of thorium is substantial (~15%) as one-mole oxygen atom takes away one mole of thorium which is 15 times heavier.

5.2 *Chloride Route*

Choice of magnesium as the reducing agent is also available if thorium chloride is used as the starting material. This process is similar to Kroll's process, which is also used for preparing zirconium, titanium, etc. However, thorium chloride is highly hygroscopic. Removal of moisture may result in its conversion to $ThOCl_2$, which does contain oxygen and thereby defeats the purpose of going for the chloride route [12]. The product is thorium sponge, which quickly picks up oxygen.

5.3 *Fluoride Route*

ThF_4 is reduced by calcium in the presence of zinc. Th–Zn eutectic and CaF_2 are liquid at reaction temperatures; hence, the product is formed in massive form (i.e. consistent solid without pores). Zinc is distilled off from Th–Zn. No experience in India about this route is available.

6 Fuel Fabrication Aspects

6.1 *Fuel Options for Water-Cooled Reactors Working in Thorium-Based Fuel Cycle*

Initially, it was anticipated that thorium will be required to supplement uranium at a very early stage of Indian nuclear power programme. Considerable thorium fuel-burning experience was thus generated globally in various types of reactors which include BWRs, PWRs, gas-cooled HTRs and MSRs. Recently, a lot of studies have been conducted in research laboratories and universities for thorium utilization in advanced reactors in many countries, e.g. India, Russia, USA, China, France, UK, Canada. The IAEA–INPRO programme has also identified a number of thorium-based fuel cycle options which include different fuel options for water-cooled reactors such as (Th-LEU) MOX, (Th–Pu) MOX and (Th-^{233}U) MOX.

6.2 *Fabrication of (Th–LEU) MOX*

Manufacturing technology of (Th-LEU) MOX is almost similar to that of the conventional powder pellet process used for enriched UO_2 [13]. Figure 14 shows the general flow sheet for fabrication of thoria-based mixed oxide fuel. Initial milling of thoria powder is carried out to change the platelet morphology so as to enhance the compatibility of the powder. The milled powder is mixed with UO_2 along with additives,

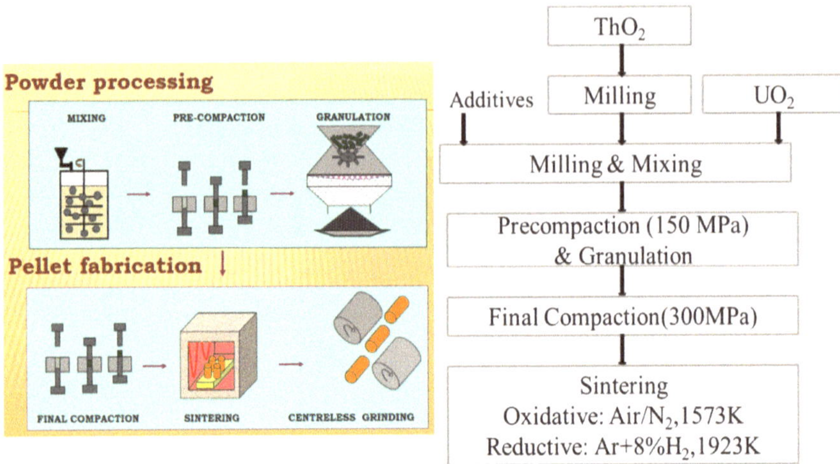

Fig. 14 Powder metallurgy process for fuel fabrication

e.g. polyethylene glycol (PEG) and oleic acid (OA) followed by pre-compaction at 150 MPa and granulation. The additives reduce interparticle friction and also friction between die and particles. Final compaction is carried out at 300 MPa followed by sintering either in oxidative or in reducing atmosphere.

6.2.1 Major Issues in (Th–LEU) MOX Fuel Fabrication

Criticality Issues: Commercial LWR fuel fabrication plants are licensed to handle up to 5% ^{235}U enrichment. (Th-LEU) MOX fuels require handling of higher ^{235}U enrichments (~20%). Fabrication plants will have to be designed to take care of more stringent criticality requirements (till the two oxide powders are mixed with each other). This may limit the size of equipment and may require adoption of batch-type mode instead of continuous mode for fuel fabrication, especially for some of the wet processes.

Health Physics Issues: Higher radiation shielding is required for handling (Th-LEU) MOX fuel. Steel or lead shielding of few cm thick has to be provided for a various equipment depending on the age of thorium being handled. It is well known that daughter products of ^{228}Th are γ emitters; e.g. ^{212}Bi emits 0.7–1.8 meV, and ^{208}Tl emits 2.6 meV gamma (Fig. 15).

Higher ^{234}U Concentration in LEU: Because of the presence of higher ^{234}U concentration in LEU as compared to Nat. U (Table 4), the activity is substantially higher in LEU, which needs to be handled during fabrication of the fuel.

Thoron Management: Fabrication plant ventilation system has to take care of ^{220}Rn by providing features for delay time prior to release. Additional steps to be considered

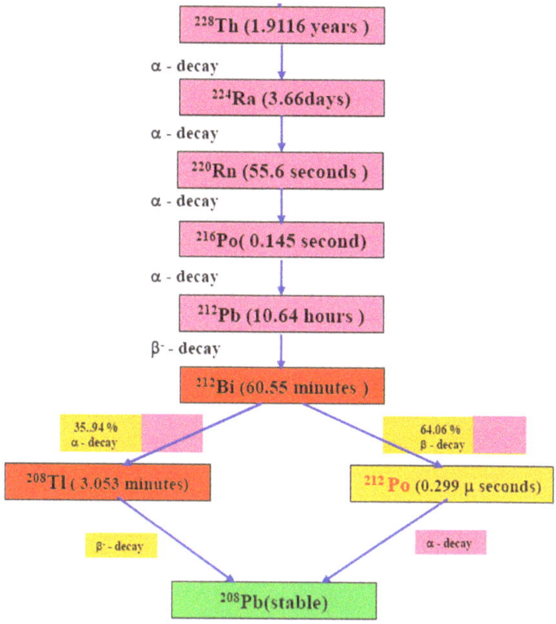

Fig. 15 Decay chain of Th-228

Table 4 Specific activity comparison between LEU and Nat. U [13]

Isotope		^{234}U	^{235}U	^{238}U
Specific activity (Bq/mg)		231,300	80.011	12.445
wt%	Nat. U	0.0053	0.711	99.284
	LEU	0.193	19.75	80.057
Total activity (Bq/g)	Nat. U	25,280		
	LEU	471,926 (almost 18 times that of Nat. U)		

in fabrication processes are conditioning of as-received ThO_2 powder and mixing and milling of the UO_2 and ThO_2 powders in an attritor to obtain desired homogeneity. Figure 16a, b shows morphology of as-received ThO_2 and UO_2 powder, respectively [13]. The morphology of ThO_2 powder is platelet type whereas that of UO_2 is regular. The platelet morphology is not suitable for good compactibility. The platelet morphology of ThO_2 powder is broken by milling process (Fig. 16c).

Sintering Behaviour of (Th-LEU) MOX Fuel: Thoria has higher melting point and thus requires higher sintering temperature and the use of dopants like MgO or Nb_2O_5 to obtain the high pellet density (Table 5).

Recycling and Recovery of Process Rejects: Recycling of process rejects is difficult due to stable nature of thoria. Dry method can be used for recycling limited quantities (~10%). Oxidative–reductive processing of the sintered rejects is carried out followed by milling and mixing with fresh powders (Table 6).

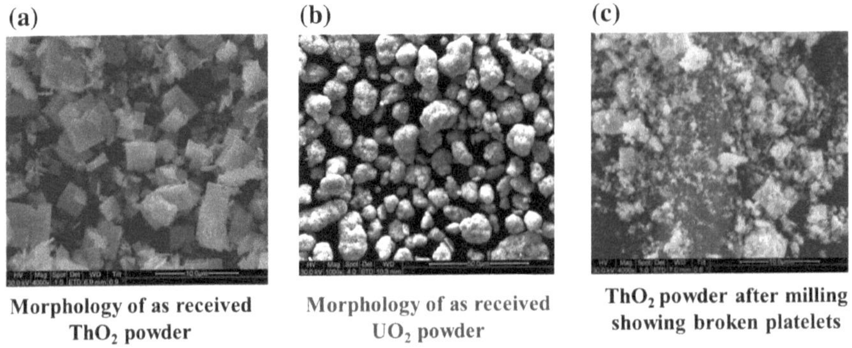

Fig. 16 Morphology of ThO_2 and UO_2 powder

Table 5 Sintering behaviour of (Th-LEU) MOX fuel [13]

Composition	Dopant	Sintering temperature (°C)	Sintered density (% T.D)
ThO_2-20% UO_2	–	1600	80
	0.05% MgO	1700	92
	0.05% MgO + 0.05% Nb_2O_5	1700	94

Table 6 Recycling and recovery of process rejects of (Th-LEU) MOX fuel [13]

Composition	% fraction of rejected material added with fresh powder	Sintered density, (% T.D)
ThO_2 + 13 w% UO_2 (with 0.5 w% Nb_2O_5)	10%	95%
	>20%	Pellet cracking is observed

Wet Recycling Aspects: Wet recycling method will have to be adopted for processing large quantities of process rejects. The following aspects need to be considered during wet recycling:

- Recycling of fabrication rejects requires dissolution of the chemically stable thoria matrix in higher concentrated HNO_3 at high temperatures and addition of fluoride. Studies carried out at BARC have shown that with 8M nitric acid containing 0.010–0.015M fluoride catalyst under boiling conditions dissolves the pellets in 48–72 h [14]. Recovery of thorium by oxalate precipitation and uranium by conventional ADU process is one of the methods. During trials with recovered ThO_2 and UO_2 powders at BARC, pellets of acceptable quality have been fabricated.
- Laboratory scale studies indicate the feasibility of adopting this process. Some of the aspects which have to be taken into consideration are
  - Long dissolution time

- Criticality aspects
- Corrosion of dissolution tank material.

6.3 Fuel Fabrication Aspects of (Th–Pu) MOX

Conventional powder pellet process inside glove boxes similar to that used for (U-Pu) MOX can be adopted for (Th–Pu) MOX fuel fabrication [13] as shown in Fig. 17.

The issues identified for fabrication of (Th-LEU) fuel are almost the same for that of (Th–Pu)-based fuel. In addition, fuel fabrication using glove box facilities restricts the size of manufacturing equipment, which results in reduction of throughput. This restriction is a greater concern as the (Th–Pu) MOX fuel tonnage requirements for water-cooled reactors are considerably high. For the (U-Pu)-based fast reactors, where the tonnages of fuel requirements are lower, this does not affect the overall economics significantly.

6.4 Fuel Fabrication Aspects of (Th–^{233}U) MOX

The major issues in fabrication of (Th–^{233}U) MOX are the radiological aspect associated with recycling of ^{233}U [13]. The concentration of ^{232}U in the uranium obtained from reprocessing of thorium fuels is in the order of a few hundred ppm (typically about 500 ppm from PHWR) [13]. ^{228}Th in recycled uranium is in the order of ppm level compared to ppb level in natural thorium. This requires a fully shielded facility (Table 7) for handling hard gamma being emitted from Th-228.

Steps followed for the fabrication of thoria-based MOX fuel is given below [13]:

(a) Sol-gel Microsphere Pelletization:

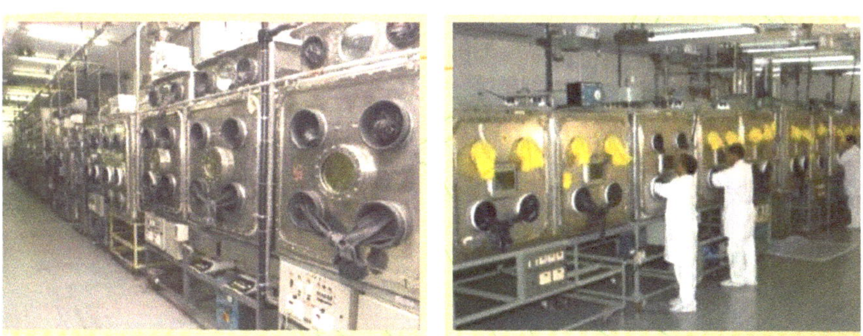

Fig. 17 Picture of glove box for handling (Th–Pu) MOX fuel

Table 7 Shielding requirement for handling of (Th-^{233}U) MOX fuel [13]

Fuel type	Technique	Type of facility
(Th-^{233}U) MOX	Powder pellet	Fully shielded facility
	Sol-gel microsphere pelletization	Fully shielded facility
	Pellet impregnation	Partially shielded facility
	Coated agglomerate pelletization	Partially shielded facility
(Th-LEU) MOX	Powder pellet	Conventional UO_2
(Th–Pu) MOX	Powder pellet	Glove box

Fig. 18 Flow sheet for fabrication of sol-gel microsphere pelletization

Figure 18 shows the flow sheet developed for fabrication of thoria-based MOX fuel. Nitrate solution of Thoria and UO_2 is mixed along with NH_4NO_3, and gel microspheres are prepared. Homogeneous microspheres thus prepared are washed and calcined. The microspheres are subjected to compaction and sintered to achieve high density.

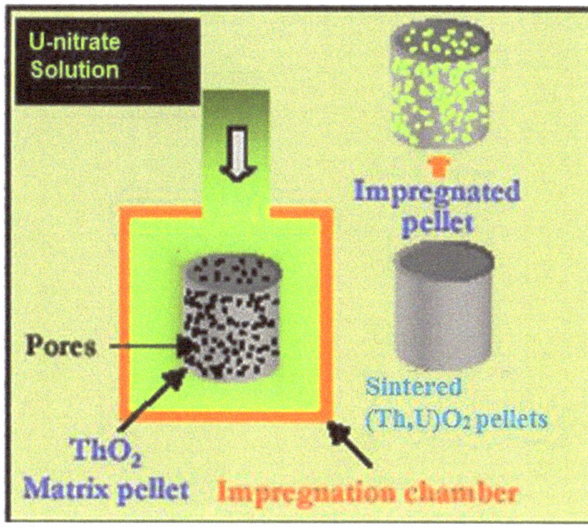

Fig. 19 Pellet impregnation technique showing impregnation of uranium nitrate solution in low-density pellet

(b) Pellet Impregnation Technique:

Low-density pellets (70% TD) are fabricated and subjected to impregnation followed by drying and sintering. A schematic of this process is shown in Fig. 19.

Following steps are followed for this process

- Fabrication of partially sintered low-density ThO_2 pellets in unshielded facility by powder—pellet route
- Impregnation with uranyl nitrate solution in shielded facility followed by drying and sintering.

(c) Microsphere Impregnation Technique:

Porous microspheres prepared by sol-gel technique are subjected to impregnation by uranyl nitrate solution. Thoria microspheres are prepared in an unshielded facility and subjected to impregnation by uranyl nitrate solution in a shielded facility (Fig. 20). The microspheres are then calcined and compacted followed by sintering at high temperature.

(d) Coated Agglomeration Pelletization (CAP) Process:

The fabrication flow sheet of (Th, U)O_2 pellets by CAP process is shown in Fig. 21. ThO_2 powder is mixed with water in a mixer and then subjected to extruder and spherodiser to obtain spheroids of regular morphology. These steps of the process are carried out in a non-shielded facility. The spheroids thus obtained are coated with uranium in a shielded facility followed by compaction and sintering to obtain high-density pellets. Figure 22 shows thoria and thoria-based products obtained from different steps of CAP process.

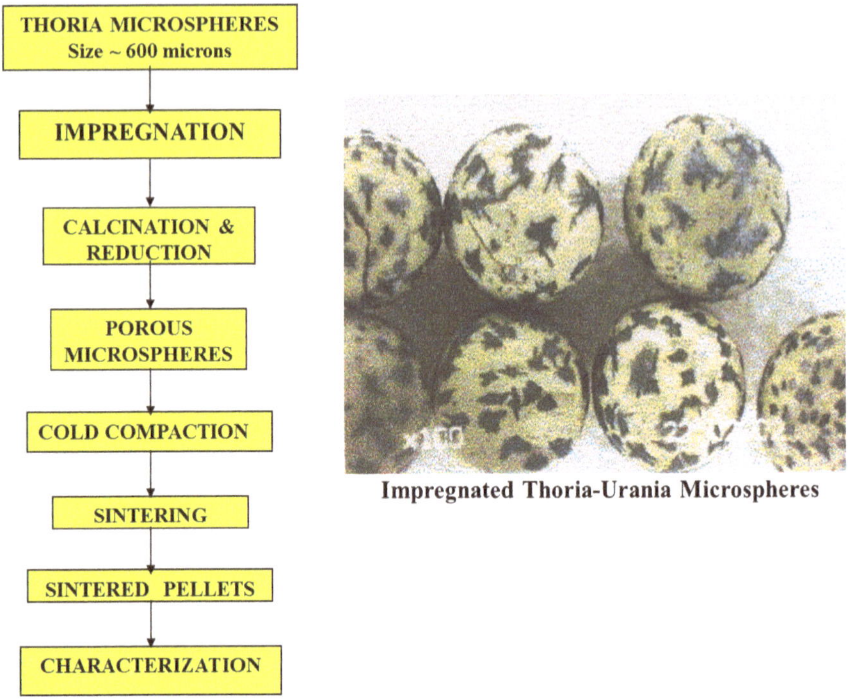

Fig. 20 Flow sheet of microsphere impregnation technique

A summary of different processes discussed above is given in Fig. 23. The steps which require shielding have been shown in different colours.

7 Measurements of Thorium Properties

The thermo-physical properties of thoria-based fuel are very important for burning the thoria-based fuel in nuclear reactor. These properties vary not only with temperature but also with burnup. Several experiments have been carried out in BARC [15] to determine important properties of thoria-based fuel which include melting point, density, thermal expansion and thermal conductivity. In addition, modelling and experiments have been carried out for thermal expansion of fuel required for fuel design for finding the extent of thermal stresses [16].

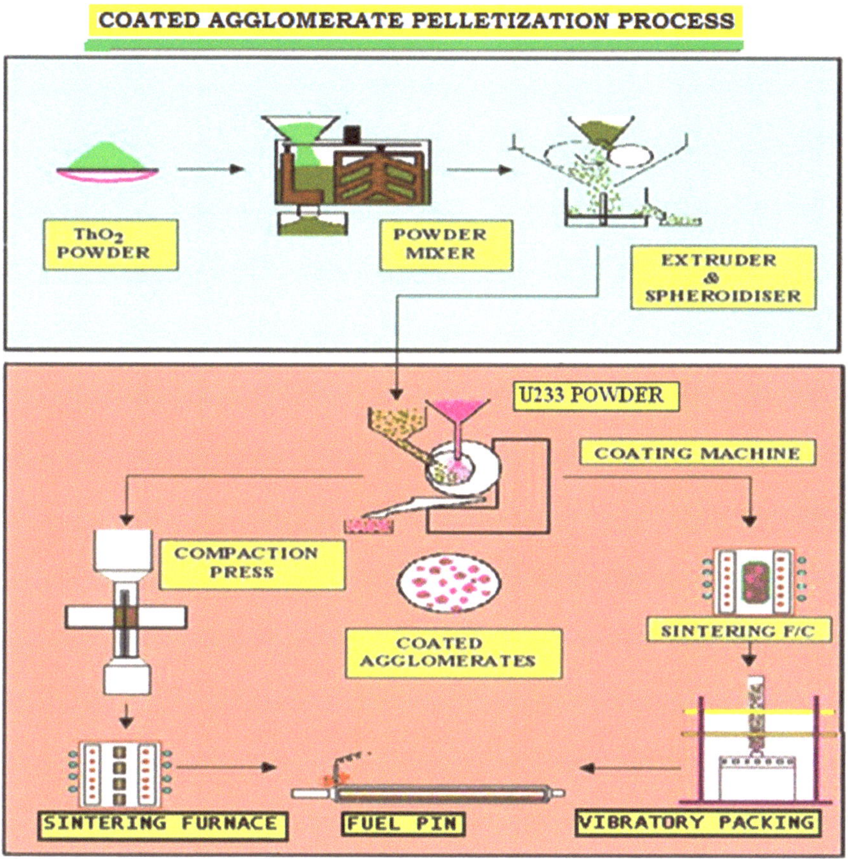

Fig. 21 Flow sheet of coated agglomerate pelletization process

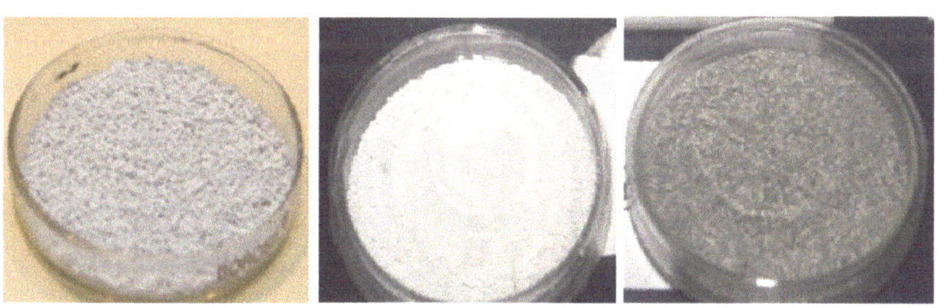

(a) ThO_2 extrude, (b) ThO_2spheroids (c) ThO_2 agglomerates coated with U_3O_8.

Fig. 22 Thoria and thoria-based products obtained from different steps of CAP process

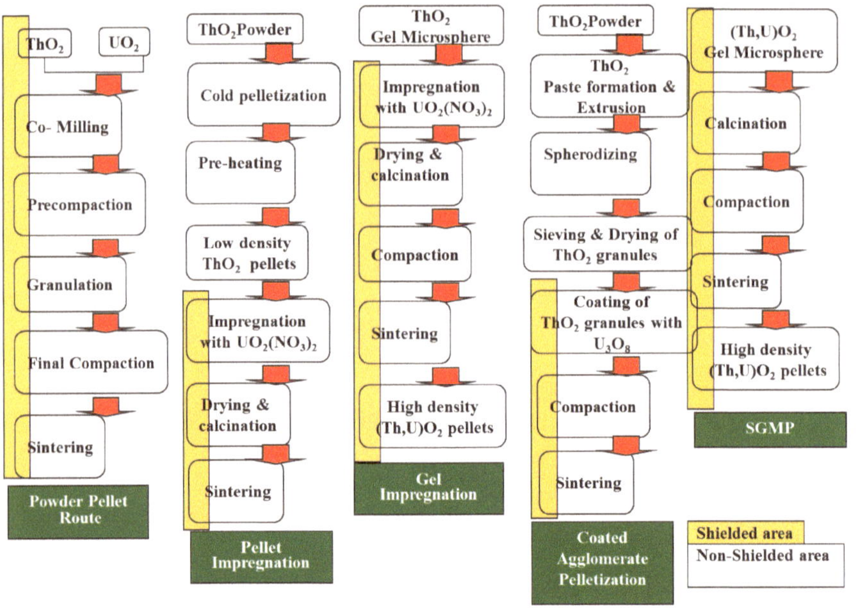

Fig. 23 Flow sheets of different processes

8 Thorium Irradiation Experience in Indian Nuclear Reactors

Thorium as the fuel has been used in nuclear reactors since the 1960s worldwide (Table 8). Flux flattening in PHWRs was carried out by using thoria bundles. U-233 fuel has been used in the research reactor KAMINI in India. The Shippingport is the first reactor which demonstrated breeding of 233-U with a conversion/breeding ratio ~1.03%.

8.1 Irradiation Experience of Thorium in India

India has a steady programme for utilization of thorium as part of the requirement for its three-stage programme. To this end, research activities have been initiated in every aspect of the thorium fuel technology.

- CIRUS: Thoria rods (J-rods) were irradiated in CIRUS research reactor in the reflector region to produce U-233. The J-rods were having about ~2–3 ppm of U-232 [17].

Table 8 Thorium-based reactors worldwide [18]

Reactor	Type	Power	Fuel	Operation period
Pebble bed AVR, Julich, Germany	HTGR	15 MWe	Th + HEU	1967–1988
Dragon, Winfrith, UK	HTGR	20 MWth	Th + HEU	1964–1973
Peach Bottom, USA	HTGR	110 MWth	Th + HEU	1967–1974
KAMINI, India	MTR thermal	30 kWth	^{233}U-Al alloy	1996–till date
SUSPOP/KSTR, the Netherlands	Aqueous homogenous suspension	1 MWth	Th + HEU	1974–1977
MSRE ORNL, USA	Molten fluorides	7.5 MWth		1964–1969
THTR, Germany	HTGR	300 Mwe	Th + HEU (pebble type)	1983–1989
Fort St. Vrain, USA	HTGR	330 Mwe	Th + HEU (prism type)	1976–1989
Shippingport, USA	PWR	100 Mwe	Th + ^{235}U/Th + Pu	1977–1982

- PURNIMA-II (1984–1986): Handling experience of U-233 was obtained in PURNIMA-II facility, wherein experiments were performed with uranyl nitrate solution containing U-233 reflected by BeO blocks [19].
- PURNIMA-III (1990–93): Important physics parameters for design of KAMINI reactor with U-233 as fuel were generated in PURNIMA-III facility by conducting several experiments with ^{233}U-Al dispersion fuel in the form of plates [19]. In this facility, BeO canned with Zircaloy was used as reflector and void and temperature coefficient of reactivities were measured.
- KAMINI: This reactor was commissioned in IGCAR at Kalpakkam in 1996 with U-233 as fuel in the form of plates of U-Al alloy [19]. It is a small reactor with 30 kWth power and has very high flux to power ratio. The reactor serves for neutron radiography of fuel of fast breeder test reactor (FBTR).
- Thorium for flux flattening in PHWR: In the fresh core of 220 MWe Indian PHWRs with all natural uranium fuel bundles, there was a limit on the maximum power that could be obtained from this reactor due to flux peaking with the initial core. As a result, there was a delay of ~100 FPD of operation to reach full power. This problem was overcome by flattening the flux using thoria bundles without any loss of burnup in UO_2 fuel and without compromising the worth of shutdown systems [20]. Figure 24 shows the loading pattern of thoria fuel bundles in the reactor core. These were accomplished first in 1991 at KAPS-1, subsequently in KAPS-2, KGS-1 and KGS-2, and RAPP-3 and RAPP-4.

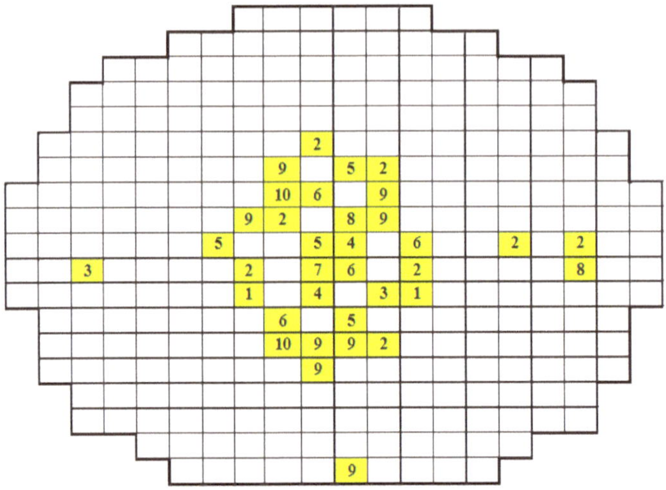

Fig. 24 Initial fuel loading with 35 thoria bundles in PHWR [8]

8.2 U-233 Production in Fast Breeder Test Reactor

For the production of U-233, thoria subassemblies (SAs) were loaded beyond the active core beginning from the sixth ring to twelfth ring in FBTR [21]. In Table 9, the annual production rate of U-233 is given. The first row in the table indicates the amount of U-233 generated in each of the rings, starting from the sixth ring and ending in the twelfth ring. The subsequent rows correspond to the U-233 production when blanket SAs are loaded from the seventh, eighth and ninth to twelfth ring, respectively. In the inner rings, though the total number of subassemblies is low, the production is higher as the neutron flux is higher. The presence of thoria blanket subassemblies in one ring shields the assemblies in the subsequent rings. In shifting the beginning of the blanket region from the sixth ring to the ninth ring, the production rate is nearly halved. Moreover, an advantage of the inner ring of blanket subassemblies is that the concentration of U-233 is higher; that is, less number of subassemblies has to be reprocessed to acquire the same amount of U-233.

Table 9 Production of U-233 (kg) in various rings in 1 year [21]

6	7	8	9	10	11	12	Total
1.9787	1.71	1.4749	1.2375	1.0151	0.8274	0.7172	8.9608
	1.9306	1.5609	1.2963	1.0618	0.8673	0.7527	7.4696
		1.7583	1.3702	1.117	0.9135	0.7957	5.9547
			1.5135	1.1701	0.954	0.8312	4.4688

Table 10 Production of U-232 (ppm) in various rings in 1 year [21]

6	7	8	9	10	11	12
55.94	29.32	15.63	8.61	4.88	2.86	1.69
	26.43	13.84	7.37	4.07	2.36	1.35
		12.11	6.56	3.49	1.96	1.10
			5.71	3.19	1.73	0.98

In Table 10, respective U-232 content (in ppm) in U-233 for the results presented in Table 9 is given. The U-232 content is lower in the outer rings. The faster spectrum of FBTR results in lesser U-232 production.

The U-233 yield (in gram) per subassembly per full-power year (FPY) of operation is shown in Fig. 25 for various blanket configurations [21]. The blue bars indicate the yield in different rings if blanket subassemblies (BSAs) are loaded from the sixth ring to the twelfth ring. Similarly, a red bar indicates the yield if thoria BSAs are loaded from the seventh ring. Yellow- and violet-coloured bars correspond to loading BSAs from the eighth and ninth rings, respectively. The production rate varies from 60 g/FPY for the sixth ring subassemblies to 15 g/FPY for the twelfth ring subassemblies. In terms of the total mass, FBTR can generate 9.8 kg/FPY of U-233 if blanket region extends from the sixth to twelfth rings. It is worth mentioning that the peak U-232 content is limited to 240 ppm in uranium in the sixth ring blanket SAs and 6 ppm in the twelfth ring BSAs after irradiated for five FPYs. Depending on the configuration, the U-232 content averaged over the blanket region is calculated to be as low as 20 ppm.

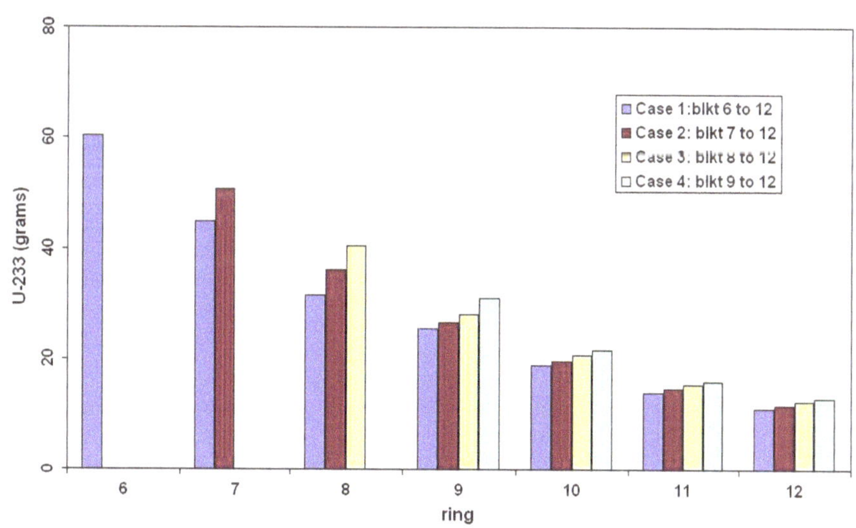

Fig. 25 U-233 production per full-power year in different cases [21]

Table 11 Net U233 production [21]

Blanket region beginning ring	Total number of BSAs	U-233 (kg)	Number of additional fuel SAs required
6	378	46	5
7	342	38.5	3
8	300	30.7	1
9	252	23.2	–

The maximum U-233 that can be generated in FBTR is dependent on how long the BSAs can reside in the core. Residence time of the BSAs is limited by their peak LHR (limited to 137 W/cm) and heat deposition in the nearby stainless steel reflector subassemblies. Based on these criteria, the residence time for the BSAs in any configuration will be in excess of five FPYs. At the end of five FPYs, the U-232 content (in ppm) becomes five times that present at the end of one FPY. In spite of this, it is seen that loading from the eighth ring still leads to U-232 content being limited to 60 ppm. The net U-233 generated in the blanket region of FBTR at the end of five FPYs of operation and the number of additional fuel subassemblies required are given in Table 11.

9 Post Irradiation Examination (PIE)

Thoria-based fuel clusters, AC-6 and BC-8, were fabricated, irradiated and examined in Trombay [22].

9.1 *PIE of AC-6 Cluster*

The AC-6 cluster contains six pins of ThO_2 + 4% PuO_2 irradiated up to 18,500 MWd/Te HM burnup for 2 years. Details of PIE included both non-destructive tests and destructive tests as given in Table 12. Details of the irradiation conditions are given in Table 13. Calibration was carried out by Cs-137 and Co-60 sources. The spectra obtained showed the presence of Cs-137, Cs-134, Eu-154, Tl-208, etc. [22], with higher counts of Cs-137 near the top and bottom portion of the elements, which could be due to migration of some caesium to these regions since they are relatively cooler.

Table 12 Details of tests conducted for PIE [22]

Non-destructive testing	Destructive testing
• Visual examination • Diameter measurement • Leak testing (liquid nitrogen and alcohol leak test) • Ultrasonic testing • Eddy current testing • Gamma scanning	• Fission gas analysis (released fission gases) • Metallographic examination – Optical metallography – SEM studies – α and β-γ autoradiography

Table 13 Details of test conditions for PIE [22]

Thermal neutron flux	Temperature of the coolant	Pressure of the coolant	Peak burnup	Peak rating
5×10^{13} n/cm^2/s	240 °C	105 kg/cm^2	18,500 (MWd/Te)	40 (kW/m)

9.2 PIE of BC-8 Cluster [22]

In BC-8 cluster, two tiers (6 + 6 mixed pins) were irradiated successfully to ~10,000 MWd/Te HM burnup. The Cs137 counts of these fuel pins were found to increase towards the reactor centre and follow the flux profile in the reactor as shown in Fig. 26. Average diameter of the fuel pins from BC-8 cluster, before and after irradiation, is shown in Fig. 27 and is found to be changed. The fission gas release during irradiation is shown in Table 14 and found to be very small [22].

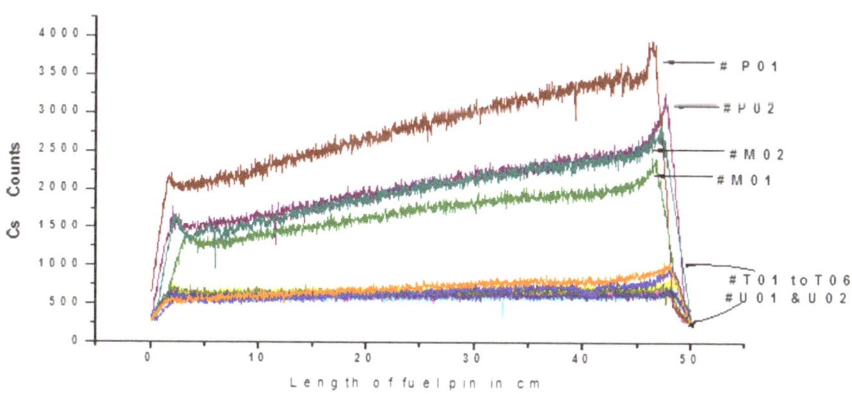

Fig. 26 Cs counts along the length of pin in BC-8 cluster [22]

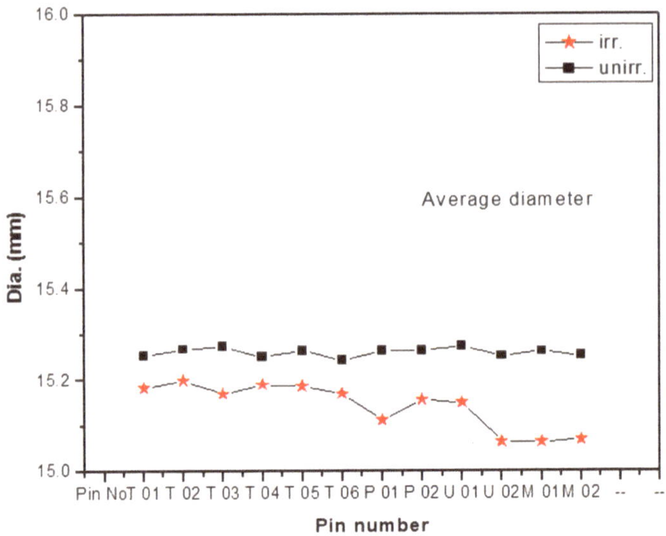

Fig. 27 Measured diameter of fuel pins before and after irradiation [22]

Table 14 Fission gas release during irradiation

Pin nos.	Fuel composition	Burnup (GWd/t)	Fission gas generated (cc)	Internal pressure (atm)	Volume of fission gases at STP (cc)	% FGR
M-01	$(U\text{-}3\%Pu)O_2$	6.9	150	1.33	0.07	0.05
M-02	$(U\text{-}3\%Pu)O_2$	8.0	182	1.59	0.03	0.02
P-01	$(Th\text{-}6.75\%Pu)O_2$	10.8	217	2.87	0.75	0.35
P-02	$(Th\text{-}6.75\%Pu)O_2$	8.1	163	1.52	0.87	0.54
U-01	UO_2	2.5	58	1.99	0.02	0.03
U-02	UO_2	2.9	67	1.40	0.02	0.03

10 Utilization of Thorium in Indian NPPs

10.1 Utilization of Thorium in PHWRs

As seen before, India has a rich experience of burning thorium in PHWRs. In addition, a lot of research were conducted for studying the feasibility of introducing thorium in Indian PHWRs with different options such as: at core level wherein pure thoria bundles can be placed at desired locations in the core, at channel level wherein some channels can be fuelled with enriched U or Pu while thorium is put in separate channels (seed and blanket type), at bundle level wherein fissile-enriched uranium

Table 15 Reduction in uranium requirement in PHWR deploying thorium [23]

	OTC—all natural uranium	OTC-3% enriched uranium + Th	Recycled U + Th + external fissile feed 4 gm/kg
Burnup (MWd/T)	7200	60,000	25,000
Annual natural uranium requirement (tons) per Gwe	170	120	39
Annual fissile material in discharge fuel (kg) per Gwe	600 kg plutonium	210 kgU^{233}	

or plutonium can be introduced in some fuel pins and fertile thorium in other pins, and at pin level wherein thorium is mixed in all the fuel pins.

Option 1: Th with Enriched U^{235} in PHWR
Studies [23] revealed that use of 30 wt% of U^{235} (10% enriched) mixed with Th in all fuel pins results in a burnup of ~60,000 MWd/T with an annual saving of nearly 50 tons (30%) of natural uranium per GWe compared to normal all natural uranium. This results in annual consumption of ~600 kg of U^{235}, and production of ~200 kg of U^{233} for a 1 GWe capacity plant.

Option 2: (Th + U^{233}) MOX in Inner Three Rings and (Th + Pu) MOX in Outer Most Ring in PHWR Fuel Bundle
In this case, 3% U^{233} with thorium is put in the inner three rings to reduce U^{233} requirement and 1.5 wt% Pu with thorium is put in the outer most ring [23]. A burnup of nearly 20,000 MWd/T is achieved with U^{233} in self-sustaining mode [23]. Plutonium is completely burnt out (plutonium burner) which needs to be supplied externally.

Option 3: Targeting Larger Saving of Natural Uranium
One GWe yr requires ~170 tons (Table 15) of natural uranium in PHWRs. To have a larger saving of uranium, in the first cycle, thorium can be burnt with 2.4% U^{235} and subsequent cycles with reprocessed uranium + Th + external fissile feed.

10.2 *Utilization of Thorium in Advanced Reactors*

Apart from PHWRs, India has launched a dedicated R&D programme for utilization of thorium for large-scale power production in advanced nuclear reactors such as AHWR, HTRs and MSRs.

AHWR
AHWR is an advanced reactor with 300 MWe capacity, employing the established pressure tube technology of heavy water reactors and several passive safety systems

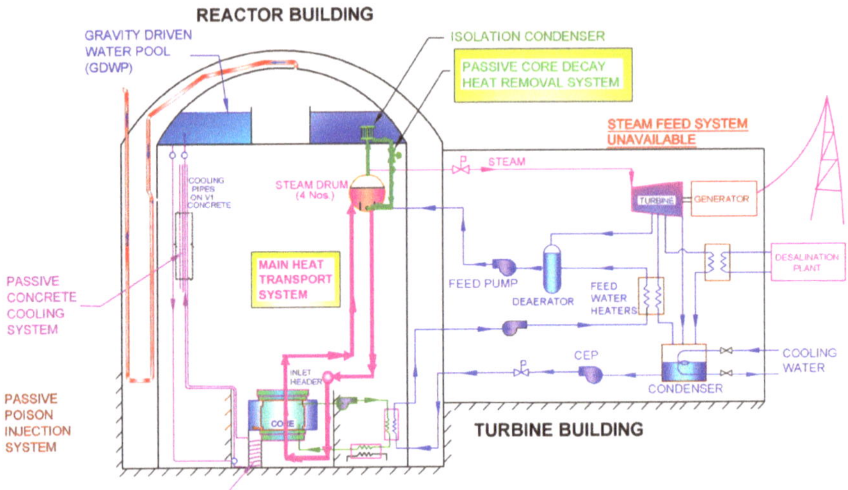

Fig. 28 General arrangement of AHWR

so as to have almost no impact in public domain due to any accident. Unlike the PHWRs, this reactor uses boiling light water as coolant in vertical channels. Heavy water is used as moderator similar to that of PHWRs. Another major objective of this reactor is to produce a significantly large fraction of power from thorium using a combination of $(U^{233}\text{-Th})O_2$ and $(Pu\text{–}Th)O_2$ fuel, while U-233 remains inside the core in a self-sustained mode, i.e. without any external feed in equilibrium cycle. Figure 28 shows a conceptual design of the reactor systems.

The fuel cluster of AHWR consists of 54 fuel pins with 12 pins in the inner ring, 18 pins in the intermediate ring and 24 pins in the outer ring. The fuel pins in the outer ring have $(Th\text{–}Pu)O_2$ fuel, and the fuel pins in the inner and intermediate rings have $(Th\text{-}^{233}U)O_2$ fuel [24] as shown in Fig. 29.

The fission heat from the core is removed by thermosyphon not only during normal operation of the reactor but also during all transients and accidents; this enhances the passive safety of the reactor. In addition, AHWR design employs several other passive systems such as isolation condenser for removal of decay heat during SBO conditions (Fig. 30), passive injection of ECCS water during LOCA, cooling of moderator and endshield by passive means in case of SBO, passive cooling of containment. To mitigate an extended SBO conditions, a large pool of water (8000 m^3 capacity) is located inside the primary containment, which can remove decay heat by passive means for more than 100 days.

The AHWR design has the flexibility to burn not only Pu-based thorium oxide fuel (AHWR reference core) but also LEU-based thorium oxide fuel [8]. Both types of fuels have favourable neutronic properties as given in Table 16. The reactor is designed to produce 60% power from thorium at a burnup of 40 GWd/Te in AHWR with closed fuel cycle for reference core, and the power fraction from $^{233}U/Th$ in the AHWR-LEU at discharge burnup of 60 GWd/Te is about 38% [8].

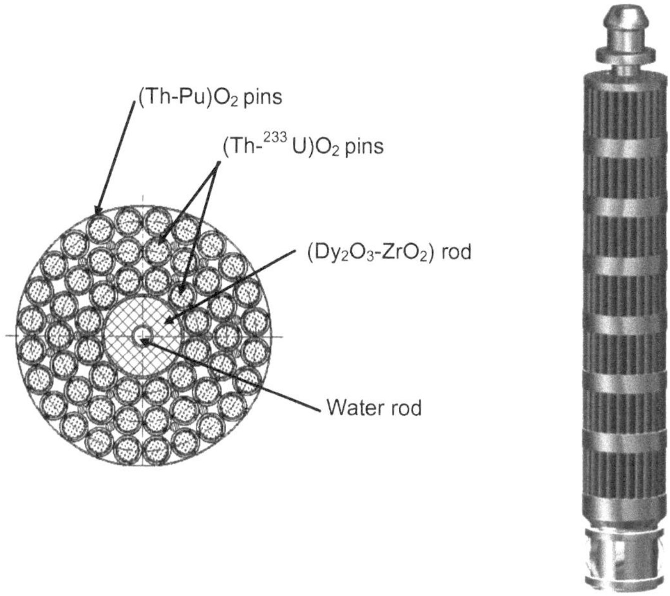

Fig. 29 AHWR fuel cluster assembly

For AHWR-LEU, the annual requirement of thorium is about 4.19 Tons and that of uranium is about 1.17 tons which contains 230 kg of fissile U^{235}. Annual discharge fuel contains 63 kg of U^{233}, 16 kg of plutonium and 22 kg of U^{235}. The presence of U^{232} makes it proliferation resistance, and the presence of U^{238} makes U^{233} denatured [25].

AHWR-LEU also provides a better utilization of natural uranium (Fig. 31) as a result of a significant fraction of the energy extracted by fission of U-233 converted in situ from the thorium fertile host as compared to PHWRs and modern LWRs. With high burnup possible today, LEU–thorium fuel can lead to better/comparable utilization of mined uranium [9].

In addition to the above-mentioned passive safety systems, Post Fukushima, mitigatory systems against core meltdown accidents have been incorporated in AHWR design which include containment filtered vent system for ensuring containment integrity and no activity release into the environment, core catcher for cooling corium and PARs for hydrogen mitigation

Indian High Temperature Reactor Programme

India is having a well-defined high-temperature reactor programme. Three reactors, viz. CHTR, IHTR and IMSBR, are being designed under the programme.

The Indian CHTR is a 100 kWth reactor utilizing thorium fuel with U-233 as fissile feed. The coolant is lead-bismuth eutectic alloy, and moderator is beryllium oxide [18]. The life of the core is 15 effective full-power years. In addition to burning thorium, CHTR design has several passive safety features. The coolant outlet

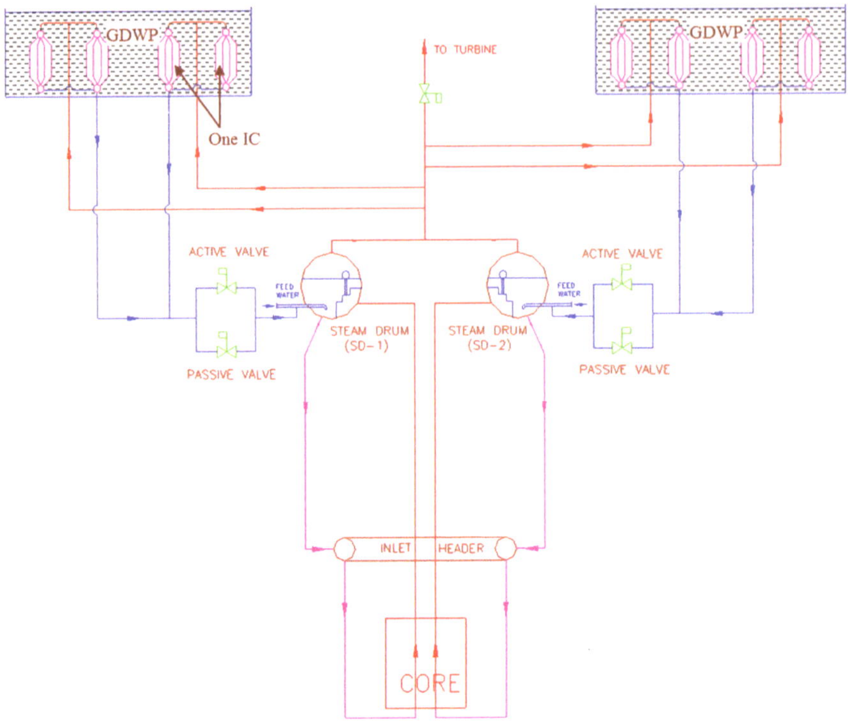

Fig. 30 Schematic of passive decay heat removal system of AHWR

temperature from the core is 1000 °C which enables not only hydrogen production from water but also other process applications. Table 17 provides important design parameters of CHTR. Schematic of the reactor is given in Fig. 32.

The Indian High Temperature Reactor is a 600 MWth reactor with 1000 °C temperatures. It is a pebble bed reactor, designed to operate with TRISO-coated fuel particles and molten salt coolant as coolant. The objective is to produce hydrogen, electricity and desalination water [18]. Natural circulation of coolant for reactor heat removal under normal operation is the mode of heat removal. Figure 33 shows the details of the reactor systems.

The Indian MSBR is designed with the following objectives [18]: apart from using thorium in a self-sustained ^{233}U-Th equilibrium cycle, it has several inherent safety features. To meet India's large energy demand, it is envisaged to deploy several such reactors in the third stage of India's nuclear power programme. In addition to reduction in use of fissile materials in the reactor, it is envisaged to have minimum waste production and avoid graphite and beryllium and beryllium-based salts.

Two design concepts are being considered for possible core configuration of IMSBR [26], i.e. loop type in which a smaller core with fuel and coolant salts as

Table 16 Neutronic characteristics of AHWR design [8]

Parameters	AHWR reference	AHWR -LEU
Fuel type/no. of pins in		
Ring 1 (10 fuel pins)	(Th, Pu) MOX/6% Pu	(Th, LEU) MOX/30% LEU
Ring 1 (2 fuel pins)	(Th, Pu) MOX/6% Pu and 2% Gd	(Th, LEU) MOX/30% LEU and 5% Gd
Ring 2 (18 fuel pins)	(Th, Pu) MOX/3.75% Pu	(Th, LEU) MOX/24% LEU
Ring 3 (24 fuel pins, axial gradation)	(Th, U) MOX Bottom: 3.7% U (80% fissile) Top: 3.3% U (80% fissile)	(Th, LEU) MOX Bottom: 18% LEU Top: 14% LEU
No. of control rods (AR/RR/SR), worth, mk	24 (8 each) 10.8/11.1/10.5	24 (8 each) 12.2/12.8/9.4
RR worth in nominal configuration, mk	6.4	6.8
Shutdown system-1, worth(mk)/no. of SOR Worth when 2 max. worth rods not available, mk	−69.4 (37 SOR) −49.2 (35 SOR)	−78.8 (45 SOR) −56.0 (43 SOR)
Power from thorium	61%	38%
Reactivity coefficients:		
Fuel temperature, $\Delta k/k/°K$	-21.2×10^{-6}	-25.5×10^{-6}
Channel temperature, $\Delta k/k/°K$	$+16.1 \times 10^{-6}$	-29.9×10^{-6}
Void coefficient (0.45 to 0.0 g/cc), $\Delta k/k/\%$ void	-120.0×10^{-6}	-25.0×10^{-6}
Moderator temperature, $\Delta k/k/°K$	$+29.6 \times 10^{-6}$	-17.4×10^{-6}

one salt flowing out of the core in external circuits and coming back. The second is a pool type in which the core is submerged in a pool of molten fuel/coolant salt.

Loop-type design has the following features [26]:

- Fuel salt: LiF (77.6%)-ThF_4 (19.7%)-$^{233}UF_4$ (2.7%)
- Blanket salt: LiF (77.6%)-ThF_4 (22.4%)
- Core dimension (m): 2 × 2
- Power: 1900 MW_{th}/850 MW_e
- Power density: 300 W_{th}/cm^3
- Breeding ratio: 1.09
- Initial fissile inventory: 3.8 T (4.27 T/GW_e)
- Forced circulation.

Table 18 shows physics design details of a loop-type design configuration [8]. The design details of pool-type concept are given below [8]:

- Fuel salt: LiF (77.6%)-ThF_4 (19.7%)-$^{233}UF_4$ (2.7%)

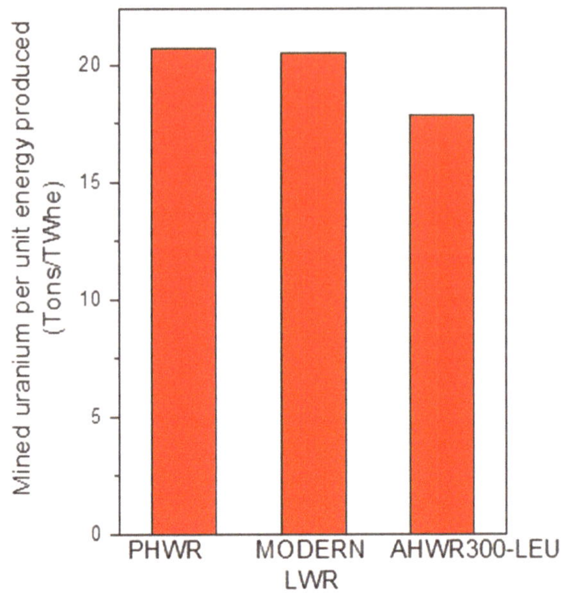

Fig. 31 Comparison of mined uranium consumption for different thermal reactors

- Blanket salt: LiF (77.6%)-ThF_4 (22.4%)
- Core dimension (m): 2 m × 2 m
- Power: 1900 MW_{th}/850 MW_e
- Power density: 300 W_{th}/cm^3
- Breeding ratio: 1.1
- Initial fissile inventory: 5.4 T (6.1 T/GW_e)
- Natural circulation.

Table 19 shows physics design details of a pool-type design configuration [8].

11 Accelerator-Driven Systems (ADS) for Thorium Utilization

In 1995, Prof Carlo Rubia proposed an "Accelerator-Driven subcritical reactor System (ADS)" (Fig. 34) for utilization of thorium for nuclear energy generation [27] which can also be used for incineration of minor actinides and transmutation of radioactive fission products [28]. This has generated a lot of interest in ADS. ADS is a new type of reactor which produces power even though it remains subcritical throughout its life. Subcritical reactors produce fewer neutrons by fission than are lost by absorption and leakage, and require an external supply of neutrons to maintain a constant reactor power. This external neutron supply comes from the interaction of a high-energy proton beam with a heavy nucleus such as lead through spallation

Table 17 Major design and operating characteristics of CHTR [18]

Attributes	Design parameters
Reactor power	100 kWth
Core configuration	Vertical, prismatic block type
Fuel	U-233 and thorium-based TRISO-coated fuel particles shaped into fuel compacts with graphite matrix
Fuel enrichment by 233 U	33.75%
Refuelling interval	15 effective full-power years
Fuel burnup	≈68,000 MWd/t of heavy metal
Moderator	BeO
Reflector	Partly BeO and partly graphite
Coolant	Molten Pb–Bi eutectic alloy (44.5% Pb and 55.5% Bi)
Mode of core heat removal	Natural circulation of coolant
Coolant flow rate through core	6.7 kg/s
Coolant inlet temperature	900 °C
Coolant outlet temperature	1000 °C
Loop height	1.4 m (actual length of the fuel tube)
Core diameter	1.27 m
Core height	1.0 m (height of the fuelled part and axial reflectors)
Primary shutdown system	Mechanical shut-off rods
Secondary shutdown system	Out of core axial movement of movable BeO reflector blocks

Table 18 Physics design details of loop-type design

Composition: $LiF\text{-}ThF_4\text{-}UF_4$	He (% mol) (%)	k_{eff}	ICR_fuel	ICR_blanket	ICR_total
77.6%–19.9%–2.5%	0	0.9975	0.988	0.092	1.08
77.6%–19.9%–2.5%	0.5	1.0006	0.979	0.086	1.065
77.6%–19.7%–2.7%	0	1.0428	0.998	0.093	1.091
77.6%–19.7%–2.7%	0.5	1.0447	0.989	0.089	1.078
77.6%–19.4%–3.0%	0	1.1024	1.011	0.097	1.108

Table 19 Physics design details of pool-type design

Composition: $LiF\text{-}ThF_4\text{-}UF_4$	He (% mol)	k_{eff}	ICR (fuel)	ICR (blanket)	ICR (total)
77.6%–19.9%–2.5%	0	0.9852	0.962	0.150	1.112
77.6%–19.7%–2.7%	0	1.0288	0.972	0.153	1.125
77.6%–19.4%–3.0%	0	1.0872	0.988	0.155	1.143
77.6%–19.9%–2.5%	0.5	0.9872	0.953	0.142	1.095

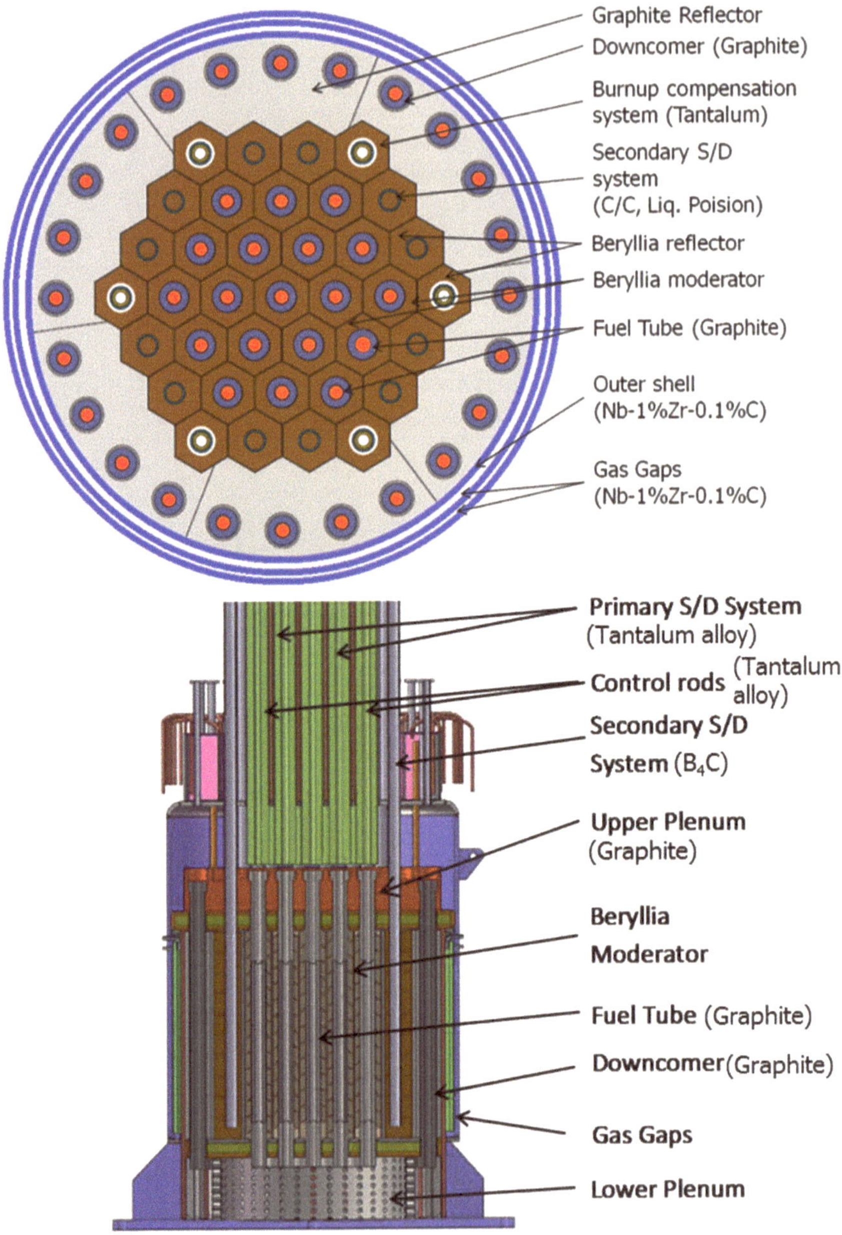

Fig. 32 Schematic of CHTR

Fig. 33 Details of IHTR [18]

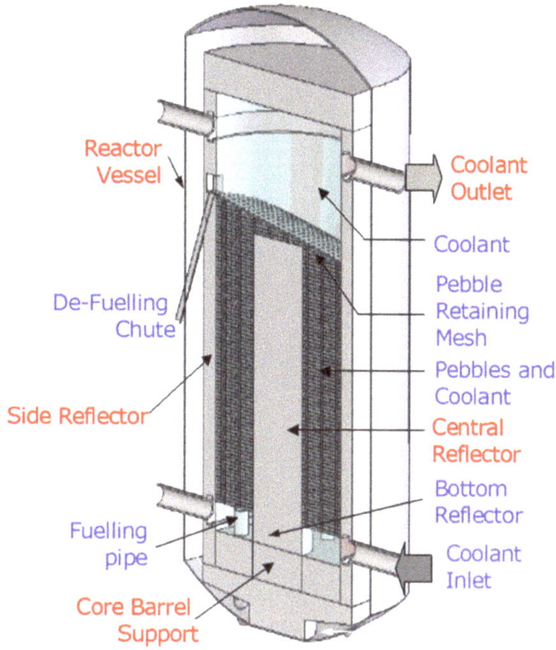

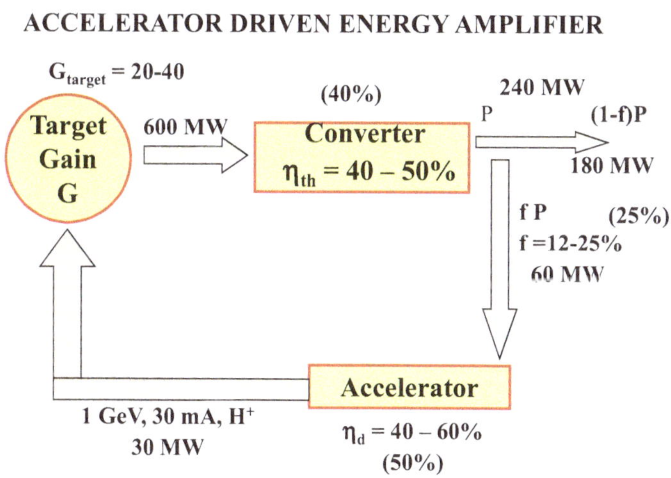

Fig. 34 Schematic of accelerator-driven energy amplifier

reaction. The power level in an ADS is higher for stronger external sources (higher accelerator currents and suitable beam energies) and for reactors which have $k_{eff} \sim 1$.

India has large resources of thorium which forms an important part of its thorium utilization for future nuclear energy generation [29]. In ADS, the accelerator delivers

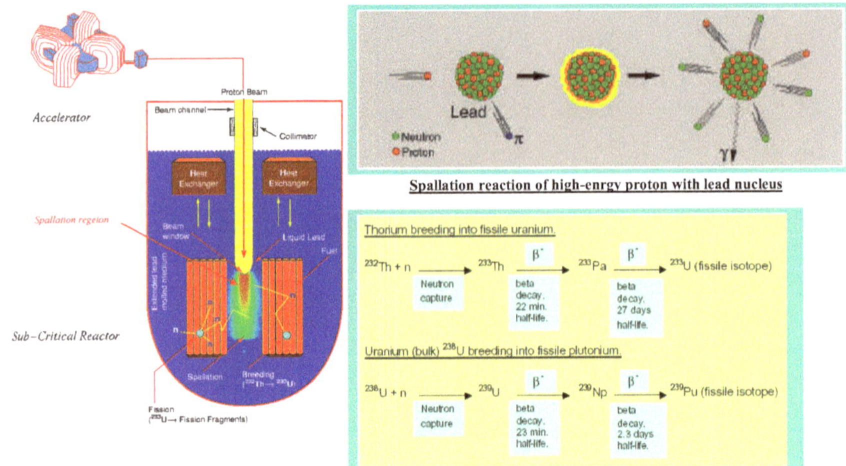

Fig. 35 Schematic of accelerator-driven system

additional neutrons over and above those coming from fission. Moreover, long-term reactivity changes due to burnup are not controlled using parasitic absorber rods. The ADS is, therefore, expected to possess superior breeding characteristics as compared to critical reactors. Being a new type of reactor, the ADS requires the development of several technologies related to high-power accelerators, removal of the intense heat generated by the interaction of the high-power proton beam with the target, subcritical reactor and associated materials development.

An ADS consists mainly of three components, namely a high-intensity proton accelerator, a spallation target and a subcritical reactor. The high-intensity proton accelerator is coupled to the subcritical reactor through spallation target as shown in Fig. 35. The practical realization of ADS requires the development of a high-energy (~1 GeV) and high-current (few 10 s mA) proton accelerator to produce the intense spallation neutron source required to drive the subcritical reactor. In order to provide uninterrupted beam power to the spallation target, over long period of time accelerator needs to be stable, reliable and rugged. The efficiency for conversion of electric power to beam power should be as high as possible (>40%). Accelerators using superconducting structures could meet these requirements very easily. The beam loss in the accelerator must be less than 1 W/m (~1 nA/m) so that hands-on maintenance of the accelerator components can be safely done. In order to avoid undesirable thermal shocks to the fuel elements, the operation of accelerator should operate in nearly CW mode. In high-current proton accelerators, the dominance of space charge effects at lower energies (<10–20 MeV) causes beam loss and also initiates oscillatory particle motions that appear later as a beam halo in the high-energy sections. The beam halo is the dominant particle loss mechanism resulting in severe activation of components in the linac and is closely connected to the emittance mismatch in the low-energy sections of the accelerating structure. For these reasons,

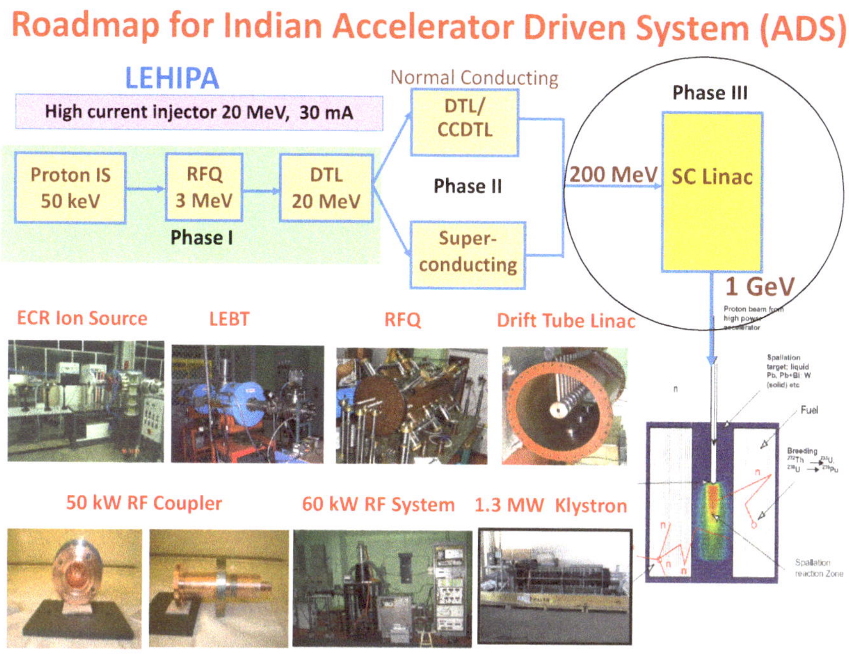

Fig. 36 Road map for accelerator development for Indian ADS

it is very important for beam to smoothly move across segments in the low-energy injector sections. In view of this, it is envisioned that 1 GeV accelerator for ADS be pursued in three phases, namely 20 MeV, 200 MeV and 1 GeV [30, 31]. The most challenging part of this CW proton accelerator is the development of the low-energy injector, typically 10–20 MeV. Therefore, BARC has initiated the development of a low-energy (20 MeV) high-intensity (30 mA) proton accelerator called LEHIPA [32] as the front-end injector of the 1 GeV accelerator for the ADS programme. In LEHIPA, the proton beams extracted at 50 keV from the 2.45 GHz ECR ion source are injected into a 3 MeV radio frequency quadrupole (RFQ) [33] and a 20 MeV drift tube linac (DTL) [34]. The low-energy beam transport (LEBT) and medium-energy beam transport (MEBT) lines match the beam from the ion source to RFQ and from RFQ to DTL, respectively. The operating frequency of LEHIPA is 352.2 MHz which was chosen due to the availability of power sources. The main criterion for the design of the linac is to have minimum beam loss (<1 W/m). The layout of the 1 GeV accelerator for Indian ADS (including 20 MeV LEHIPA) is shown in Fig. 36. The LEHIPA is under commissioning process which will be a unique facility for beam studies at low energies.

In Indian ADS programme, lead-bismuth eutectic (LBE) is being used for spallation target for which both window and windowless options have been considered. Experimental studies related to the development of the spallation target and materials are also pursued at BARC [35]. Computer codes have been developed for thermal hydraulics for LBE target simulations. Experimental loops for validation of thermal hydraulics codes and corrosion studies on window materials have been built and will be operational very soon. Initially, studies are being done using 30 MeV proton beams from a high-current cyclotron at Kolkata.

It is proposed to use one-way coupled reactor system in Indian ADS programme, which allows considerable reduction in the beam current [36]. In this system, two subcritical cores are used. The inner core is subcritical fast reactor with thermal neutron absorber liner surrounded by a gap, and outer core is subcritical thermal reactor. The neutrons leaking from inner core can go to thermal core and get multiplied. However, neutron from thermal reactor cannot go to inner core due to absorber liner that is why it is called one-way coupled system.

12 Reprocessing Experiences in Thorium Fuel Cycle in India

12.1 Experience in Research Reactor-Irradiated Thoria Reprocessing

Irradiation of thorium and thoria in research reactor was started in 1960 in India. Separation of ^{233}U from irradiated thoria was achieved in the year 1970 at pilot plant scale employing THOREX process [37]. Solvent extraction-based process flow sheets were developed and deployed for the selective recovery of ^{233}U. Since recovery of ^{233}U was important during initial period, lower percentage of TBP (5% in odourless kerosene) was adopted. The pilot plant at BARC essentially comprised of, a dissolver cell for rod charging, decladding and dissolution operation and a series of glove boxes for solvent extraction, product concentration, precipitation, ion-exchange purification, etc. Since aluminium was the clad material, chemical decladding using alkali was adopted. The thoria was then dissolved in nitric acid containing sodium fluoride and aluminium nitrate solution. The complete dissolution could be achieved in about 12 h, and the resultant solution had Th concentration of ~200 g/L with an acidity of ~4M having ~0.12% ^{233}U. As the irradiation was carried out in research reactor having low flux, activities due to fission products and ^{232}U content were low. Therefore, entire operation comprising of separation, concentration and ion-exchange purification of ^{233}U was carried out sequentially arranged inside glove boxes.

The uranium product was then concentrated and purified by anion-exchange method. For carrying out separation by anion exchange, it was necessary to have uranium in chloride medium free from nitrate ions. Since concentrated uranium solution contained a large amount of nitrate, uranium and thorium in the solution were first precipitated with ammonia, and the precipitate washed free of nitrate ions and then dissolved in concentrated HCl. The solution was conditioned to ~8M HCl and loaded on a preconditioned anion-exchange column. The uranium-loaded column was washed using 8M hydrochloric acid for removing thorium. The ^{233}U was then eluted in 0.5M HNO_3. The eluted product contained pure ^{233}U with a few ppm of Th contamination. The product solution of ^{233}U was then precipitated with ammonia solution and filtered and washed to remove ammonium salts. Ammonium diuranate thus obtained was transferred to silica muffle and ignited at 850 °C to get U_3O_8. This was followed by setting up of another pilot facility at Kalpakkam to validate the flow sheet employed in the first facility. These facilities had provided good experience to develop THOREX process to a more robust process. ^{233}U recovered from the initial campaigns was used for fuelling KAMINI reactor.

The recovery of ^{233}U from irradiated thoria was followed by several developmental studies. The major areas covered were: (a) removal of bulk thorium from ^{233}U product, (b) final purification of the ^{233}U, (c) testing of the equipment, (d) thoria-based fuel dissolution studies, (e) material development for thorium fuel cycle and (f) waste management issues.

Summary on some of these areas is described below.

12.1.1 Studies to Improve the Decontamination of ^{233}U Product

During the operation of initial facilities, ^{233}U product often contained significant quantity of thorium requiring additional steps for purification. Solvent extraction data under varying experimental conditions were generated [38]. The feed used in the studies covered Th concentration range from 50 to 250 g/L, U from 0.50 to 15 g/L and acidity from 1.00 to 4.00M HNO_3. These data were used for evaluation of a number of theoretical stages for the given flow sheet conditions. Based on the results, two separation schemes were proposed. Each scheme was tested using air-pulsed mixer-settlers containing 20 stages. The feed acidity was maintained at 4.00M during extraction, and 2.00M HNO_3 was used as scrub. During experiments, 12–14 stages were employed for uranium extraction and 6–8 stages for scrubbing the thorium. The feed used in these experiments contained 200 g/L Th with 0.12% U. The flow rates of aqueous feed, organic and scrub for these two schemes were maintained at 1:2:0.6 and 1:1:0.3, respectively. Under such experimental conditions, it was possible to get uranium-loaded organic phase containing only ~5–10% Th. To improve the product purity with respect to Th, more number of scrub stages were required. Counter-current experiments using laboratory scale mixer-settler units were also conducted using lower percentage of TBP (2 and 3% in *n*-dodecane). These experiments, carried out under different feed conditions, provided the possibility of Th removal from uranium-bearing organic phase. Results showed good decontamination factor with respect to thorium [39].

12.1.2 Tail-End Purification of ^{233}U

(i) Ion Exchange in Acetic Acid

In the pilot plant at BARC, the tail-end purification of ^{233}U from Th was carried out by anion-exchange method from chloride medium in lead-shielded glove boxes. This method had certain disadvantages like corrosion of the process equipment, poor decontamination factor with respect to iron and gassing of the resin column, but gave excellent decontamination with respect to Th. As an alternate to chloride, acetate medium was also considered as uranium is known to form anionic complexes in acetate medium [40]. Feed preparation in this method is little cumbersome. While preparing the feed solution in acetate medium, thorium and uranium are first precipitated with NH_4OH. The precipitate is washed with water to make it free from NO_3^- ions. The amount of water taken should be at least 200 mL per gram of thorium present. Acetic acid (~50% solution) is then added slowly till the solution acquires a pH of about 2.5. The free acetic acid concentration at this stage is found in between 2 and 3 M. Uranium forms three predominant species in aqueous acetic acid medium. In 0.50 M acetic acid and above, it exists mainly as $UO_2(OAC)_3^-$. When $Th(OH)_4$ comes in contact with concentrated acetic acid, most of it gets converted to tetra acetate which has low solubility in concentrated acetic acid. The K_d values for uranium in Dowex 1×4 (acetate form) decrease with increase in free acetate concentration. The presence of thorium has no effect on uranium loading. Column is loaded to 30–60% capacity with respect to uranium and washed with 2.00M acetic acid to remove Th. Uranium is finally eluted quantitatively using 0.50M HNO_3. The thorium contamination in the product was found in the range of 50–150 ppm.

(ii) Cation Exchange in Nitric Acid

A method involving cation-exchange resin was also studied [41]. The method involves preferential exchange of Th^{4+} ions over UO^{2+} on the resin column from a feed solution containing thorium and uranium at 0.50M HNO_3. This yields an effluent stream almost free from Th containing >99% ^{233}U. Separation of thorium from uranium is found satisfactory when feed and elution acidities are chosen appropriately. Purity of uranium was found better when the thorium was loaded to only ~60% of the resin capacity. Cation exchanger, Dowex 50W×8, was found more suitable for this purpose. The feed solutions used in these experiments had thorium concentration of ~1.00 g/L at 0.50M HNO_3 with varying concentrations of uranium in the range of 0.30–3.00 g/L. Depending upon the thorium loading on the column, dilute nitric acid (1–2M) was used as eluent. When loading was ~60% of the resin capacity by Th^{4+} ions, 1M HNO_3 was chosen for elution of uranium. However, when loading was less than 30%, the elution was preferred with 1M HNO_3 which required larger volumes, resulting in a product with low concentration. Elution with 2M HNO_3 generated a product with relatively higher in uranium concentration. The ^{233}U product solution containing low nitric acid from the solvent extraction cycle was found suitable as feed for the final purification by ion exchange. It was also observed that processing rates depend on Th content in the feed solution that varies from batch to batch. After preferential washing out uranium, bulk of the Th from the ion-exchange column was eluted using 4.00M HNO_3. Removal of thorium from the column with nitric acid alone was not very simple which was solved by the use of Dowex 50W× 4 having less cross-linking [42].

(iii) Oxalate Precipitation

A laboratory scale precipitation method was developed for the purification of uranium from thorium [43]. The two-step precipitation method involves the precipitation of thorium as oxalate in the first step from a feed solution containing uranium and thorium at 1.0M HNO_3. Under experimental conditions, uranium remains in the supernatant. Th as thorium oxalate precipitate is separated by filtration. In the second step, uranium from supernatant is precipitated as ammonium diuranate using ammonia solution as precipitant. In order to meet varying feed conditions, the method was tested for feed solution containing varying acidity and thorium in the range of 1–6.00M HNO_3 and 1–50 g/L, respectively. Parameters like feed acidities, U and Th concentrations, excess oxalic acid concentrations in the supernatant, precipitation temperatures, precipitate wash volumes were optimized to obtain near quantitative recovery of uranium and thorium with less than 50 ppm uranium losses to ammonium diuranate filtrate. Results indicate that the optimum concentration of nitric acid in the feed should be about 1.00M, and concentration of uranium should not exceed 10 g/L. However, acidities up to 5M HNO_3 could be tolerated in the feed. The free oxalic acid concentration should be restricted to 0.1M. The distribution patterns of different fission products and stainless steel corrosion products during various steps of this process are also studied. Separated thorium oxalate could either be converted to oxide or be recycled back to process after dissolution in nitric acid. The supernatant could be treated appropriately prior to its final disposal. The uranium oxide obtained had good purity. Laboratory experiments with 25 and 100 g lot uranium yielded products confirming the suitability of the method. The technique was deployed in the THOREX campaigns later at Kalpakkam. The process was further improved by adding one more precipitation step for decontamination of uranium product with respect to iron impurity [44].

12.1.3 Equipment Qualification

Another pilot plant operated in this domain at IGCAR, Kalpakkam, to test the equipment systems that were developed for fast reactor fuel reprocessing [45]. In the system, necessary instrumentation was provided for remote operation and maintenance. The flow sheet employed was similar to that used earlier. The final purification was carried out by improved precipitation technique described above. The performance of equipment with respect to capacity and efficiency was quite satisfactory. The decontamination factors with respect to fission products and thorium were also found satisfactory. The equipment operated and maintained remotely. Considerable experience was gained in handling radioactivity, remote maintenance and trouble shooting in radioactive environment. Efforts to avoid pumps in radioactive environment by using alternative techniques were also successful. Electrically operated systems also could be avoided in remotely maintained radioactive cell. Feedback from the experience of using remote-handling equipment helped to refine design for future programme.

12.1.4 Thoria Dissolution Studies

Numerous investigations of thoria dissolution in HF-HNO_3 mixtures had also been made with and without aluminium nitrate with thoria pellets of different preparative histories. The study conducted in the laboratory was restricted to two dissolvent mixtures 13 and 8M HNO_3 with 0.1M $Al(NO_3)_3$ and two concentrations of HF, 0.03 and 0.05M. In subsequent work, NaF was tried instead of HF since it is easier to use and is reported to have enhanced dissolution rate. During chop leach process, because of the presence of fluoride ions, some dissolution of Zircaloy fines can also be expected. The aim of the study was to establish the lowest acidity and fluoride ion concentration for dissolving thoria fuels at an acceptable rate with a view of minimizing the feed adjustment for subsequent solvent extraction cycle in THOREX process and also to study the rate of corrosion of stainless steel and Zircaloy in different dissolvent mixtures. Thoria used for study was sintered pellets (1680 °C), green pellets and MgO-doped (0.5–2.5%) sintered pellets. In order to study uniform corrosion, intergranular corrosion, weldments and stress corrosion of SS304L, SS coupons of standard along with coupons of Zircaloy tube cladding were placed in the dissolvent mixture during dissolution [46]. The sintered pellets with 95% TD were used. The vessels were heated to the boiling temperature by electrothermal mantle, which produced agitation throughout the solution while boiling. The volume of the dissolvent was 250 mL. Approximately, 75 g of thoria was added to have a thorium concentration of ~1.00–1.10M. Appropriate quantities of HF/NaF/$Al(NO_3)_3$ were added to make dissolvent of desired strength. The temperature of refluxing was about 110–115 °C. Samples were drawn at regular intervals to monitor progress of dissolution. The dissolution studies along with magnesia-doped pellets were conducted in an aqueous medium containing 8.00M HNO_3 and 0.03M HF. More than 95% of the dissolution was completed in 13M HNO_3 +0.03M HF+0.1M $Al(NO_3)_3$ mixture in 92 h, while the time taken for the same in 8M HNO_3 +0.03M HF + 0.1 M $Al(NO_3)_3$ mixture was 172 h. In a comparison study conducted with 0.03M of NaF and 0.03M HF, the dissolution rate in the presence of NaF was found to be marginally higher initially as compared to HF and levelled out at 82% for both after 40 h of refluxing. But total time taken for complete dissolution was same in both the cases. Studies on dissolution using HF and NaF at a little higher concentration of 0.05M showed marginally improved dissolution rate. Studies with magnesia-doped sintered pellets were conducted in dissolvent containing 8.00M HNO_3 +0.03M HF+0.01M $Al(NO_3)_3$. It was found that the dissolution behaviour was significantly influenced by the presence of magnesia in the thoria pellet. Lower content helped the dissolution, whereas higher content reduced the dissolution rate. About ~1.5% MgO was found to be the optimum addition. The corrosion rates of plain SS are welded and stressed, and ring coupons were also found within the acceptable limits. In all the cases, the corrosion rate of Zircaloy cladding tube was measured. In the absence of $Al(NO_3)_3$ and in the presence of Zircaloy tube, the corrosion rate of SS was significantly higher. This showed that the concentration of Zr in solution was not enough to complex free fluoride ions, and therefore the presence of aluminium nitrate is essential to minimize the corrosion of SS vessel. A sintered (Th-4% Pu)O_2 pellet (sintered at 1680 °C,

Table 20 Thorium lean raffinate from reprocessing

Parameter	Value	Parameter	Value
Volume	18,700 L	^{137}Cs	6.34 mCi/L
Acidity	2.1M HNO_3	^{90}Sr	7.5 mCi/L
α	9×10^{-3} mCi/L	^{125}Sb	9.13×10^{-3} mCi/L
β	23.25 mCi/L	^{106}Ru	2.92×10^{-3} mCi/L
Th	0.245 g/L	F^-	106 mg/L
Al	1.97 g/L	Fe	126.65 mg/L

~94% TD) weighing 12.93 g was dissolved in 50 mL of solvent mixture 13M HNO_3 + 0.03M HF + 0.1M $Al(NO_3)_3$ [47]. During dissolution, pre-weighed coupons of SS and Zircaloy-2 of known dimension were also immersed. Samples were drawn after every 4 h of continuous refluxing at 115 °C. Thorium and plutonium contents were estimated. After complete dissolution of the pellet, coupons were washed, dried and weighed for calculating corrosion rate. Results revealed that the dissolution rate was >50% in first 4 h reaching around 86% after 16 h of refluxing for both Pu and Th. Later on, dissolution rates slow down and it takes 32 h more for 95% dissolution for Th. Around the same time, 90% of plutonium was observed to be dissolved. This was in general agreement with the dissolution behaviour of sintered thoria pellets reported earlier where the rate of dissolution of thorium slowed down with increasing hours of reflux and thorium concentration in the dissolved solution. The fine residue left behind was treated for further 8 h after which the plutonium dissolution reached around 94%. However, quantitative dissolution of the pellet was complete in ~44 h. The corrosion rate of SS was calculated and found well within acceptable limits (~14 mpy). The final solution had zirconium concentration of about 0.002M which was calculated by weight loss method. Addition of $Al(NO_3)_3$ in thoria dissolvent has undesirable effects in the subsequent vitrification process of high-level waste which is generally carried out at a temperature range of ~900–1000 °C. Besides, highly corrosive nature of fluoride ion, lowering its concentration in the dissolution reaction, was advantageous in reducing the corrosion of dissolver and other plant vessels and equipment. Detailed studies have been carried out [48], wherein $Al(NO_3)_3$ may not be required to inhibit the corrosion rate when lower concentration of fluoride is added in the presence of stronger complexing metal ions, viz. Zr(IV) and Th(IV) which are already present in the system.

12.1.5 Engineering Facilities for Uranium–Thorium Separation

Later in the year 2002, an engineering facility, Uranium–Thorium Separation Facility (UTSF), was designed and commissioned at Trombay for the processing and recovery of ^{233}U. The necessary modifications from pilot plants, Trombay and IGCAR, were incorporated in this facility. Decladding operation was similar to that adopted earlier using alkali. Thoria dissolvent, i.e. 13M HNO_3 containing 0.1M $Al(NO_3)_3$ and 0.03M NaF, was used for dissolution. Unlike previous campaigns, the feed after dissolution was found to have acidity in the range of about 5.00–6.00M HNO_3 with comparatively lower Th and ^{233}U. Process flow sheet using 3% TBP in *n*-dodecane was deployed for preferential extraction of ^{233}U with lesser co-extraction of thorium in the organic phase. Specially designed combined airlift mixer-settler unit (CALMSU) was chosen as solvent extraction contactors. The mixer-settler unit had 12 stages for extraction (2 modules), 18 stages for scrubbing (3 modules) and 12 stages for stripping (2 modules). In order to get better decontamination with respect to thorium, more number of stages were provided for scrubbing thorium from organic phase using nitric acid. Using this single-cycle modified THOREX process flow sheet, 56 irradiated thoria rods were processed in 8 batches. In the reconversion laboratory, a pre-concentration step using ammonia as precipitant was also adopted to concentrate Th and U product received from the plant. This step gave a concentration factor of about 10. Subsequently, sequential precipitation developed in-house was adopted for reconversion operation. Th from uranium in concentrated solution was removed as oxalate in the first step using oxalic acid as precipitant. The supernatant containing Fe and ^{233}U was subjected to carbonate precipitation under boiling conditions to remove Fe if any in the second step. Uranyl carbonate was finally subjected to ammonia precipitation step for uranium recovery. In this campaign, which lasted for ~2 years, about 800 g ^{233}U of very good isotopic purity ($^{232}U < 3$ ppm) was recovered. This generated about 10 m^3 of THOREX raffinate. To recover thorium from such raffinate stream, detailed laboratory studies [49] were undertaken to understand the physicochemical parameters such as solubility of thorium at different concentrations of nitric acid, limiting concentration of thorium in the organic phase from Th bearing feed solution at ~5.5M HNO_3 using TBP (30–42% in *n*-dodecane) to avoid third phase. Extraction and stripping of thorium by TBP in *n*-dodecane were carried out using simulated raffinate initially, and based on those results counter-current extraction and stripping studies were carried out for thorium recovery. The study established the scheme of extracting thorium from the THOREX raffinate using 38% TBP in *n*-dodecane. The limiting organic concentration of thorium under the experimental conditions was found to be well above 30 g/L. Thus, an organic to aqueous phase ratio of 3:1 was employed for extraction. Counter-current extraction, scrubbing and stripping under various experimental conditions provide useful data for designing the flow sheet on an engineering scale operation. Based on the above studies, a scheme was proposed [50] and deployed for the recovery of thorium from THOREX raffinate generated during the recovery of ^{233}U from research reactor-irradiated thoria rods at UTSF, Trombay, employing TBP (38% in *n*-dodecane). The product was subsequently stored as thorium nitrate in one of the process tanks at 2M HNO_3. About 18 m^3 thorium lean raffinate (Table 20) was transferred to Waste Management Plant, Trombay, for further treatment.

12.2 Reprocessing of AHWR Spent Fuel

Discharged fuel from AHWR is expected to contain fissile materials, viz. U^{233} and plutonium in the range of ~2–4% of thorium. Closing the fuel cycle for AHWR is a real challenge to separation scientists and technologists as it involves the separation of three components, viz. plutonium, inbred uranium and thorium from bulk of radioactive fission products. Distribution data under different experimental conditions played a very important role in designing solvent extraction-based process flow sheets for reprocessing spent fuels from reactors. Among the various solvents proposed for the three-component separation, TBP was found to be most promising extractant. Based on the experience from THOREX process, a flow sheet employing TBP (5% in *n*-dodecane) was developed for three-component separation in our reprocessing laboratory. Initial data from batch equilibration studies from nitric acid medium were found to be promising for separating U, Pu and Th. Solvent extraction parameters such as effect of nitric acid concentration, aqueous to organic phase ratios, nitric acid concentration for scrubbing of Th, optimization of partitioning agent and its composition, behaviour of fission products were studied extensively which are described below.

12.2.1 Batch Equilibration Studies

The distribution ratios of U, Pu and Th in batch mode were determined initially both from pure nitric acid solution containing Al^{3+} and F^- as well as from simulated feed solution. The composition of simulated feed solution used is given in Table 21.

Results of distribution ratio as carried out from aqueous nitrate medium as well as from simulated feed solution in the presence of fission products are given in Table 22. The distribution ratio values signify that even in the presence of 100 g/L Th, significant extraction of U is achievable in 5% TBP in *n*-dodecane as extractant and though D_{Pu} values are slightly lower, extraction can be achieved using higher organic to aqueous phase ratio in counter-current mode. From the table, it is also evident that the presence of fission products or metal ions from structural materials mainly corrosion products do not affect the D_{An} values to any significant extent.

Though the distribution ratio of Th is much less, significant Th extraction in the organic phase is expected due to the presence of bulk concentrations of Th (100 g/L) in the feed solution. To scrub out the co-extracted Th from the organic phase, scrubbing with different concentrations of nitric acid was studied. The optimized scrub acid

Table 21 Composition of simulated feed solution of AHWR spent fuel (average discharge burnup of 43 Gwd/Te after 5 y cooling)

Elements	Concentration (mg/L)	Elements	Concentration (mg/L)	Elements	Concentration (mg/L)
Pb	0.0174	Y	63.03	Cs	317.3
Ac	8.55E-05	Zr	436.9	Ba	213.67
Th	100.00[a]	Nb	6.24 E-4	La	157.05
Pa	0.67735	Mo	327.99	Ce	325.85
U[1]	1.474[a]	Tc	94.33	Pr	151.70
Np	0.283	Ru	280.98	Nd	503.2
Pu[1]	0.725[a]	Rh	38.03	Pm	3.06
Am	0.08[a]	Pd	276.7	Sm	95.4
Cm	0.016[a]	Ag	13.35	Eu	14.1
Ge	0.16	Cd	22.22	Gd	39.5
As	0.064	In	0.132	Tb	0.558
Se	9.86	Sn	9.48	Dy	0.256
Rb	55.44	Sb	1.81	Ho	0.053
Sr	108.97	Te	79.38	Er	0.0187

[a]In g/L, U and Pu were added appropriately so as to get their conc. to 2 g/L in feed

Table 22 Distribution ratio of U, Pu and Th extractant: TBP (5% in *n*-dodecane). Phase ratio of organic to aqueous = 2, time of contact = 10 min

Actinides	Distribution ratio from	
	Simulated feed	HNO_3 medium
UO_2^{2+}	2.57	2.88
Pu^{4+}	1.12	1.15
Th^{4+}	<0.1	<0.1

concentration of 3.00M HNO_3 was found to reduce the co-extracted Th in the organic phase to <200 mg/L with insignificant loss of U and Pu.

To select a reagent suitable for reductive partitioning of plutonium from uranium, various reducing agents, viz. acetohydroxamic acid, hydroxyl amine nitrate (HAN), ascorbic acid, tertiary butyl hydroquinone, ferrous ammonium sulphate, were studied [51]. Reagent mixture of 0.30M HAN + 0.20M hydrazine nitrate (HN) + 0.60M HNO_3 was found to be most suitable. HN in the partitioning agent was added as scavenger for HNO_2. Performance of indigenously synthesized HAN was compared with that received from imported sources and was found similar. The possibility of re-extraction of uranium released in the aqueous phase during reductive partitioning was also studied by increasing the acidity of the aqueous phase to 3.5M HNO_3 using 5% TBP in *n*-dodecane as wash. Extraction of uranium was found quantitative without loss of Pu(III) from the aqueous phase. Stripping of U from Pu lean

Table 23 Decontamination factors for U, Pu and Th in product streams after first cycle. Organic to aqueous phase ratio = 2, contact time = 10 min

FPs	DFs		
	U	Pu	Th
^{137}Cs	8.89×10^5	6.09×10^5	5.97×10^4
^{144}Ce	$>7.30 \times 10^5$	$>5.07 \times 10^5$	5.07×10^4
^{106}Ru	1.22×10^6	8.36×10^5	9.11×10^4
^{90}Sr	1.94×10^6	1.96×10^6	2.07×10^5

organic phase was finally carried out by 0.01M HNO_3, and quantitative stripping was observed at aqueous to organic phase ratio of (1:3). Raffinate phase from extraction experiment was used for Th recovery studies using 38% TBP in *n*-dodecane. Results were identical with those observed using THOREX raffinate of UTSF origin [49].

Overall decontamination factors (DFs) of different actinides with respect to individual fission products in final product streams after each single cycle of operation (extraction, partitioning, U-stripping, Th extraction and stripping) are reported in Table 23. It is evident from the results that very high decontamination factors are achievable even after a single cycle of operation. Further improvements in the DF value can be achieved using conventional purification cycles as in PUREX and THOREX process.

Under optimized conditions, counter-current experiments were carried out inside specially designed glove boxes employing laboratory scale mixer-settler units to demonstrate the separation. The flow sheet was validated using simulated waste using simulated AHWR spent fuel solution with reference burnup of 43 GWd/Te after 5y cooling.

12.2.2 Counter-current Studies

Extraction studies in counter-current mode were carried out in glove boxes using feed solution containing uranium and plutonium each at 2.00 g/L level having 100.00 g/L that is 3.00M HNO_3. The details of the experiments are given elsewhere [51]. All the steps, viz. extraction and scrubbing, partitioning of Pu, stripping of U, extraction of Th from raffinate of extraction run and Th stripping, were carried out. Figure 37 shows the scheme used for extraction cum scrubbing experiment inside the glove box. The mixer-settler unit used during the run is shown in Fig. 38.

Stage-wise concentration profile for U, Pu and Th at steady-state conditions in organic phase is shown in Fig. 39.

Results from these experiments indicated quantitative extraction of uranium and plutonium with negligible loss in the raffinate phase under the experimental conditions. Thorium contamination of about 206 mg/L was observed in the organic phase. The composite organic phase generated from the extraction cum scrub run had the concentration of uranium, plutonium and thorium of 1.40 g/L, 0.87 g/L and 200 mg/L,

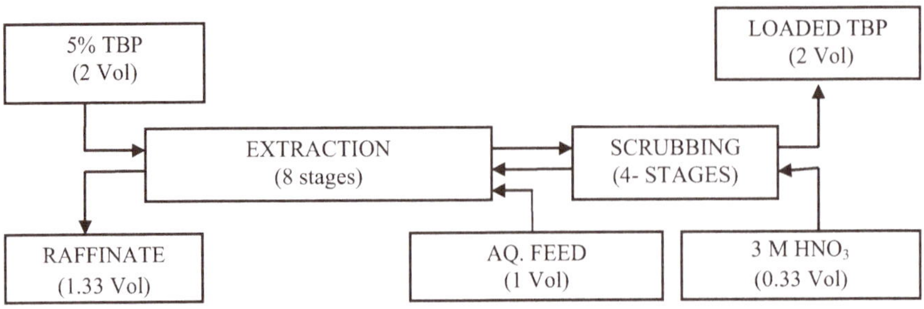

Fig. 37 Scheme used for counter-current extraction cum scrub experiment

Fig. 38 Mixer-settler unit housed inside glove box

Fig. 39 Concentration profile of U, Pu and Th during extraction cum scrub cycle

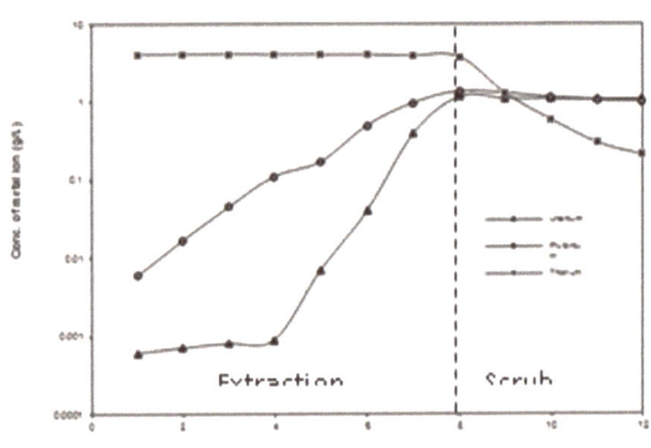

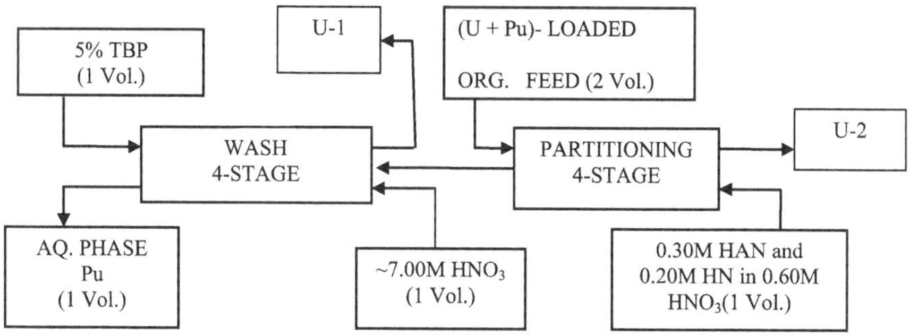

Fig. 40 Scheme used for partitioning using HAN

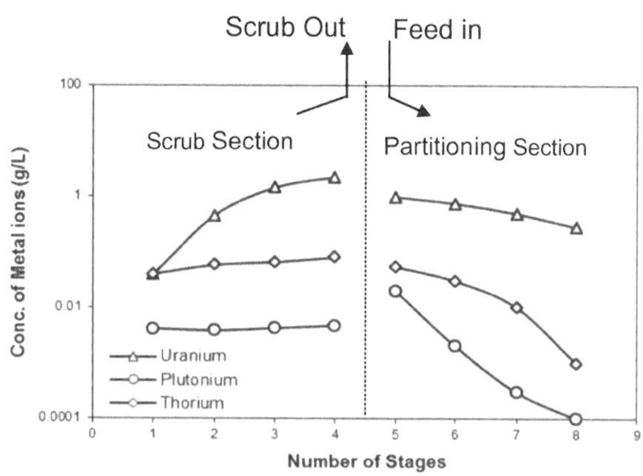

Fig. 41 Stage-wise concentration profile of U, Pu and Th in organic phase in partitioning step

respectively, with 0.09M acidity which was partitioned using a partitioning agent of 0.30M HAN + 0.20M HN + 0.60M HNO_3. Nitrate concentration in aqueous phase was increased by addition of 7.00M HNO_3 in the fifth stage. Consequently, the acidity of the aqueous phase was increased to ~3.50M and thus enabled the re-extraction of U to the organic phase. The reason behind release of U into aqueous phase during partitioning is low nitrate concentration in aqueous phase. The organic stream (U-1), obtained by washing the aqueous phase containing Pu(III) by 5% TBP in *n*-dodecane during partitioning, was found to have major portion of uranium compared to Pu lean organic stream (U-2) coming out of the partitioning stage and collected separately. The scheme used for partitioning is shown in Fig. 40, and stage-wise concentration profile of U, Pu and Th in organic phases under steady-state condition is shown in Fig. 41.

Stripping of uranium from composite organic phase (U-1 + U-2) using 0.01M HNO_3 showed quantitative stripping of uranium at an aqueous to organic phase ratio of 1:3 (concentrations of uranium at different stages during stripping are shown in Fig. 42). As the concentration of Th in the raffinate phase is similar to that observed in THOREX raffinate of Uranium–Thorium Separation Facility, Trombay, the flow

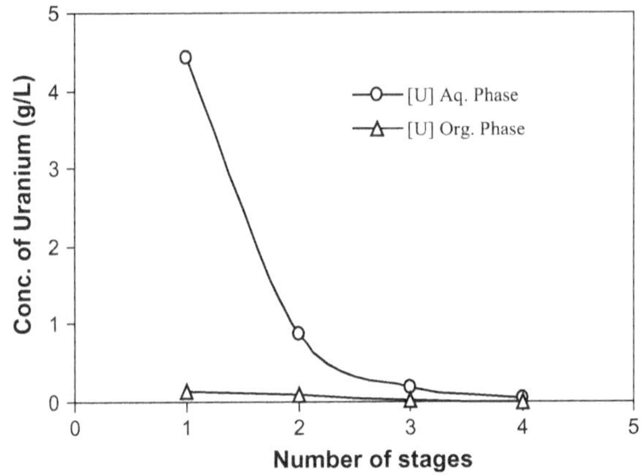

Fig. 42 Stage-wise profile for U-stripping

Table 24 Analysis of three product streams

Concentration (g/L)		
Th-stream	Pu-stream	U-stream
Th: 8.00	Pu: 0.87	U: 4.35
U: 0.001	U: 0.01	Pu: 0.004
Pu: 0.0042	Th: 0.18	Th: 0.109

sheet used earlier using 38% TBP in *n*-dodecane can be adopted for the recovery of thorium. The analysis of the three product streams obtained using the proposed flow sheet is given in Table 24.

As stated above, the purity of individual components with respect to other actinides can be improved to the desired level using conventional techniques such as solvent extraction, precipitation and ion exchange. During the course of the above experiments, neither third-phase formation nor crud formation was encountered at any stage of the runs. The flow sheet was validated using simulated AHWR spent fuel solution with reference burnup of 43 GWd/Te after 5y cooling. All the schemes, viz. extraction cum scrubbing, partitioning, U-stripping, extraction and stripping of Th, were carried out under optimized conditions. Schematic flow sheet of AHWR spent fuel employing 5% TBP in *n*-dodecane is depicted in Fig. 43.

Recovery of plutonium and uranium was found to be quantitative. No crud formation, third-phase formation or choking problem was observed during the entire experiment. The uptake of fission products in the organic phase was found to be negligible except Ru which was detected at very low concentration.

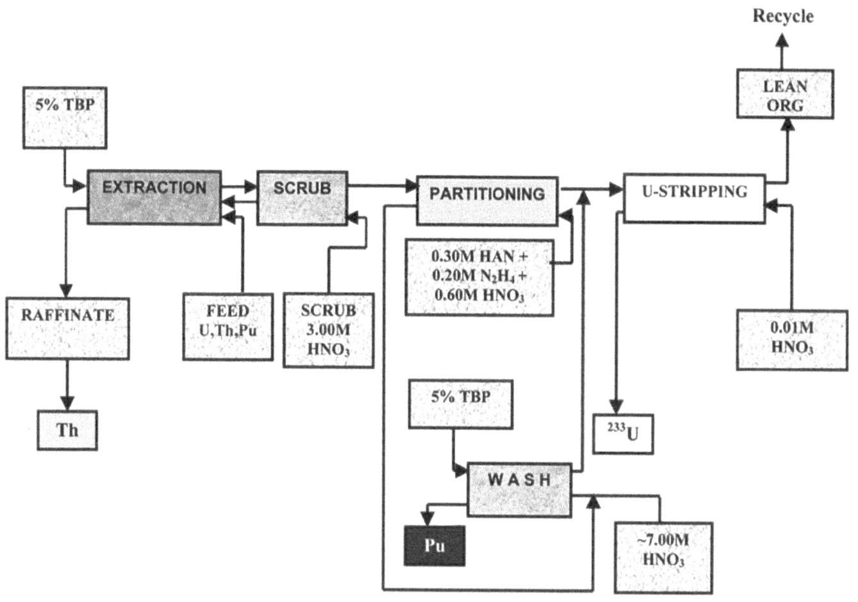

Fig. 43 Process flow sheet for three-component separation of spent AHWR fuel

12.3 Processing of Thoria Irradiated in Power Reactors

To process PHWR-irradiated thoria fuel, an engineering facility, Power Reactor Thorium Reprocessing Facility, Trombay, was built which was based on in-house R&D and experience from UTSF, Trombay. The facility was commissioned on January 2015 and operated successfully giving more confidence in thoria processing. In this facility, two batches (~50 kg per batch) of power reactor-irradiated thoria bundles were processed. Irradiated thoria bundle end-plate cutting was carried out by laser, and fuel pins were dismantled. Subsequently, chopping of the fuel pins was carried out by single pin mechanical chopper. Thoria was dissolved in dissolvent solution containing appropriate concentration of HNO_3, $Al(NO_3)_3$ and NaF. Dissolved solution was subjected to 5% TBP in *n*-dodecane-based THOREX process employing CALMIX mixer-settler units. After scrubbing thorium from organic phase by HNO_3, the uranium was stripped in aqueous phase using dilute nitric acid (0.01M). This product was passed through cation-exchange resin column to remove traces of thorium present in uranium product. Catalytic reduction of uranium was carried out using Adams catalyst in the purified product solution. U(IV) thus obtained was precipitated as $U(C_2O_4)_2$ and calcined to uranium oxide. Isotopic content of uranium in both the batches was carried out and found to be about 165 and 285 ppm, respectively.

Thus, India has gained enough confidence in thorium processing technology at the back end of nuclear fuel cycle.

13 Challenges Ahead

In Th-U233 Fuel Cycle

- Creation of commercially viable technology suitable for fuel cycle of Th-U233 owing to high activity of U 232 embedded in Uranium 233
- Sustainability of nuclear energy itself
- Waste management particularly the long-lived ones
- Increasing complexity of systems to meet the moving safety/regulatory requirements
- Associated increasing costs are important challenges to assure sustainability of the nuclear technology itself.

References

1. Human Development Report 2015 Access to Energy and Human Development, UNDP (2015)
2. http://niti.gov.in/writereaddata/files/document_publication/Niti_annual_report-2014-15.pdf
3. Energy Statistics 2012 Central Statistics Office, Ministry of Statistics and Programme Implementation, Government of India (2012)
4. http://pib.nic.in/newsite/PrintRelease.aspx?relid=174832 (2017)
5. http://www.npcil.nic.in/main/allprojectoperationdisplay.aspx
6. I. Pioro, *Handbook of Generation IV Nuclear Reactors* (Woodhead Publishing). eBook ISBN: 9780081001622, Hardcover ISBN: 9780081001493 Imprint
7. S.A. Bhardwaj, Indian Nuclear Power Programme—The Role of Thorium and The Challenges Ahead, A Plenary Talk, NCPTh-2014, 22–24 Dec 2014, NUB, Anushaktinagar, Mumbai, India
8. P.D. Krishnani, Physics Considerations for Utilisation of Thorium in Power Reactors, Lecture, ThEC-15, 12–15 Oct 2015, NUB, Anushaktinagar, Mumbai, India
9. A. Kakodkar, Opportunities with Thorium in Global Context, A Plenary Talk, NCPTh-2014, 22–24 Dec 2014, NUB, Anushaktinagar, Mumbai, India
10. A. Kakodkar, Harnessing Thorium for Nuclear Power: Challenges Ahead, A Plenary Talk, ThEC-15, 12–15 Oct 2015, NUB, Anushaktinagar, Mumbai, India
11. C.K. Asnani, R.N. Patra, Thorium 'Mining, Extraction and Purification'—Indian Context, Lecture, NCPTh-2014, 22–24 Dec 2014, NUB, Anushaktinagar, Mumbai, India
12. G.K. Dey, Determination of Thermal Expansion & Thermal Conductivity of $(Th,Ce)O_2$ MOX: A Combined Theoretical & Experimental Approach, Lecture, NCPTh-2014, 22–24 Dec 2014, NUB, Anushaktinagar, Mumbai, India
13. A. Kumar, Challenges in Manufacturing of Thorium based Fuels, Lecture, NCPTh-2014, 22–24 Dec 2014, NUB, Anushaktinagar, Mumbai, India
14. C. Srinivas, M.A. Rao, S.B. Shinde, S.A. Khot, BARC Internal Report—BARC/2015/I/014—Wet Recycling Process for Reject $(Th,U)O_2$ Fuel
15. C.G.S. Pillai, P. Raj, Thermal conductivity of ThO_2 and $Th_{0.98}U_{0.02}O_2$. J. Nucl. Mater. **277**, 116–119 (2000)
16. D. Jain et al., Thermal diffusivity and thermal conductivity of thoria–lanthana solid solutions up to 10 mol.% $LaO_{1.5}$. J. Nucl. Mater. **353**, 35–41 (2006)
17. T.K. Basu, M. Srinivasan, Thorium Fuel Cycle Development Activities in India, M 1990 BARC-1532 (1990)
18. P.K. Vijayan, A. Basak, V. Shivakumar, Advantages and Challenges of Thorium Based Power Reactors, Lecture, NCPTh-2014, 22–24 Dec 2014, NUB, Anushaktinagar, Mumbai, India

19. M. Srinivasan, K. Chandramoleshwar, C.S. Pasupathy, K.K. Rasheed, K. Subba Rao, U-233 fuelled experimental reactors in India—PURNIMA-II and KAMINI, in *Proceedings of 4th International Conference on Emerging Nuclear Systems*, Madrid, Spain, 30 June–4 July 1986
20. K. Balakrishnan, K. Kakodkar, Optimization of the initial fuel loading of the Indian PHWR with thorium bundles for achieving full power. Ann. Nucl. Energy **21**, 1–9 (1993)
21. G. Srinivasan, Irradiation of Thorium in FBTR, Lecture, ThEC-15, 12–15 Oct 2015, NUB, Anushaktinagar, Mumbai, India
22. J.L. Singh, Post Irradiation Examination of Thoria Mixed Oxide Fuels in Indian Hot Cells, Lecture, NCPTh-2014, 22–24 Dec 2014, NUB, Anushaktinagar, Mumbai, India
23. S. Banerjee, H.P. Gupta, Nuclear Power from Thorium: Different Options, A Plenary Talk, NCPTh-2014, 22–24 Dec 2014, NUB, Anushaktinagar, Mumbai, India
24. R.K. Sinha, A. Kakodkar, Design and development of the AHWR—the Indian thorium fuelled innovative nuclear reactor. Nucl. Eng. Des. **236**, 683–700 (2006)
25. A. Thakur, et al., Fuel cycle flexibility in advanced heavy water reactor (AHWR) with the use of Th-LEU fuel, in *International Conference on Future of Heavy Water Reactors (HWR-Future)*, Ottwa, Canada (2011)
26. P.K. Vijayan, et al., Conceptual design of Indian molten salt breeder reactor. PRAMANA J. Phys. **85**(3) (2015)
27. F. Carmini, R. Klapisch, J.P. Revol, J.A. Rubio, C. Rubia, CERN AT/93–47 (ET) C. Rubia, et al., CERN/AT/95-44 (ET), CERN/LHC/96-0 (EET), CERN/AT/95-53 (ET), CERN/LHC/97-01 (EET)
28. C.D. Bowman, et al., Nucl. Instrum. Methods **A320**, 336 (1992); C.D. Bowman, Ann. Rev. Nucl. Part. Sci. **48**, 505 (1998)
29. S.S. Kapoor, BARC Report BARC/2001/R/004
30. P. Singh, S.V.L.S. Rao, R. Pande, T. Basak, S. Roy, M. Aslam, P. Jain, S.C.L. Srivastava, R. Kumar, P.K. Nema, S. Kailas, V.C. Sahni, Pramana J. Phys. **68**, 331 (2007); R. Pande, S. Roy, S.V.L. S Rao, P. Singh, S. Kailas, Pramana J. Phys. **78**, 247 (2012)
31. P. Singh, Thorium Utilization in ADS, Lecture, NCPTh-2014, 22–24 Dec 2014, NUB, Anushaktinagar, Mumbai, India
32. P. Singh, *Proceedings of Indian Particle Accelerator Conference (InPAC2013)*, VECC, Kolkata (Nov 2013)
33. S.V.L.S. Rao, P. Singh, Pramana J. Phys. **74**, 247 (2010)
34. S. Roy, R. Pande, S.V.L.S. Rao, S. Krishnagopal, P. Singh. J. Instrum. **10**, 11008 (2015)
35. P. Satyamurthy, L.M. Gantayet, A.K. Ray, Pramana J. Phys. **68**, 343 (2007)
36. S.B. Degweker et al. Ann. Nucl. Energy **26**, 123 (1999); O.V. Shvedov et al. IAEA-TECDOC 985, D4.1, 313 (1987)
37. N. Srinivasan, M.N. Nadkarni, G.R. Balasubramaniam, R.T. Chitnis, H.R. Siddiqui, Pilot Plant for the Separation of ^{233}U at Trombay, BARC-643 (1970)
38. G.R. Balasubramaniam, R.T. Chitnis, A. Ramanujam, M. Venkatesan, Laboratory Studies on the Recovery of Uranium 233 from Irradiated Thorium by Solvent Extraction Using 5% TBP Shell Sol-T as Solvent, BARC report-940 (1977)
39. R.R. Chitnis, P.S. Dhami, A. Ramanujam, V.P. Kansara, Recovery of ^{233}U from Irradiated Thorium Fuel in Thorex Process Using 2%TBP as Extractant, BARC Report BARC/2001/E/030
40. A. Mukherjee; R.T. Kulkarni, S.G. Rege, P.V. Achuthan, Anion exchange separation of uranium from thorium for the tail-end purification of ^{233}U in Thorex process, in *Radiochemistry and Radiation Chemistry Symposium*, Pune, India, 7–11 Dec, Paper No. SC-12 (1982)
41. R.T. Chitnis, K.G. Rajappan, S.V. Kumar, M.N. Nadkarni, Cation Exchange Separation of Uranium from Thorium, BARC Report-BARC-1003 (1979)
42. P.V. Achuthan, C. Janardanan, N. Vijayakumar, P.V.E. Kutty, A. Mukherjee, A. Ramanujam, Studies on the Separation of Thorium and Uranium on Variously Cross Linked Dowex 50W Resins, BARC Report BARC/1993/E/09(1993)
43. A. Ramanujam, P.S. Dhami, V. Gopalakrishnan, A. Mukherjee, R.K. Dhumwad, Separation and Purification of Uranium Product from Thorium in Thorex Process by Precipitation Technique, BARC-1496 (1989)

44. P. Govindan, A. Palamalai, K.S. Vijayan, M. Raja, S. Parthasarathy, S.V. Mohan, R.V. Subba Rao, Purification of ^{233}U from thorium and iron in the reprocessing of irradiated thorium oxide rods. J. Radioanal. Nucl. Chem. **246**(4), 412 (2000)
45. G.R. Balasubramaniam, Reprocessing of irradiated thorium—Indian experience, in M. Srinivasan, I. Kimura (eds.) *Thorium Utilisation*, Proceedings of Indo-Japan Seminar on Thorium Utilisation: 65: Mumbai, India (1990)
46. P. K. Vijayan, S.S. Shinde, U. Jambunathan, A. Ramanujam, Studies on the dissolution of sintered thoria pellets in $HNO_3 + Al(NO_3)_3$ mixture containing HF and NaF, INSAC-2000, vol. 2, pp. 148 (2000)
47. N.M. Sinalkar, C. Janardanan, K. Vijayan, S.S. Shinde, U. Jambunathan, A. Ramanujam, Studies on the dissolution characteristics of sintered thoria (ThO_2-4%PuO_2 pellet, in *Nuclear and Radiochemistry Symposium*, 2001, University of Pune, 7–10 Feb, p. 210, (2001)
48. C. Srinivas, V. Yalmali, A.S. Pente, P.K. Wattal, S.D. Misra, Thoria/Thoria-Urania Dissolution Studies for Reprocessing Application, BARC/2012/E/007
49. P.S. Dhami, P.W. Naik, N.L. Dudwadkar, P. Jagasia, R. Kannan, P.V. Achuthan, A. Dakshinamooorthy, S.K. Munshi, P.K. Dey, Studies on the Physico-Chemical Parameters of Extraction and Stripping of Thorium nitrate from THOREX Raffinate Using 38%TBP in *n*-dodecane, BARC/2008/E/010 (2008)
50. P.S. Dhami, P. Jagasia, S. Panja, P.V. Achuthan, S.C. Tripathi, S.K. Munshi, P.K. Dey, Studies on the development of a flow-sheet for AHWR spent fuel reprocessing using TBP. Sepn. Sci. Tech. **45**, 1147 (2010)
51. P.S. Dhami, P. Jagasia, S. Panja, P.W. Naik, P.V. Achuthan, S.C. Tripathi, S.K. Munshi, P.K. Dey, Validation of the flow-sheet proposed for reprocessing of AHWR spent fuel: counter-current studies using TBP. Desalination Water Treat. **38**, 184 (2012)

Sustainability and the Role of Thorium in Our Future Energy System

Björn Stigson

Abstract How will 9 billion people live well within the limits of the planet? Demand for electricity will go up by 40% over the next 25 years and all of that is in the emerging economies. We are inevitably facing a carbon and pollution constrained world in 2050. Therefore, it is clear that when we look at the future energy mix, we need clean, safe, and scalable energy. Nuclear with thorium and renewable energy will play important roles in this mix. Competition, the fact that the markets will demand low carbon and pollution solutions, will play a major role in the climate agenda going forward. Indeed, the transformation to a sustainable world presents the biggest business opportunity the world has seen in its modern industry. The companies that see it, understand it, and invest in it will see enormous opportunities. The companies that don't understand and don't invest won't be around in the future. No part of society can create a sustainable world on their own. Therefore, we need new partnerships between governments, business, civil society, and academia, where each part is delivering on their specific responsibilities and roles. The World Business Council for Sustainable Development (WBCSD) was created for global business to have a neutral platform, which was not the individual company but a platform for like-minded that want to drive change in a certain direction. The nucleus for such a platform for thorium is the International Thorium Energy Organization (IThEO).

Keywords Sustainability · World energy mix · Energy systems analysis

1 Our Future World—A Growth Story

World's growth is driven by a few strong trends:

Population growth. Over the next decades, we are going to add one-third more people on this planet and they are going to be born and raised in emerging economies. By 2050, 85% of the world's population will be in what we today call emerging

B. Stigson (✉)
Stigson and Partners AB, Duvnäs, Sweden
e-mail: bjorn@stigsonpartners.com

A. K. Nayak and B. R. Sehgal (eds.), *Thorium—Energy for the Future*,
https://doi.org/10.1007/978-981-13-2658-5_4

economies. Fifteen percentage of the population is the old world; Europe, North America, Japan, and Australia.

Shift in global economic power. The growth of population in emerging economies and the growing middle class are driving faster economic growth in emerging economies. Already in 2025, we expect that 60% of the world's GDP is in the emerging economies.

Increased urbanization. We are about 3 billion people living in cities today. We expect that by 2060, there will be 6 billion people living in cities. These new cities will be built in Asia and Africa. If we build them with the type of urban infrastructure we have today, we will not have a sustainable world in 2050. This is one very important issue for the future.

Aging population. In some countries, the population is actually decreasing. Germany and Japan are the typical examples. The focus in emerging economies is understandably to improve quality of life and reduce poverty. That requires economic growth and the use of resources, not the least energy. The energy demand in the OECD countries is going to stand still for the next 40 years and all the growth in energy demand is coming from the emerging economies.

These simple but strong trends lead to a resource and carbon-constrained world. Inevitably, it is a world that will have to go through a transition to sustainability. The economy and society will change. The pressure on the environment will grow. We will have to find a way to govern this transformation. Here, we are struggling in different ways both nationally and globally.

2 The Green Race Is On

Since a few years, there is what I have come to call the "green race" going on, a green competition. Which countries and companies will become the leading suppliers of the resource efficient, low polluting product system solutions that the world markets will inevitably be asking for? If you want to be a leading exporter or a global company, you will have to provide those types of solutions. If you can't, you will not be an exporter. If you want to be such a leading country or company, you will have to transform your domestic market so that you build up a demand in your home country. This then becomes the source of competence and scale as a platform for exports.

If we look at some countries for major trends, we see that:

China has taken the lead. They are investing far more than anyone else in transforming their economy. This transformation is a major part of their current and their next five-year plan. I have been an advisor to the Chinese State Council on these matters for more than 16 years and gotten to know how they think. Their investments in renewables are alone larger than that of the next 10 countries put together. There are a very strong awareness and determination in the Chinese leadership to change the way the society and economy work. Last year PM Li Keqiang said, "we are going to declare war against our inefficient and unsustainable model of growth and way of life." They are also seeking international cooperation and here it is interesting to note that there is a bilateral agreement between Germany and China on innovation that was signed last autumn.

Japan, the neighbor to China, is the world's most energy efficient economy. They have a good technology platform for green solutions, good global companies and are pushing for innovation. What has also happened with their shutdown of nuclear is that they have become the world's largest importer of natural gas; thirty percentage of global supply is going to Japan. Even for an economy like Japan's, it is becoming too costly and therefore they are considering reopening nuclear.

USA is world's most innovative economy but they have, as we know, a political deadlock in Washington. You can't get innovations to really penetrate the economy if you don't have the support of the right type of political framework. Since that is not happening, they are falling behind China in the green race. A lot is happening decentralized in states and cities. What is interesting is that Silicon Valley is now starting to make major investments in renewable and new nuclear technologies. The big thing in the USA is the shale gas revolution, which has made the use of coal go down. This means that the USA is going to meet its requirements for reducing CO_2 emissions.

EU has historically been the global leader in green technologies export with 30% global market share, of which half is Germany. In my view, however, the transformation of the internal market is too slow, leading to EU losing global market share. There are two big things going on in Europe. One is the dependence on energy imports from Russia and how that is being handled in the present situation and then it is the transformation of the German energy system, which has created a lot of issues. Germany is looking at sustainability as a core element of their industrial strategy and they have major investment programs in research. They now have a 5-year research program, which has a dedicated budget for sustainability on 1 billion euros per year into four areas: resource efficiency, climate and energy, quality of life, and education. When the Chinese PM signed their contract, he said "Germany has the technology, we have the markets. It's a great partnership going forward." The two world-leading exporters are joining forces on a number of fronts.

India clearly needs major investments both to provide electricity to the people who lack it and also to improve reliability of supply. A major increase of energy efficiency is also needed. Since China has a program to reduce emissions, India is likely on track to become the world's largest emitter of CO_2. That will create a lot of political pressure on India going forward.

3 Transforming to a Low Carbon Economy—The Role of a Facilitating Platform

Energy systems on the scale of a country like India are very rigid systems; it takes a very long time to change them.

The World Business Council for Sustainable Development, where I was the president for 18 years, is a coalition of the world's 200 largest companies and their chief executives dedicated to deal with the issues of sustainable development for global business. These 200 companies have a value on the stock exchanges of 8.000 billion US dollars. The organization also has national organizations in a regional network in 60 countries with one in India.

In 2008, it becomes clear that global business needed to develop its thinking about what the world will really look like going forward. A world that will be carbon and pollution constrained in 2050. How will 9 billion people live well within the limits of the planet? And if we are going to share resources more equitable than we have until now, what will that look like?

WBCSD got 30 of its members engaged for about 2 years and they developed a pathway to 2050. They divided society and the economy into 9 different elements from soft issues like peoples values to hard issues like materials, mobility, buildings, and energy. They looked at what a sustainable world would look like in 2050 for each element, and what transformations we would have to go through to arrive at this sustainable world. They concluded that from a technical perspective this is probably doable, as well as from global business perspective. They identified that the real big issue is whether there is a political will to go through these transformations that will be very disruptive in society; transformations that will not only create winners but also create both winners and losers. Is there political will and capacity to drive this type of change? This became and still is a very substantial discussion that we have in global business. If we look at the climate negotiations, it is clear that there are difficulties for global society to solve these types of issues.

Systems issues were another observation that was made. Energy, as an example, is connected to water, which is connected to climate, which is connected to food, which is connected to land use and so forth. The understanding of systems solutions in society is very limited. The International Institute for Applied Systems Analysis, the head of which I am an advisor to, sees a huge need for learning how different systems interact if we are going to solve these issues. In many ways, both in the political world and in the academic world, we act very much like in silos and are very poor in understanding the cross-sectoral issues.

The major finding for business was that this transformation to sustainability is the biggest business opportunity it has ever seen in the history of modern industry. The companies that see it, understand it, and invest in it will see enormous opportunities. The companies that don't understand and don't invest won't be around in the future.

4 The Energy and Climate Landscape

Fast changing energy landscape, more distributed electricity generation, electrification of transportation will probably go much quicker than what people think. Solutions for electricity storage are being developed that will be very important for distributed electricity generation, smart grids.

Interestingly, it is the IT companies like IBM, Cisco, Google, and Apple that are seen by the traditional energy equipment suppliers as their competitors in smart electricity systems. Electricity utilities' roles have changed with the growing energy supply generated from renewable energy sources and they have lost customer base. A good example is Germany, but also the rest of Europe has been affected in particular. Between 2008 and 2013, the 20 biggest utilities in Europe lost 500 billion euros in market value. Actually, more has been lost on the utilities than on banks in the financial crisis in Europe. Accompanying this development is the issue of where should the base load electricity come from and the question of shale gas.

When we look at the future energy mix, it is clear that we need clean, safe, and scalable energy. Nuclear with thorium and renewable energy will play important roles. Looking at the energy system, we see that the oil industry probably is facing a similar situation as the utilities, where their traditional linear business model is coming under pressure. However, the power sector has taken many actions though it has lost tremendous value and should today be more broadly defined than the traditional electricity utilities. Demand for electricity will go up by 40% over the next 25 years and all of that is in the emerging economies; there is no increase in the OECD countries.

The export of energy-intensive goods like steel, aluminum, chemicals, paper, and so on is changing. The shale gas revolution in the USA has made electricity much cheaper in the USA than in the EU and Japan. The IEA is projecting that over the next 25 years the EU and Japan will lose one-third of their energy-intensive export of goods partly to the USA but also to emerging economies.

The IEA has been pushing for energy efficiency. It has concluded that about 60% of emission reductions must come from energy efficiency in the consumption phase, not in the production phase, both as individuals and industrially across all parts of society.

What will affect the climate agenda going forward? The awareness of the issue is growing partly as an effect of the Paris Climate Agreement and this will have an impact. And so will the relative price reduction for renewable energy production (wind and solar) versus in particular fossil fuels. However, competition, the fact that the markets will demand low carbon and pollution solutions will play a major role. Energy efficiency will play a role to lower emissions. A price on carbon will drive change in the energy mix. There will be a lot of technology innovations with, e.g., electrification of vehicles, electricity storage, new nuclear, and CO_2 management. We don't believe that carbon capture and storage will save the coal industry; it is a way too complicated and therefore expensive. However, if we could change CO_2 from being a pollutant into being a raw material by splitting the molecule, it could change the whole dynamics of climate change. We are not there yet but today we see interesting research going on in many parts of the world to make it possible.

5 Need for Partnerships and Collaboration

No part of society can create a sustainable world on their own. Governments can't do it on their own, and markets are too slow to drive transformational changes. Markets are very good at driving change in very well-defined areas but they are not good at driving the type of change sustainability requires. Therefore, we need new partnerships between governments, business, civil society, and academia, where each part is delivering on their specific responsibilities and roles.

Governments must create the regulatory framework that can stimulate actions toward resilient societies.

Business must innovate and implement actions for a resource efficient low pollution world.

Civil society has a crucial role to support the difficult trade-offs between conflicting priorities that governments and business will be facing.

Academia must educate leaders that can lead society toward a resilient world and stimulate science and innovation of new solutions.

In most parts of the world, neither governments nor big businesses have the trust of the public. Therefore, we will not succeed if governments or big businesses push these changes onto the public. If we are going to get public acceptance for the types of disruptive transformational changes that are required, we must have a close cooperation with civil society.

WBCSD was created for global business to have a neutral platform, which was not the individual company but a platform for like-minded that want to drive change in a certain direction. Therefore, if thorium is going to be successful, a platform that is not only the industry that is developing the energy systems is needed. The nucleus for such a platform for thorium is the International Thorium Energy Organization (IThEO). It can have the constituency that is broad enough from different interests, which will create a level of confidence in the people the field is addressing. These types of facilitators are needed to drive change because without them it is very difficult to gain acceptance. Through WBCSD we managed to create trust for big businesses in what we wanted to do. The field of Thorium Energy can do the same.

What India is aiming to do with thorium is warmly welcomed as a carbon and pollution constrained energy source. It can play an important role among many other things such as energy efficiency, reliability of supply, grid improvements, and so forth. Thorium can play a role but can't get us there alone—whole energy systems analysis is needed for that. Germany had not done their energy system analysis when they took their decision to radically change their energy system, and have ended up with many unintended consequences and issues as a result. The recommendation to India is to think long and hard about what the Indian energy system can and should look like, do the systems analysis, and then decide where the points are where it is necessary to drive change. Thorium will certainly play a role in the future Indian energy system.

6 Conclusions

How will 9 billion people live well within the limits of the planet? Demand for electricity will go up by 40% over the next 25 years and all of that is in the emerging economies. We are inevitably facing a carbon and pollution constrained world in 2050. Therefore, it is clear that when we look at the future energy mix, we need clean, safe, and scalable energy. Nuclear with thorium and renewable energy will play important roles in this mix.

Competition, the fact that the markets will demand low carbon and pollution solutions will play a major role in the climate agenda going forward. Indeed, the transformation to a sustainable world presents the biggest business opportunity the world has seen in its modern industry. The companies that see it, understand it, and invest in it will see enormous opportunities. The companies that don't understand and don't invest won't be around in the future.

No part of society can create a sustainable world on their own. Therefore, we need new partnerships between governments, business, civil society, and academia, where each part is delivering on their specific responsibilities and roles.

The World Business Council for Sustainable Development (WBCSD) was created for global business to have a neutral platform, which was not the individual company but a platform for like-minded that want to drive change in a certain direction. The nucleus for such a platform for thorium is the International Thorium Energy Organization (IThEO).

Acknowledgements The author would like to thank the Government of India and IThEO for the opportunity to deliver this message to the international Thorium Energy community, ThEC15 in Mumbai was indeed a great experience.

Technology Considerations for Deployment of Thorium Power Reactors

Matthias Krause

Abstract Deployment of any nuclear reactors, in particular innovative designs, requires a healthy combination of design and safety concept maturity, design/concept validation, integration into a national energy policy, cooperation between vendor/operator/regulator bodies, and finally a sound economic/financing plan and perhaps most importantly for long-term success, supportive government policy and acceptance by an educated public. This paper briefly summarizes these aspects as they relate to unique challenges and opportunities for a potential thorium-based water-cooled power reactor program and suggests areas for technical collaboration. Based on past efforts and current knowledge, water-cooled reactors are capable of accommodating thorium-containing or even thorium-based fuel designs, known technology challenges can be mastered and there are no "showstoppers" in a gradual transition to a "complementary" thorium/uranium-based nuclear energy program, while there remain significant obstacles to full Th/^{233}U implementation. The potential of thorium-based fuel options being a "better" or "superior" fuel compared to U/Pu-based fuels is clear. However, there are a few remaining areas that need further development work to generate knowledge and reduce uncertainties in existing physical parameters and implementation concepts. These are summarized and suggestions for international collaboration are made to help a gradual near-term introduction of Th fuel into the cores of current or advanced water-cooled power reactors. In light of the current lack of clear economic or overwhelming scientific advantages, perhaps the main argument for starting a thorium fuel cycle implementation is sociopolitical, namely greater public acceptance stemming from the fact that (1) Th does not carry the military connection of U/Pu and (2) the waste from any Th-based fuel cycle will have less long-lived minor actinides.

Keywords Thorium · Water-cooled reactors · Fuel cycle
Renewable nuclear energy

M. Krause (✉)
International Atomic Energy Agency, 1400 Vienna, Austria
e-mail: M.Krause@iaea.org

A. K. Nayak and B. R. Sehgal (eds.), *Thorium—Energy for the Future*,
https://doi.org/10.1007/978-981-13-2658-5_5

1 Introduction and Background

Thorium as a fuel in a nuclear reactor has been considered, researched, and proposed for many decades in most reactor types, in fact since the beginning of peaceful uses of nuclear energy. Its inherent advantages over uranium are:

- Greater abundance and easier mining (no natural radioactivity), currently a "waste product" in various industrial mining activities;
- In principle can achieve a fully self-sustainable breeding fuel cycle;
- Higher thermal conductivity and higher melting temperature, thus cooler fuel centerline temperature and larger margins to fuel melting in severe accidents. More suited as a "no-melt fuel";
- Chemically stable (has only one oxidation state), thus would not oxidize during defected fuel operation or during transients that may induce fuel cladding defects or in a waste repository;
- Fission produces much less minor actinides (including Pu), significantly reducing waste radiotoxicity beyond ~100 years;

There are also a few inherent disadvantages:

- It needs fissionable material added at the beginning of the reactor operation (^{235}U or ^{239}Pu).
- A lower delayed neutron fraction requires faster reactor regulation and neutronic trip shutdown response due to reduced reactivity worth of control and shutdown devices. Reactivity coefficients are also negatively affected. However, these effects are comparable to MOX fuel.
- Chemically stable, making the dissolution process during reprocessing more difficult.
- The high-energy gamma emitters in ^{232}U decay chain necessitate more expensive, remotely operated, shielded reprocessing facilities; the presence of ^{220}Rn also requires a more complicated ventilation system and the presence of ^{228}Th in the "excess" thorium requires shielded secure storage for ~20 years.
- Beta decay produces long-lived radionuclides (including ^{231}Pa) that contribute to waste radiotoxicity.

Other, external, factors that have changed over the past decades and could change again in the coming years could make thorium as a reactor fuel more or less attractive. These include

- Economics: Fuel cost, recycling cost, relative magnitude of the fuel cycle cost components as a fraction of overall cost of nuclear power generation (including capital and financing costs). The increased reprocessing costs may be offset by benefits resulting from higher fuel burnup and longer fuel cycle lengths;
- Resource availability and fuel cost and/or security, both global and local, including for competing technologies such as fossil;

- Government (national) policy and public opinion relating to nuclear energy, which in combination determines a country's willingness to invest in technology, in particular valuation of CO_2 emissions reduction in setting national energy mix goals.

Then there are more subtle technology considerations related to fuel manufacturing, reactor operation, and waste management that are unique to thorium (and different from uranium), which are discussed in the following sections. The recent OECD/NEA Report on the *Introduction of Thorium in the Nuclear Fuel Cycle* [1] states: "Improving the efficiency of utilization of mineral resources (whether it is uranium or thorium) while reducing ultimate waste streams are among the major challenges that the nuclear energy industry must address if nuclear energy is to develop significantly and become a sustainable source of energy for the long term," and "the use of thorium has been considered as an option for nuclear fuel since the advent of nuclear energy." Thorium cannot be a real alternative to uranium, however, at least at the initial stage, because it has no fissile isotope. Thorium does, nonetheless, generate the fissile nuclide ^{233}U by neutron capture in a nuclear reactor. ^{233}U is an excellent fissile nuclide, with advantages over the ^{235}U or ^{239}Pu used in current reactors. This inherent "neutronic attractiveness" of ^{233}U, particularly in thermal neutron spectrum reactors, makes the use of thorium scientifically interesting, despite the multiple challenges yet to be resolved technically and industrially. The thorium option has never been fully discarded, and "the thorium fuel cycle" has, albeit with fluctuating intensity, continuously been studied worldwide."

These challenges, together with the inherent, economic, social, and technology factors affecting in general the choice of nuclear and in particular of a nuclear fuel cycle, lead to the conclusion that a gradual and symbiotic transition from the current pure-U/Pu fuel cycle to a combined Th/U/Pu fuel cycle, of which various forms are possible, can offer an optimal solution. It would allow the flexibility of continuing with the transition, remaining at a certain future status-quo, or returning to a pure-U/Pu cycle, whichever is consistent with the nuclear energy strategy, be it national or regional.

2 Reactor Deployment Considerations

Any advanced concept, whether it is fuel or other major reactor technology, needs to have a solid basis rooted in sufficient R&D, analysis, testing, and, ideally, prototyping, before it can be put forward for deployment. Much research has been completed on various thorium fuel options in the past, to a large degree current analysis tools can be used to analyze thorium behavior during normal operation, extended operation (high burnup) and accident conditions (however, more validation is required before a solid safety case can be presented for licensing) and in-reactor irradiations and post-irradiation examinations have shown very promising results. The potential of thorium-based fuel options being a "better" or "superior" fuel compared to U/Pu-based fuels is clear. However, there are a few remaining areas that need further development work to generate knowledge and reduce uncertainties in existing parameters and concepts [1, 2].

In a homogeneous core concept, the fuel material itself consists of a fissile/fertile mix, while in a heterogeneous core the fissile and breeding fuel regions (pins, bundles, or assemblies) are physically separated and only neutronically coupled (also referred to as seed/blanket configuration). The technical challenges are different whether considering a heterogeneous or homogeneous core implementation. These can be summarized into two broad areas:

- Fuel manufacturing challenges (e.g., ensuring consistent composition without hot spots, optimization of additives to facilitate sintering of high-density pellets), material properties, high-burnup properties and behavior, and reprocessing schemes for ThO_2 containing significant amounts of UO_2 or PuO_2. This is not an issue in a heterogeneous core.
- Effective radiation shielding during the reprocessing of used fuel and worker internal contamination monitoring in the absence of ^{239}Pu (since there is no ^{239}Pu in irradiated thorium fuel, marker candidates for radiotoxicity are likely to be ^{232}U and/or ^{228}Th depending on cooling time. Dosimetry procedures and sensitive methods may need to be further developed and established). The latter applies only to the Th reprocessing branch for a heterogeneous core.

In addition to the above generic issues, which could be categorized as "knowledge gaps," some reactor technology concepts should be further investigated, or optimized, taking into consideration realistic and specific implementation options, i.e., "detailed design gaps." This can be done through detailed analytical studies, which have certainly been done for reactor concepts such as the AHWR (India) and the SCWR (Canada), but are not published to the author's knowledge:

- Detailed physics/thermal hydraulics and analysis, both steady and transient, of heterogeneous fuel assemblies to determine to what degree local power/coolant flow matching is required in order to achieve the target burnup, breeding ratio, and percentage energy from Th, while maintaining all safety/trip margins and regulatory/operational limits. Since in a heterogeneous configuration, the seed and blanket fuel channels or assemblies may have significantly different powers that further vary with time in the refueling cycle, either flow matching or derating (compared to operation with uranium fuel) is needed.
- Determination of fuel material, fuel/cladding interaction, and fission gas release relevant to the fresh-to-high-burnup region of homogeneous core (mixed Th/U/Pu) fuel, including any microstructure effects. This is needed to verify/validate predictive models and assumptions.
- Feasibility of recycling such homogeneous Th/U/Pu fuel, because a once-through cycle would likely not yield the expected results of fuel cycle sustainability and waste MA reduction [3].
- Options of on-site reprocessing/refreshing should be investigated in light of the higher shielding requirements for both core concepts.

The above apply to any reactor type and are part of the vendors' design process. In order to support individual efforts and increase the visibility of such efforts and their underlying concepts and R&D, some areas are perhaps suited for international collaborations. Suggested topics include:

- What are Th-specific reactor design considerations in a LWR/BWR/PHWR? How could a U → Th transition be designed into an evolutionary reactor concept or what should be "modified" to make it suitable/optimized for thorium? For example, is static or dynamic (changing over time or burnup) channel flow/power matching needed/possible/desired.
- An objective, unbiased investigation into which currently deployed reactor type is most suitable (CANDU, PWR, BWR, etc.) for near-term introduction of thorium fuel, including the effects on fuel handling/refueling schemes, outage management, reactor control/shutdown requirements, and overall power/efficiency.
- What are realistic targets of % energy from Th in a commercial power reactor? This could be investigated in detail using refueling codes, possibly integrating thermal hydraulics codes, and ideally confirmed by ever-increasing use of Th in a suitable power reactor with a "mixed core," first as an additive to U fuel, then as a continually more significant core load (dedicated fuel assemblies for Th introduction—mixed, but heterogeneous fuel assemblies—and finally, a homogeneous Th-based core). Since such an introduction into an existing power reactor would most likely require some derating, the process should be financially supported by a national or international governmental group.

In order to maximize the chances of success for such a pilot transition scenario from U/Pu to Th fuel cycle in an existing water-cooled reactor, cooperation among nuclear countries is absolutely necessary and a formal internationally financed program would be ideal.

3 Conclusions

Large-scale implementation/deployment of thorium fuel is unlikely to happen based on current or foreseeable economic, technological, waste management, or safety/security/safeguards drivers. Therefore, the significant licensing work which would need to be undertaken for the implementation of thorium fuels and the R&D required for closing remaining technology gaps is economically not justified at present. However, the same can be said for fast breeder reactors and GenIV concepts.

The 2011 NEA report on *Trends towards Sustainability in the Nuclear Fuel Cycle* [4] concludes that "the successful large-scale reactor technology demonstration efforts conducted in the past suggest that there should not be insurmountable technical obstacles preventing the use of thorium fuel and its fuel cycle in existing and evolutionary L/HWRs. However, the industrial infrastructure, research, design and licensing data are not in place to allow a rapid deployment of thorium fuels in current reactors in the short term." While progress has been made in the past five years, notably in India, the statement remains valid.

Therefore, a long-term vision and steady effort, including international collaborations on thorium issues (technological, economical, and sociopolitical), are required to make a gradual transition possible that could eventually be called "Renewable Nuclear Energy."

References

1. OECD/NEA, *Introduction of Thorium in the Nuclear Fuel Cycle*. NEA Report No. 7224 (2015)
2. IAEA, *Role of Thorium to Supplement Fuel Cycles of Future Nuclear Energy Systems*. IAEA Nuclear Energy Series No. NF-T-2.4 (2012)
3. IAEA, *Thorium Fuel Cycle-Potential Benefits and Challenges*. IAEA-TECDOC-1450 (2005)
4. OECD/NEA, *Trends Towards Sustainability in the Nuclear Fuel Cycle*. NEA Report No. 6980 (2011)

LFTR: In Search of the Ideal Pathway to Thorium Utilization—Development Program and Current Status

Benjamin Soon

Abstract Thorium has gained substantial attention as a potential energy source that could rival and eventually replace fossil fuels as humanity's primary energy source. This could not have come at a more opportune time as concerns about global climate change from CO_2 emissions and the specter of finite fossil fuel resources create serious challenges for the continuation of our advanced industrial societies, which are reliant on readily available and affordable energy. Thorium also potentially represents the catalyst with which the nuclear industry could reinvent itself and finally gain widespread public acceptance. There are many opinions on how to utilize thorium as a fuel, but the question of what constitutes an 'ideal' pathway has mostly been underemphasized. Many specific characteristics of the thorium fuel cycle can differ significantly depending on the conditions and methodologies of utilization; characteristics such as safety, efficiency, waste profile and volume, and fissile protection can vary greatly according to reactor design and utilization philosophy. With thorium, we have been given an opportunity to start over, a blank slate. Therefore, in imagining the 'Thorium Economy' to come, it behooves the scientific and engineering communities to consider the most 'elegant' solution physically possible—what constitutes the 'ideal' and is it possible to reconcile it with what is both economically and technically practical? The characteristics desired of an 'ideal' nuclear reactor, in the areas of safety, efficiency, economy, and sustainability, and the five key design choices that could enable such a reactor will be discussed. This will be followed by an overview of the liquid fluoride thorium reactor, a two-fluid molten salt reactor currently under development by Flibe Energy in the USA. LFTR is a direct descendant of the MSRE, which was developed at Oak Ridge National Laboratory (ORNL) in the USA during the 1960–70s. The rationales and scientific concepts for the choices made on the design of the LFTR and how it could potentially achieve many of the desired characteristics of the 'ideal' nuclear system will be the focus of this article.

Keywords LFTR · Thorium · Flibe · Technology development

B. Soon (✉)
Flibe Energy Inc., 5952, Huntsville, AL 35814, USA
e-mail: benjamin.soon@flibe-energy.com

A. K. Nayak and B. R. Sehgal (eds.), *Thorium—Energy for the Future*,
https://doi.org/10.1007/978-981-13-2658-5_6

1 Introduction

Energy is the master commodity. Since the dawn of the industrial age, we have used fossil fuels as our primary energy source, and all modern economic activity can be measured in units of energy, be it from the direct combustion of liquid fuels or electricity. However, the use of fossil fuels comes at a cost, both in terms of emissions of CO_2 and other pollutants, and in terms of social and political costs.

Thorium has gained substantial attention as a potential energy source that could rival and eventually replace fossil fuels as humanity's primary energy source. This could not have come at a more opportune time as concerns about global climate change from CO_2 emissions and the approaching specter of finite fossil fuel resources create serious challenges for the continuation of our advanced industrial societies, which are reliant on readily available and affordable energy. Thorium also potentially represents the catalyst with which the nuclear industry could reinvent itself and finally gain widespread public acceptance. There are many opinions on how to utilize thorium as a fuel, but the question of what constitutes an 'ideal' pathway has mostly been underemphasized. Many specific characteristics of the thorium fuel cycle can differ significantly depending on the conditions and methodologies of utilization; characteristics such as safety, efficiency, waste profile and volume, and fissile protection can vary greatly according to reactor design and utilization philosophy.

With thorium, we have been given an opportunity to start over, a blank slate. Therefore, in imagining the 'Thorium Economy' to come, it behooves the scientific and engineering communities to consider the most 'elegant' solution physically possible—what constitutes the 'ideal' and is it possible to reconcile it with what is both economically and technically practical?

In this article, I will discuss the characteristics desired of an 'ideal' nuclear reactor, in the areas of safety, efficiency, economy, and sustainability, and the five key design choices that could enable such a reactor. This will be followed by an overview of the liquid fluoride thorium reactor, a two-fluid molten salt reactor currently under development by Flibe Energy in the USA. LFTR is a direct descendant of the MSRE, which was developed at Oak Ridge National Laboratory (ORNL) in the USA during the 1960–70s.

The rationales and scientific foundations for the choices made on the design of the LFTR will allow it to potentially achieve many, if not all, of the desired characteristics of the 'ideal' nuclear energy system.

2 The Problem with Nuclear Energy

In any given year, energy takes one of the top three spots on the list of urgent issues facing humanity alongside famine and clean water availability. When considering the 'problem' of energy, there are many theories on how to solve it. But when one runs the numbers and considers the hard realities of engineering and economics, none are more compelling or attractive than nuclear energy.

The energies of the atomic nucleus are millions of times greater than those released from chemical combustion. It does not emit any pollutants and has shown itself to be the safest form of energy. Yet, this seemingly obvious choice has not come into its own as our primary energy source.

The main problem is that public perception and politics are just as essential as technical and economic arguments when considering the whole picture; there must be acceptance among a society immensely paranoid about radiation; and that reality leads us to accept that we must fundamentally adjust the way we harness the power of the atom. This point was driven home by the public reaction, and continued backlash, to the events at Fukushima. Regardless of our opinion as engineers and scientists regarding the actual situation, this is the reality we face. How we deal with this will be discussed later in this article.

Putting aside perceptual problems, I believe that all engineers share one driving passion; we seek the elegant solution. And sadly—even discounting the issue of public perceptions—despite all the good that traditional nuclear technology has done, in the context of large-scale civilian power generation, it could hardly be considered the most elegant of solutions.

3 Thorium as a Resource

If one were to strip away all the extraneous arguments from the 'problem' of energy, it can all be boiled down to a single factor, 'limited resources.' After all, competition—and the potential for conflict—only occurs over a limited resource.

Thorium—for all intents and purposes—represents a virtually unlimited energy resource. One does not need to be an economist to appreciate the significance of this. In my opinion, any effort to develop thorium as a fuel on the strength of its abundant nature is already laudable.

Furthermore, we know that thorium if properly implemented, in the right machine, has the potential for far more than simple resource abundance.

3.1 What Is the Most Elegant Solution?

So, what would the ideal modality of utilizing thorium look like? How would we create a machine that satisfies the public concern for safety, while meeting the need for a better, more efficient system based on a truly sustainable end-to-end supply chain?

I have said we should be in the business of seeking elegant solutions. Thus, we should always seek to first consider what the ideal pathway is, before resorting to compromises.

To do this, we must define what is desirable and reconcile it with what is both possible from nature and practical in an engineering sense. It makes no sense to come up with what one might consider 'the ultimate energy source,' but one which is practically and economically unviable to build, operate, and/or sustain.

3.2 First Principles and the Four Major Desirable Characteristics

All nuclear systems are basically comprised of three essential sections. Though common to most energy systems, the design choices surrounding these sections substantially affect the characteristics of the resulting reactor.

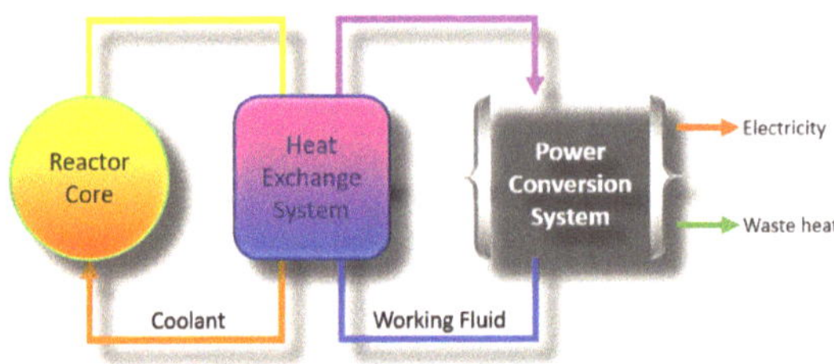

There are four major characteristics we would desire to maximize in an ideal system: (1) **Safety**, (2) **Efficiency**, (3) **Economy**, and (4) **Sustainability**.

As with any good engineering practice, we should first attempt to achieve these characteristics with the bare minimum and only add on additional sections if it is absolutely necessary. This is engineering 101; the simpler the machine, the less expensive, and prone to failure it will be.

3.3 Inherent Versus Engineered

From a practical engineering standpoint, wherever possible, we would want to make choices that lead to desirable characteristics that are physically Inherent to the reactor, rather than those that we need to engineer solutions for.

This is an important point because we know that any engineered solution is vulnerable to failure no matter how miniscule the chance is—even a statistical insignificance of $P > 0.005$ is a nonzero probability—but an inherent characteristic is immune to failure because there was no problem to engineer a solution for in the first place!

While this is contingent on whether such an approach is physically or chemically possible in the first place, where such solutions exist, and are not excessively difficult to implement, such options should be pursued first with all due diligence.

3.4 Safety

The first and most important characteristic that would be desired from an ideal nuclear system would be one with the best possible safety characteristics. This is essential because it is likely the key to overcoming public aversion to nuclear energy in general.

Deterministic vs Probabilistic Safety
Elimination of fuel/core damage risk
Elimination of dispersion mechanisms
Minimization of source terms

If it were possible, we should aim to design a reactor that is capable of operating under a deterministic safety regime, rather than a probabilistic one, which has been the case all this time with legacy nuclear technology. This means that if it were possible to entirely eliminate accident scenarios such as fuel melt, explosive dispersion, and any other radiological contaminant propagation mechanisms, it should be necessarily pursued as a priority.

3.5 Efficiency

Next, we would desire to have a system that had the maximum possible efficiency. Some of the key factors to maximizing efficiency lie in finding the best way to implement the following characteristics.

- High fuel utilization;
- High thermal conversion efficiency;
- High volumetric heat capacity (in the coolant);
- Minimal mechanical complexity;
- Control simplicity.

3.6 Economy

Thirdly, there must be an economic case for the technology. If it cannot be made, then no matter how brilliant the technology, the chances of widespread adoption is close to, if not exactly, zero. Like in the case of efficiency, the overall design choices should favor the following.

- Operational simplicity;
- Minimize fuel costs;
- Modular fabrication and construction (standardization);
- System simplicity.

3.7 Sustainability

Finally, the total ecology of the reactor should be sustainable in terms of overall supply chain and complete cradle-to-grave life cycle. This includes decommissioning and how one would dispose of spent components such as reactor vessels, fuel.

The factors which should be considered in making design choices are as follows.

- Low waste volume;
- Minimize/eliminate TRU production (especially plutonium);
- Fuel efficiency;
- Supply chain sustainability (incl. political);
- Low overall environment impact (incl. resource extraction and fabrication).

4 The Five Key Design Choices

With the above-desired characteristics and their underlying considerations in mind, if we examine the system commonalities of reactor concepts, one quickly realizes that the pathway to achieving these system properties lie within five key design choices, many of which have overlapping influences on each of the four desirable characteristics and they are as shown.

- Fuel material
- Burner vs Breeder
- Solid vs Liquid Fuel
- Thermal vs Fast (Neutron Spectrum)
- Coolant Type

4.1 Fuel Material

The choice of fuel material is probably the most crucial design choice of any energy system. This decision substantially impacts, even determining the availability of downstream design choice options, hence it must be the first and most important consideration. Thorium is the most ideal fuel choice due to—among other things—its vast abundance as a natural resource.

More significantly, when implemented correctly, the thorium fuel cycle has the potential to eliminate the production of plutonium from the waste stream, a massive advantage in terms of sustainability and political acceptability.

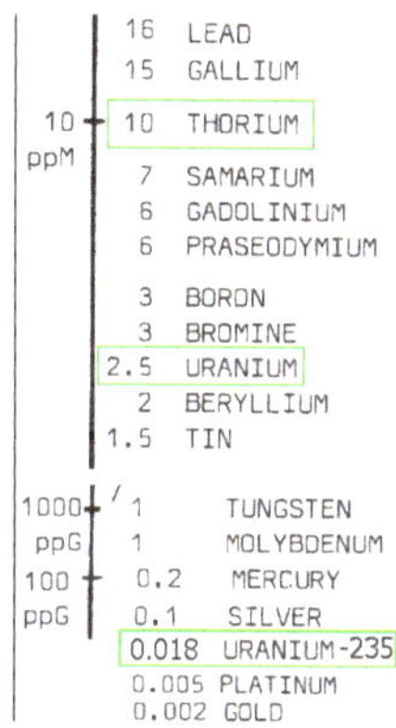

source: energyfromthorium.com

4.2 Burner Versus Breeder

The next important question is whether we want to use this fuel in a breeder or a burner? Setting aside any professional bias or any present technical/economic debate, it is generally accepted that for nuclear energy to be truly sustainable, a more advantageous use of thorium as a nuclear fuel would be in a breeder. The reason for this is the low natural fissile resource availability, and breeders are the only way to increase, or at the least, maintain our global fissile inventory. As opposed to constantly burning the only—ever-reducing—natural fissile we have; U-235, which is almost as rare as platinum.

Only Breeders offer a chance for truly sustainable nuclear energy

It should be noted that it is only possible to breed with the thorium or plutonium fuel cycle, so if U-235 (the current fuel cycle in traditional reactors) were chosen previously as the fuel, breeding would not be possible. Furthermore, only thorium can breed in the thermal spectrum, the advantages of which will be discussed shortly.

4.3 Fast Versus Thermal (Neutron Spectrum)

Next, which neutron spectrum is preferable for a practical reactor?

Fast Reactors have the advantage of being able to breed with any fuel cycle and potentially perform a specialist role in the disposal of legacy spent fuel. Advocates of the fast spectrum argue that it provides the best breeding ratios and require no moderator, thus having a simpler core structure.

However, fast reactors require significantly more fissile inventory and a much more complex control system. Not to mention that the significantly harder neutron flux will lead to greater core structural component degradation and lower lifetimes. The lack of a moderator, while having the advantage of allowing simpler core structures, sacrifices a significant safety characteristic inherent to them, which increases the complexity of the control and safety system necessary to operate fast reactors. The net result is that fast reactors are, by nature, very 'twitchy' and costly machines to build and operate. Over the past 40 years, there has been very little indication that they would ever be economically viable.

Thermal reactors on the other hand require much less fissile inventory and have much simpler control system requirements, and the presence of a moderator gives thermal reactors many inherent physical properties that lend toward stronger safety and control stability characteristics.

The disadvantage of thermal reactors is of course that you cannot breed using traditional uranium fuels (U-235 and plutonium fuel cycles), but since we are interested in the thorium fuel cycle, this is no impediment to us. Hence, all other things being equal, and we weren't trying to accomplish any specialized roles, I will argue that a thermal spectrum reactor is preferable choice in this respect. It is potentially safer, and cheaper, and easier to control.

4.4 Solid Versus Liquid Fuel (Fuel Form Factor)

Next, we talk about the fuel form; this is very important because it ultimately determines the resistance to damage and maximum potential efficiency of both the reactor structure and its fuel.

Liquid fuels can be continuously recycled and have neutron poisons removed. Because they are fluid and self-homogenizing, they are impervious to the issue of fuel element damage over time. They do not require expensive fuel fabrication and allow for extremely high fuel utilization rates. Furthermore, a liquid expands and contracts far more readily than solids, possessing an extremely strong negative thermal coefficient of reactivity and self-regulating nature; a very attractive inherent physical safety characteristic.

Solid fuels are the basis for current technology and we are familiar with its processes. However, solid fuels require complex and expensive fuel fabrication. They inevitably become damaged by the neutron flux and neutron poison build up in a reactor core and will need to be replaced before completely burning up its fissile content. Not to mention that reprocessing solid fuels, as is necessary for breeders, is much more complicated.

But more than that, solid fuels are and will always remain the primary safety vulnerability in nuclear reactors that use them. Since they are immobile, it will always be necessary to engineer emergency systems bring cooling to the fuel, and we know that engineered systems are vulnerable to failure, regardless of whether they are passive or active, so what's left is to determine the consequence of such failure and its acceptability. Figure 1 shows the basic principle and mechanics of a loss-of-coolant-accident such as that which happened at Fukushima in 2011.

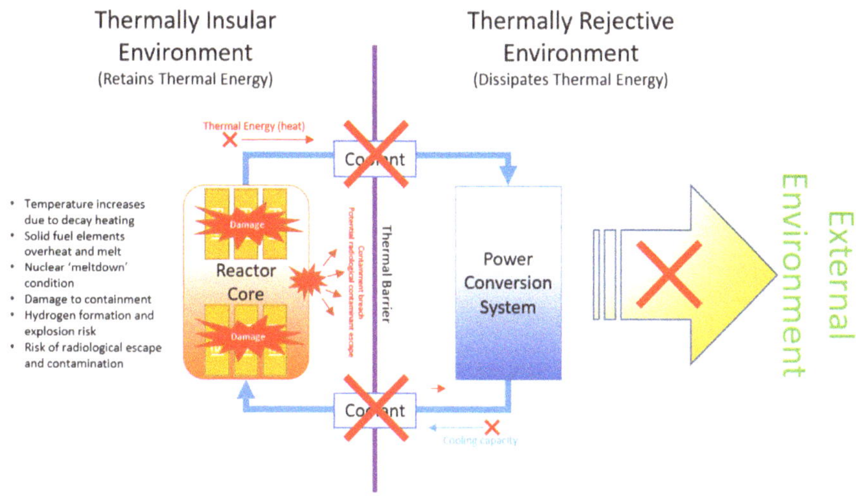

Fig. 1 Basic principle of loss-of-coolant-accident

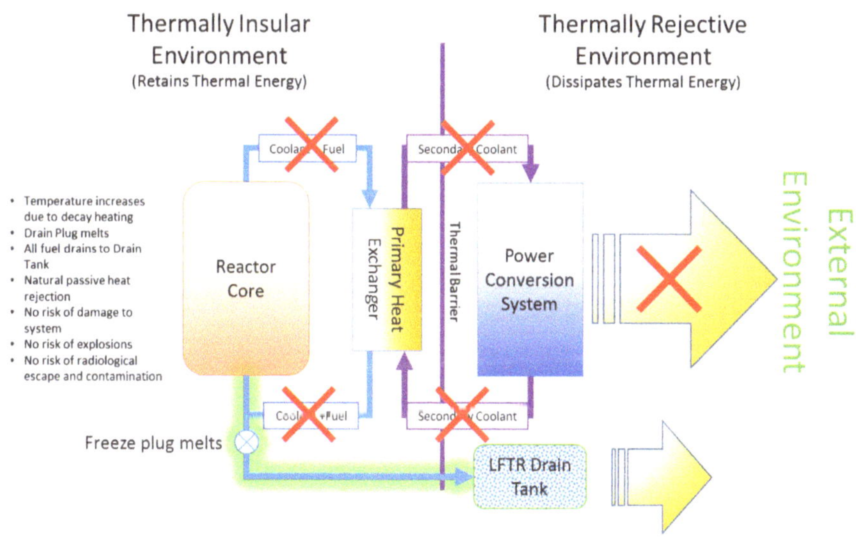

Fig. 2 Liquid-fueled systems are invulnerable to fuel melt

In contrast, a significant safety advantage of liquid fuels is that they are mobile and can be moved from the core to an environment that maximizes the passive rejection of decay heat in lieu of complex, fail-vulnerable engineered coolant systems, as shown in (Fig. 2). So, in this respect, a liquid fuel is preferable.

4.5 Coolant Type

Finally, we need to talk about the coolant, because the properties of the primary coolant essentially determine the reactor's size, and ease of manufacturing of components, and play a critical role in the overall safety characteristics of the system.

We would wish for the greatest margin of safety paired with the most efficient option in a coolant. With lower pressures we get significantly better safety, and with higher temperatures we get greater efficiency.

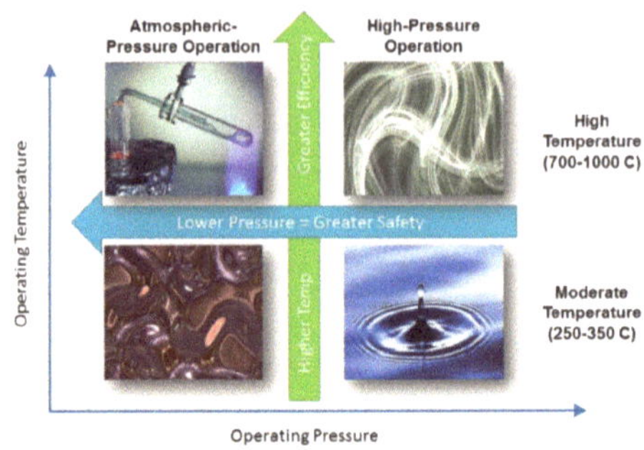

We can see from this 2 × 2 temperature and pressure matrix of the commonly available coolants that only salts can reach high temperatures yet operate at low pressures, making them ideal candidates for the transport of heat in a nuclear reactor core. But molten salts are more than just a coolant; it can also be a fuel carrier.

Fluoride salts, as a class of materials, embody many advantages. Most importantly, they are chemically compatible with the thorium fuel cycle and physically compatible for use as a liquid fuel medium.

They also possess numerous highly desirable nuclear, mechanical, and chemical characteristics ideal for use in a nuclear reactor.

- Extremely Stable
- V. Large liquid temp. range (~1000°C)
- Impervious to radiation damage
- Freezes at ~400°C
- Can be used as fuel carrier

The properties of molten fluoride salts make them ideal candidates for the transport of thermal energy in a nuclear reactor core.

5 The Ideal Scenario

To summarize, when we examine what a reactor that could maximize the four desirable characteristics might look like through the logical lens of the five primary design choices, it becomes naturally apparent that it would likely be

A **Liquid**, **Fluoride** salt, **Thorium** fueled, **Thermal** spectrum, **Breeder** Reactor.

The question is, is it possible to create such a reactor in a practical way? Fortuitously, the answer is, yes.

5.1 The Molten Salt Reactor

It is precisely this type of reactor that some of the greatest names in the annuals of science had envisioned would power the future back in the mid-twentieth century.

They called this reactor a 'molten salt reactor' or MSR. And this is the core enabling technology for the LFTR.

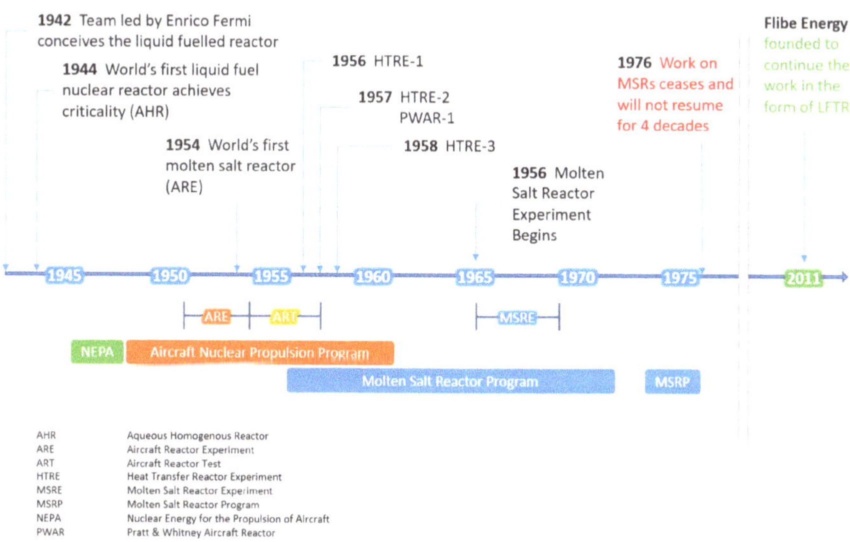

We can see from the above historical timeline that the technology has seen significant development work, right up to the point the MSR program was finally canceled in 1974 and the technology fell into obscurity for the next four decades. The MSRE was by itself a very successful exercise in engineering excellence. The experiment demonstrated the viability of the concept and its practical implementation.

While the reason for the program's cancellation is an important question, it is also one that is long and very complicated in the telling, one that would require a separate, dedicated article to explain.

Nonetheless, it is sufficient to know for now that the MSR's demise was a political decision, and the underlying reasons of which were almost entirely bureaucratic in nature; **there was no technical justification why it should have been abandoned**.

6 The Liquid Fluoride Thorium Reactor

The LFTR is the modernized and improved iteration of the original two-fluid MSR designed by Weinberg's team, incorporating the already proven core concept with modern materials and chemical processes that further increase efficiency, safety, and sustainability. Significantly, it is designed specifically to use the pure thorium fuel cycle.

6.1 Demonstrated Technology

LFTR is a direct descendant of the MSRE. The core technology enabling this reactor was already successfully demonstrated. But it is more than simply the MSRE's spiritual successor; it is the system optimized to use thorium as fuel that the Alvin Weinberg—who ironically was also one of the inventors of the pressurized light water reactor that we currently use—had intended to create.

Alvin Weinberg

Kirk Sorensen, principal designer of the LFTR and the world's foremost authority on the history of thorium research, rediscovered Weinberg's work in the early 2000s and started the thorium energy movement. He has worked ever since to realize Weinberg's dream of a world set free of energy constraints.

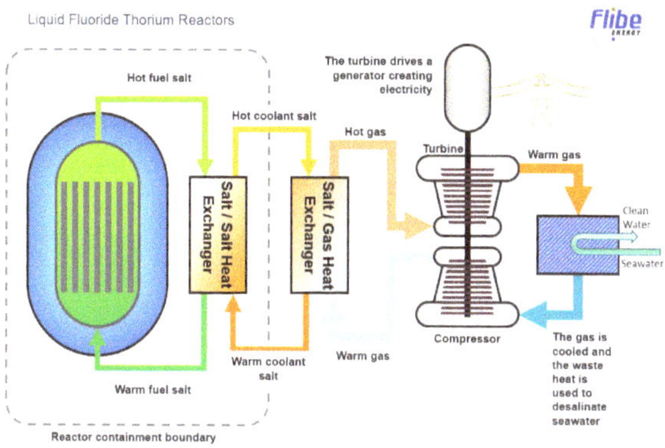

How an LFTR works is by circulating a naturally self-homogenizing fuel coolant mix called a 'fuel salt' in what is called the primary loop which passes through a graphite-moderated core and primary heat exchanger.

Fission occurs when the fuel salt enters the core and achieves the configuration for criticality, generating thermal energy from fission reactions that heat up the fuel salt. Upon exit from the core, fission promptly ceases as the fuel is naturally removed from the moderator, causing it to depart from the critical configuration.

The thermal energy is then transferred to a secondary coolant loop which ultimately transfers it to a closed cycle gas turbine PCS for upwards of 45% conversion efficiency!

This system can be cooled by air giving it substantial siting flexibility, but if it were located next to a body of salt water, it could use the enormous amount of clean, high-quality waste thermal energy to desalinate water, that would have otherwise been discarded to the environment. This would be of tremendous value and importance in some regions where water is just as important as electricity.

6.2 Advanced Power Conversion Advantages

Another important advantage of an LFTR is its ability to couple to an advanced gas turbine power conversion system.

This can lead not only to improvements in electrical generation efficiency but also to reductions in size and capital costs as you can see from the juxtaposition of turbine sizes compared to a reference 250 MW steam turbine.

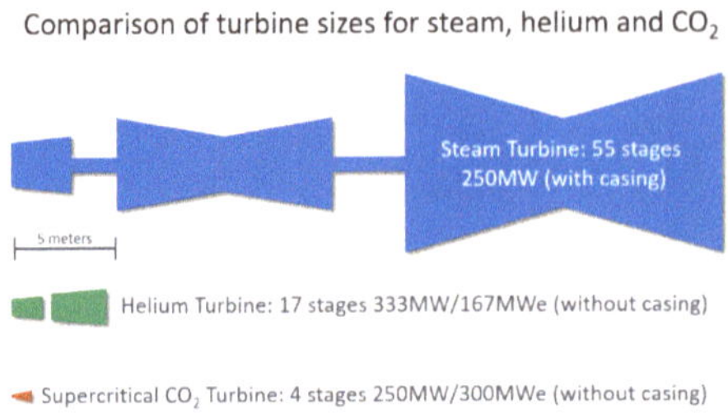

6.3 Modularity

There has also been a far greater interest in small modular reactors, with the goal of building reactor components in a factory environment and reducing site preparation time.

Creating modular reactors with liquid fluoride technology is much easier because the reactors do not operate at high pressure and can be shipped and returned unfueled.

6.4 Physical Inherent Safety

Having a liquid fuel enables a remarkable inherent passive safety feature to be implemented that can solve perhaps the most vexing problem in reactor safety.

For many years, engineers have been concerned primarily with how to cool the solid fuel of a reactor in the event of an accident to prevent a meltdown.

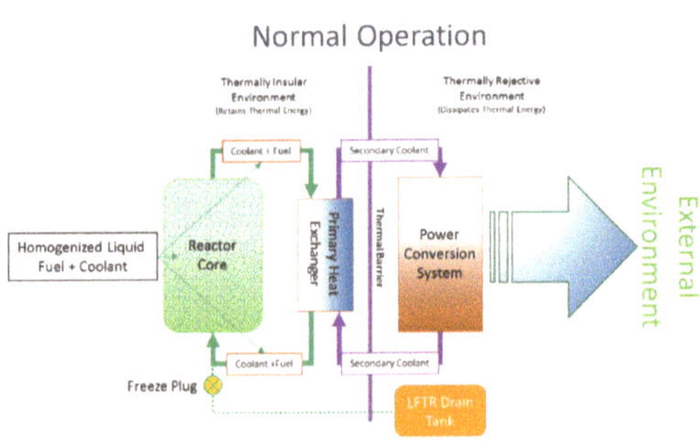

The use of a liquid fluoride fuel enables a simple and remarkable solution, based on the melting temperature of the salt, which is about 400 °C.

The reactor is fitted with a drain line that is kept plugged by a slug of frozen salt, which is kept frozen by an active flow of coolant over the outside of the drain line.

In the event of a complete loss of power, the salt plug would melt, causing the fuel salt in the reactor to drain through the line into a dedicated tank called a drain tank. This tank is configured to maximize the passive rejection of decay heat to the environment.

This enables the reactor to dispense with a multitude of emergency core cooling systems that are required in solid-fueled reactors, particularly those that operate at high pressure.

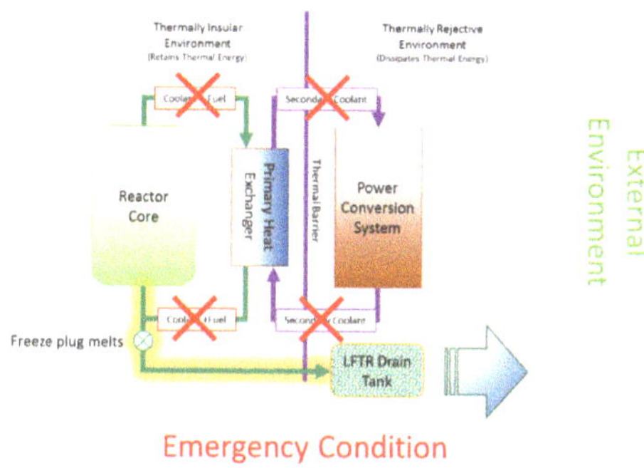

This remarkable safety feature is one of the most compelling arguments for consideration of the liquid fuel approach.

If you recall, earlier I implicated that the elegant solution is one where you minimize the number of problems that one needs to engineer solutions for. I'm sure that already you can see how the LFTR inherently avoids all the basic problems that necessitate highly complex control and safety systems in traditional and solid-fueled reactor technologies.

6.5 *How LFTR Would Utilize Thorium*

Thorium would be used in an LFTR by generating a fluoride salt and inserting it into a blanket, where it would absorb neutrons and transmute to fissile U-233, which would then be inserted into the core to repeat the cycle.

6.6 *Advantages of Thorium in LFTR*

Significantly, only when used in an LFTR will thorium fuel have the following properties:

High fissile protection characteristics (aka strong proliferation resistance) via the unavoidable evolution of U-232 in LFTR fuel. Contrary to some assertions, it is impossible to breed or separate LFTR fuel without U-232 contamination, **unless extremely extensive, expensive, and very easily detectable activities** (that will be eminently and quickly detected by international monitors) are undertaken. Furthermore, substantial modification to the LFTR will be required that would immediately reveal obvious nefarious intention.

And with the LFTR design, there is an additional safeguard that is to be implemented with international monitors like the IAEA known as 'Ionium spiking' that will further enhance this already very strong defense.

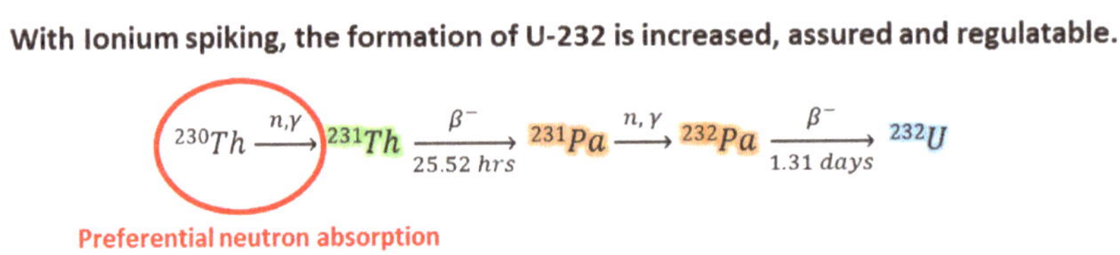

But even if we were to assume the hypothetical case where a determined actor with the resources, technical know-how, and requisite stealth were to attempt to divert an LFTR's fuel for weapons; one of the key characteristics of the LFTR is that removal of any strategically useful amounts of U-233 would cause it to shut down permanently, and instantly flag the would-be proliferator state or actor to international monitors. Furthermore, even if heavily modified, the LFTR's maximum breeding ratio under perfect conditions is only 1.07; it would take a very long time to 'skim' enough material to make a credible nuclear threat without causing a shutdown scenario. There are much easier and cheaper ways to make nuclear weapons, LFTR is not one of them and will never be.

The LFTR would have **Ultra-low waste production**, both in volume and in HLW content, thanks to LFTR's 90% + fuel utilization rate. With the potential to completely eliminate plutonium production if properly implemented.

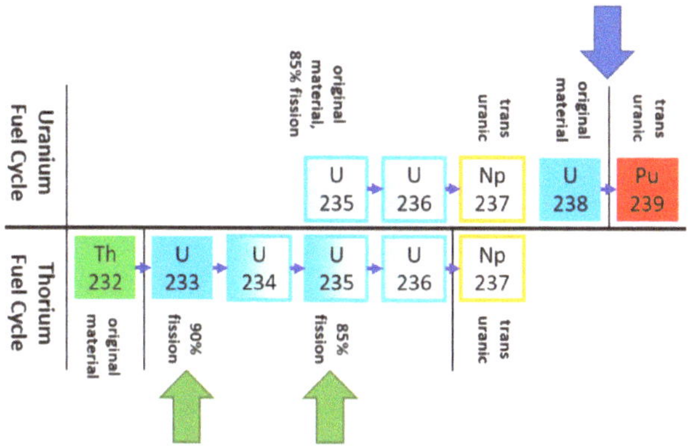

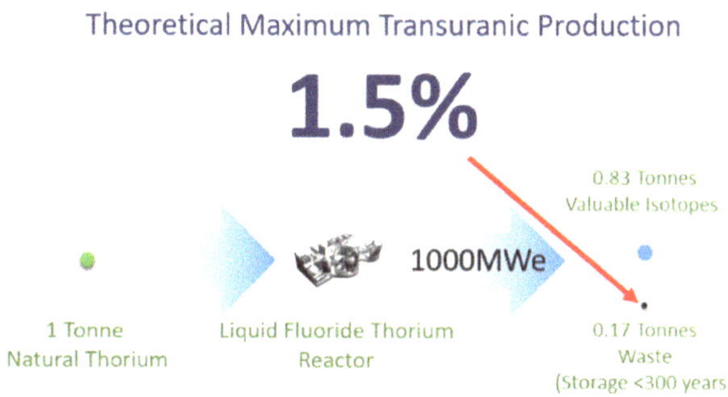

It has **technical zero safety radius** outside the reactor building (though whether it's practical to implement is a separate issue). This is because the low-pressure liquid fluoride thorium fuel system is invulnerable to accidents like fuel melt and explosive dispersion.

Invulnerable to Fuel Melt accidents
No explosive dispersion mechanisms
Fuel Immobilization in accident scenario
Minimal radiological propagation risk

7 Conclusion

We are in exciting times; few have been given the opportunity to potentially affect the future of our society in such a profound way.

The LFTR concept presents a highly optimized way of utilizing thorium as fuel, and the data we have thus far certainly suggests that it may very well be the 'elegant solution' to nuclear energy that we've long sought.

Part II
Fuel Cycle

Technology Assessment of Near-Term Open-Cycle Thorium-Fuelled Nuclear Energy Systems

W. J. Nuttall, S. F. Ashley, R. A. Fenner, P. D. Krishnani and G. T. Parks

Abstract As part of the RCUK–India civil nuclear research collaboration, British and Indian researchers have assessed the merits and disadvantages of, and potential for, open-cycle thorium–uranium-fuelled (Th–U-fuelled) nuclear energy systems. The research centred on fuel cycle modelling and life-cycle assessment of three Th–U-fuelled nuclear energy systems and compared these to a reference uranium-fuelled nuclear energy system, all operating with open nuclear fuel cycles. The results indicate that thorium-based fuels offer little benefit over conventional uranium-fuelled approaches for open-cycle nuclear energy production. This chapter provides an overview on the project and stresses overarching conclusions.

Keywords Thorium · Nuclear fuel cycle · Life-cycle analysis
Technology assessment

Nomenclature

ADSR	Accelerator-driven subcritical reactor
AHWR	Advanced heavy water reactor (DAE India funded)
CANDU	A widely adopted type of pressurised heavy water reactor developed in Canada
EPR	European pressurised reactor (AREVA)
FC	Fuel cycle
GHG	Greenhouse gas

W. J. Nuttall (✉)
School of Engineering and Innovation, The Open University, Milton Keynes MK7 6AA, UK
e-mail: william.nuttall@open.ac.uk

S. F. Ashley · R. A. Fenner · G. T. Parks
Department of Engineering, University of Cambridge, Cambridge CB2 1PZ, UK

P. D. Krishnani
Reactor Physics Design Division, Bhabha Atomic Research Centre,
Trombay, Mumbai 400085, India

A. K. Nayak and B. R. Sehgal (eds.), *Thorium—Energy for the Future*,
https://doi.org/10.1007/978-981-13-2658-5_7

GT-MHR	Gas turbine modular helium reactor (General Atomics)
LCA	Life-cycle analysis
LCOE	Levelised cost of electricity
MOX	Mixed oxide (nuclear fuel)
MSR	Molten salt reactor (various concepts)
PR	Proliferation resistance
PWR	Pressurised water reactor
SNF	Spent nuclear fuel

1 Introduction

As part of the RCUK–India civil nuclear research collaboration, British and Indian researchers have assessed the merits and disadvantages of, and potential for, open-cycle thorium–uranium-fuelled (Th–U-fuelled) nuclear energy systems. An overview of the systems chosen and the rationale for choosing these systems is presented in Ref. [1]. The work centred on assessing various performance indicators spanning material flows, waste composition, economics, emissions and proliferation resistance. These performance indicators were determined by using the UK National Nuclear Laboratory's fuel cycle modelling code ORION [2, 3] and also the life-cycle analysis code GaBi [4]. In addition, background issues surrounding the proliferation resistance of Th–U-based fuels were presented at the UK PONI 2012 Conference [5]. The purpose of the paper is to provide a summary of the project. To serve this end, the paper refers to some example results from component studies [3, 4] and stresses the overarching project conclusions.

2 Scope of the Research

Most considerations of thorium-fuelled nuclear energy focus on closed nuclear fuel cycles. That is cycles in which spent reactor fuel is reprocessed to yield fissile material for further useful electricity generation. The intention motivating the research was to understand whether thorium might have a useful role to play in open nuclear fuel cycles in which spent nuclear fuel is sent for disposal as radioactive waste without reprocessing. The motivation of the research has been described at greater length elsewhere [1].

It would not have been practical to seek to assess all conceivable technologies that could utilise thorium in open nuclear fuel cycles. Hence, three Th–U-fuelled reactor systems using uranium enriched to ~20% ^{235}U were assessed and compared to a reference technology. These were: AREVA's European pressurised reactor (EPR); India's advanced heavy water reactor (AHWR); and General Atomics' gas turbine modular helium reactor (GT-MHR). The reference technology was a uranium-fuelled EPR. These three Th–U-fuelled reactor systems are considered to be "near-term" technologies, in so far as one (the EPR) is under construction, another is ready for deployment (AHWR), and the third (GT-MHR) appears to be based entirely on proven technology—albeit with some engineering challenges remaining.

Accelerator-driven subcritical reactors (ADSRs) were excluded from this study on the grounds that significant reliability challenges lie in the path of commercial deployment of this technology for power production [6]. Another technology with potentially strong synergies with thorium fuel cycles is the concept of the molten salt reactor (MSR). MSR concepts were also considered out of scope, as (i) significant engineering challenges lie ahead for this technology, and (ii) the recirculation of liquid fuel/coolant shapes and constrains the fuel cycle towards approaches that have more in common with conventional closed fuel cycles than conventional open-cycles.

This work considers a strong definition of the "open" nuclear fuel cycle; i.e. no separated fissile materials (such as separated plutonium) are available from previous reprocessing. As such, the reactor systems are initiated and operated using enriched uranium. This restriction implies that potential technologies to address the disposal options for the UK's surplus separated civil plutonium are out of scope. An example of one of the out-of-scope technologies is the CANMOX approach (from CANDU Energy), whereby UK separated plutonium might be used to manufacture a Th–Pu mixed oxide (MOX) fuel suitable for use in the company's EC-6 reactor [7].

Although this study focusses on technologies that have a high technological readiness level, certain advances (most notably in fuel cladding materials) would be required. Advanced fuel cladding materials (i.e. silicon carbide and TRISO) are assumed for the Th–U-fuelled EPR and Th–U-fuelled GT-MHR, respectively. It is well established, see for instance Ref. [8], that the increased burn-up afforded by advanced cladding materials enables greater ^{233}U breeding and thus improves the performance of thorium cycles. Changes in the neutron spectrum, in particular hardening of the spectrum by reducing moderation, can also be exploited in the proposed future reactor systems to alter the breeding or burning characteristics of thorium-based fuels [9]. Further developments in either of these areas of technology could change the conclusions reached here about the potential of open thorium fuel cycles.

3 Methodology

To assess the three open-cycle Th–U-fuelled reactor systems, various analyses were performed. The bulk of the analysis is derived from the UK National Nuclear Laboratory's ORION fuel cycle modelling code [2]. The code performs inventory analysis to determine the throughput of material throughout a number of facilities in the nuclear fuel cycle, including storage buffers (that can represent the mine, mill and deep geological repositories), fuel fabrication facilities, reactors and, when applicable, reprocessing facilities. To model the isotopic inventories within a reactor, ORION requires shielded cross sections which are dependent on burn-up. These are produced by post-processing the results from deterministic or Monte Carlo-based neutronic assessments of the reactor core. In addition, radiotoxicity and decay heat calculations can be performed from the generated radionuclide inventories. Further details on the ORION fuel cycle modelling code and the reference data and reactor physics codes used to generate the burn-up-dependent shielded cross sections for the reactor technologies are contained in Ref. [3]. In addition, the results from the ORION fuel cycle modelling code provided the necessary inputs for the economic modelling (both levelised nuclear fuel cycle costs and levelised cost of electricity (LCOE) values, using SimLab) and proliferation resistance assessment (using the National Nuclear Laboratory's proliferation resistance tool). The results are also detailed in Ref. [3] and a brief summary of these results is presented in Sect. 4.

Life-cycle environmental emissions associated with thorium production and use in open Th–U-based nuclear fuel cycles were also investigated [4]. A comparative life-cycle analysis (LCA) was performed for the three Th–U-based systems using GaBi4 and the EcoInvent database [4]. The LCA considered the construction, operation and decommissioning of each of the reactor systems including all of the associated facilities of an open fuel cycle. This included the development of LCA models to describe the extraction of thorium from monazitic beach sands and for the production of heavy water. Life-cycle impacts per kWh (or per kg for thorium dioxide production) were obtained and are summarised in Sect. 4. Further details on this methodology, the underlying data used in this analysis and all life-cycle impact analysis results are detailed in Ref. [4].

4 Results

The full results have been published previously in Refs. [3, 4] and are summarised in Table 1.

To summarise, the three Th–U-fuelled nuclear energy systems involved a larger amount of separative work capacity than the equivalent benchmark U-fuelled system,

Table 1 Performance indicator values per kWh electricity generated [3, 4]. Green (italic) denotes better performance (whereas red (underlined) denotes poorer performance) than the benchmark uranium-fuelled EPR

Reactor	U_3O_8 (mg)	Enrich. (μSWU)	U fuel (mg)	SNF (nm^3)	Radiotox. (Sv)*	Heat (μW)*	FC Cost (¢)	LCOE (¢)	CO_2(eq) (g)	PR score
EPR UO_2	18.4	14.1	1.76	3.77	0.39	2.11	0.77	12.1	6.60	7.06
EPR UO_2/ThO_2	<u>25.4</u>	<u>24.3</u>	*0.68*	*3.03*	*0.35*	<u>2.21</u>	<u>0.90</u>	<u>12.2</u>	<u>6.86</u>	*7.22*
AHWR UO_2/ThO_2	<u>22.0</u>	<u>21.2</u>	*0.51*	<u>4.55</u>	*0.33*	<u>2.47</u>	<u>0.99</u>	<u>13.7</u>	<u>13.2</u>	*8.68*
GT-MHR UO_2/ThO_2	<u>52.6</u>	<u>50.8</u>	*1.21*	<u>32.3</u>	*0.28*	*1.52*	<u>2.37</u>	<u>15.7</u>	<u>10.7</u>	<u>6.40</u>

* Values for radiotoxicity and decay heat are taken 50 years after the fuel has been discharged.

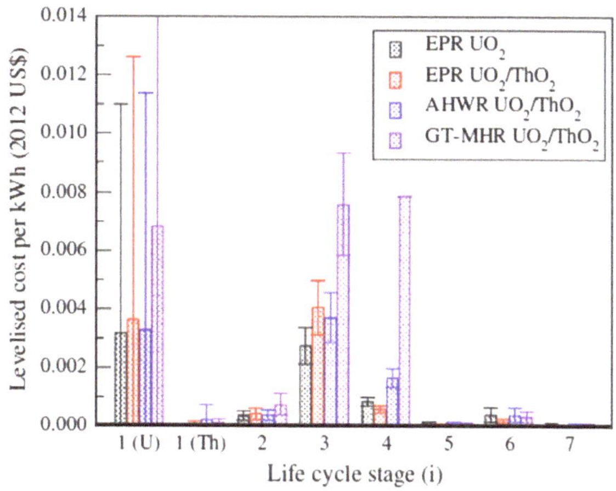

Fig. 1 Levelised nuclear fuel cycle costs per kWh generated [3]. The life-cycle stages (i) are described in the text. If the figure is monochrome: each set of four bars follows the sequence in the key

with higher levelised fuel cycle (FC) costs and with a greater LCOE. While a decrease of approximately 6% is seen in the amount of uranium ore per kWh needed by the Th–U-fuelled AHWR when compared to the U-fuelled EPR reference, the two other Th–U-fuelled systems needed more uranium ore per kWh than the U-fuelled EPR reference. Insignificant advantages and disadvantages were seen for both the quantity and the properties of the spent nuclear fuel (SNF) produced by the various nuclear energy systems investigated. Two of the Th–U-fuelled systems revealed some improvement regarding the proliferation resistance (PR) of the SNF produced. While open-cycle uranium-fuelled pressurised water reactors (PWRs), such as the EPR, can be considered to be highly resistant to proliferation, some improvement could be achieved by using a thorium–uranium MOX fuel in an EPR [3]. Similarly good scores are also achieved by Th–U-fuelled AHWR [3].

In Fig. 1, indicative results from the most challenging aspect of this research, a comparison of the fuel cycle costs per kWh of electricity generated considering all seven stages of the open nuclear fuel cycle, are presented. These seven stages are:

1. Producing milled ore (U_3O_8 or ThO_2);
2. Converting U_3O_8 to UF_6;
3. Uranium enrichment;
4. Nuclear fuel fabrication;
5. Fuel transport;
6. Interim SNF storage;
7. Final SNF waste disposal.

Bringing together the aspects illustrated in Fig. 1, and having deployed a Monte Carlo analysis, the mean LCOE (and uncertainty to one standard deviation) was found to be 121 ± 16 US$/MWh for the reference U-fuelled EPR, 122 ± 17 US$/MWh for the Th–U-fuelled EPR, 137 ± 18 US$/MWh for the Th–U-fuelled AHWR, and 157 ± 14 US$/MWh for the Th–U-fuelled GT-MHR. Noting the error bars and the difficulties inherent in such an analysis, it is apparent that thorium does not represent

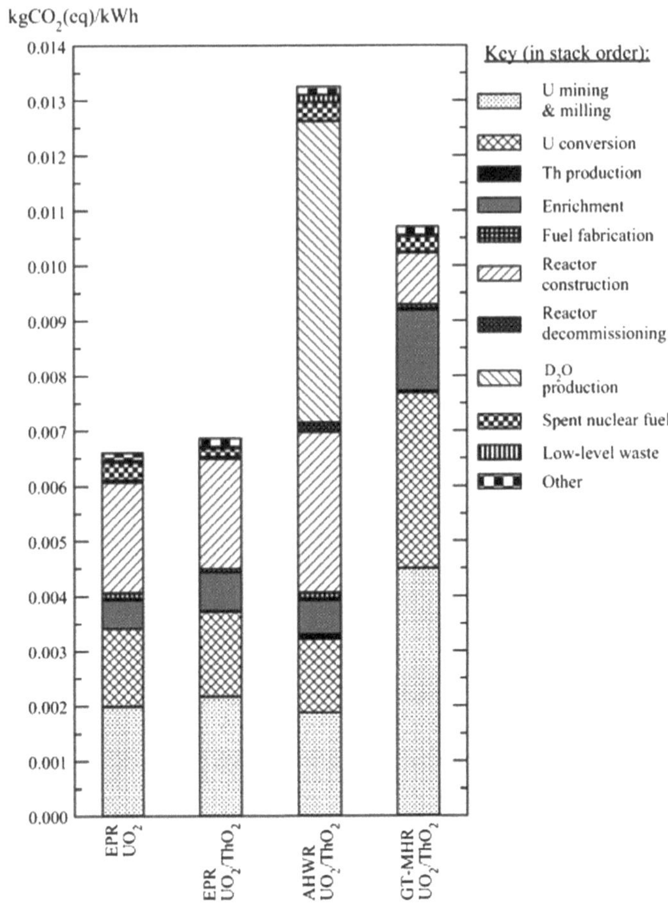

Fig. 2 Greenhouse gas impact per kWh of electricity generated for each of the four systems studied. For details, see Ref. [4]

an economically attractive fuelling option for these systems operating with an open nuclear fuel cycle.

The results of the life-cycle impact analysis highlight that the reference U-fuelled EPR has the lowest greenhouse gas (GHG), ozone depletion and eutrophication emissions per kWh generated; this is largely a consequence of the second-lowest requirement for uranium ore per kWh generated. The GHG result is shown in Fig. 2. The results demonstrate that the greatest overall contributor to environmentally harmful emissions is the requirement for mined or recovered uranium (and thorium) ore, albeit with the possible exception of nuclear energy systems requiring heavy water (see Ref. [4] for a detailed description and breakdown of the various emissions). Thorium obtained from monazitic beach sands is associated with lower overall damaging emissions than uranium (either conventionally mined or recovered from in situ leaching) when considered in a like-for-like comparison of mining and recovery methods. If thorium were to become a viable nuclear fuel, we would expect monazitic beach sands (and equivalent placer deposits) to be utilised first and preferentially, despite such resources only forming 30% of the overall known thorium ore deposits. Generally, and unsurprisingly, for the four nuclear energy technologies considered, the range of CO_2(eq) emissions per kWh generated (6.60–13.2 g CO_2(eq)/kWh) appears to be low when compared to the emissions generated by the dominant fossil fuel combusting electricity generation technologies.

5 Conclusions

The analysis shown here indicates that thorium fuel offers little benefit over conventional uranium-fuelled approaches for open-cycle nuclear energy production limited to the low-enriched uranium standard of 20% ^{235}U. Hence, short- to medium-term interest in thorium should be restricted to those countries with an interest in spent nuclear fuel reprocessing or with a need to reduce inventories of fissile material (such as separated plutonium).

Acknowledgements The work was funded by the Engineering and Physical Sciences Research Council (UK) under grant no. EP/I018425/1. We are most grateful to all those who worked to establish the RCUK–India civil nuclear collaboration. We also wish to note our gratitude to all those listed as co-authors and in the acknowledgements sections of the papers where our individual studies are more fully described.

References

1. S.F. Ashley, R.A. Fenner, W.J. Nuttall, G.T. Parks, Open cycle thorium–uranium-fuelled nuclear energy systems. Proc. ICE Energy **166**(2), 74–81 (2013)
2. R. Gregg, C. Grove, Analysis of the UK nuclear fission roadmap using the ORION fuel cycle modelling code, in *Proceedings of IChemE Nuclear Fuel Cycle Conference*, IChemE, Manchester, UK, 2012
3. S.F. Ashley, B.A. Lindley, G.T. Parks, W.J. Nuttall, R. Gregg, K.W. Hesketh, U. Kannan, P.D. Krishnani, B. Singh, A. Thakur, M. Cowper, A. Talamo, Fuel cycle modelling of open cycle thorium-fuelled nuclear energy systems. Ann. Nucl. Energy **69**, 314–330 (2014)
4. S.F. Ashley, R.A. Fenner, W.J. Nuttall, G.T. Parks, Life-cycle impacts from novel thorium–uranium-fuelled nuclear energy systems. Energy Convers. Manag. **101**, 136–150 (2015)
5. S.F. Ashley, W.J. Nuttall, G.T. Parks, A. Worrall, On the proliferation resistance of thorium–uranium nuclear fuel, in *Proceedings of UK PONI Conference*, 2012
6. S.J. Steer, W.J. Nuttall, G.T. Parks, L.V.N. Gonçalves, Predicting the contractual cost of unplanned shutdowns of power stations: an accelerator-driven subcritical reactor case study. Electr. Power Syst. Res. **81**, 1662–1671 (2011)
7. CANDU Energy, *EC6 and the CANMOX Project,* corporate brochure available at: http://www.candu.com/site/media/Parent/CANDU%20brochure-CANMOX-0905.pdf no date

8. E. Shwageraus, H. Feinroth, Potential of silicon carbide cladding to extend burnup of Pu-Th mixed oxide fuel, in *Proceedings of American Nuclear Society (ANS) Annual Meeting*, Hollywood, Florida, USA, pp. 658–660, 2011
9. B.A. Lindley, C. Fiorina, F. Franceschini, E.J. Lahoda, G.T. Parks, Thorium breeder and burner fuel cycles in reduced-moderation LWRs compared to fast reactors. Prog. Nucl. Energy **77**, 107–123 (2014)

Sorption of Protactinium (V) on Silica Colloid and the Effect of Humic Acid

Madhusudan Ghosh and Rakesh Verma

Abstract Protactinium (V) is strongly sorbed on silica colloids in the pH range of 3–9 in $NaClO_4$ medium. The sorption of Pa (V) is lower (70–80%) below pH 3 and above pH 9. There is no significant change in the sorption behavior of Pa (V) on silica colloid in the presence of humic acid between pH 3 and 9. However, there is a small reduction in the sorption below pH 2 and above pH 10. To elucidate the sorption mechanism the charge of Pa (V) species in different pH medium was determined by ion-exchange technique using ^{231}Pa tracer. The average charge on the Pa (V) species in dilute $HClO_4$ (0.1–3.0 M) medium was 1+. Although neutral species predominated between pH 6 and 11, negatively charged species started forming in higher pH and were predominant above pH 11. The zeta potential of silica, humic acid and silica-humic acid systems were also measured. The point of zero charge (PZC) of silica was about pH 2, and the magnitude of the zeta potential increased with increase in the pH. The zeta potential of humic acid was negative between the pH 1 and 12, and the magnitude increased with increase in the pH. The zeta potential of silica–humic acid combined system in the pH range of 1–12 was negative, whereas the magnitude was between that of silica and humic acid. The strong sorption of Pa (V) on silica colloid in the pH range 3–9 was assigned to surface complexation, whereas the reduction in sorption (below pH 3 and above pH 9) was attributed to electrostatic repulsion, as a fraction of Pa (V) exists as cationic species at pH $\leq$ 2 and anionic species above pH 9. The reduction in the sorption of Pa (V) below pH 2 and above pH 10 in the presence of humic acid was assigned to reduced magnitude of the zeta potential of silica in the presence of humic acid and complexation of Pa (V) with humic acid, respectively. Comparison of the results of sorption of Pa (V) on silica with that of Nb (V) showed the chemical similarity between the two metal ions.

Keywords Silica colloid · Protactinium · Humic acid · Sorption
Radioactive waste

M. Ghosh (✉) · R. Verma
Analytical Chemistry Division, Bhabha Atomic Research Centre, Mumbai 400085, India
e-mail: ghoshm@barc.gov.in

R. Verma
e-mail: rvermabarc@gmail.com

A. K. Nayak and B. R. Sehgal (eds.), *Thorium—Energy for the Future*,
https://doi.org/10.1007/978-981-13-2658-5_8

1 Introduction

^{231}Pa, an alpha-emitting isotope with a relatively long half-life (32,760 year), is formed during irradiation of thorium in a thermal nuclear reactor for breeding 'fissile' isotope ^{233}U [1, 2] as shown below.

$$^{232}_{90}\mathrm{Th} \xrightarrow{(n,2n)} {}^{231}_{90}\mathrm{Th} \xrightarrow{\beta^-} {}^{231}_{91}\mathrm{Pa}$$

Uranium (^{233}U) and thorium are separated from fission products by thorium–uranium extraction (THOREX) process which is based on solvent extraction using tributyl phosphate. During separation by THOREX process [3, 4], ^{231}Pa along with fission products remains in the aqueous phase and forms part of the high-level radioactive waste. ^{231}Pa could have long-term radiological impact that needs to be suitably resolved [5, 6].

It is of major concern if high-level radioactive waste stored underground [7, 8] comes in contact with flowing groundwater; subsequently, radionuclides may leach out and migrate [9–11]. The rate of radionuclide transport will depend upon the distribution of radionuclide between aquifer solid and groundwater apart from the groundwater velocity. It is expected that elements with extremely low aqueous solubility and strong sorption on aquifer solid will be relatively immobile. However, the migration of plutonium, that has low solubility, is known to occur [12]. A three-phase model of contaminant transport, known as colloid-facilitated transport [13], was proposed to explain the phenomenon [14, 15]. The three-phase system consists of stationary aquifer solid, the mobile aqueous phase, and non aqueous mobile colloids.

The key processes in colloid-facilitated radionuclide transport in natural groundwater systems are (i) colloid production, stability, and filtration; (ii) sorption and desorption processes of radionuclides onto colloids and colloids onto geologic materials; and (iii) colloid transport. The genesis and stabilization of groundwater colloids, the factors controlling their transport through porous media, and the importance of major pathways and mechanism to predict the role of colloids in subsurface transport of contaminants have been reported [15].

Sorption of radionuclides on colloidal surfaces may be due to electrostatic forces, ion exchange, surface reactions, and co-precipitation, and the rate and magnitude of sorption may depend upon the concentration of radionuclide and colloid, pH and ionic strength of solution, and natural complexing species present in the solution. Oxides, especially those of Si, Al, and Fe, occur as colloids in groundwater. Various aspects of colloids that facilitated the transport of radionuclides in natural groundwater systems have been reviewed [16], and one of the key areas of current research is the factors affecting the sorption of radionuclides onto colloidal surfaces.

Pa (V) has a great tendency to hydrolyze and polymerize in aqueous media forming hydrophilic colloidal system with strong adsorption properties [17, 18]. Pa (V) is a likely candidate for colloid-facilitated transport, owing to its extremely low solubility and tendency to form colloids. Knowledge of the extent of association of protactinium with natural colloids is essential for understanding its colloid-facilitated transport. The interaction of natural and anthropogenic radionuclides with components of natural water has been surveyed [19, 20], and the sorption of niobium [21] and uranium [22] on hydrous silica has been reported.

The most important natural ligands found in groundwater are humic acid. The presence of humic acid may affect the sorption and migration of radionuclides in the groundwater [23, 24]. It is reported that the anthropogenic actinides are about ten times more concentrated in humic acid fraction than in the overall topsoil [25]. The interaction of humic acid with radionuclides increases the mean residence time of actinides in mobile phase and reduces the sorption of radionuclides to the stationary phase. Humic acid can alter the oxidation state of the metal ions. Humic acid reduces U (VI) to U (IV) and Np (V) to Np (IV) [26].

In this paper, we report the sorption behavior of Pa (V) on silica colloids, studied using ^{231}Pa radiotracer, as a function of pH.[1] The sorption experiments were carried out in a solution of 0.05 M sodium perchlorate ($NaClO_4$) in the pH range of 1–12. Effect of humic acid on the sorption of Pa (V) on silica was also studied. The charge of Pa (V) species in perchloric acid ($HClO_4$) medium was determined by ion-exchange technique using ^{231}Pa radiotracer. The probable mechanism of sorption of Pa (V) on silica and effect of humic acid is discussed in terms of the zeta potential and the species of Pa (V) that exists in the solution. The sorption of Nb (V) on silica was also studied using radiotracer (^{94}Nb and ^{95}Nb) and compared with that of Pa (V) to study their chemical similarity.

2 Materials and Methods

2.1 *Materials*

The chemicals used for the study are colloidal silica (AEROSIL 200, commercial grade), anion exchanger AG 1-X8 (Bio-Rad, 50–100 mesh), cation exchanger AG 50W-X8 (Bio-Rad, 50–100 mesh), $NaClO_4$ (Sigma Aldrich), and humic acid–sodium salt (Sigma Aldrich).

2.2 *Characterization of Colloidal Silica*

2.2.1 X-Ray Diffraction (XRD) Measurements

Philips X-ray diffractometer (PW1710 model) with Ni-filtered Cu-K_α radiation was used to record the XRD pattern of silica.

[1]Preliminary results of this work were presented in the ThEC15: International thorium energy conference: gateway to thorium energy; Mumbai (India); 2015.

2.2.2 Surface Area Measurement

Specific surface area measurement was carried out using an indigenous surface area analyzer (BARC) using Brunauer–Emmett–Teller (BET) method. About 25 mg of accurately weighed silica was heated electrically at about 110 °C in a glass sample cell for nearly 4 h in continuously flowing mixture of nitrogen and helium. The sample was then cooled to liquid nitrogen temperature in a continuously flowing mixture of nitrogen and helium. The adsorbed nitrogen was desorbed by bringing the sample to room temperature. The changes in the nitrogen concentration of the flowing gas stream, consequent to the adsorption and desorption, were monitored by a thermal conductivity bridge. The desorption signal was electronically integrated from which the surface area of the sample was computed.

2.2.3 Particle Size Measurement

Silica colloid was added to deionized water, ultrasonicated (Retsch GmbH, Germany) (1 min, 35 kHz, power = 140 W), and transferred to a quartz cuvette (universal 'dip' cell). It was assumed that the soft conditions of ultrasonication would not have altered particle size. Malvern 4800 Autosizer with a 7132 digital correlator was used for dynamic light scattering (DLS) measurements. The light source employed was He–Ne laser (632.8 nm) with maximum power of 15 mW. CONTIN method was used to analyze the correlation functions.

2.2.4 Zeta Potential Measurement

The mobility of charged particles in liquid medium under fixed applied electric field depends on its surface charge, and this principle utilized to measure the zeta potential. The method for the zeta potential measurement has been reported [21], and the procedure was described. Malvern Zetasizer Nano Z (UK) was used for the zeta potential measurements of silica (2 g L^{-1}), humic acid (5 mg L^{-1}), and silica (2 g L^{-1})–humic acid (5 mg L^{-1}) combined systems. He–Ne laser (632.8 nm, 4 mW) light source was used for the measurements. Universal 'dip' cell with 10-mm light pathway was used for the measurement. The experiments were performed at ambient condition (25 ± 1) °C and at fixed ionic strength of 0.05 M $NaClO_4$. The pH was adjusted between 1 and 12 by using dilute solutions of $HClO_4$ and NaOH. The pH was measured by glass electrode (Oakton, pH700, USA). Ultrasonication (1 min, 35 kHz, power = 140 W) of sample solution was carried out before the zeta potential measurements to avoid aggregation, and it was assumed that the soft conditions of ultrasonication would not have altered particle size or species.

2.3 *Humic Acid: Purification and Characterization*

2.3.1 Purification

Humic acid was purified [26] before its use, and the purification procedure is described briefly. Sodium salt of humic acid (~10 g) was dissolved in 100 mL of 0.1 M NaOH and centrifuged at 3000 rpm for 30 min. Supernatant was decanted, and the precipitate containing humin and clay minerals was discarded. Solid NaF (0.3 g for 1 g of humic acid) was added to humic acid solution and stirred overnight with N_2 bubbling to dissolve silicate impurities. Humic acid was precipitated with HCl solution at pH 1, redissolved in 0.1 M NaOH, and reprecipitated. The humic acid precipitate was washed with 0.1 M HCl to remove Na^+ ions and was followed by washing with deionized water. Purified humic acid was air dried and stored in a desiccator.

2.3.2 Elemental Analysis

As-received and purified sodium salts of humic acid were characterized by energy dispersive X-ray spectrometry (EDS) (Quanta, Model: 3D FEG-I, Switzerland). Hydrogen and nitrogen concentrations in humic acid were determined by CHNS analyzer (Euro Vector EA3000EA-IRMS elemental analyzer, Italy).

2.3.3 Proton Exchange Capacity

The total exchange capacity was determined by Baryta method [27], and the procedure is described briefly. About 80 mg of accurately weighed purified humic acid was equilibrated with excess 0.1 M $Ba(OH)_2$ solution (50 mL) on a mechanical shaker for about 24 h with occasional N_2 gas bubbling to remove CO_2. After equilibration, the solution was filtered and the precipitate was washed with deionized water. Filtrate and washings were combined, and the solution was titrated using 0.5 M standard HCl.

Capacity corresponding to –COOH group was determined by equilibrating purified humic acid (~80 mg) with 50 mL of 0.1 M $Ca(OOCCH_3)_2$ solutions for 24 h. After equilibration, the solution was filtered and the precipitate was washed with deionized water. The filtrate and washings were combined, and the solution was titrated using 0.5 M standard NaOH.

Capacity of as-received humic acid was also determined using the above procedure.

2.4 Purification of ^{231}Pa Radiotracer

For preparing radiotracer, an old stock of ^{231}Pa [28] in organic medium was extracted by 0.5 M HF solution. A white gelatinous precipitate appeared on addition of alkali indicating the presence of impurities. Energy dispersive X-ray fluorescence analysis of the precipitate showed the presence of Ti and Zr. Hence, ^{231}Pa was purified by anion exchange method [29]. The unpurified ^{231}Pa solution was loaded on anion exchanger AG 1-X8 (50–100 mesh, Cl^- form, Bio-Rad) filled in a polypropylene column. 9 M HCl + 0.004 M HF was used as both loading and eluting medium. The eluted fractions containing ^{231}Pa were neutralized with NH_4OH. On evaporation of the solution on sand bath, white crystals of NH_4Cl appeared which was removed by heating the solution in a platinum dish using Bunsen burner. The residue was taken in HNO_3, and few drops of HF were added. Concentration of ^{231}Pa in the final solution was 9.5×10^{-6} M.

All sorption experiments were carried out in polypropylene containers (Tarson). The sorption experiments were carried out after removing HF by repeated evaporation of aliquot of ^{231}Pa radiotracer solution with HNO_3.

2.5 Sorption Experiments

2.5.1 Sorption of Humic Acid on Silica

Calibration plot was obtained in the concentration range 0.5–15 mg L^{-1} of purified humic acid in 10 mL of 0.05 M $NaClO_4$ solutions. The solutions were equilibrated for about 30 min on a mechanical shaker before absorbance measurement at 257 nm using UV–Vis spectrophotometer (V-530, Jasco).

Humic acid (15 mg L^{-1}) was added to previously prepared silica (2 g L^{-1}) solution in 0.05 M of $NaClO_4$. The pH was adjusted between 1 and 12, and the solutions were equilibrated for 48 h on a mechanical shaker. Phase separation was carried out by centrifugation at 10,000 rpm (Hermle Labotechnik GmbH Z383K, Germany) for 1 h. The extent of separation was checked by measuring silica in supernatant by DLS technique. The absorbance of the supernatant was measured at 257 nm, and concentration of humic acid in supernatant was obtained from the calibration plot.

In a separate experiment, colloidal silica suspension (2 g L^{-1}) was prepared in acidic (pH 1), neutral (pH 7), and alkaline (pH 11) aqueous media, equilibrated on a mechanical shaker and centrifuged. The supernatant was analyzed by inductively coupled plasma optical emission spectrometry and contained 2 mg L^{-1} (pH 7), 14 mg L^{-1} (pH 1), and 120 mg L^{-1} (pH 11) of silica.

2.5.2 Sorption of ^{231}Pa on Silica

Sorption studies were carried out in 0.05 M $NaClO_4$ medium in the pH range of 1–12 by equilibrating ^{231}Pa (~2.9×10^{-9} M) with silica (2 g L^{-1}) for 48 h on a mechanical shaker. Total volume of the solution was 25 mL. Phase separation was carried out using high-speed centrifuge (10,000 rpm), and the completeness of separation was checked by measuring silica in supernatant by DLS technique.

Supernatant (10 mL) was taken for measurement of gamma ray (283.69 keV, relative abundance 1.7%) of ^{231}Pa (counting time 60,000 s) using high-purity germanium (HPGe) detector (EG&G ORTEC) coupled with 8 k channel analyzer (relative efficiency: 45%; resolution: 1.9 keV at 1132 keV of ^{60}Co). The percentage sorption was calculated by comparing the net counts of the supernatant with that of the standard solution as shown in Eq. 1.

$$\text{Sorption}\,(\%) = (A_0 - A)/A_0 \times 100 \tag{1}$$

where A_0 = net counts of the standard solution and A = net counts of the supernatant.

Identical experiment was carried out to study the sorption of niobium (1.12×10^{-9}) mL^{-1} on silica using ^{94}Nb and ^{95}Nb radiotracers [21].

2.5.3 Sorption of ^{231}Pa on Silica in the Presence of Humic Acid

The sorption of ^{231}Pa on silica in the presence of humic acid (5 mg L^{-1}) was measured using the procedure described in Sect. 2.5.2 under identical conditions.

2.6 Determination of Charge on Pa (V) Species

A very simple technique based on ion exchange has been described by Marhol, Strickland, and Welch [30–32] for determining the charge of an ion in aqueous solutions below 10^{-6} M using radiotracer. About 100 μL of ^{231}Pa solution (0.061 μg mL^{-1}) and 1 g of mono-functional strong cation exchanger resin (AG 50W-X8) were added to 10 mL of $HClO_4$ solutions of desired molarities (0.1–3 M). The resin was equilibrated for 2 h using a mechanical shaker. To determine the charge on Pa (V) in neutral and basic media (10^{-8} to 0.5 M NaOH), ^{231}Pa tracer was equilibrated with 1 g of anion exchanger resin (AG 1-X8). The count rate of 283.69 keV gamma ray of ^{231}Pa in the solution and resin were measured using HPGe detector, and the distribution ratios (D) of Pa were determined using Eq. 2.

$$D = A_{\text{resin}}/A_{\text{solution}} \quad (2)$$

where A_{resin} and A_{solution} are count rates of resin (g^{-1}) and solution (mL^{-1}) at equilibrium.

The overall net charge on Pa (V) ion was determined from the slope of Log (D) versus Log (M) plot.

3 Results and Discussion

3.1 Characterization of Silica and Humic Acid

X-ray diffraction pattern indicated the amorphous nature of silica. BET surface measurements and DLS showed that the specific surface area and particle size were $(160 \pm 20)\ m^2\ g^{-1}$ and 160–310 nm, respectively [21].

EDS spectra of unpurified humic acid indicated the presence of appreciable amount of the impurities like Al, Si along with Na. The other impurities like Fe, K, Ca, Ti were also found. After purification, the relative contents of C, S increased while that of O decreased and concentration of other impurities decreased significantly (Table 1). The atomic ratios of H/C, N/C, S/C did not change substantially, establishing that there was no major chemical alternation and modification of humic acid molecule during the purification process. The large decrease in O/C ratio after purification indicated the removal of oxygen-rich impurities such as clay and other minerals.

The proton exchange capacities of as-received and purified humic acid were (4.1 ± 0.1) meq g^{-1} and (5.4 ± 0.2) meq g^{-1}, respectively. The capacity of purified humic acid corresponding to –COOH group was (3.8 ± 0.1) meq g^{-1}, and hence, the exchange capacity corresponds to phenolic –OH group which was about 1.6 meq g^{-1}.

The point of zero charge (PZC) of silica was about pH 2, and the magnitude of the zeta potential increased with increase in the pH. The zeta potential of humic acid was negative between pH 1 and 12, and the magnitude increased with increase in the pH. The zeta potential of silica–humic acid combined system in the pH range of 1–12 was negative, whereas the magnitude was between that of silica and humic acid.

Table 1 Elemental composition of Na salt of humic acid (as-received) and purified humic acid

Element	As-received (atom %)	Purified (atom %)
H	3.1 (±0.05)	4.7 (±0.05)
C	36.2 (±0.3)	57.9 (±0.94)
N	0.4 (±0.03)	0.7 (±0.03)
O	50.2 (±0.4)	37.91 (±0.3)
Na	9.2 (±0.2)	0.41 (±0.2)
Al	1.7 (±0.2)	0.06 (±0.03)
Si	1.7 (±0.2)	0.02 (±0.01)
S	0.3 (±0.1)	0.35 (±0.07)
K	0.2 (±0.1)	–
Ca	0.1 (±0.1)	–
Ti	0.2 (±0.02)	0.07 (±0.02)
Fe	0.3 (±0.1)	0.17 (±0.06)
(H/C)	0.086	0.078
(N/C)	0.011	0.012
(O/C)	1.387	0.720
(S/C)	0.015	0.013

3.2 Determination of Charge on Pa (V) Species

For the determination of the ionic charges of an anion and cation in aqueous solution, mono-functional ion-exchange resins with strongly acidic and basic groups are used, respectively [31]. The overall net charge on Pa (V) ion, determined from the slope of Log (D) versus Log (M) plot (Fig. 1) in the dilute perchloric acid medium (0.1–3.0 M) was about 1.12+. In the pH range, Log (D) versus Log (M) plot (Fig. 2) has two distinct slopes, one between pH 6 and 11 and another above pH 11. Between pH 6 and 11, the charge on Pa (V) ion was 0.1−, which indicated that the major fraction of protactinium was present as neutral species and negatively charged species started forming only at higher pH. Above pH 11, the charge on Pa (V) ion was ~0.8−, suggesting that the protactinium predominantly existed as singly negatively charged species. It is difficult to explicitly identify Pa (V) species from the net average charge as more than one species is likely to exist simultaneously in the solution. The above findings are consistent with the literature [33, 34].

3.3 Sorption Studies

In order to model the migration of Pa (V) under aqueous environmental conditions, it is essential to measure the degree of sorption of Pa (V) on the colloids and the effect

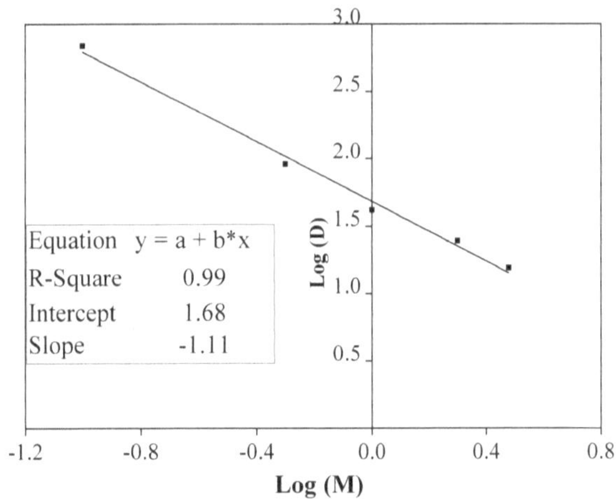

Fig. 1 Distribution ratio (D) of Pa versus similarity (M) of $HClO_4$

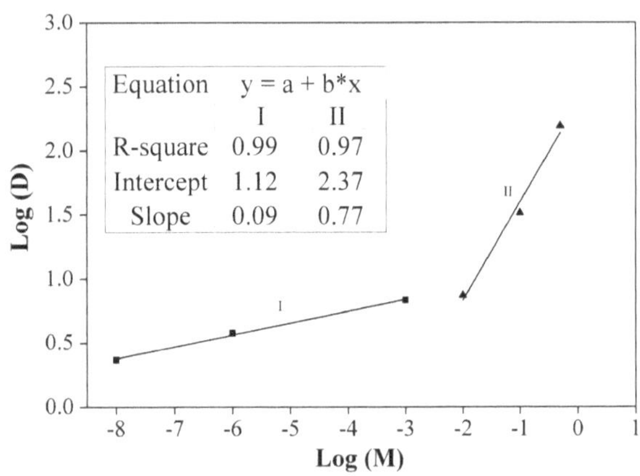

Fig. 2 Distribution coefficient (D) of Pa versus concentration of OH^- ion

of naturally occurring ligand, humic acid on the sorption. In addition, understanding of the chemical species of Pa (V) and nature of interactions with natural colloids are also required.

3.3.1 Sorption of Humic Acid on Silica

pH-dependent sorption of purified humic acid (2 mg L^{-1}) on silica (2 g L^{-1}) was carried out in 0.05 M $NaClO_4$ solution, and the results are presented in Fig. 3. The sorption of humic acid decreased from 77.8% at pH 1 to 9.8% at pH 12. Point of zero charge (PZC) of silica was at ~pH 2, whereas the zeta potential of humic acid is negative even at pH 1, which explains the high sorption of humic acid on silica at low pH. Above pH 2, the zeta potential of both silica and humic acid is negative and the magnitude increases with increasing pH. Therefore, electrostatic repulsion

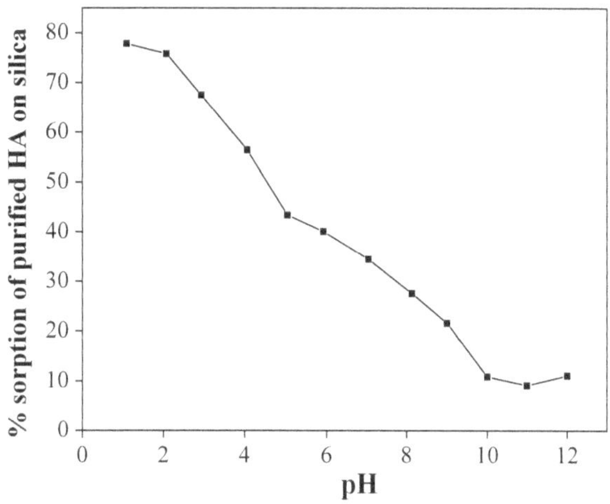

Fig. 3 pH-dependent sorption of purified humic acid on silica

reduces the sorption of humic acid on silica above pH 2. Even though there was a reduction in the sorption with increasing pH, some amount of humic acid was still sorbed on silica, which is not fully understood [35, 36]. Sorption of humic acid on silica decreased with increasing pH and showed an upward trend beyond pH 10 which was not significant.

During the present work, polypropylene tubes were used instead of glass to prevent the loss of protactinium. It was found that the sorption of protactinium onto polypropylene tube was (2–5)% depending upon pH of the medium.

3.3.2 Sorption of Protactinium on Silica

The sorption of Pa (V) on silica from 0.1 to 8 M $HClO_4$ solutions has been reported, and it was found that the sorption decreased from about 25% to less than 1% [37]. The sorption of Pa (V) on silica in the pH range has not been reported in the literature. The sorption of Pa (V) on silica as a function of pH is shown in Fig. 4. The sorption behavior can be explained in terms of ionic species of Pa (V) in aqueous medium and the zeta potential of silica. The species of Pa (V) that exist in non-complexing media (0.1 M $Na/HClO_4$) are (a) $PaO(OH)^{2+}$ between pH 0.3 and 2, (b) $PaO(OH)_2^+$ or PaO_2^+ between pH 2 and 5, (c) $Pa(OH)_5$ or $PaO_2(OH)$ between pH 5 and 9, and (d) $Pa(OH)_6^-$ above pH 9 [33, 34]. Zeta potential of silica is positive $\leq$ pH 2, and Pa (V) also exists as doubly positive charged species resulting in negligible sorption because of electrostatic repulsion. Between pH 3 and 5, the zeta potential of silica is negative and Pa (V) exists as mono-positive species which leads to enhanced sorption. Between pH 5 and 9, quantitative sorption can be explained in terms of surface complexation.

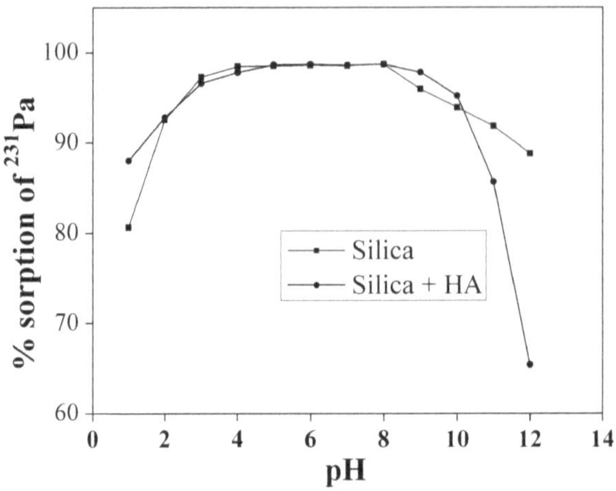

Fig. 4 Effect of pH and humic acid on sorption of ^{231}Pa on silica

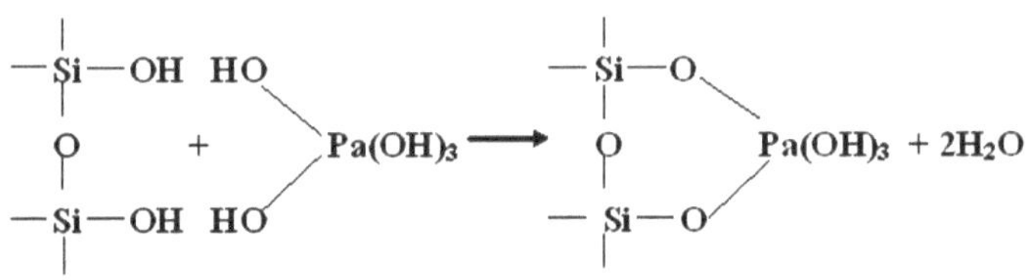

Above pH 9, Pa (V) exists as negatively charged species and the zeta potential of silica is highly negative, and the electrostatic repulsion causes the decreased sorption.

3.3.3 Effect of Humic Acid on Sorption of Protactinium on Silica

In the presence of humic acid, the sorption behavior of Pa (V) between pH 3 and 9 is similar to that in the absence of humic acid (Fig. 4). The lower sorption below pH 2 can be explained in terms of reduced magnitude of the zeta potential of silica in the presence of humic acid. Above pH 9, the functional groups of humic acid are fully ionized and the carboxylate group complexes with Pa (V), which results in the decreased sorption. It is known that Pa (V) ions form anionic complexes with carboxylic acid group [33, 38]. However, complexation of Pa (V) with humic acid has not been reported in the literature.

The sorption behavior of Pa (V) on silica without humic acid is similar to niobium (Fig. 5), suggesting the chemical similarity between the two metal ions. It is known that protactinium and niobium are chemical analogs under certain conditions [33, 39, 40].

Conventional absorption spectroscopy is not suitable for the determination of speciation of Pa (V) in solution as the solubility of Pa (V) is very much lower than the detection limit of the technique. In addition, it is known that the speciation of

Fig. 5 Effect of pH on the sorption of Nb and Pa on silica

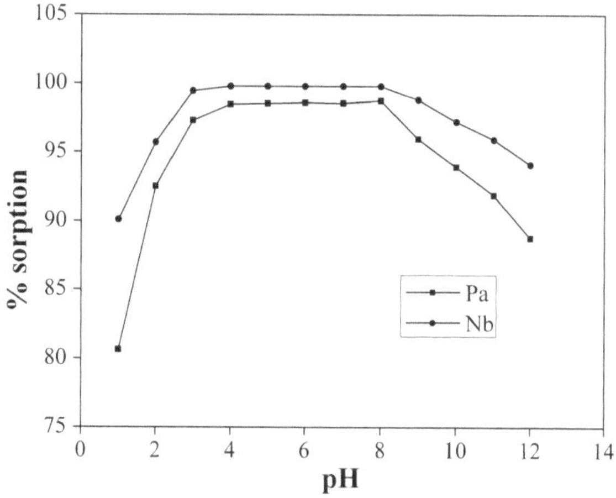

a hydrolyzable metal ion is solely governed by its concentration and the pH of the solution. Hence, inferences drawn by carrying out spectroscopic studies at higher concentration of Pa (V) ion will not be valid at very low concentrations. Our earlier attempt to study silica–humic acid, silica–niobium, and silica–humic acid–niobium system by FT-IR and Raman spectroscopy did not provide any useful information where the concentrations of niobium and humic acid were 1.12×10^{-8} M and 5×10^{-3} g L^{-1}, respectively. X-ray absorption spectroscopy using synchrotron radiation has been used to study the speciation of actinides [41, 42]. However, at present we do not have the facility to handle protactinium at synchrotron.

4 Conclusion

Pa (V) is strongly associated with colloidal silica in the pH range of 1–12. The sorption of Pa (V) on colloidal silica in the pH range 3–9 is due to surface complexation, whereas the nature of speciation of Pa (V) ions in the solution leads to the reduction in sorption below pH 3 and above pH 9. The sorption of Pa (V) on silica in the presence of humic acid is similar but slightly lower to that in its absence. The presence of humic acid lowers the sorption of Pa (V) on colloidal silica below pH 2 and above pH 9 due to the reduced magnitude of the zeta potential of silica in the presence of humic acid and complexation of Pa (V) with humic acid, respectively. The sorption behavior of Pa (V) and Nb (V) on silica showed the chemical similarity between the two metal ions.

The results of this study will be useful for modeling migration behavior of protactinium in natural aquatic system as silica Pa (V) pseudocolloid may lead to colloid-facilitated transport. An investigation into the sorption mechanism of protactinium using chemical methods is underway.

References

1. Organization for economic cooperation and development, in *Introduction of Thorium in the nuclear fuel cycle: Short-to long-term considerations*, NEA No. 7224 (2015)
2. Organization for economic cooperation and development, in *Perspectives on the Use of Thorium in the Nuclear Fuel Cycle*, NEA No. 7228 (2015)
3. J.R. Oliver, *Thorex Process: Third Uranium Cycle Studies*. Oak Ridge National Laboratory, U.S. Atomic Energy Commission, ORN L-2473 (1958)
4. R.H. Rainey, J.G. Moore, *Laboratory Development of the Acid Thorex Process for Recovery of Consolidated Edison Thorium Reactor Fuel*. Oak Ridge National Laboratory, U.S. Atomic Energy Commission, ORN L-3155 (1962)
5. *Thorium Fuel Cycle: Potential Benefits and Challenges*. IAEA-TECDOC-1450 (2005)
6. M. Srinivasan, K. Subba Rao, M.V. Dingankar, The actinide waste problem in perspective, in *Proceedings of the Indo-Japan seminar on Thorium Utilisation*, Bombay, India, 10–13 Dec, pp. 184–196 (1990)
7. Organization for economic cooperation and development (OECD), in *Geological Disposal of Radioactive Waste: An Overview of the Current Status of Understanding and Development*. Paris. EUR 9130 (1984)
8. R.C. Ewing, R.A. Whittleston, B.W.D. Yardley, Geological disposal of nuclear waste: a primer. Elements **12**, 233–237 (2016)
9. H. Basu, R.K. Singhal, M.V. Pimple, A. Kumar, A.V.R. Reddy, Association and migration of uranium and thorium with silica colloidal particles in saturated subsurface zone. J. Radioanal. Nucl. Chem. **303**, 2283–2290 (2015)
10. I. Plecas, S. Dimovic, Leaching kinetics of Mn and Sr radionuclides fixed in cement-vermiculite composition. J. Radioanal. Nucl. Chem. **264**(3), 687 (2005)
11. I. Plecas, Leaching study in process of solidification of radionuclide 60 °C in Concrete. Int. J. Environ. Eng. Sci. **2F**, 193 (2011)
12. A.B. Kersting, D.W. Efurd, D.L. Finnegan, D.J. Rokop, D.K. Smith, J.L. Thompson, Migration of plutonium at Nevada test site. Nature **397**, 56–59 (1999)
13. R. Kretzschmar, T. Schafer, Metal retention and transport on colloidal particles in the environment. Elements **1**, 205–210 (2005)
14. B.D. Honeyman, Colloidal culprits in contamination. Nature **397**, 23–24 (1999)
15. J.F. McCarthy, J.M. Zachara, Subsurface transport of contaminants. Environ. Sci. Technol. **23**, 496–502 (1989)
16. F.P. Bertetti, Colloid-facilitated transport of radionuclides in natural groundwater systems literature review. U.S. Nuclear Regulatory Commission Contract NRC–02–02–012 (2006)
17. H.W. Kirby, The radiochemistry of protactinium, National academy of science. Nuclear Science Series, NAS-NS-3016 (1959)
18. A. Chetham-Strode, O.L. Keller, Proceeding of protactinium chemistry symposium. U.S.A, REPORT TID-7675, pp. 189–193 (1963)
19. K.H. Lieser, Radionuclides in geosphere: sources, reactions in natural waters and interaction with solids. Radiochim. Acta **70**(71), 355–376 (1995)
20. H. Nitsche, Synchrotron X-ray absorption spectroscopy: a new tool for actinide and lanthanide speciation in solids and solution. J. Alloys Compd. **223**, 274 (1995)
21. M. Ghosh, P.S. Remya Devi, R. Verma, A.V.R. Reddy, Sorption of niobium on colloidal silica and the effect of humic acid. J. Radioanal. Nucl. Chem. **306**, 147–153 (2015)
22. P.N. Pathak, G.R. Choppin, Kinetics and thermodynamics of uranium (VI) sorption on hydrous silica. Radiochim. Acta **95**, 507–512 (2007)
23. S. Sachs, G. Bernhard, Influence of humic acids on the actinide migration in the environment: suitable humic acid model substances and their application in studies with uranium—a review. J. Radioanal. Nucl. Chem. **290**, 17–29 (2011)
24. H. Geckeis, T. Stumpf, Institute for nuclear waste disposal. Annual Report. KIT-SR 7559 (2009)
25. G. Buckau, Humic substances in performance assessment of nuclear waste disposal: actinide and iodine migration in the far field. 3rd Technical progress report, FZKA 7070 (2005)

26. M. Samadfam, Y. Nitisa, S. Sato, H. Ohashi, Complexation thermodynamics of Sr(II) and humic Acid. Radiochim. Acta **73**, 211–216 (1996)
27. F.J. Stevenson, *Humus Chemistry: Genesis, Composition, Reactions* (Wiley, New York, 1982)
28. S. Sethi, P. Anupama, L.M. Gantayet, Pilot plant experience of recovery of rare, natural protactinium from the insoluble muck: installation and operation. TSM-203, CHEMCON, Chennai (2001)
29. K.A. Kraus, G.E. Moore, Anion exchange studies. III. Protactinium in some HCl-HF mixtures: separation of niobium, tantalum and protactinium. J. Am. Chem. Soc. **73**(6), 2900–2902 (1951)
30. M. Marhol, Ion exchanger in analytical chemistry, their properties and use in inorganic chemistry, vol. XIV (1982), p. 57
31. J.D.H. Strickland, Use of Ion exchanger resin to determine charge of an ion in aqueous solution. Nature **169**, 620 (1952)
32. G.A. Welch, The Protactinyl ion. Nature **172**, 458 (1953)
33. E.S. Pal'shin, B.F. Myasoedov, A.V. Davydov, *Analytical Chemistry of Protactinium* (Ann Arbor-Humphery Science Publishers, London, 1970)
34. B. Fourest, The hydrolysis of protactinium (V) studied by capillary diffusion. J. Solution Chem. **33**(8), 957–973 (2004)
35. S. Kumar, B.S. Tomar, S. Ramanathan, V.K. Manchanda, Effect of humic acid on cesium sorption on silica colloids. Radiochim. Acta **94**, 369–373 (2006)
36. M. Samadfam, T. Jintoku, H. Ohashi, T. Mitsugashira, M. Hara, Y. Suzuki, Effects of humic acid on the sorption of Am(III) and Cm(III) on kaolinite. Radiochim. Acta **88**, 717–721 (2000)
37. Y. Inoue, O. Tochiyama, S. Hamashima, Removal of ^{233}Pa from ^{237}Np using manganese dioxide or silica gel. Int. J. Appl. Radiat. Isotopes **29**, 561–565 (1978)
38. M. Mendes, S. Hamadi, C.L. Naour, J. Roques, A. Jeanson, C.D. Auwer, P. Moisy, S. Topin, J. Aupials, C. Hennig, M.D. Giandomenico, Themodynamical and structural study of Protactinium (V) oxalate complexes in solution. Inorg. Chem. **49**, 9962–9971 (2010)
39. A. Chetham-storde, O.L. Keller, Physico-Chimie du Protactinium. Centre Nat. de la Rech. Scient, Pub. No. 154, Paris (1966), pp. 189–193
40. F. Monroy-Guzman, Anion exchange behaviour of Zr, Hf, Nb, Ta and Pa as homologues of Rf and Db in Fluoride Medium. J. Mex. Chem. Soc. **54**(1), 24–33 (2010)
41. R.A. Mark, S. Lynda, X-Ray absorption spectroscopy of the actinides. (2006). https://doi.org/10.1007/1-4020-3598-5_28
42. R.J. Silva, H. Nitsche, Actinide environmental chemistry. Radiochim. Acta **70**(71), 377–396 (1995)

Thorium Utilization in Fast Breeder Reactors and in Cross-progeny Fuel Cycles

Bal Raj Sehgal

Abstract Utilization of thorium fuel in fast breeder reactors has a distinct advantage, namely the reduction in the positive sodium void coefficient, which has very favorable implications for the licensing of the liquid metal fast breeder reactors (LMFBRs). However, the breeding ratio obtained is lower than that obtained when the U-238-based fuels are employed. Thorium-fueled fast reactors breed large quantities of U-233, which can be employed in advanced reactors and thermal reactors moderated by water and heavy water to increase fuel utilization by $\approx 20\%$; thereby extracting a greater amount of energy per mined ton of uranium ore. The cross-progeny fuel cycles indeed provide excellent fuel utilization performance. Thorium-based metal alloy fuel in LMFBR applications would eliminate the breeding gain penalty incurred by employing the thorium-based oxide fuel. However, there is not an extensive database in the world for the lengthy irradiations of thorium metal fuel in LMFBRs. Thorium metal and oxide fuels, in general, should have better properties and stability than the uranium metal and oxide fuels. Thorium fuel cycles have to be closed since the benefit is obtained only when the U-233 is used. India is the only country in the world, which has extensive facilities for reprocessing of irradiated uranium- and thorium-based fuels, in thermal reactors moderated by light and heavy water and in 500 MWe LMFBRs. The cross-progeny fuel cycles would be a natural vision to pursue India. This paper was written in 1982 and presented at the US–Japan Seminar on thorium fuel cycle held in October 1982. The calculations performed and the results quoted in this paper are of that vintage. However, the cross-sectional data for Th and other materials has not changed significantly since that time. The same holds for the methodologies in computer codes, diffusion theory, and the other methodologies employed in this paper, versus those in computer codes currently in use. This is a review paper incorporating results from several papers from different authors, and it is being submitted to remind the community that with the introduction of GEN IV LMFBRs, other possibilities for thorium utilization could spring forth and should be studied further and in more depth.

B. R. Sehgal (✉)
Nuclear Power Safety Division, Royal Institute of Technology (KTH), Alba Nova University Center, 10691 Stockholm, Sweden
e-mail: balraj@kth.se

A. K. Nayak and B. R. Sehgal (eds.), *Thorium—Energy for the Future*,
https://doi.org/10.1007/978-981-13-2658-5_9

Keyword Thorium utilization · Fast breeder reactor · Cross-progeny fuel cycles

1 Introduction

Thorium utilization in thermal reactors and advanced convertor reactors (ACRs) has been of considerable interest since the start of the nuclear power era. This interest stems from the fact that in the thermal spectrum the 233U (bred from 232Th) is a superior (higher η value) fissile material than the naturally occurring 235U, and it may be possible to build a thermal breeder reactor with the 233U enriched fuel. Many studies [1] have been performed, and some experimental and demonstration reactors with thorium-based fuels have been built. Principal among them are (a) the Indian Point 1 pressurized water reactor (PWR), (b) the Peach Bottom and Fort Saint Vrain high temperature gas reactors (HTGRs), (c) the light-water breeder reactor (LWBR) at Shippingport, and (d) the molten salt reactor experiment (MSRE) at Oak Ridge. Unfortunately, none of these reactors graduated to truly commercial deployment, and large-scale utilization of thorium-based fuels in thermal and advanced reactors has not matured yet. Studies, on use of thorium in a spectrum-shift reactor [2] and in D_2O moderated and cooled PWRs [3], have also been reported, and the results look promising. However, an experience base to reliably assess the economic feasibility of large-scale utilization of thorium-based fuels in thermal and advanced reactors is not yet at hand.

Utilization of thorium in fast reactors has been studied off and on over the years. A study [4] by the author (in collaboration with others) investigated the fuel cycle and the safety parameter characteristics with thorium-based fuels in fast reactors. Studies on gas-cooled fast reactors (GCFRs) with ThO_2 blankets [5] have been reported. However, there is probably a very small irradiation experience with thorium-based fuels in fast reactors.

There are two major features that distinguish thorium-based fuels in fast reactors from the conventional 238U-based fuels. These are (a) the lower breeding potential and (b) the reduction in the positive sodium void coefficient of reactivity. The former feature, a consequence of the lower fast fission bonus from 232Th than that from 238U, has traditionally inhibited the consideration of thorium for applications in the fast reactors. The latter feature has gained more attention lately due to the enhanced margins obtained against energetic core disruptions during the initiating phase of the beyond-the-design basis events postulated for the LMFBRs.

Symbiotic (or cross-progeny) fuel cycles have been proposed; the use of thorium in this context is attractive since it replaces the primary resource of naturally occurring uranium fuel. In fact, the start of the thorium fuel cycle is only possible with fissile fuel (Pu or 235U) from the uranium fuel cycle. An attractive cross-progeny fuel cycle is shown in Fig. 1, in which uranium is employed in the already existing LWRs to produce 239Pu, which is then combined with thorium and burned in the fast breeder reactors (FBRs). The 233U produced in the FBRs can be employed in either HTGRs or ACRs to sustain a self-generation 232Th–233U fuel cycle or as a replacement

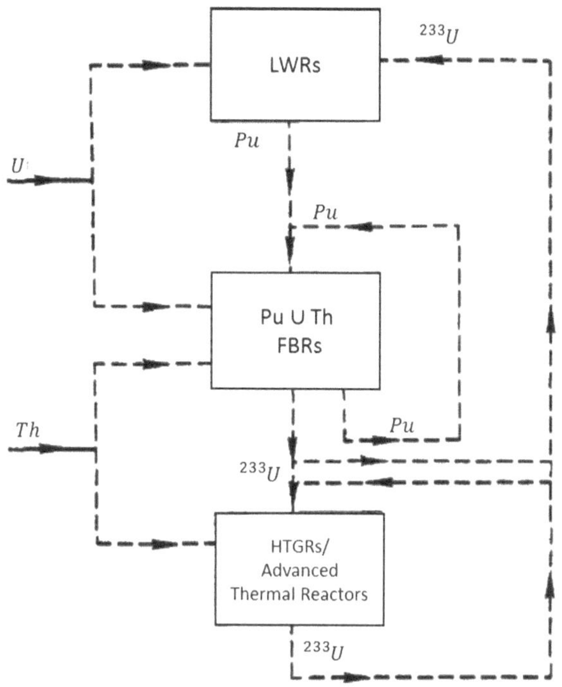

Fig. 1 A typical symbiotic fuel cycle including thorium fuel FBRs

for 235U in the LWRs. Many options become possible when both the uranium- and thorium-based fuels are employed. If the HTGRs and the ACRs do not come into being due to the insufficiency of resources, a firm resolve, or market success, the 233U fuel can be employed in standard LWRs to increase the conversion ratio and decrease the apparent fissile material consumption by ~20%. Thus, the breeding gain lost with the use of thorium in fast breeder reactors may be recovered when 233U is used either in the LWRs or in ACRs or in HTGRs.

The cross progeny fuel cycles require the closure of both of the uranium and thorium fuel cycles so that Pu and 233U are available. In fact, the benefit of thorium utilization is obtained only when 233U is separated and used subsequently in other or the same reactor system. The uranium fuel cycle as employed currently in its once-through mode in LWRs has to be closed before the fast reactors can be introduced into the nuclear power economy.

The current situation regarding nuclear power development is different in different countries. Clearly, the environment in USA is not particularly conducive to the entry of thorium as a participant in the nuclear power generation scenarios. The environment may be different in other countries where thorium utilization may be of greater significance due to the lack or paucity of indigenous fossil and uranium fuel resources and a greater commitment to efficient generation of electricity through nuclear power. It is clear, however, that the use of thorium as a supplement to the uranium fuel will stretch the available uranium supply considerably. Alternatively, with a given supply base of uranium, much larger electricity growth rates can be sustained [6].

Another issue of some significance, which has to be faced (head-on) when thorium utilization is envisaged, is that of diversion-resistance of (necessarily) closed fuel cycles. The introduction of fast reactors, ACRs, thorium fuel all depends upon the utilization of the bred material. The 233U produced in thorium-based fuel cycles has been suggested [7] as a vehicle for achieving greater diversion-resistance in the fuel cycles. This is achieved by denaturing 233U by mixing it with 238U; and given the high gamma activity of the daughters of the 232U contaminant (bred along with 233U from 232Th), the mixture would be highly self-protective. The combination of perceived greater diversion-resistance and improved fuel utilization, due to the utilization of 233U, is the main impetus for the suggested introduction of thorium-based ACRs in Ref. [7]. A similar diversion-resistant fuel cycle for LMFBRs has been suggested [8], and an international variant of the ACR concept has been proposed [9]. The fact remains, however, that closure of both thorium and uranium fuel cycles is required, albeit a very dirty and messy one. This would entail order of magnitude higher expenditures on fuel reprocessing and fabrication, in addition to perhaps a much larger risk of significant occupational exposures. It is also becoming clear that diversion-resistance of fuel cycles cannot be exactly quantified, and the exclusion of the utilization of the technologies of FBRs and ACRs based solely on proliferation concerns is not justified. The employment of improved safeguard measures, fuel accounting controls, and secured sites should remove any remaining concerns regarding this issue.

2 Thorium-Based Fuels for Fast Breeder Reactors (FBRs)

Thorium can be employed in the form of mixed oxide, mixed carbide, and metal alloy fuel. Thorium oxide mixed with plutonium oxide appears to be a viable fuel since it is similar in physical properties to the mixed uranium–plutonium oxide fuel currently in use in most of liquid metal fast breeder reactors (LMFBRs). The mixed thorium oxide has a higher melting point, slightly higher thermal conductivity, but lower density than the mixed uranium oxide fuel. Thorium–uranium oxide has been used as fuel for the Indian Point LWR, and post-irradiation examinations showed that it behaved very well. The thermal and irradiation performance of the mixed ThO_2 fuels should, therefore, be very similar to that of the UO_2–PuO_2 fuel, for which an extensive database exists. Similar conclusions can perhaps be stated about the similarity of the thermal and irradiation performance of the mixed thorium carbide to that of the mixed uranium carbide fuels, for which a small database exists.

Thorium in the form of metal alloys (Th–U, Th–Pu, Th–U–Pu) was studied [10] on a laboratory scale some years ago. Some key material and thermal property data, e.g., phase diagrams, fabrication potential, thermal conductivity, density, were established. Some small-sample irradiation testing [11–13] was performed to establish swelling rates, and one transient-heating-to-failure experiment [14] was performed.

Results of these limited investigations (highlighted in Ref. [4]) indicated that thorium metal alloys have reasonably high melting temperatures, limited swelling rates, and relatively mild failure modes. These results also indicated that the thorium metal alloys have significantly better material, thermal, and irradiation performance than the uranium metal fuel employed for the EBR-II reactor, which achieved consistently very high burnup. The candidacy of the sodium-bonded thorium metal alloys as fuel for LMFBR applications can be considered as attractive.

3 Performance of Thorium-Based Fuels in a LMFBR Core

In this section, we will review the application of thorium-based fuels in the core of a representative LMFBR. Both the thorium mixed oxide and metal alloy fuels are considered, and the calculated results for the important fuel cycle and safety parameters are compared with those obtained with the PuO_2–UO_2 fuel. Most of the information reported here in Table 1 is from Ref. [4]; other published information was consulted, and the veracity of the information in Ref. [4] was confirmed.

The basic system considered for the evaluation of the performance was a cylinder with a core radius of 1.4 m and a height of 0.93 m, i.e., a core volume of 5630.1. The core was divided into three radial regions, namely inner, control, and outer cores with outer radii, respectively, of 0.92, 1.06, and 1.4 m. The core was enclosed by radial and axial blankets. The axial blankets have the same volume fractions as the core, except that the fissile material in the fuel is replaced by the fertile material. Axial and radial reflectors enclose the blankets. A gap of 0.1 m, in height, is provided above the fuel height, i.e., between the core and the upper axial blanket for the collection of the fission gases.

The thorium-based fuels considered were:

the three oxides: ThO_2–235UO_2; ThO_2–233UO_2, and ThO_2–239PuO_2 and
the metal alloys: Th–235U; Th–233U–239Pu, and Th–233U Pu with LWR generated plutonium containing 58.9% 239Pu, 21.4% 240Pu, 14.2% 241Pu, and 5.5% 242Pu. In these alloys, the content of the plutonium was =8.8 wt%. The radial blanket compositions were also varied, being either ThO_2, UO_2, Th metal, or U metal.

The fuel cycle calculations were performed with an eight group cross-sectional set collapsed from a 29 group cross-sectional set, which was obtained from the ENDF/B-III data through the MC2 [15] code. The average broad-group cross sections for the core region were generated at critical buckling and at zero buckling for the others. The cross sections for the 232Th and 238U were determined at 700 K and 1200 K to obtain the Doppler reactivity coefficients. The core region volume fractions were: 46% coolant, 17% clad and structure, and 37% fuel. The smear density for thorium metal alloy fuel rods was taken as 76% to allow swelling of the fuel with burnup into open radial space within the clad as is done in the Mark II fuel of EBR-II. The smear density for the oxide fuel rod was taken as 85%. The blanket region

Table 1 Results of fuel cycle calculations total power =2500 MW (th)

	Case no.	Core fuel material	Radial blanket material	Core volume (l)	Cycle length (days)	Total core fissile mass (kg)	Fissile prod[b]/fissile abs	Extra fissile material prod. (kg/cycle)				%ΔK for Na[c] voiding whole core	Doppler[c] coefficient whole core
								^{233}U	^{235}U	^{239}Pu	^{241}Pu		
Equilibrium core	1	ThO_2–$^{235}UO_2$	ThO_2	6795	223	3920.5	0.84	477.1	−588.5	…	…	…	…
	2	ThO_2–$^{233}UO_2$	ThO_2	6795	225	2569.2	1.05	38.7	…	…	…	…	…
	3	ThO_2–$^{239}PuO_2$	UO_2	6795	228	2958.7	1.15	434.3	…	−341.6	3.8	…	…
	4	Th–^{235}U	Th	5630	262	4010.0	0.96	584.3	−619.4	…	…	…	…
	5	Th–^{233}U	Th	5630	261	2645.3	1.14	108.0	…	…	…	…	…
	6	Th–^{233}U	U	5630	265	2584.7	1.21	−44.1	…	210.1	…	…	…
	7	Th–^{233}U–^{239}Pu	U	5630	280	2695.7	1.31	335.8	…	−104.9	1.7	…	…
	8	Th–^{233}U–Pu[d]	U	5630	271	2502.6	1.14	266.8	…	31.3	−48.9	…	…
	9	UO_2–$^{239}PuO_2$	UO_2	6795	256	2950.8	1.23	…	−31.3	165.1	7.6	…	…
First core	1	ThO_2–$^{235}UO_2$	ThO_2	6795	223	4202.0	0.8.	563.5	−685.8	…	…	−1.89	−0.0078[c]
	2	ThO_2–$^{233}UO_2$	ThO_2	6795	226	2611.7	1.08	59.8	…	…	…	−1.98	−0.0107
	3	ThO_2–$^{239}PuO_2$	UO_2	6795	228	3084.7	1.21	587.1	…	−460.7	2.1	0.62	−0.0093

(continued)

Table 1 (continued)

Case no.	Core fuel material	Radial blanket material	Core volume (l)	Cycle length (days)	Total core fissile mass (kg)	Fissile prod[b]/fissile abs	Extra fissile material prod. (kg/cycle)				%ΔK for Na[c] voiding whole core	Doppler[c] coefficient whole core
							^{233}U	^{235}U	^{239}Pu	^{241}Pu		
4	Th–^{235}U	Th	5630	262	4384.7	0.94	707.5	−755.3	…	…	−2.52	-0.0068[c]
5	Th–^{233}U	Th	5630	261	2738.5	1.17	134.4	…	…	…	−2.58	−0.0086
6	Th–^{233}U	U	5630	265	2674.3	1.22	−37.7	…	216.3	…	−2.25	−0.0087
7	Th–^{233}U–^{239}Pu	U	5630	280	28174	1.36	458.6	…	−191.6	1.0	−0.52	−0.0084
8	Th–^{233}U–Pu[d]	U	5630	271	260.6	1.39	358.8	…	−11.54	−66.4	−0.73	−0.0089
9	UO_2–$^{239}PuO_2$	UO_2	6795	256	2988.2	1.30	…	−40.7	211.5	3.3	3.25	…

[a]Beginning of cycle values
[b]Middle of cycle values
[c]These values apply to the start of the first core cycle
[d]Contains light-water reactor-generated Pu
[e]Values are those of Th only

volume fractions were 70% fertile material, 15% structure, and 15% coolant, and the reflectors contained 85 vol.% stainless steel and 15 vol.% coolant.

The calculations were performed in R-Z geometry with the REBUS [16] code, and the results are presented in Table 1 for both the equilibrium and the first core. The reference PuO_2–UO_2 core is case 9, and the comparison depicts the penalty incurred in breeding ratio with the ThO_2-based fuels. The penalty disappears when the metal alloys are employed as the fuel material. The 235U-fueled (both oxide and metal alloy) cores lead to a convertor, instead of a breeder reactor. In these calculations, the core fuel is discharged at 9 atom % average burnup after three cycles and the radial blanket after eight cycles. The cycle lengths obtained in this manner were then employed for the first core calculations. It is seen that the equilibrium cycle lengths for the metal alloy fuels are a little bit longer than those for the reference PuO_2–UO_2 fuel. The PuO_2–ThO_2 fuel has a shorter cycle length than the reference fuel. The breeding ratios for the equilibrium cycle are somewhat lower than those for the first core cycle, due to the fission product build up, except for the 235U-fueled systems where the 233U produced helps in increasing the breeding (conversion) ratio. The fissile inventories for the three metal alloy-fueled cores are approximately the same as they are for the two oxide fuels, with the latter group requiring slightly more material.

The sodium void Δk and Doppler coefficients for the starting composition of the first core cycle were determined with the CITATION code [17] and are also shown in Table 1. It is clear that the most outstanding feature of the thorium-fueled cores is the very marked reduction in the positive reactivity coefficient for sodium voiding from the core. In fact, except for the PuO_2–ThO_2-fueled core, the whole-core voiding coefficient is negative. The calculations also show a sufficiently high Doppler coefficient. The values obtained for these coefficients play key roles in strengthening the arguments against the occurrence of energetic core dispersal during the initiating phase of the hypothetical accidents considered in licensing appraisals, e.g., the loss of flow and the transient overpower without scram events.

The sodium void worth distribution calculations were also performed for some of these reactors with diffusion perturbation theory. The radial distributions along the axial midplane and core upper edge and the axial distributions along the core center and core outer edge are compared in Figs. 2 and 3 for some of the cases in Table 1. The low magnitude of sodium void reactivity is exhibited by the Th–233U metal alloy core. The Th–233U–239Pu-fueled core exhibits approximately one-third the sodium void reactivity change as does the PuO_2 core. It is also seen that Th–233U-fueled core (operating exclusively on the thorium fuel cycle) has negative reactivity worth for sodium voiding in a very large region of the core and only when the voiding takes place in a region less than 1 m radius that a positive reactivity addition could occur. The whole-core voiding reactivity change is negative for an all thorium cycle metal alloy-fueled core.

The results of the calculations shown in Table 1 can be examined for deducing the effect of the blankets on the fuel cycle. Comparison of cases 5 and 6 shows that employing the U metal blanket instead of the Th metal blanket for the Th–233U metal alloy core increases the breeding gain; however, the substitution results in

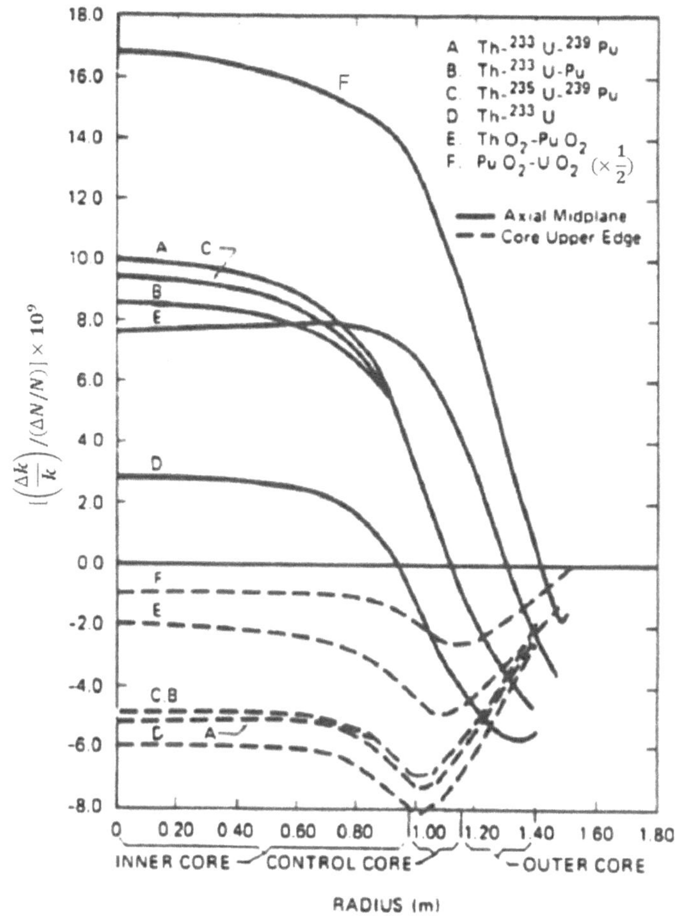

Fig. 2 Radial distribution of sodium void worth

extra 239Pu produced instead of extra 233U, and it may not be what is desired. The blanket fertile material does not greatly affect the reactor physics parameters, except for the bred material formed during the cycle, which in turn determines the type of the cross-progeny fuel cycles that can be started or sustained.

4 Performance of Thorium-Based Fuels in a GCFR

Thorium-based fuels can also be employed in a GCFR. The GCFR-proposed fuel element design is very similar to that of a LMFBR, except that the fission products produced are vented to a purification system. The GCFR cores are generally larger in

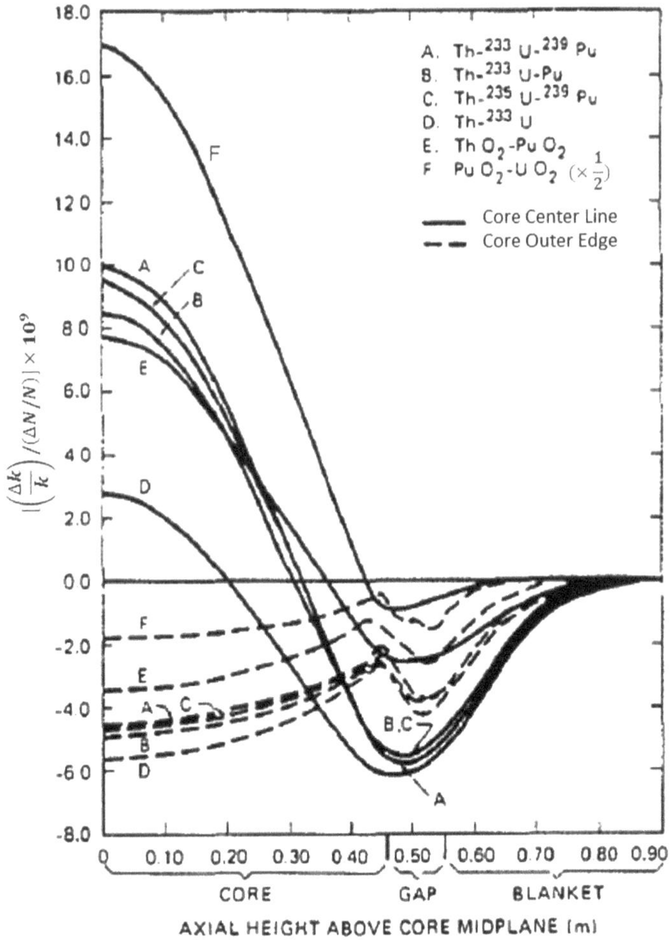

Fig. 3 Axial distribution of sodium void worth

size than the LMFBR cores since the power density is substantially less than that in the LMFBR core. A proposed 1200 MWe design has a core height of ~1.35 m, core radius of ~1.6 m, with core volume of ~11,000 L. The axial blanket is of ~0.60 m height at each end, and there are two rows of radial blanket. The core has four enrichment zones for keeping the radial power distribution as flat as possible. The fuel assembly contains 331 rods of ~0.76 cm diameter. The radial blanket design is similar to that of the LMFBR.

The proposed fuel for the GCFR core is a mixed oxide, i.e., a mixture of PuO_2 and ThO_2. The proposed blanket fertile fuel may contain UO_2 or ThO_2. A completely thorium-based fuel cycle would burn the 233U produced in a self-sufficient (self-generation) mode, and only the extra 233U produced could be employed in starting or sustaining a thermal reactor in the cross-progeny mode. The main impetus for using

Table 2 Results of comparative study between thorium and uranium blanket in a 1200-MWe reference mixed oxide GCFR

Parameter	Uranium blanket	Thorium blanket	Thorium core and blanket
Average core enrichment	14.3	15.0	15.6
Core peak to average power	1.33	1.39	1.51
Internal core conversion ratio	0.82	0.79	0.71
Breeding ratio	1.48	1.46	1.27
Compound doubling time, year[a]	9	10	17
Excess fissile material	$^{239}Pu + ^{241}Pu$	^{233}U	^{233}U
Production, kg/year	460	410	240
Levelized fuel costs, mill/kWh[b]	1.1[c]	0.4–0.6[b]	…

[a] At 80% load factor, 1-year out-of-pile time 1*X*% losses
[b] At fuel cycle parameters from Ref. [3]
[c] Pu at $10/g
[d] ^{233}U at $14–25/g

thorium-based fuels in a GCFR is to produce 233U and not to reduce the coolant void coefficient since the latter is very small anyway.

A set of calculated results reported in Ref. [5] are shown in Table 2, where the comparison is made between an all thorium (232Th–233UO_2) core and thorium oxide in the radial blanket and the plutonium–uranium oxide-fueled core with either a UO_2 or a ThO_2 blanket. It is seen that the breeding ratio does not change significantly when uranium is replaced by thorium in the radial blanket. The thorium oxide blanket on the PuO_2–UO_2 core produces a large quantity of the 233U. The all thorium-based breeder has a lower breeding ratio and produces a smaller quantity of excess 233U for future deployment. The GCFR can produce relatively large quantities of 233U in the blankets, while burning large quantities of 239Pu.

5 Performance of ^{233}U-Fueled LWRs, HTGRs, and ACRs

As indicated in Fig. 1, the cross-progeny fuel cycles will employ the 233U produced in the thorium-fueled fast breeders either in the currently available LWRs or in the HTGRs and ACRs. The use of 233U fuel in LWRs on the basis of straight replacement of 235U, i.e., 233U-enriched uranium, reduces the total fissile inventory considerably. The results of calculations reported in Ref. [18] show that for a 30,000 MWe/metric ton burnup, the starting enrichment of 233U for a PWR will be 2.44% instead of the

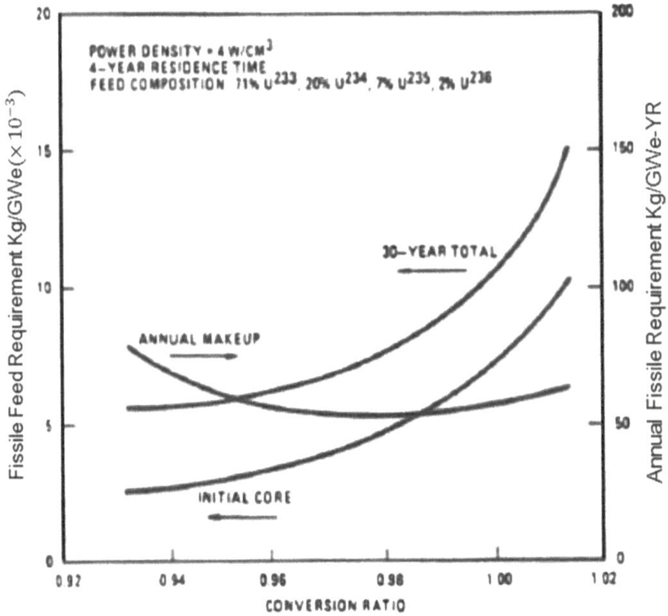

Fig. 4 Fissile requirements for near-breeder HTGR

3% required with 235U fuel. The integral conversion ratio for 233U-fueled core is ~10% higher than that for the 235U-fueled core. If the 233U is mixed with Th in the form of oxide fuel and employed in unmodified PWRs, the initial enrichment is ~3%; however, the conversion ratio is ~0.75, i.e., = 25% higher than that available currently from the slightly enriched UO_2-fueled PWRs.

The HTGR has been designed to employ thorium fuel, and therefore, replacement of 235U by 233U will be a natural extension of the concept. It is possible to construct a HTGR to have a conversion ratio slightly greater than 1.0. However, as is common in most ACRs [19], increasing the conversion ratio requires higher initial inventory. The best possible use of the 233U resource occurs when the design conversion ratio is closer to 0.9. This is shown in Fig. 4 reproduced from Ref. [20]. It is seen that the initial core inventory for the HTGR design with a conversion ratio greater than 1.0 is ~3 times that with a conversion ratio close to 0.9. The annual makeup increases with lower conversion ratios; however, the gain made in the initial inventory is enough to make the 30-year fuel requirements smaller for the lower conversion ratio design.

There are several ACR designs which can employ 233U to advantage. The principal one among them was the LWBR at Shippingport which has demonstrated breeding with the 233U–232ThO_2 fuel in a light-water-moderated and -cooled system. The LWBR concept aimed at using a core design providing strict neutron economy to achieve breeding. A number of special design features had to be incorporated in the core: (a) a seed-blanket design with variable 233U content, (b) a high metal/water ratio (between 1.5 and 2.3 compared to 0.8 for a PWR), and (c) reactor control by movable fuel or fertile rods in addition to the safety shutdown rods made of neutron poisons. The reactor also used high 233U enrichment, low average fuel burnup, and power density of approximately one-half of that in a PWR. The LWBR did not produce as much power from a given vessel as does a present-day PWR.

The CANDU reactor is another possible ACR option for the use of 233U produced in thorium-fueled LMFBRs. The CANDU core with 233U feed and ThO_2 fertile material can reach sufficiently high conversion ratios that the fuel cycle becomes self-sufficient [21]. The CANDU system has had excellent operating experience over the years with net plant factors of above 80%. The capital costs of CANDU system are estimated to be greater than those of the PWRs. However, the continuous refueling scheme reduces the fuel cycle costs [22].

Other thermal or near-thermal advanced reactors have been proposed, e.g., the molten salt breeder [23], the spectrum-shift control reactor, in each of which the 233U–Th fuel cycle produces very high conversion ratios and is very resource-efficient.

6 Special Features of Thorium Fuel Cycles

Thorium fuel cycles in both thermal and fast reactors have their own special features. One is the buildup of 232U in the irradiated fuel, whose decay begins a long chain of nuclides emitting gamma rays between 0.4 and 2.6 meV. Remote handling of the recycled fuel is, therefore, required, and the fuel reprocessing and refabricating become complicated. This, however, will also be required for the plutonium recycle fuel. Another feature is the buildup of 233Pa which decays to 233U with a 27 days' half-life. The shutdown reactivity requirement is therefore increased. Finally, the delayed neutron fraction β of 233U–232Th fuel is less than that for the 239Pu–238U fuel. This will affect the reactor period and the rate of power increase below prompt critical. More stringent operational controls, e.g., lower worth for control rods and slower withdrawal rates may be required.

7 Thermal Performance of Thorium-Based Fuels in FBRs, LWRs, HTGRs, ACRs

Thermal performance of thorium mixed oxide fuels has been determined for FBRs [4], LWRs, HTGR, and the various ACRs. There is essentially no difference between the thermal and irradiation performance of the ThO_2- and UO_2-based fuels.

The thorium metal alloy fuel is suitable for application in LMFBRs, where its thermal performance should be satisfactory and indeed better than that of the Th mixed oxide fuel. The swelling of thorium metal alloy fuel is less than that for the U metal alloy fuel, and it can be contained successfully, as is the U metal fuel in the EBR-II Mark II fuel element. The Thorium metal alloy fuel has not had any operational experience so far.

8 Summary and Conclusions

The application of thorium-based fuels in fast breeder reactors and in cross-progeny fuel cycles has been discussed. The fuel of choice for LMFBR application should be thorium metal alloy fuel, since its breeding and thermal performance are as good or better than that with the PuO_2–UO_2 fuel. The major advantage of employing thorium-based fuels in LMFBR is that of the reduction in the sodium void coefficient in a uniform geometry core. If one considers, that for the early LMFBR plants, the PuO_2–UO_2 fuel is replaced by the PuO_2–ThO_2 fuel; the calculated results for a 1000 MWe core show, that for whole-core voiding, the sodium void coefficient could be reduced from the level of ~10$ to that of 2$, without making any changes in the core geometry. Thus, measures such as introduction of internal blankets (heterogeneous core designs), reduction of core height increase axial leakages. Seed-blanket-type conceptual designs [24] required for reducing the magnitude of the positive sodium void coefficient for a PuO_2–UO_2-fueled core may not be required. These measures generally imply much higher initial fissile loadings and several operational problems for which engineering solutions have to be sought. The radially heterogeneous designs appear to be an effective way to reduce the magnitude of the positive sodium void reactivity coefficient in large cores; however, they have very high fissile loadings and the swing in the rate of heat generation with operational time, in the internal blankets during the burnup cycle which has to be provided for.

The application of thorium-based fuels in GCFR should be with thorium oxide in the radial blanket, since the GCFR does not have the large positive coolant void coefficient of reactivity. The breeding ratio of a GCFR is high, and the fertile ThO_2 blanket will produce copious amounts of 233U, which can either be recycled in GCFR or employed in a cross-progeny mode. The use of 233U, produced either from a LMFBR or a GCFR fueled with thorium, in a thermal or near-thermal reactor will lead to highly efficient (possibly self-sufficient) fuel cycles. The ACRs perform especially well with the U fuel, and the desirable choice would be the HTGRs and CANDU reactors for service respectively in USA and in Canada. Even the current generation LWRs will benefit from the use of 233U as fuel. The major issue to be settled before the application of thorium is possible in any reactor concept is that of closure of the fuel cycle. Without it, thorium application will lead to greater need of fissile material.

Another aspect of the current situation in USA and other parts of the world is the extraordinarily large stockpile of Pu collected from the military and the commercial LWR operations. When the thorium reprocessing is well established, the large quantities of Pu available should be utilized and the thorium-fueled LMFBRs could serve as transmuters of Pu to 233U. A case can be made for such an application of Pu and thorium since the penalty incurred in using thorium in FBRs can be recovered when 233U is employed in the LWRs. The LMFBRs and LWRs will be perhaps the most feasible combination with the use of thorium in LMFBRs, and the utilization of reprocessed 233U in LWRs and cross-progeny cycles would be of considerable overall benefit.

Acknowledgements The author acknowledges the work performed by several authors whose results, published in various journals and proceedings of meeting, have been quoted here. The objective of this paper was to provide a review on the subject of this paper.

References

1. P.R. Kasten et al., Assessment of Thorium Fuel cycles in Power Reactors. ORNL/TM-5565 (Jan 1977)
2. R.A. Matzie, G.P. Menzel, Conceptual Design of a Large Spectral-Shift-Controlled Reactor. CEND-377 (Aug 1979)
3. C.S. Yang, B.R. Sehgal, Physics of thorium-based fuel cycles in a D_2O cooled PWR. Trans. Am. Nucl. Soc. **32**, 794 (1979)
4. B.R. Sehgal, J.A. Naser, C.L. Lin, W.B. Loewenstein, Thorium-based fuels in fast breeder reactors. Nucl. Technol. **35**, 635 (1977)
5. R.J. Cerbone, N. Tsoulfanidis, Thorium utilization in gas-cooled fast breeder reactors. Trans. Am. Nucl. Soc. **22**, 703 (1975)
6. E.L. Zebroski, B.R. Sehgal, Advanced reactor development goals and near term and long term opportunities for development. Trans. Am. Nucl. Soc. **24**(13) (1976)
7. F. Von Hippel, H.A. Feiveson, R.H. Williams, An evolutionary strategy for nuclear energy—a safer economic alternative to the breeder and fission power: an evolutionary strategy. Sci. Mag. **203** (Jan 1978)
8. M. Levenson, E.L. Zebroski, A Fast Breeder System Concept—A Diversion Resistant Fuel Cycle, in *5th Energy Technology Conference*, Washington, D.C. (27 Feb 1978)
9. D.H. Ligon, R.H. Brogli, International symbiosis—the role of thorium and the breeders. Nucl. Technol. **48**, 261 (1980)
10. B. Blumenthal et al., Thorium-Uranium-Plutonium Alloys as Potential Fast Reactor Fuels. Part I: Thorium-Uranium-Plutonium Phase Diagram. ANL-7258, Argonne National Laboratory (Sept 1968)
11. B. Blumenthal et al., Thorium-Uranium-Plutonium Alloys as Potential Fast Reactor Fuels, Part II: Properties and Irradiation Behavior of Thorium-Uranium-Plutonium Alloys, ANL-7259, Argonne National laboratory (Oct 1969)
12. J.A. Horak et al., The Effects of Irradiation of Some Binary Alloys of Thorium-Plutonium and Zirconium-Plutonium. ANL-6428, Argonne National Laboratory (July 1962)
13. J.H. Kittel et al., Effects of Irradiation on Thorium and Thorium-Uranium Alloys. ANL-5674, Argonne National Laboratory (Apr 1963)
14. C.E. Dickerman et al., Behavior of Th-20 wt.% U fast reactor fuel under transient heating to failure. Nucl. Appl. **3**, 9 (1967)
15. B.J. Toppel, A.L. Rago, D.M. O'Shea, MC2, A Code to Calculate Multigroup Cross Sections. ANL-7318, Argonne National Laboratory (1967)
16. A.P. Olson, A User's Manual for the Reactor Burnup System. REBUS-2, FRA-TM-62, Argonne National Laboratory (Mar 1974)
17. T.B. Fowler, D.R. Vondy, G.W. Cunningham, Nuclear Reactor Core Analysis Code: CITATION, ORNL-TM-2496, Rev. 2. Oak Ridge National Laboratory (1971)
18. N.L. Shapiro, J.R. Rec, R.A. Matzie, Assessment of Thorium Fuel Cycles in Pressurized Water Reactors. Electric Power Research Institute, EPRI NP-359 (Feb 1979)

19. B.R. Sehgal, C. Braun, A. Adamantiades, Role of Advanced Reactors in Electricity Generation, Paper 80-WA/NE-7. American Society of Mechanical Engineers, New York
20. G.W. Shirley, R.H. Brogli, HTGR near breeder cycles. Trans. Am. Nucl. Soc. **28**(339) (June 1978)
21. E. Critoph et al., Prospects for Self Sufficient Equilibrium Thorium Fuel Cycles in CANDU Reactors. AECL Report 5501 (1976)
22. J.A.L. Robertson, The CANDU reactor system: an appropriate technology. Science **199**, 657 (1978)
23. J.R. Engel, H.T. Kerr, E.J. Allen, Nuclear characteristics of a 1000 MW(e) molten-salt breeder reactor. Trans. Am. Nucl. Soc. **27**, 705 (1975)
24. B.R. Sehgal, C.L. Lin, E.L. Fuller, A liquid-metal fast breeder reactor core design with seed-blanket modular assemblies. Nucl. Technol. **57**, 149 (1982)

Electrolytic Reduction of Uranium Oxide in Molten Fluoride Baths in Small Electrolytic Cells

Nagaraj Alangi, Jaya Mukherjee and P. Anupama

Abstract Use of molten fluoride salt towards electrolytic reduction of actinides and lanthanides by molten salt electrolysis is of interest for problems related to metallic nuclear fuels. Electrolytic reduction of uranium oxide for producing uranium metal has also assumed importance because of the introduction of AVLIS in the nuclear fuel cycle. Feasibility of electrolytic reduction of uranium oxide in molten fluoride melts has been demonstrated in laboratory scale electrolytic cells. Experiments were performed to study the effect of variation of size of the cell, anode and cathode configurations, liquid cathode materials, anode materials and different electrolytes. Reproducible feasibility results with different forms of uranium oxide and liquid cathodes have been demonstrated.

Keywords Electrolytic reduction · Molten fluoride · Uranium oxide
Liquid cathode

1 Introduction

Electrolytic reduction of uranium oxide to uranium for producing uranium metal has assumed importance because of the introduction of metallic nuclear fuels and atomic vapour laser isotope separation (AVLIS) in the Indian nuclear fuel cycle.

A process with the following considerations, for the uranium feed metal preparation, can be advantageously integrated to the nuclear fuel cycle.

a. Minimal number of steps or a single step process
b. The least effluent and minimal radioactive waste generation
c. Process amenable for operation in a radiochemical environment
d. Low or nil potential for criticality hazard from the neutronic accident point of view
e. Process capable of accepting any of following feed material directly

N. Alangi (✉) · J. Mukherjee · P. Anupama
Laser and Plasma Technology Division, Bhabha Atomic Research Centre, Mumbai 400085, India
e-mail: nagaraj@barc.gov.in

A. K. Nayak and B. R. Sehgal (eds.), *Thorium—Energy for the Future*,
https://doi.org/10.1007/978-981-13-2658-5_10

a. Oxides (UO_2, UO_3, U_3O_8 etc.)
b. Partially oxidized metal
c. Recycled bath.

f. Amenable for continuous or semi-continuous operation

In the electrolytic reduction process above the melting point of uranium, the uranium oxide dissolves in the molten fluoride. The electrolyte is composed of fluorides of Li, Ba, Ca, etc., which provides the desired electrochemical window of operation. Addition of UF_4 enhances dissolution of UO_2 and its decomposition to U^{4+} and O^{2-} ions. The tetravalent uranium ions migrate to the cathode and get converted uranium metal at the cathode. Liquid uranium is deposited in the cell operating above the melting point of uranium. Oxygen ion migrates to the graphite anode and gets converted to CO and leaves the cell.

Uranium metal production by electrolysis of UO_2 above the melting point of uranium with UF_4 as additive in LiF–BaF_2 base electrolyte has been reported by various authors since 1958. Niedrach et al. [1–5] published several reports describing a process for the electrolytic reduction of UO_2, UO_3 and U_3O_8 to uranium metal. Henrie and coworkers [6] studied the preparation of liquid uranium, by direct electrolysis of UO_2 dissolved in 35 wt% UF_4, 5 wt% LiF and 60 wt% BaF_2 melts. Piper et al. [7–12] and Stevenson [13] reported investigations on the electrolytic production of uranium metal from UO_2-carbon consumable anode pellets by mixing UO_2 powders and coal tar pitch in BaF_2 65 wt%—LiF 10 wt%—UF_4 25 wt% electrolyte. Haas et al. [14] verified the work that was reported in UF_4 based electrolytes by conducting the experiments already reported by others with more data in the areas of optimization of current efficiencies, current densities, bath composition, composition, generation of cell gases and solubility of feed materials. Takeshi Shimada et al. [15, 16] reported direct electrolysis of uranium oxide with LiF–BaF_2 and CaF_2–BaF_2 based salts with 25 wt% UF_4 solvent.

The objective of the study was to identify a cell configuration that leads to production of uranium with LiF–BaF_2-25wt% UF_4 and utilize the cell to study the effect of the source of uranium in the form of UO_2 powder, UO_2+C pellets or sintered UO_2 pellets and also different liquid cathodes to produce uranium alloy as product.

2 Materials Used and Electrolyte Bath Preparation

LiF and BaF_2 were sourced from M/s. Sigma Aldrich with CAS No. 7789-24-4 and 7787-32-8, respectively. They were of reagent grade with minimum assay of 99%. UF_4 was procured from Uranium Extraction Division, BARC which was of nuclear grade and had a minimum assay of 99%. Nuclear grade UO_2 powder and sintered UO_2 pellets were obtained from Nuclear Fuel Complex, Hyderabad and were of the same specifications as that used for reactor applications. All the salts were dried in vacuum at 1073 K for 3 h in a graphite crucible and then transferred into the desiccators, as stock.

Cathodic and anodic materials were chosen such that the overpotential was minimized and they were compatible with the cathodic and anodic products, respectively. High-density graphite (density > 1.95 g/cc) components were used in all the experiments for the crucible and the anode compartment. For the cathode compartment application, electrically insulating hexagonal boron nitride was used which has good corrosion resistance to liquid uranium, U alloys and also electrolytes of $LiF–BaF_2$ with UF_4. All the structural materials in the high-temperature zone, namely, the radiation shields, the anode and cathode connector rods which were not in contact with the electrolyte, were made of either graphite or molybdenum. The connectors for the anode and cathode that were in contact with the electrolyte were made of high-density graphite.

2.1 Salt Preparation

The electrolyte mixture was prepared by weighing and mixing the individual pre-dried salts of LiF, BaF_2 and UF_4 in the desired proportions in a mortar and pestle and transferred to a 1200 cc graphite crucible. The conditioning procedure of the salt mixture involved in first equilibrating the electrolyte mixture at 673 K for 24 h and for another 24 h at 1473 K while a continuous stream of high-purity argon gas having < 0.1 ppm of oxygen provided an inert atmosphere.

2.2 Preparation of UO_2+C Pellets

Two types of UO_2+C pellets were prepared and used in the experiments. In the first type of UO_2+C pellets, stoichiometric amount of carbon required for the reduction of UO_2 was added along with 1 wt% zinc behenate as the binder and 1 wt% naphthalene as the lubricant. Powder mix was prepared in ~50 g batches and pellets of 13 mm diameter, ~8 mm height each weighing ~4 g were prepared by cold pelletization at $5T/cm^2$ load. The green pellets were heated at 673 K in vacuum of 1E-2 m bar to remove the binder and lubricant, further heated to 1573 K in high-purity argon gas atmosphere at a rate of 100 K/h soaked for 4 h and then cooled to room temperature. In the second type of pelletization, carbon powder was mixed with stoichiometric amount of UO_2, ~ 2–3 wt% biomass tar (obtained from pyrolysis of municipal waste) was added as binder. Detailed chemical analysis of the tar was not conducted as the composition depended on the nature of the municipal waste. Pellets of 13 mm diameter, ~8 mm height were prepared at $5T/cm^2$ load and weighed ~3.8 g each. The pellets were heated to 1573 K at 100 K/h in argon gas and soaked for 4 h for graphitization of tar.

No shrinkage of the pellets was observed after heating; therefore, separate porosity measurement was not conducted. The density of UO_2+C pellets prepared by the above-mentioned process varied between 3.5 and 3.6 g/cc. The sintered pellets remained intact during handling and assembling of the electrolytic cell.

3 Experimental Set-up

The set-up for electrolysis experiments consisted of a muffle to provide an inert gas cover and a heating system to achieve the desired temperature. The material selection was based on thermodynamic considerations, functionality, properties of the materials, cost and ease of fabrication. A recrystallized alumina tube of 100 mm inner diameter, 115 mm outer diameter and 800 mm length, closed at one end, was used as muffle. Water-cooled stainless steel flanges with Viton O ring seals were used on the open end of the alumina tube. The flange had openings for gas inlet and outlet connections and fitted with leak tight feed-throughs for electrode connections (to anode and cathode) and thermocouple. The inert gas purge line tubes and sheathed thermocouples were taken through feed-throughs with Viton O ring seals. The anode and cathode electrical connecting rods were taken through Teflon seals; both the anode and cathode connecting rods could be independently lifted or lowered to achieve the desired immersion depth in the molten salt bath. Where the crucible was used as an electrode, the inter-electrode distance was maintained by the extent of insertion into the crucible, which was calibrated prior to the experiment. The alumina tube was heated by a 6 kW resistively heated furnace controlled by a PID controller within the range of ± 3 K. A pack of radiation shields of 98 mm diameter was used to prevent heat losses and to maintain a uniform temperature zone; the pack of shields was supported from the flange by 3 mm diameter Inconel 600 rods. A high-density graphite crucible (different sizes in different experiments) was used to contain the salt mixture. The crucible containing the electrolyte rested on a perforated alumina plate of 98 mm diameter and 5 mm thick, which in turn rested on the floor of the alumina tube. All the anode compartments to hold UO_2 powder, UO_2 pellet or UO_2+C pellet were made out of high-density graphite. This anode compartment, which contained the powder, had a provision to fasten an 8 mm diameter threaded rod. A graphite rod of 10 mm diameter with M8 threading was connected to the anode compartment on one end and the other end of the rod was connected to a graphite adapter, which was fastened to a 6 mm diameter molybdenum rod, as shown in the schematic of the experimental set-up in Fig. 1. A graphite rod was used to provide electrical connection to the cathode chamber, which had a provision to connect a 6 mm diameter molybdenum rod similar to that for the anode. The desired cathode current density was attained by providing an insulating BN liner to the graphite crucible, exposing only the desired area of the graphite surface. All the other high-temperature structural members were made of either graphite or molybdenum.

High-purity argon with <0.1 ppm of oxygen was purged continuously during the experiment, with the oxygen concentration in the inlet gas being monitored continu-

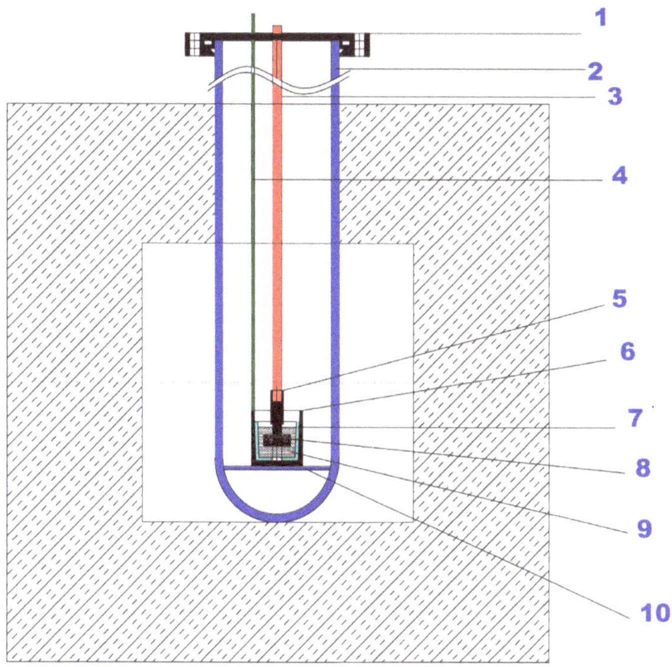

1. Water cooled stainless steel flanges
2. Recrystallized alumina tube
3. Molybdenum rod anode connector
4. Nickel cathode connection
5. Graphite powder holder
6. Graphite crucible cathode
7. BN insulating liner
8. Uranium oxide powder
9. Electroyte
10. Alumina base plate

Fig. 1 Schematic of the electrolysis set-up

ously by an online Systech 800 Series oxygen analyzer. The furnace was housed inside a fume hood and the purge gas released into the fume hood was passed through HEPA filters and a water scrubber to remove radioactive particulate and fluoride vapour, respectively before exhausting to the environment. A photograph of the experimental set-up in the fume hood is shown in Fig. 2.

4 Experimental Variables for Electrolysis of UO_2 to U Metal

The effects of the following were studied in this work: (1) variation of size of the cell, (2) anode and cathode configurations, (3) liquid cathode materials, (4) anode materials and (5) different electrolytes. Four cell sizes S-80, S-350, S700 and S-1400 were used in the complete experimental plan, which had electrolyte holding

1 Fumehood
2 Furnace inside fumehood
3 Thermocouple
4 Experimental setup with alumina tube
5 Fumehood sash
6 Control panel for furnace

Fig. 2 Photograph of the experimental set-up for electrolytic reduction inside the fume hood

capacity of 80, 350, 700 and 1400 g, respectively. Different anode and cathode design configurations were used in each cell size. The configurations of the set-ups are shown in Figs. 3 and 4. The cathode area in different cell designs varied from 0.785 to 11.43 cm^2. Similarly, the polarity of the electrolyte container was either anodic, cathodic or electrically floating.

In the electrolysis experiments with S-80 and S-350 cell sizes, four different anode designs were used: A bucket type anode as shown in Fig. 5a to hold ~5 g of UO_2, a perforated anode cylinder to hold ~10 g UO_2 as shown in Fig. 5b, floating BN powder holder for ~25 g UO_2 as shown in Fig. 5c and a graphite powder holder anode for ~25 g of UO_2 as shown in Fig. 5d. The experiments were conducted with inter-electrode distance varying from 10 mm to 30 mm. The features of all the cell configurations used in this study are compiled in Table 1.

Given the severity of the operational conditions imposed due to the melting point of uranium, the high-temperature chemical reactivity of uranium metal with the electrolyte and the container material, and also the corrosion of the container by the molten electrolyte, there was no scope to vary a large number of parameters. With the limited solubility of UO_2, ~2 wt% in LiF–BaF_2-25 wt% UF_4 and the size of the cell chosen for the experiments for the feasibility studies, continuous addition of UO_2 was not considered. Thus, each experiment was a batch process. As the primary objective of the study was to demonstrate the feasibility of the process of uranium oxide to uranium metal conversion, no separate effort was directed towards the determination and improvement of the cathode and anode current efficiencies.

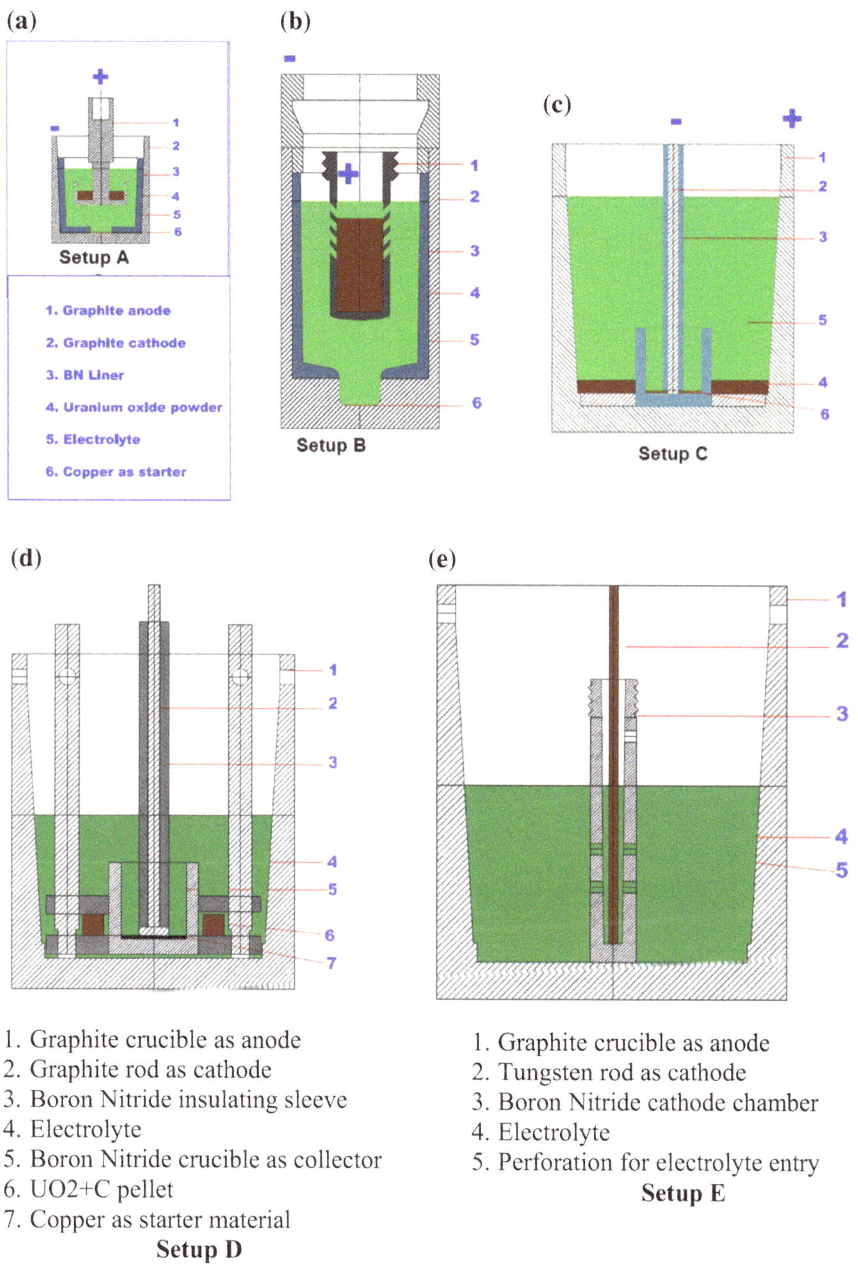

Fig. 3 Schematic of set-up A, B, C, D and E the for electrolysis experiments

Fig. 4 Schematic of set-up F the for electrolysis experiments

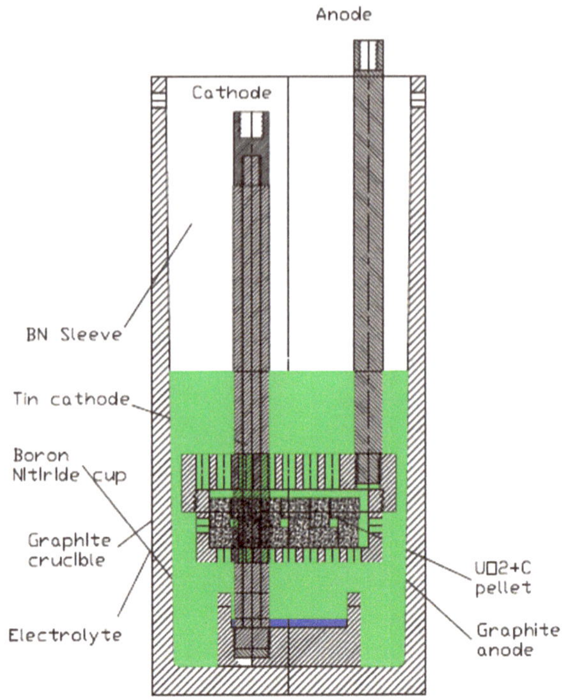

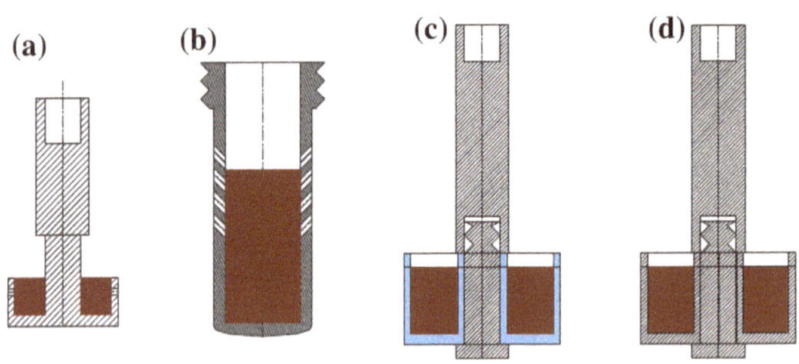

Fig. 5 Different anode designs used in the electrolysis studies in S80 and S350 electrolytic cells

The experimental procedure involved in equilibrating the electrolyte and uranium oxide source (Sintered UO_2 pellets, UO_2+C pellets, UO_2+C+biomass tar pellets and UO_2 powder), with the electrolyte after pre-treatment at 673 K and 1473 K for 24 h at each temperature. Several electrolysis experiments were conducted in LiF:BaF_2(1:1)-25 wt% UF_4 salt mixtures in different electrolytic cell designs to arrive at the cell configuration which results in uranium deposition. The electrolysis was conducted at 3.5 V with a cathode current density of ~2A/cm^2 for durations between 200 and 300 min. Off-gas analysis, which provides information on the nature of the anodic reaction, was not included in the study. The post-electrolysis step involved cooling

Table 1 Comparison of different electrolytic cell designs

	Set-up A	Set-up B	Set-up C	Set-up D	Set-up E	Set-up F
Cell size	S80	S350	S700	S700	S700	S1400
Capacity, g	80 g	350 g	700 g	700 g	700 g	1400 g
Graphite crucible dimensions	47 mm OD, 42 mm ID, 50 mm height	58 mm OD, 50 mm ID, 100 mm height	95 mm OD, 85 mm ID, 110 mm height	95 mm OD, 85 mm ID, 110 mm height	95 mm OD, 85 mm ID, 110 mm height	95 mm OD, 85 mm ID, 210 mm height
Polarity of crucible	Cathodic	Cathodic	Anodic	Anodic	Cathodic	Electrically floating
Anode	Graphite bucket with UO_2	UO_2 in graphite cylindrical bucket with perforations	Graphite crucible with UO_2 at the bottom	Graphite crucible, pellet holder and UO_2+C	Graphite crucible	UO_2+C in closed graphite compart-ment with perforations
Cathode	Graphite crucible	Graphite crucible	Copper in contact with graphite rod	Copper in contact with graphite rod	2.5 mm Φ tungsten rod	Liquid copper or tin
Cathode area, cm^2	0.785 cm^2	3.18 cm^2	1.25 cm^2	1.25 cm^2	2.1 cm^2	11.43 cm^2
Source of uranium	UO_2 powder	UO_2 powder	UO_2 powder	UO_2+C pellets	UF_4 only	UO_2+C pellets

the contents to room temperature. The contents of the crucible, the cathode or the anode compartments were examined and samples were taken for analysis. In a few of the experiments, the crucible along with the boron nitride liner was cut across the diameter to study stratification in the electrolyte and preferential settling of uranium oxide.

5 Results and Discussion

Although the results were encouraging for the smaller cells (S-80, S-350 and S-700), electrolysis experiments did not yield any recoverable uranium. There was no coalesced liquid uranium obtained after these experiments. Molten uranium being the desired product, dispersed, re-oxidized, dissolved uranium in the electrolyte that is not recoverable, was not considered as the product. In all the experiments with S-80 and S-350 cells, a mixture of UO_2 and electrolyte was always observed at the bottom of the crucible owing to settling of higher density of UO_2 in the electrolyte. In the small cells, due to the proximity of the anolyte and catholyte, the transfer of U^{3+} which was generated at cathode by re-dissolution of the deposited uranium, to the anode

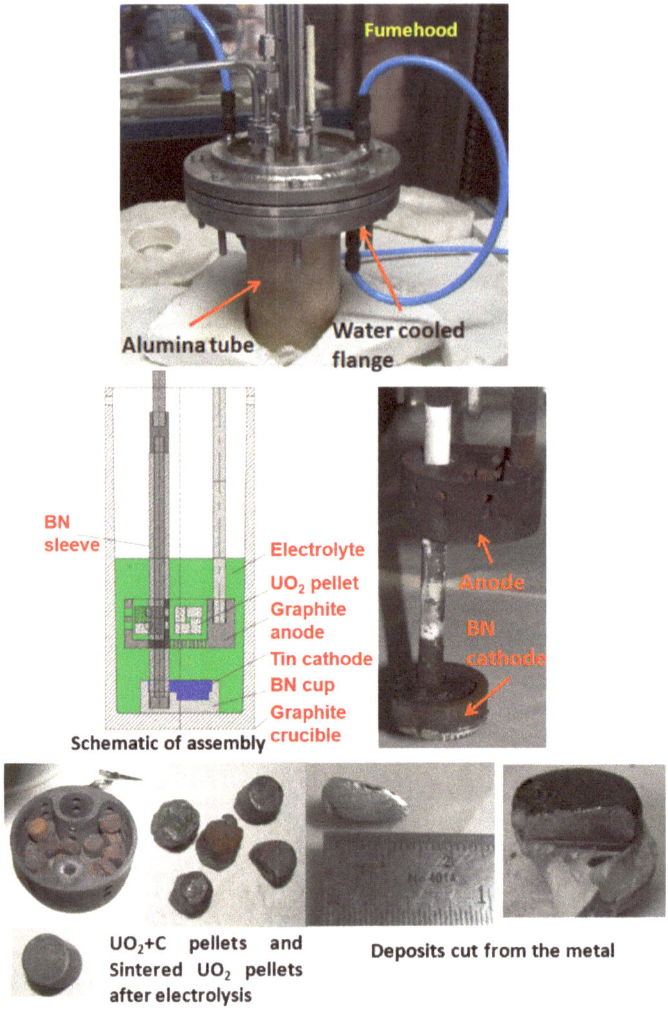

Fig. 6 Photographs of components of electrolysis studies

region was faster. Near the anode, U^{3+} was oxidized to U^{4+}, thereby setting up a cyclic reaction which resulted in a large parasitic current. This was confirmed in the cells of S-350 type by the appearance of green colour near anode indicating greater abundance of U^{4+} and brown or red colour at the cathode indicating predominant presence of U^{3+}. In case of S700 cells with both UO_2 powder and UO_2+C pellets, electrolysis did not yield any coalesced uranium that could be collected. However, the contents of the cathode chamber in the BN crucible were found to have higher concentration of ~28% uranium as against 18.9% uranium in the electrolyte. This indicated that uranium ions were migrating towards cathode and reducing to uranium, but quickly

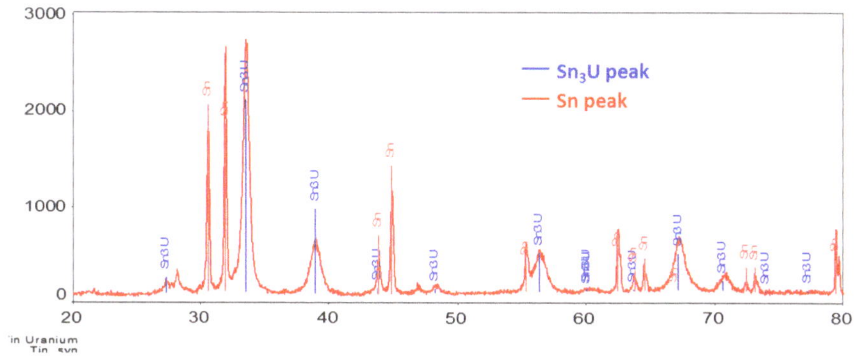

Fig. 7 Typical XRD analysis of metal confirming the formation of USn_3

redissolving the product into the electrolyte. The colour near the cathode, which was brown or red, indicated presence of UF_3, supporting the inference of re-dissolution.

Hence the efforts were directed towards use of a liquid cathode (set-up F) of higher quantity which alloys or traps and acts as an immediate sink to the uranium produced, thereby reducing its activity in alloy form. Tin was used as the liquid cathode in the experiments based on its thermodynamic suitability with the electrolyte and container materials. In the experiments with UO_2+C pellets, USn_3 compound was formed with tin metal, which was confirmed by XRD on a sample from tin mass. The total uranium in the tin cathode mass was ~6.8 g. Formation of USn_3 compound indicates that production of coalesced uranium metal required a good combination of favourable factors, namely, availability of dissolved UO_2, formation and coalescence of uranium and reduction of re-dissolution after electrolysis. The absence of any one factor might lead to either poor or no deposition of uranium. The anode current densities are limited by the onset of anode effect; hence, a high ratio of anode to cathode area is desirable. In the current experiment, cathode area was 11.4 cm^2 and the anode area was ~110 cm^2; hence, no anode effects were observed even when the cell operated at 2A/cm^2 cathode current density. In fact, it posed no problem for trials even higher cathode current densities.

With this arrangement, the effect of different forms of UO_2 was further investigated. UO_2 powder and sintered UO_2 pellets were used in separate experiments in LiF–BaF_2-25 wt% UF_4 electrolyte with the identical configuration and anode designs shown in Fig. 6a, b, respectively. The electrolysis conditions were maintained identical to those of experiments with UO_2+C pellets. In these experiments, USn_3 formation was again confirmed by XRD. It was expected that depolarization of the anode due to the presence of carbon along with UO_2 in UO_2+C pellets will help the electrolysis. However, there was no change in the potential to maintain a cathode current density of 2A/cm^2, either with UO_2 powder or with sintered UO_2 pellets compared to the potential with UO_2+C pellets. The rate of dissolution of UO_2 pellets might have been sufficient to maintain the required cathode current density, and hence no anode effect was observed. One could conclude that the dissolution rate of uranium oxide from UO_2 powder, sintered UO_2 pellets or UO_2+C pellets did not have any effect on the electrolysis. Photographs of the components used in the electrolysis are shown in Fig. 6. Process repeatability in LiF:BaF_2:25%UF_4 electrolyte was established. Typical XRD pattern of the USn_3 cathode product is shown in Fig. 7.

6 Conclusions

The technical feasibility of a process for electrolytic reduction of uranium oxide to uranium-tin alloy in LiF–BaF_2-25 wt% UF_4 electrolytes at 1473 K has been demonstrated in laboratory size cells. Given the operating conditions, a continuous process for the production of uranium should be possible. Coalescence of uranium metal was a major cause for the poor production of metal in all the small cell experiments. In the electrolysis experiments, the form of uranium oxide, namely UO_2 powder, UO_2+C pellets or UO_2 pellets, was not found to have influence on the formation of uranium. Also anode effect never occurred in any of the electrolysis runs as sufficient dissolved uranium oxide must have been available and the anode current density was one order less than the cathode current density. Possibly the $U^{4+} \leftrightarrow U^{3+}$ cyclic reaction led to depolarization of anode. The current efficiencies were found to be in the range of 7–8% in all the experiments. Although very high-cathode current efficiencies have not been obtained, sufficient metal has been produced to indicate that kilogram level production is possible. Higher efficiencies should be possible in large size electrolytic cells than that were obtained in laboratory cells.

References

1. L.W. Niedarch, B.E. Dearing, The preparation of uranium metal by the electrolytic reduction of its oxides. J. Electrochem. Soc. **105**(6), 353–358 (1958)
2. L.W. Niedarch, New approach to electrolytic uranium. Nucleonics **16**(1), 64–65 (1958)
3. L.W. Niedarch, A.C. Glamm, Uranium purification by electrorefining. J. Electrochem. Soc. **103**(9), 521–528 (1958)
4. L.W. Niedarch, B.E. Dearing, Electrowinning of uranium from its oxides. *Knolls Atomic Power Laboratory*, Rep No. KAPL-1761 (1957)
5. L.W. Niedarch, A.C. Schafer, Electrowinning of uranium from its oxides II A preliminary engineering evaluation. *Knolls Atomic Power Laboratory*, Rep No KAPL-1668 (1957)
6. T.A. Henrie, Electrowinning rare-earth and uranium metals from their oxides. J. Metals **16**, 978–981 (1964)
7. R.D. Piper, R.F. Leifield, Electro reduction of uranium oxides to massive uranium metal. *Mallinckrodt Chemical Works*, Rep No. MCW-1447 (1960)
8. R.D. Piper, R.F. Leifield, Electrolytic reduction of uranium metal from uranium oxides. I & EC Process Design Dev. **1**(3), 208–212 (1962)
9. R.D. Piper, E.C Gay, Process development quarterly progress report. *Mallinckrodt Chemical Works*, Rep No. MCW-1479 (1962)

10. R.D. Piper, Process development quarterly progress report. *Mallinckrodt Chemical Works*, Rep No. MCW-1480 (1963)
11. R.D. Piper, J.D. Vie, J.N. Reyland, R.J. Kornfeld, D.E. Treadway, Methods for tapping molten uranium from an electrolytic cell. *Mallinckrodt Chemical Works*, Rep No. MCW-1481 (1963)
12. R.D. Piper, Laboratory report of the uranium Electroreduction process. *Mallinckrodt Chemical Works*, Rep No. MCW-1500 (1966)
13. J.W. Stevenson, Electrolytic uranium project: Cell operation, terminal progress report. *Mallinckrodt Chemical Works*, Rep No. MCW-1514 (1966)
14. P.A. Haas, P.W. Adcock, A.C. Coroneos, D.E. Hendrix, Small cell experiments for electrolytic reduction of uranium oxides to uranium metal using fluoride salts. Metall. Mat. Trans. B. **25**, 505–518 (1994)
15. Takeshi Shimda, Nobuo Tedzuka, Yuzo Shimizu, Masanobu Miyake, Electrochemical reduction of uranium oxide in molten fluoride mixture. J. Alloy. Compd. **204**, 1–4 (1994)
16. Takeshi Shimda, Nobuo Tedzuka, Yuzo Shimizu, Masanobu Miyake, Electrode reactions of Uranium during electrolysis of its oxide in a molten fluoride bath. J. Alloy. Compd. **206**, 249–253 (1994)

Evolution of Actinides in ThO_2 Radial Blanket of Prototype Fast Breeder Reactor

Abhitab Bachchan, K. Devan, G. Pandikumar and P. Puthiyavinayagam

Abstract The third stage of India's nuclear program envisages utilization of thorium through Th–U233 cycle in view of the better abundance and relative merits of thorium. For early introduction of Thorium into the nuclear energy system, several R&D programs have started to find the best possible route of thorium utilization. Toward this, efforts were made to assess the feasibility of Th–U cycle in a fast spectrum reactor like prototype fast breeder reactor (PFBR). The effect on core neutronic parameters and actinide evolution with the replacement of depleted UO_2 in the PFBR blanket SA with thorium oxide has been studied using 3D diffusion code FARCOB. Study shows that by the introduction of thorium blanket, core excess reactivity is coming down by ~535 pcm, and core breeding ratio is slightly lower than that with conventional oxide blanket. The distribution of region wise power production is slightly changed. Power from radial blanket is reduced from 3 to 2% while the core-1 power is increased from 49 to 50%. The estimated U^{233} production is 7.6, 11.5, and 14.1 kg/t with 180, 360, and 540 days of irradiation, respectively. In addition, U-232 production is also estimated using ORIGEN-2 code. The amount of U-232 per radial blanket SA is about 165 ppm at discharge burnup of 3500 GWd/t.

Keywords Thorium · PFBR · Actinide buildup · Neutronic parameter
One group cross section

1 Introduction

Thorium-based fuels ensure long-term sustainability of nuclear power in India because of its large abundance. Towards this, thorium utilization through U^{233}–Th cycle is necessary to be studied in various reactor systems [1]. The three-stage program of India targets utilization of thorium in the second-stage fast breeder reactor

A. Bachchan (✉) · K. Devan · G. Pandikumar · P. Puthiyavinayagam
Reactor Design Group, Indira Gandhi Centre for Atomic Research, HBNI, Kalpakkam 603102, Tamil Nadu, India
e-mail: abslindia@igcar.gov.in

A. K. Nayak and B. R. Sehgal (eds.), *Thorium—Energy for the Future*,
https://doi.org/10.1007/978-981-13-2658-5_11

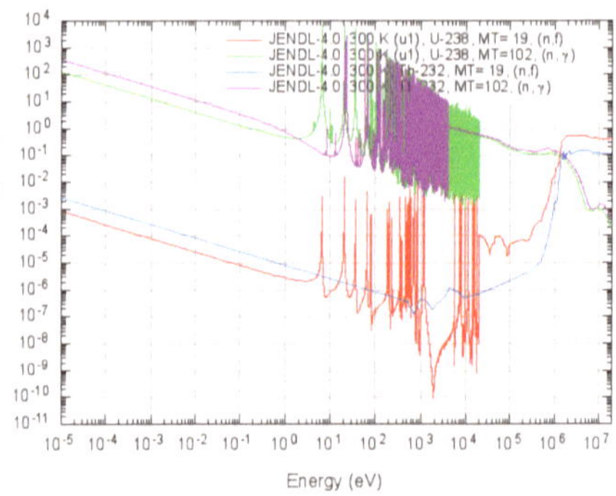

Fig. 1 Capture and fission cross-section for U-238 and Th-232

and third stage with advanced heavy water reactor (AHWR). The irradiation of thorium has been made in a research facility to demonstrate the challenging task of reprocessing and fuel fabrication [2]. The radiological problem in Th–U cycle is associated with U-232(~69 years), whose daughter products are hard gamma emitters with longer half-life. For example, one of the daughter products of U-232 like Tl^{208} is a hard gamma emitter (E_{γ}: 2.6 MeV) which has very small half-life. The increase in build-up of Tl-208 results in increased dose rate in the fuel. Another concern of Th–U fuel cycle is the high toxicity potential of Th-229 [3]. Thorium utilization has been investigated in the European Fast Reactor EFR and CAPRA cores [3]. The role of liquid metal cooled fast reactors (LMFBRs) in the utilization of thorium resources in India had been investigated by Ramanna and Lee [4]. The main focus of the present study is to see the thorium utilization in fast spectrum reactor like PFBR. Presently, PFBR uses depleted UO_2 as the blanket material and later it may be replaced with thorium oxide (ThO_2). The capture and fission cross section for Th-232, and U-238 is shown in Fig. 1. The changes in core neutronic parameters and actinide evolution in PFBR with ThO_2 blanket is studied using FARCOB and ORIGEN-2. The salient observations of this study are reported in this paper.

2 Details of PFBR Core

The prototype fast breeder reactor (PFBR) core has two enriched fuel zones of mixed oxide (PuO_2–UO_2) fuel. The active core consists of 85 fuel sub-assemblies (FSA) of low enrichment (20.7% PuO_2) in the central region followed by 96 FSA of higher enrichments (27.7% PuO_2). The radial and axial blankets are of depleted UO_2 [5]. The fuel SAs are surrounded by 117 radial blankets (RB) which are arranged in two rows and followed by 138 steel reflectors, one row of inner B_4C shielding assemblies, in-vessel storage, and shielding assemblies. The active core contains 12 absorber rods (AR) for reactivity management and shutdown. Out of the 12 ARs, nine are

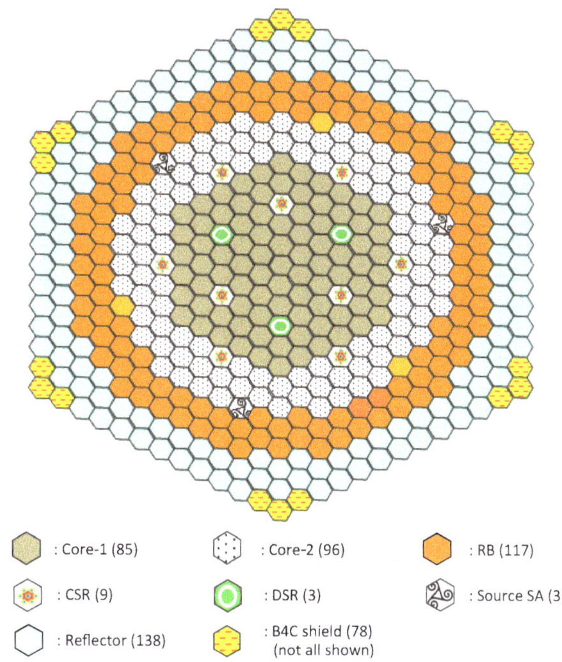

Fig. 2 Equilibrium core of PFBR

designated as the control and safety rods (CSRs) and the other three as the diverse safety rods (DSRs). The core also contains three neutron source assemblies placed at 120° symmetry in between core-2 and RB assemblies. The core layout used for the neutronics calculations is given in Fig. 2.

All FSAs are made of 217 helium-bonded fuel pins. Each fuel pin has 1000 mm column of fuel, 300-mm-long upper and lower axial blankets and upper (230 mm) and lower (750 mm) fission gas plenum. RB assembly contains 61 blanket pins of diameter 143 mm. Active length of radial blanket is 1600 mm, and the total mass of thorium in a blanket SA is about 112 kg at the beginning of cycle. Fuel and blanket pins are encased in a 20% cold worked D9 steel cladding and are separated from each other by a helical spacer wire. The thorium blanket SA design (like pin dia, pin pitch, material fraction) is assumed to be the same as that of the reference UO_2 blanket SA.

3 Method of Analysis

The neutronics calculations based on 3D diffusion theory have been performed using FARCOB code [6] to assess the core neutronics parameters. Neutronics calculations are performed in 26 groups using homogeneous model and the adjusted Russian multigroup library called ABBN-93 [7]. It is to be noted that this cross-section library has been adjusted for medium-sized sodium-cooled fast reactor cores using MOX

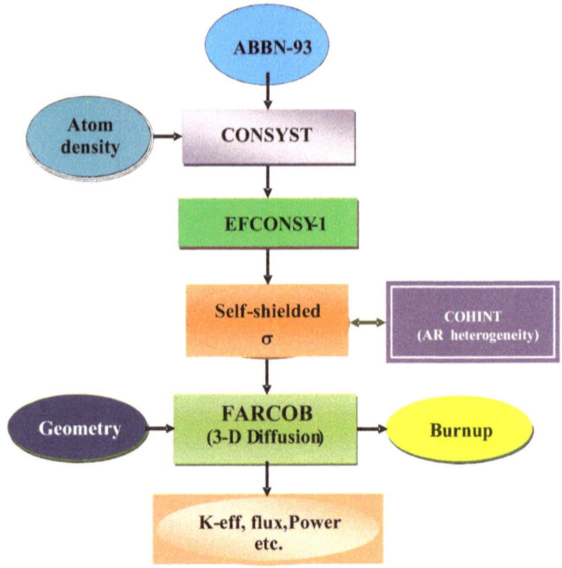

Fig. 3 FARCOB code system

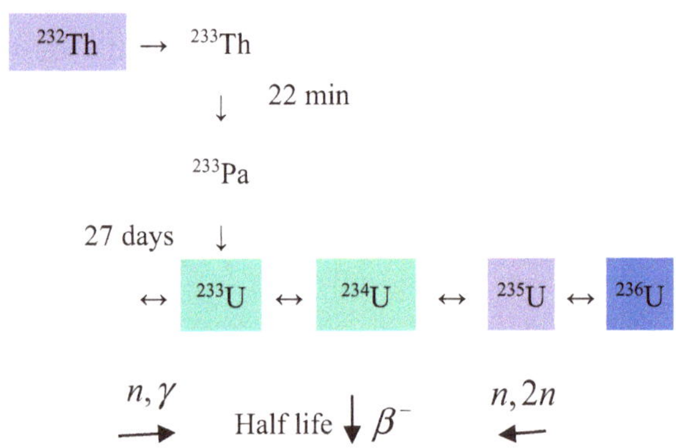

Fig. 4 Actinides chain for Th–U cycles

fuels. Isotopic concentration is calculated using the code ATOMIX [8]. Heterogeneity effect is taken into account while preparing the self-shielded cross sections for fuel, blanket, and absorber rods [9]. For the absorber rods, the transport equation is solved by considering 2D cylindrical model using the first flight collision probability method (using COHINT code [10]) to obtain the fluxes for cross-section homogenization. A flowchart showing various codes/modules and cross-section libraries in the FARCOB code systems is given in Fig. 3. The actinides chain used in FARCOB to study the Th–U cycles is given in Fig. 4 [11].

For U-232 production estimation, ORIGEN2 [12] has been used. ORIGEN2 works based on a point source model, and on the summation approach, to estimate isotopic inventories. The code solves the Bateman equation for estimating the production and depletion of nuclide. The complex reactor physics details including geometry, self-shielding are assumed to be fed through the effective spectrum averaged reaction cross-section data that pertains to the reactor under study. The code is associated with several databases of nuclear data for different types of reactors. The data suitable for PFBR has already been included in the system. In addition to cross sections, the database consists of decay data and fission product yield data.

4 Results and Discussions

4.1 Nuclear Data

Table 1 gives the PFBR radial flux spectrum averaged one group self-shielded cross sections of Th-232, U-233, U-238, and Pu-239. It shows that there is significant difference in fission cross sections of fertile nuclides, viz., Th-232 and U-238. The fission cross section of Th-232 is approximately four times lower compared to U-238. The capture cross sections are almost similar. Due to lower values of fission cross section and average number of neutrons released per fission (ν), the value ofη Th-232 is about six times smaller than that of U-238.

4.2 Influence on Core Parameters by with Thoria Blanket

The first change observed with ThO_2 blanket with respect to the UO_2 is the reduction of core excess reactivity by ~535 pcm. This is due to the enhanced absorption in thorium compared to uranium (Table 1). Larger capture to fission ratio of Th^{232} compared to U^{238} makes thorium systems less neutron economic. Table 2 shows that power production in RB comes down from 3 to 2% which leads to higher power production from the fuel for obtaining the same core power. It is also observed that there is small reduction in the breeding ratio (~0.02%).

Table 1 Comparison of nuclear data corresponding to ThO_2 and depleted UO_2 radial blankets

RB Material	Nuclide	One group cross sections[a]		ν	$\frac{\sigma_c}{\sigma_f}$	η
		$\sigma_{capture}$	$\sigma_{fission}$			
Thorium	Th232	0.43	0.0048	2.36	89	0.027
	U233	0.41	3.6012	2.50	0.12	2.24
Uranium	U238	0.38	0.0218	2.77	17.4	0.16
	Pu239	1.03	2.2830	2.89	0.36	2.12

[a]PFBR radial blanket neutron flux spectrum averaged self-shielded cross sections

Table 2 Impact of ThO_2 blanket on power production in fresh core

Region	Region wise power production (%)	
	With UO_2 blanket	With ThO_2 blanket
Core-1	49	50
Core-2	45	45
LAB	1	1
UAB	1	1
RB	3	2

4.3 Actinide Build-up in the Radial Blankets

Figure 5 shows the build-up of actinides (kg/t) in the radial blanket for continuous irradiation. The estimated U^{233} production is about 8, 12, and 14 kg/t in 180, 360, and 540 days of irradiation. Figure 5 also shows that Pa^{233} is initially builds-up and comes down while U-233 builds-up continuously. The production of U^{234}, U^{235}, and fission product is small. Build-up of actinides with respect to burnup in Th–U cycles is given in Fig. 6. The estimated build-up of U^{233} in a SA with burn-up of 0.97 and 1.9 MWd/kg is about 2.4 and 3.3 kg, respectively. The available amount of U^{234} and U^{235} even with higher burn-up of 2.4 MWd/kg is only few grams.

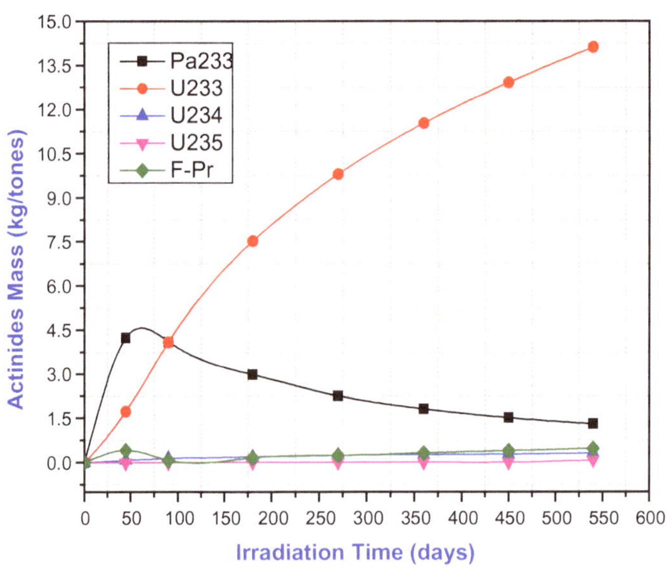

Fig. 5 Estimated mass of actinides in Th–U cycles with irradiation time

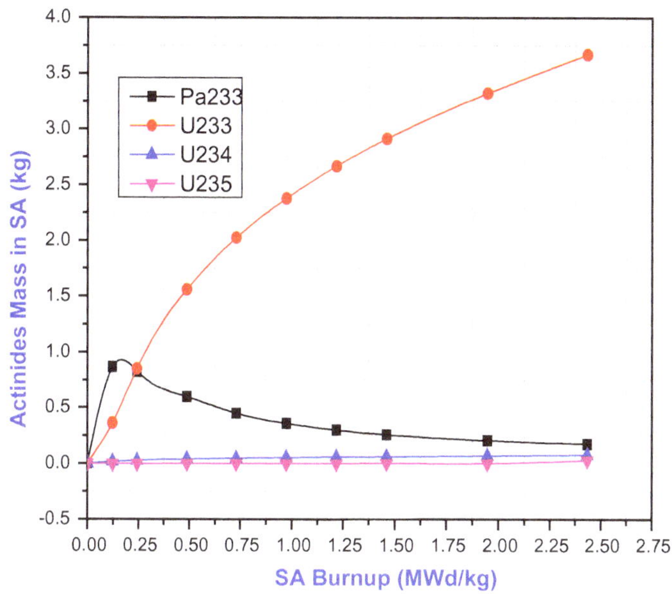

Fig. 6 Build-up of actinides with burnup

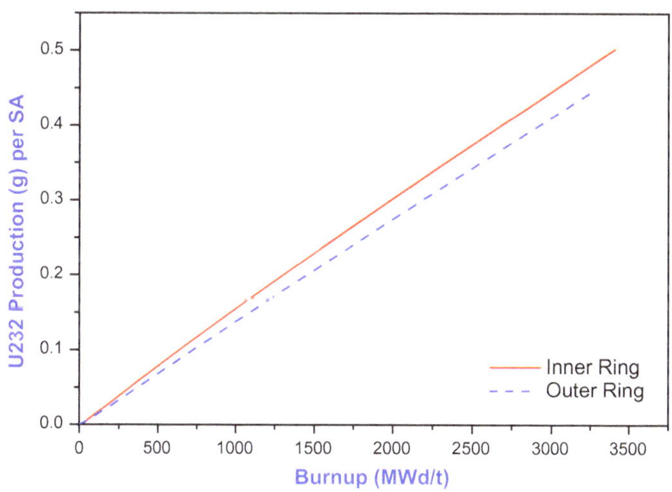

Fig. 7 Build-up of U-232 with burnup in RB

4.4 Build-up of U-232

Based on ORIGEN-2 computations, the inventory of U-232 is computed and shown in Fig. 7. The study shows that the production of U-232 is almost linear in lower burnup range. The amount of U-232 per radial blanket SA is about 0.48 g (165 ppm) at discharge burnup of 3500 MWd/t for inner ring SA. However, it can be noticed that, for the same burnup, the amount of U-232 is higher in the inner ring SA than outer ring SA.

5 Conclusions

Early introduction of thorium-based fuels in Indian nuclear power program is very essential to ensure long-term energy security. The issues associated with the Th–U fuel cycle (front-end to back-end) are being studied in thermal as well as in fast spectrum reactors. Thorium pins irradiated in CIRUS have been reprocessed and the produced U-233 is used as fuel in KAMINI (U-233 based fuel) reactor. In addition, reactors like AHWR are also designed for thorium utilization. In this context, in the present study, the U-233 production potential in fast reactor, especially in PFBR is studied using 3D diffusion code FARCOB code system and the U-232 production is estimated using ORIGEN-2.

The study shows that there is no appreciable change in core neutronic parameters with the replacement of depleted UO_2 with thorium oxide in radial blanket SA. The core excess reactivity reduces by 535 pcm, and power production in RB comes down from 3 to 2%. Due to lower values of fission cross section and average number of neutrons released per fission (ν), the value of η for Th-232 is about six times smaller than that of U-238. There is slight reduction in the core breeding with respect to the depleted UO_2 blanket (~0.02%). Estimated build-up of U-233 in a SA with burn-up of 0.97 and 1.9 MWd/kg is about 2.4 and 3.3 kg, respectively. The available amount of U-234 and U-235 even with burnup of 2.4 MWd/kg is few grams. The estimated amount of U-232 per radial blanket SA is about 165 ppm with discharge burnup ~3500 GWd/t. Studies that are more detail are under progress to know evolution of actinides including U-232 build-up.

References

1. U. Kanan, P.D. Krishnani, Energy from thorium—an Indian perspective. Sadhana **38**(5), 817–837 (2013)
2. A. Kakodkar, *Strategies for development of Thorium based power systems* (Annual Conf. Indian Nucl. Soc., Plenary talk, INSAC, 2000)
3. IAEA TECHDOC-1450, Thorium fuel cycle—potential benefits and challenges, IAEA, May 2005
4. R. Ramanna, S.M. Lee, The thorium cycle for fast reactors. Pramana-J. Phys. **27**, 129 (1986)
5. Chapter 5.1-PFBR Reactor Physics, FSAR Report-Revision-0, IGCAR, Kalpakkam (2010)
6. P. Mohanakrishnan, Development and validation of a fast reactor core burnup code—FARCOB. Ann. Nucl. Energy **35**, 158–166 (2008)
7. G.N. Manturov, *ABBN-93 Group Data Library Part-I: Nuclear Data for the Calculations of Neutron and Photon Radiation Functions, INDC (CCP)-409/L* (IAEA, Vienna, 1997)

8. K. Devan, A. Riyas, P. Mohanakrishnan, ATOMIX: A Code for Calculating atom Densities of Nuclides in a mixture, REG/RPD/CPS/24, IGCAR Report Kalpakkam (April 2003)
9. K. Devan, EFCONSY-1: An Interface between CONSYST/ABBN-93 and IGCAR Diffusion Theory Neutronics Codes, REG/RPD/NDS/90, IGCAR Report Kalpakkam (December 2000)
10. P. Mohanakrishnan, *COHINT—A Computer Code to Treat Control Rod Heterogeneity and It's Application to Fast Reactor Core Analysis, IGC-130* (IGCAR Report, Kalpakkam, 1992)
11. P. Mohanakrishnan, A. Bachchan, Introduction to Use of Computer Code FARCOB, REG/RPD/CPS/32/Rev.2, IGCAR Report Kalpakkam (September 2011)
12. A.G. Croff, *A User's Manual for the ORIGEN-2 Computer Code, Report ORNL/TM-7175* (ORNL, Oak Ridge, Tennessee, USA, 1980)

A Proposal for a First ADS Demonstrator

Jean-Pierre Revol

Abstract The international Thorium Energy Committee (iThEC), a non-profit association, under Swiss law, was founded in 2012 with the goal to promote R&D on the use of thorium in order to transmute nuclear waste and produce safe, clean and abundant energy. To meet the tremendous world energy needs, systematic R&D has to be pursued to replace fossil fuels. Nuclear energy, which is abundant, energy-intensive, produces no greenhouse gases or air pollution should be a leading candidate. iThEC is at the origin of two new initiatives: a first ADS experiment of substantial power ($\geq$1 MW) at INR Troitsk in Russia, which I will describe, and an innovative high-power superconducting cyclotron. Global cooperation is highly desirable in this domain, and I will discuss possibilities of cooperation based on the CERN experimental collaboration model.

Keywords Thorium · Accelerator-driven systems · ADS · iThEC · Nuclear waste

1 Introduction

The international Thorium Energy Committee (iThEC) [1] is a Geneva-based, non-profit association, founded under Swiss law in 2012, strongly associated with CERN. Its members, from 14 countries, include physicists, engineers, politicians and others concerned by energy issues. Several iThEC members were involved in pioneering CERN accelerator-driven experiments: FEAT [2] and TARC [3] carried out by Nobel Prize Laureate Carlo Rubbia and his team, as well as in the design, construction and exploitation of the CERN neutron test facility n_TOF [4].

iThEC is acting to promote R&D in the use of thorium in order to transmute nuclear waste and produce safe, clean and abundant energy, in particular with accelerator-driven systems (ADS). Whether countries such as Germany or Switzerland decide to

J.-P. Revol (✉)
international Thorium Energy Committee (iThEC),
Rue François-Dussaud 17, 1227 Geneva, Switzerland
e-mail: jean-pierre.revol@cern.ch

A. K. Nayak and B. R. Sehgal (eds.), *Thorium—Energy for the Future*,
https://doi.org/10.1007/978-981-13-2658-5_12

stop their nuclear program, or others such as France continue their nuclear program, or others such as India and China and several developing countries build a nuclear program, nuclear waste management is a common important issue that has not yet found a convincing solution. The additional constraint on geological repositories to make the spent fuel retrievable for future destruction requires new storage techniques to be developed. It also brings a stronger argument in favour of developing, as soon as possible, the technology needed to destroy the long-term component of nuclear waste. Therefore, it is the primary goal of iThEC to promote the development of such technology. The first task for iThEC was the organization of the ThEC13 conference, in the CERN Globe of Science and Innovation, in October 2013. This conference, attended by the major world players in thorium R&D, provided a review of the status of the field, crucial to the iThEC for the definition of its strategy.

2 iThEC's Motivations

Fossil fuels are present in finite quantities in the Earth's crust, and inevitably, they will run out sooner or later. At the rate they are used today, the timescale for exhausting these sources of energy is of the order of 50–100 years for oil and gas, and of 100–150 years for coal. If consumption continues to increase, this will occur sooner. These are short-time periods considering the implications. However, there are strong reasons for replacing fossil fuels by better fuels even sooner. Indeed, it does not make sense to burn fossil fuels till supplies have run out:

- In the first place, there is an increasing consensus that global warming due to the release of greenhouse gases issuing from human activities is a serious problem. Even if this is doubted, precaution should prevail.
- The immediate major problem associated with burning fossil fuels is air pollution. This is already very costly to society. For instance, burning coal costs Europe alone more than 40 billion Euros in annual healthcare expenses. The European Environmental Agency declared that "Air pollution poses the single largest environmental health risk in Europe today". More than half a million deaths per year in China are due to air pollution [5]. World Health Organization (WHO) estimated that one in eight of total deaths in the world is the result of exposure to air pollution. This means that the use of fossil fuels kills 10,000 people each day.
- In addition, fossil fuels could be put to much better use. For instance, instead of burning oil, it can be used in the manufacturing of plastics, rubber, tires, paints, glue, drugs, cosmetics and detergents.

In conclusion, there is every reason to stop burning fossils fuels, as soon as possible. Today fossil fuels represent 86% of the primary energy consumption [6]. The problem is that they are cheap and abundant so the current tendency is still to increase their consumption (Fig. 1). It is expected that, given a rising standard of living and a global population plateauing between 10 and 11 billion inhabitants, the world energy consumption would have to increase by a factor 3 or more in the next 100 years to

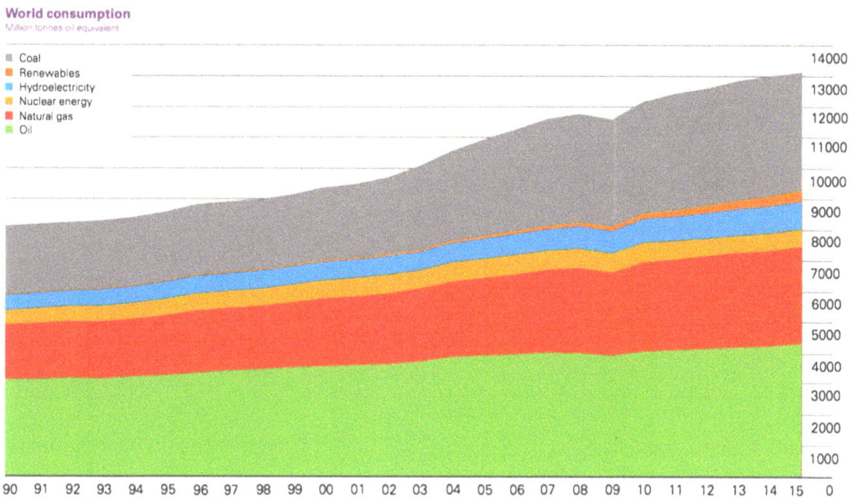

Fig. 1 World primary energy consumption from 1990 to 2015, showing that coal, gas and oil represent 86% of the world consumption, hydro 6.5%, nuclear 4.4% and renewable 2.8% [6]

keep pace with demand. Therefore, replacing fossil fuels is a formidable challenge to society that can only be met through vigorous and systematic Research and Development (R&D). As far as the energy domain is concerned, this R&D must be pursued without prejudice and must include nuclear energy. There is no reason to believe that nuclear energy could not be made acceptable to society, in terms of safety, waste management and proliferation. Its incomparable advantages are that: it is abundant and energy-intensive; it does not produce greenhouse gases or create air pollution.

This challenge may also turn out to be an opportunity, as the innovation required will drive economic growth. This is particularly true for developing countries that do not yet have a fully developed nuclear industry. These countries have the choice of copying what was done in countries with advanced nuclear industry or taking the innovative option. In this way, they would control the energy market in the future, which is clearly the intention of China.

3 Why Thorium?

Given the expected huge energy demand, any potentially important source of energy must be considered. In particular, thorium as the required technologies represents only a modest extrapolation of existing nuclear and particle accelerator technologies. As stated by Carlo Rubbia at the ThEC13 international conference, thorium constitutes "a sustainable energy resource on the human timescale" [7].

Indeed, thorium exists in abundance. There are 1.2×10^{14} tons of thorium in the Earth's crust, which is about the same as lead, and three to four times more

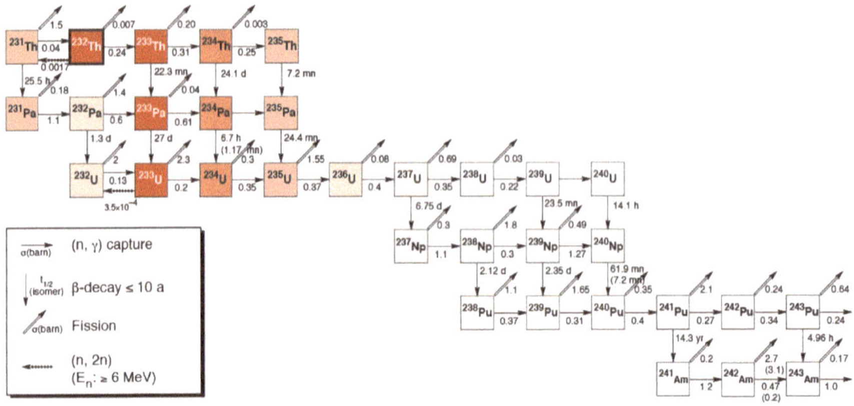

Fig. 2 Main elements produced by irradiation of thorium in a fast neutron flux. The production of TRU is strongly suppressed as it takes seven successive neutron captures to produce plutonium-239 starting from thorium-232

than uranium. Thorium used in breeding mode represents a resource potential 500 times larger than uranium-235 used in present thermal neutrons reactors. Thorium resources are broadly distributed on the Earth, and India is particularly well endowed, with the largest thorium resources [8]. Natural thorium is isotopically pure, found mostly in monazite ores, and relatively cheap as often a by-product of rare earth mining. Thorium produces smaller amounts of long-lived nuclear waste (TRU)[1] than uranium (Fig. 2). In addition, it has excellent physical properties such as higher melting point for both metallic and oxide forms, and better thermal conductivity when compared to uranium. This means that there are higher safety margins for design and operation. When used in ADS, thorium offers the most efficient method to destroy TRU through fission [9] and allows the system to be subcritical, with obvious safety advantages. Most importantly, the thorium fuel cycle has the great advantage of being proliferation resistant.

The advantage of a thorium ADS as advocated by C. Rubbia with the Energy Amplifier (EA) [9] is very clear: while for CAPRA [10] or any uranium–plutonium fast reactor the equilibrium concentration of plutonium is about 15% (Fig. 3), it is only of the order of 10^{-6} in an EA with thorium fuel. Therefore, it is easy to achieve in an EA a plutonium concentration much larger than 10^{-6} and be in efficient plutonium consumption mode. On the other hand, burning plutonium at a high rate in CAPRA requires concentrations much beyond 15%; one must exceed 40% to compete with the EA with an incineration rate, which could reach 80 kg/TWhe. Under such enrichment conditions, a uranium–plutonium fast reactor would become extremely dangerous (small bare critical mass, very little control margin, worse temperature coefficient, increased void coefficient, etc.), and the effective incineration rate would remain

[1] Transuranium elements (TRU): chemical elements with atomic numbers greater than 92 (the atomic number of uranium).

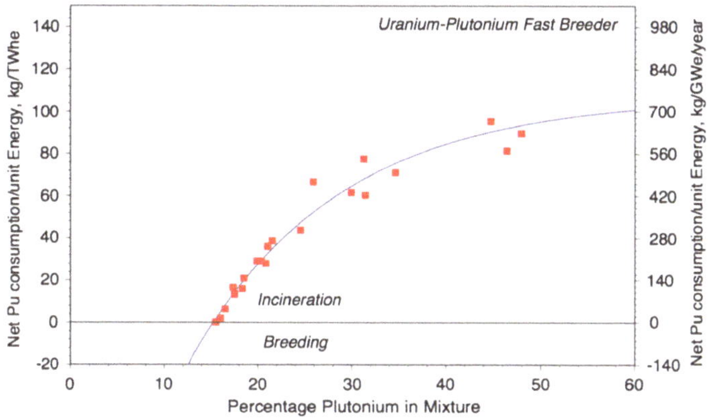

Fig. 3 Net plutonium consumption in CAPRA [10] in kg of plutonium per TWh of electric energy as a function of plutonium enrichment of the fuel mixture in percentage. (*Source* CEA unpublished)

modest because the reactivity decreases very rapidly as fission fragments build up. The situation would even be worse with high enrichment in minor actinides.

Mainly because of neutron inventory reasons, thorium cannot simply be substituted for uranium in critical reactor fuel [11]. Uranium-233 has to be bred from thorium in order to produce energy. Three basic approaches are envisaged:

1. A chain of critical reactors to first breed plutonium (in CANDU reactors). Afterwards, use the plutonium in fast sodium-cooled critical reactors around which a thorium blanket allows the breeding of uranium-233, so it can be used to manufacture fuel for advanced thermal rectors. This scheme has been adopted by India; it technically works but requires maintaining three different reactor technologies. Moreover, it does not solve the problems of nuclear waste management and sustainability.
2. Moving the fuel continuously so that there is always fresh fuel in the core. This can be done in pebble-bed reactors or in molten salt reactors, both of which have serious safety issues. In travelling wave reactors, which have yet to be developed, it is not the fuel that moves, but the neutron breeding and fission wave.
3. Using a particle accelerator to produce the extra neutrons needed to sustain a chain reaction, in so-called accelerator-driven systems. C. Rubbia demonstrated that this is the most efficient and elegant solution for using thorium [9]. This is what motivates iThEC in its efforts to promote accelerator-driven thorium systems (ADTS).

4 Accelerator-Driven Systems

Unfortunately, at the moment, developed countries do not seem to be ready to move in the direction of thorium fuel. So far, in most countries with established nuclear power generation, the development of thorium technologies has been opposed for political reasons by green parties and for economic reasons by nuclear industry, already vastly invested in the development of a new generation of fast critical uranium breeder reactors (Generation IV [12]), and which benefits from institutional funding. However, the development of ADS is not in competition with present nuclear industry programs, and it is now time to start preparing for the future.

The recent history of ADS may be characterized by four phases. The first phase, which demonstrated the basic concepts of ADS, in particular at CERN at the initiative of Carlo Rubbia, was completed in the 1990s. The FEAT experiment [2] (Fig. 4) provided a verification of the basic concepts and validated the innovative simulation code (Fig. 5) developed by Carlo Rubbia and his group. The second phase, which tested all the various elements of an ADS separately, was completed in the 2000s: a proton beam of power exceeding 1 MW at PSI; a neutron spallation source of 1 MW, MEGAPIE, also at PSI, and various developments of high-power spallation targets for SNS, ESS and EURISOL; the development of high-power superconducting accelerating cavities by the particle physics community; the EUROTRANS R&D

Fig. 4 Photograph of the FEAT experiment [2] set up at the CERN Proton Synchrotron (PS). It was a subcritical ($k_{eff} \sim 0.9$) thermal neutron system, with 3.62 tonnes of natural uranium. The PS beam was injected horizontally. The beam vacuum pipe is visible on the upper right corner of the photograph

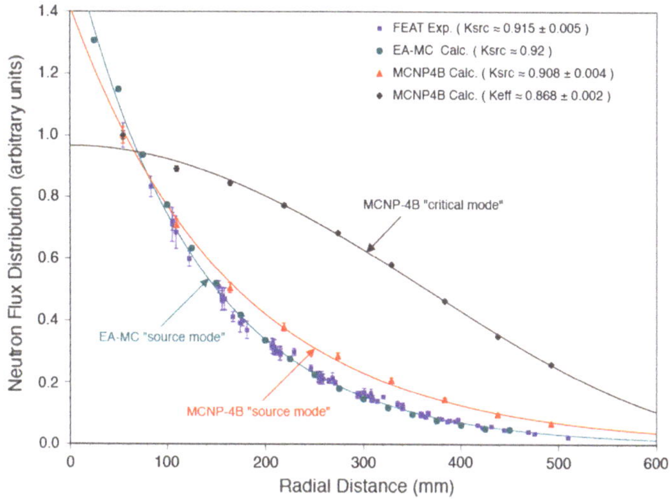

Fig. 5 Radial dependence of the neutron flux as measured in the FEAT experiment [2]. The measured points (blue squares) are compared to simulations: with the Energy Amplifier innovative code (green dots), with MCNP-4B (red triangles). The neutron flux distribution expected under critical conditions is shown to illustrate the difference between subcritical and critical systems. Note the excellent agreement with the innovative code developed by Carlo Rubbia and his group

program of the European Union on partitioning and transmutation for the uranium fuel cycle.

It would seem natural to follow these first two phases with a phase three consisting of the first coupling of a proton beam to a fast-neutron subcritical core at a significant power (≥1 MW). However, this step is still missing, more than 20 years after the first pioneering FEAT experiment. Phase three should also include the development of an accelerator optimized for industrial applications of ADS. Phase three is concentrating all the efforts of iThEC at the moment.

The good news is that there exist already two important ADS projects in the world, which are aiming directly at what could be defined as the industrial or fourth phase: the MYRRHA project [13] in Europe, which should be the flagship of ADS systems. Unfortunately, at this time it is not fully funded and the intended date of completion, 2025, will most likely be exceeded. The other important project is ADANES in China [14], which has the ambition of reaching 1000 MW of electrical power by 2032. Given their long timescales, and the large leap into industrialization, both projects would benefit from a phase three. In addition, there are many other ADS-related activities in progress, in particular in Ukraine, but still at low power and with a thermal neutron flux, based on an electron beam of 100 kW driving a subcritical core ready to be commissioned [15]. Given the importance of the energy issue, the lack of coordination and collaboration between these various efforts is regrettable.

5 iThEC—INR Proposal

In the context of phase three, iThEC is at the origin of two new initiatives: a first ADS experiment of substantial power ($\geq$1 MW) in collaboration with INR Troitsk [16], near Moscow, and a high-power superconducting cyclotron [17], in collaboration with CERN, in Geneva Switzerland.

Through a systematic review of the status of thorium R&D in the world at ThEC13, visits of the Institute of Nuclear Research RAS, Moscow (INR), and detailed discussions with INR management, iThEC became convinced of a rather unique opportunity in Russia, at the Troitsk INR laboratory. Here the existing infrastructure would allow, for the first time at significant power (a few MW), the coupling of a proton accelerator to a fast-neutron subcritical core. The accelerator exists (Fig. 6) and only needs a relatively modest refurbishment. Troitsk is already operating a neutron spallation source. A beam line towards an available experimental pit could be implemented quickly (Fig. 7). In addition, the infrastructure exists for the manipulation of radioactive materials.

It is estimated that in 5 years' time a landmark experiment could be carried out which would validate the properties of an ADTS, demonstrate safety, demonstrate the destruction of minor actinides, and test methods for the production of new radioisotopes for medicine. This would be at 5% or less of the cost of currents projects

Fig. 6 View of the 600 MeV LINAC installed in a 500 m building at INR Troitsk

Fig. 7 Photograph taken at INR Troitsk of the beam target area showing the proposed ADS pit (9) and the presently operating pulsed neutron source cell (10) and its beam line (8). Taken from [16]

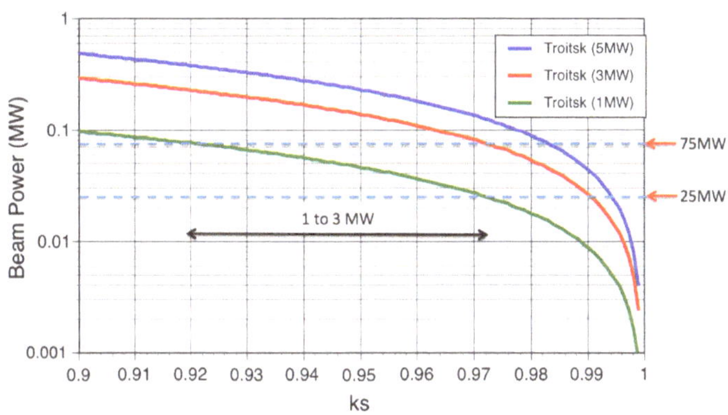

Fig. 8 Curves of constant core power in neutron multiplication coefficient k_S versus beam power in MW, for Troitsk proton beam energy of 300 MeV, showing that the required core power in excess of 1 MW can be obtained with k_S below 0.98

such as MYRRHA [13]. A road map of the project has been prepared jointly by INR management and iThEC. The Troitsk experiment would constitute in addition a welcome versatile fast neutron test facility, the sub-critical character of which, guarantees safety. Discussions with contacts in China, the Czech Republic, India, South Korea, Japan, Switzerland, Turkey, Chile, as well as with IAEA indicate that such a project, if approved, would get international support, and a collaboration on

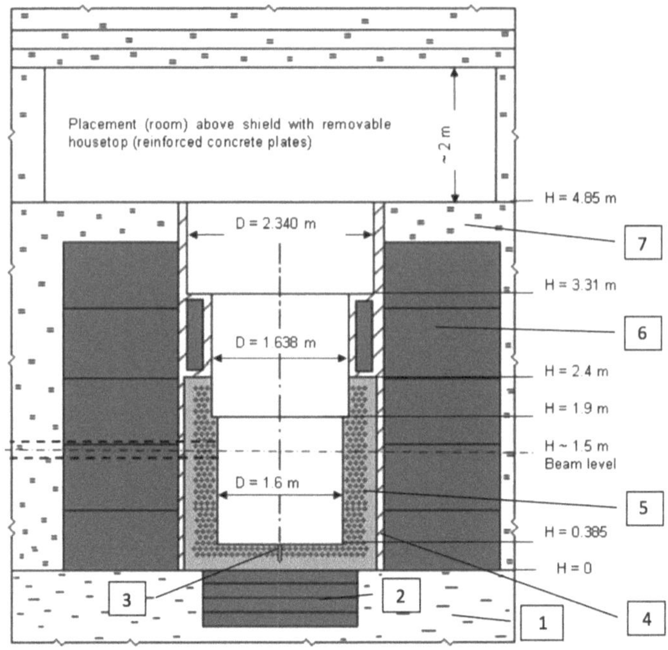

Fig. 9 Schematic of the available ADS pit, of the shield and its basic dimensions: (1) concrete base; (2) steel plates (additional protection of subsoil waters against activation); (3) radioactive water collector in case of an accident; (4) layer of serpentinite; (5) water-cooled heat shield; (6) steel shield (thickness ~ 2.5 m); (7) concrete shield (density 2.8–3.5 g/cm^3). Adapted from [16]

the CERN experimental collaboration model would be possible. Global cooperation, which iThEC is actively stimulating, is highly desirable in this domain.

The main goals of the proposed Troitsk ADS experiment are:

(1) To characterize the properties of an ADS, at substantial power (a thermal power of order 1 MW_{th} or above needed to observe temperature behaviour) and in a fast neutron flux (Destroy nuclear waste (TRU) through fission requires the use of a fast neutron flux);
(2) To demonstrate the feasibility of destroying nuclear waste efficiently (samples of various actinides will be tested);
(3) To study operational procedures and demonstrate safety;
(4) To test properties of various fuels including thorium;
(5) To test new possibilities in the production of radioisotopes of interest in industry and in medicine.

In addition, this unique fast-neutron subcritical facility could be used to test material properties, the subcritical nature of the system providing obvious safety advantages over critical systems.

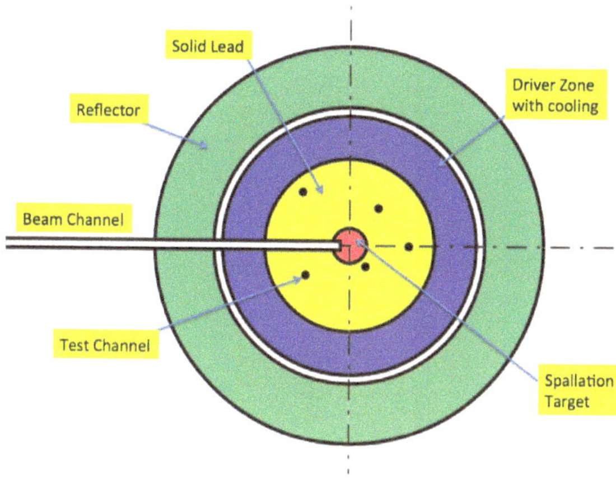

Fig. 10 Conceptual design of the subcritical core of the Troitsk ADS experiment. The area labelled "Reflector", corresponds to zone 5 on Fig. 9, which includes a water-cooled heat shield and represents a boundary condition for the design of the subcritical core, as it is already in place. The driver zone is occupied by enriched uranium fuel. The lead area has holes (test channels), which allow the introduction of neutron detectors or of samples to be irradiated, at various distances from the spallation target

The accelerator is being refurbished to reach a proton energy of 250–300 MeV and an average beam power of 25–75 kW. With such beam power, a thermal core power of 1–3 MW can be obtained with $k_s \leq 0.98$ (Fig. 8).

The ADS pit (Fig. 9) already exists and will allow the introduction of a subcritical core of outer diameter 1.6 m (Fig. 10). The proton beam is inserted horizontally, at a height of 1.1 m above the floor. iThEC and INR are presently working to simplify and optimize the core design. The next step is that ROSATOM will be preparing an engineering design that they will submit to the Russian authorities for approval.

6 iThEC Accelerator Project

iThEC is also at the origin of an accelerator project submitted to the European Union, in the framework of the Horizon 2020 FET program. This is a collaboration with CERN, PSI and leading European industrial companies to design an innovative single-stage, high-power superconducting cyclotron [17] (Fig. 11). Reliability, minimal beam losses and a much lower cost than other accelerator technologies are the main goals of the study.

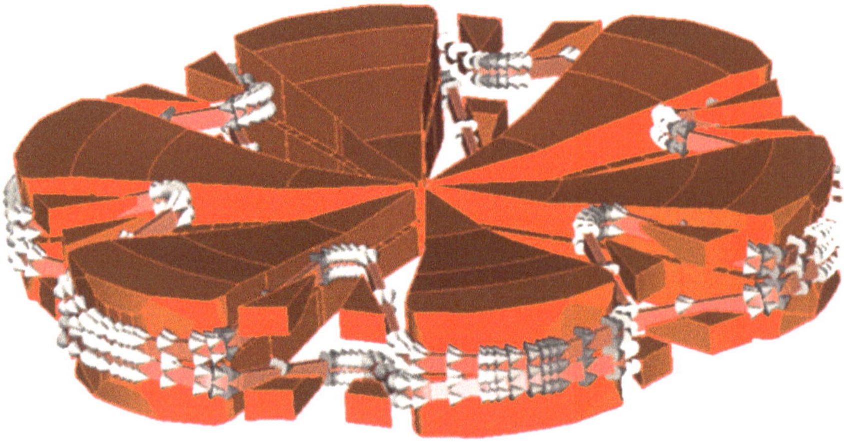

Fig. 11 3D view of the six-sector single-stage cyclotron with reversed valley B-field, considered for the Horizon 2020 FET proposal [17] (patented under the name S2CD by AIMA DEVELOPPEMENT)

7 Conclusion

In conclusion, thorium is an important energy resource that cannot be neglected. It is most important not to leave thorium ADS out of energy R&D programs. The development of thorium technologies is an opportunity for developing countries, such as India. The ADS experiment at INR Troitsk and the development in parallel of an innovative high-power superconducting cyclotron are both necessary steps towards an industrial ADS prototype and the exploitation of thorium resources. It is the priority of iThEC to obtain approval for these two projects. INR Troitsk offers a unique opportunity to advance significantly the field of ADS at a modest cost compared to all other projects and faster. As Russian authorities would like to see international support for the Troitsk experiment, iThEC and INR are seeking partners from the international scientific community.

Acknowledgements I would like to thank all my iThEC and INR colleagues and the management of INR Troitsk. Without their contributions and enthusiasm, none of the projects described here would have been possible.

References

1. iThEC's Web page: www.ithec.org
2. S. Andriamonje et al., Experimental determination of the energy generated in nuclear cascades by a high energy beam. Phys. Lett. B **348**, 697–709 (1995)

3. A. Abánades et al., Results from the TARC experiment: spallation neutron phenomenology in lead and neutron-driven nuclear transmutation by adiabatic resonance crossing. Nucl. Instrum. Methods Phys. Res. A **478**, 577–730 (2002)
4. F. Gunsing, Nuclear Data for the Thorium Fuel Cycle and Transmutation, in *Thorium Energy for the World, ThEC13 Proceedings of CERN Globe of Science and Innovation*, Geneva, Switzerland, 27–31 Oct 2013 (Springer, 2016)
5. GBD MAPS Working Group, Burden of Disease Attributable to Coal-Burning and Other Air Pollution Sources in China (Tsinghua University Report, 2016)
6. 65th edition of the BP Statistical Review of World Energy (June 8, 2016). https://www.bp.com/en/global/corporate/media/press-releases/65th-edition-of-bp-statistical-review-of-world-energy.html
7. C. Rubbia, A Future for Thorium Power?, in *Thorium Energy for the World*, in *ThEC13 Proceedings of CERN Globe of Science and Innovation*, Geneva, Switzerland, 27–31 Oct 2013 (Springer, 2016)
8. Data taken from "Uranium 2014: Resources, Production and Demand", OECD Nuclear Energy Agency and the International Atomic Energy Agency
9. C. Rubbia; et al., Conceptual Design of a Fast Neutron Operated High Power Energy Amplifier, *CERN/AT/95-44 (ET)*, 29 Sept. 1995; see also Rubbia, C. A High Gain Energy Amplifier Operated with Fast Neutrons, *AIP Conference Proceeding* 346, International Conference on ADT Technologies and Applications, Las Vegas (1994)
10. J. Rouault et al., Physics of Pu Burning in Fast Reactors: Impact on Burner Core Design, in *Proceedings of ANS Topical Meeting on Advances in Reactor Physics*, Knoxville, 11–15 Apr 1994
11. J.-P. Revol, Accelerator-Driven Systems (ADS) Physics and Motivation, in *Thorium Energy for the World, ThEC13 Proceedings of CERN Globe of Science and Innovation*, Geneva, Switzerland, October 27–31 2013 (Springer, 2016)
12. Generation IV nuclear energy systems, see: www.ne.anl.gov/research/genIV
13. H. Aït Abderrahim, MYRRHA: A Flexible and Fast Spectrum Irradiation Facility, in *Thorium Energy for the World*, ThEC13 Proceedings of CERN Globe of Science and Innovation, Geneva, Switzerland, 27–31 Oct 2013 (Springer 2016)
14. P. Luo, S.C. Wang, Z.G. Hu, H.S. Xu, W.L. Zhan, Accelerator driven sub-critical systems—a promising solution for cycling nuclear fuel, Institute of Modern Physics, Chinese Academy of Sciences, Lanzhou, 730000, China. https://doi.org/10.7693/wl20160903
15. Y. Gohar, I. Bolshinsky, D. Naberezhnev et al., Accelerator driven sub-critical facility: conceptual design development. Nucl. Instrum. Meth. Phys. Res. A. **562**, 870–874 (2006)
16. S.F. Sidorkin, A.D. Rogov, L.I. Ponomarev, E.A. Koptelov, Proposal of the ADS Research Stand Based on the Linac of the Institute for Nuclear Research of the Russian Academy of Sciences, in *Thorium Energy for the World*, ThEC13 Proceedings of CERN Globe of Science and Innovation, Geneva, Switzerland, 27–31 Oct 2013 (Springer 2016)
17. M. Conjat, J. Mandrillon, P. Mandrillon, Cyclotron Drivers for Accelerator-Driven Systems, AIMA DEVELOPPEMENT Company, in *Thorium Energy for the World*, ThEC13 Proceedings of CERN Globe of Science and Innovation, Geneva, Switzerland, 27–31 Oct 2013 (Springer 2016)

Proposal for an Experimental Fast ADS Using Th–Pu MOX Fuel for Higher Actinide Transmutation

Amar Sinha, Tushar Roy, Y. S. Kashyap, S. Bajpai and M. Shukla

Abstract An experimental program to design and study external source-driven subcritical assembly has been initiated at BARC. This program is aimed at understanding neutronics of accelerator-driven system at low-power level. In this series, an experimental thermal subcritical core BRAHMMA driven by a D-D/D-T neutron generator has been developed. Various ADS-related technique development and experiments for flux measurement, reactivity measurement using PNS techniques, noise methods, etc., have been conducted using this thermal assembly. Following the successful implementation of BRAHMMA which is thermal assembly, low-power experimental fast ADS based on Th–Pu MOX fuel has been proposed for the next stage of ADS programme to study the issues related to higher actinide transmutation using ADS. Fast ADS is an excellent tool for waste transmutation and worldwide efforts are on to study such a system on low-power scale. However, most studies such as Guinevere (Belgium) have been limited to enriched uranium and lead system, whereas from Indian point of view, thorium–plutonium fast experimental ADS system is of direct interest not only from the point of view of waste transmutation but also for thorium utilization using ADS. Design concept and issues related to a Th–Pu MOX fast subcritical assembly driven by intense D-T neutron generator for mock-up studies of waste transmuter ADS will be discussed.

Keywords ADS · BRAHMMA · Th–Pu MOX fuel · Reactivity

1 Introduction

Energy is the key to economic growth of a country. Among the various options being explored for long-term energy production, nuclear energy is being considered to

---

A. Sinha
Raja Ramanna Fellow, DAE, Mumbai, India

T. Roy (✉) · Y. S. Kashyap · S. Bajpai · M. Shukla
Technical Physics Division, Bhabha Atomic Research Centre, Mumbai, India
e-mail: tushar@barc.gov.in

A. K. Nayak and B. R. Sehgal (eds.), *Thorium—Energy for the Future*,
https://doi.org/10.1007/978-981-13-2658-5_13

be not only a long-term solution but also a carbon dioxide-free option. However, despite a very good record of operating nuclear reactors for more than 65 years, there are some apprehensions in the mind of common man about nuclear energy. One of the major concerns relates to waste produced by nuclear energy in the form of fission product and transuranic (TRU) waste which have long half-life. Among these waste nuclides, transuranic nuclides have not only large contribution in terms of radioactivity but also longest half-life extending to tens of thousands years. To allay the fears in the mind of public, technologies have to be evolved to find a long-term solution to this problem. Evolving methods of managing TRU wastes [1] is thus one of the primary goals of nuclear waste management. One of the solutions is to incinerate the long-lived TRUs by using the methods of nuclear transmutation. Nuclear transmutation converts these nuclides into nuclides which have manageable half-life. The efficiency of transmutation of these TRUs is governed by neutron economy and energy spectrum of neutrons transmuting them. A fast spectrum is ideally suited to fission TRUs to relatively low half-life fission product. However, the transuranic elements have low fraction of delayed neutrons. This creates a problem of safety and control if these transuranic are loaded in critical fast reactors. Also, the Doppler feedback for transuranic loaded fast critical reactors is low compared to conventional fast reactors owing to low fertile content. Thus, transmuting TRUs in conventional fast critical reactor has inherent safety issue.

Accelerator-driven system (ADS) has no such issue of neutron economy or safety as their power is controlled by external neutron source. ADS, being subcritical to start with, is not affected by low-delayed neutron fraction of transuranic elements or Doppler feedback. ADS has an additional advantage, particularly in Indian context, for utilization of thorium. Thorium by itself is not fissile and needs additional neutron to convert it to fissile. This puts a problem of neutron economy in using thorium in critical reactors. The spallation reaction used in ADS is a neutron-rich reaction and can easily supply additional neutrons for thorium breeding over and above those coming from fission.

Keeping this objective in mind, India has embarked upon accelerator program for ADS applications. However, besides accelerator there are major neutronics, safety, control, physics, and engineering issues which must be studied before such a concept can be realized. This is being done worldwide using a low-power system. Such low-powered systems have been designed worldwide using a variety of neutron sources such as D-T neutron source or photo-neutron source to study various physics issues of ADS. As a part of the ADS program in India, a subcritical assembly BRAHMMA (**B**eO **R**eflected **A**nd **H**DPe **M**oderated **M**ultiplying **A**ssembly) [2, 3] driven by an indigenously developed deuteron accelerator has been designed, developed, and commissioned at Bhabha Atomic Research Centre, India. This experimental facility has enabled the study of various physics parameters related to neutronic and dynamic characteristics of ADS.

BRAHMMA experimental system is first in a series of proposed subcritical assembly. Various experiments carried out at this facility will be discussed in this paper. For the next stage, a fast spectrum ADS based on Th–Pu for demonstrating waste transmutation is being designed. A preliminary design overview of fast ADS will also be discussed in this paper. The first part of the paper will highlight the role of an experimental thermal ADS for developing methods for subcriticality measurement in thermal ADS. The second part of paper will discuss next stage in the form of a conceptual proposal for fast ADS using Th–Pu MOX fuel for waste transmutation.

2 BRAHMMA—Thermal ADS

The primary strategy in making of experimental ADS system in India has been to design a modular and neutronically simple system so that various experimental studies can be done without the need of complex neutronic modeling. To do so, Indian subcritical system BRAHMMA was designed using HDPE as moderator which enabled modular design. To keep the system compact, beryllium oxide (BeO) was used as reflector. The fuel rods are arranged in a 13 × 13 square lattice with 48-mm pitch. A central cavity (144 mm × 144 mm) has been provided for inserting neutron generator. The system is described in detail in [2]. The system in the present configuration has k_{eff} of 0.89. However, due to its modular design, it is possible to increase k_{eff} of system by using enriched fuel. This subcritical system is driven by D-D or D-T neutron source in pulsed or DC mode depending upon the need of experiment (Fig. 1).

2.1 Experimental Measurements

Since the experiment channels are narrow, special miniature ^{3}He detectors have been developed [4] for experimental measurements. The miniature ^{3}He detector has

(a) (b)

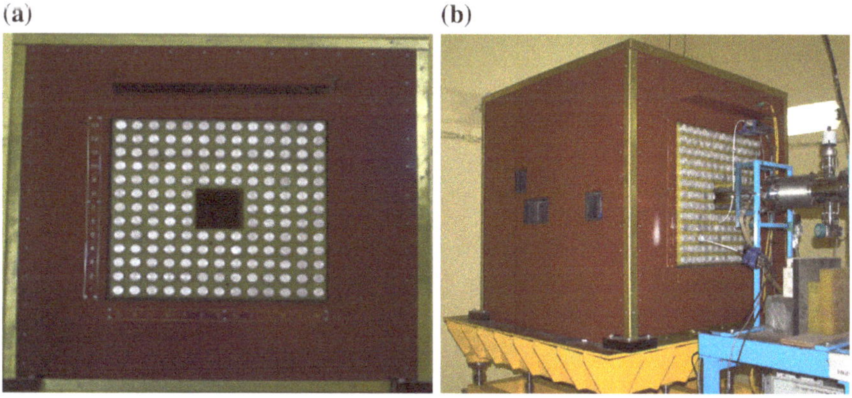

Fig. 1 **a** BRAHMMA (front view), **b** coupled to neutron generator

a diameter of 6.2 mm, active length of 70 mm, and sensitivity of 0.4 cps/nv. The experimental measurements performed have been listed below:

- Spatial flux profile—axial and radial
- Pulsed Neutron Source (PNS)
  - Area Ratio Method
  - Slope Fit Method
  - Source Jerk Method
- Bell–Glasstone spatial correction for PNS measurements
- Noise Measurements
  - Feynman-α
  - Rossi-α
- Source Multiplication

2.1.1 PNS Measurements

Pulsed neutron source (PNS) techniques [5, 6] are one of the simplest methods for reactivity measurement of a subcritical system. In PNS techniques, the detector response to a short neutron pulse is measured. Following the neutron pulse, the neutron population first rises to a peak and then dies away quickly due to the subcritical nature of the system. After a short period of time, the detector response decays much more slowly to represent the emission of delayed neutrons in the reactor. Three techniques—namely area ratio method, slope fit method, and source jerk method—have been employed to measure reactivity (ρ) and prompt neutron decay constant (α).

For area ratio and slope fit experiments, the external pulsed D-T neutron source is operated with a pulse period of 10 ms (100 Hz). The neutron flux in the subcritical assembly is recorded by the detectors from a number of repetitive neutron pulses (typically 50,000 pulses) once the delayed neutron level has reached an equilibrium value. Figure 2 shows the PNS histogram for detectors located in the reflector region.

The reactivity of subcritical system as measured by area ratio method is spatially dependent. In order to remove this spatial dependence, we need to correct measured reactivity by a spatial correction factor. For our system, we have calculated spatial correction factors by a method suggested by Bell and Glasstone [7]. This is a static simulation method which has been implemented using Monte Carlo technique [8, 9]. The spatial correction factor for area ratio method is given as

$$f = \left(\frac{\rho_{\text{cri}}}{\rho_{\text{src}}}\right) = -\left(\frac{A_{\text{d}}}{A_{\text{p}}}\right)\left(\frac{1}{\beta_{\text{eff}}}\right)\left(\frac{1-k_{\text{eff}}}{k_{\text{eff}}}\right) \tag{1}$$

where A_{p} is the prompt area, A_{d} is delayed area, ρ_{cri} is reactivity calculated by computer codes in criticality mode, and ρ_{src} is reactivity calculated by computer codes in source mode with the external neutron source. The experimental reactivity values

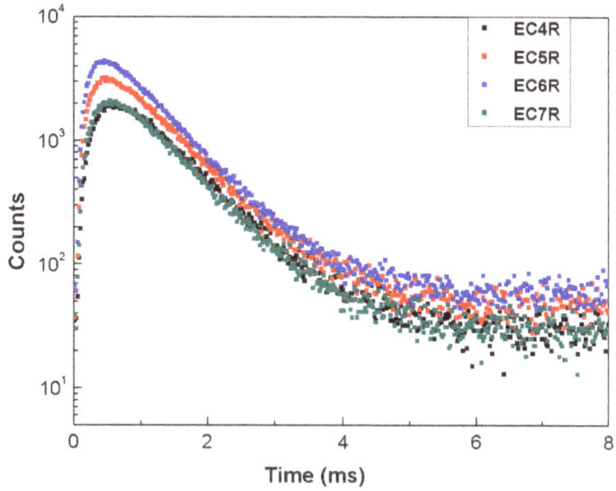

Fig. 2 PNS histogram for detectors located in reflector

measured in area ratio method are corrected by multiplying with the Bell–Glasstone correction factor.

Source jerk is another pulsed neutron technique which can be used to infer reactivity. In this technique, the subcritical system is operated at a steady flux level (say, n_0) and then the neutron source is switched off. The neutron population inside the core will attain a lower neutron level (say, n_1). This quasi-static level will decay according to the decay rate of delayed neutron precursors. Using point kinetic equation, the reactivity of the subcritical core is given by:

$$\rho_{\$} = \frac{\rho}{\beta_{\text{eff}}} = -\frac{n_0 - n_1}{n_1} \tag{2}$$

2.1.2 Noise Measurements

The reactivity of subcritical system can be measured by studying the variance to mean ratio of the neutron counts with respect to various counting gate widths. The deviation of the variance to mean value from one, known as Feynman $Y(T)$, is defined as

$$Y(T) = \frac{\text{Var}(T)}{\text{Mean}(T)} - 1 = \varepsilon \sum_{i=0}^{6} A_i \left(1 - \frac{\left(1 - \exp^{-\alpha_i t}\right)}{\alpha_i t} \right) \tag{3}$$

The $i = 0$ term corresponds to the prompt part and α_0 is the prompt decay constant. The higher order terms are for the delayed neutron precursors. The prompt decay constant is then used to obtain reactivity. Preliminary Feynman-α measurement has been carried out. The experimental Y-curve is shown in Fig. 3. The value of α_0 is obtained by fitting Eq. 3 to the measured data. The measured k_{eff} value is 0.888 ± 0.003.

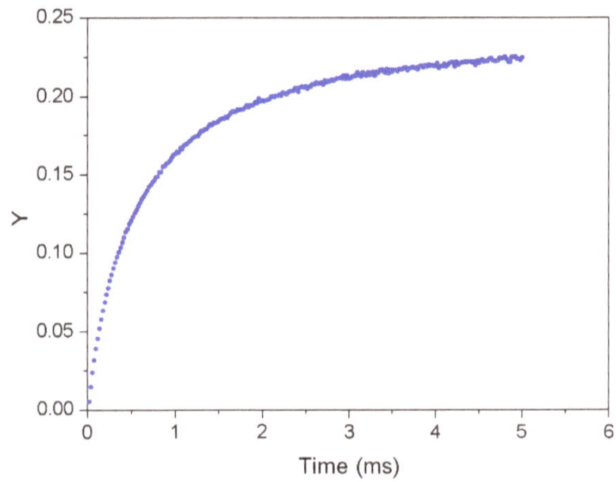

Fig. 3 Feynman Y-curve for continuous DD source

3 Proposal for Experimental Fast ADS

For the next stage of ADS programme in India, a low-power experimental fast ADS based on Th–Pu MOX fuel has been proposed to study the issues related to higher actinide transmutation using ADS. Fast ADS is an excellent tool for waste transmutation and worldwide efforts are on to study such a system on low-power scale. However, most studies such as Guinevere [10] have been limited to enriched uranium and lead system, whereas from Indian point of view, Th–Pu system is of direct interest not only for fast ADS but also as a multipurpose system to study the third stage of India nuclear program which proposes to use thorium and plutonium.

In continuation with thermal ADS programme, we propose to develop methods and tools suited for understanding behavior of fast subcritical system. In particular, we intend to study the following parameters:

- Methods of subcriticality monitoring of a fast subcritical system;
- Interpretation and validation of experimental data;
- Validation of core power/beam current relationship;
- Benchmarking and code validation activities;
- Establishing start-up and shutdown procedures;
- Safety issues of different components as well as that of the integrated system as a whole;
- Validation of dynamic behavior of an ADS in a wide range of subcritical levels, subcriticality safety margins, and thermal feedback effects.

In particular, this program will provide answers to the questions of online reactivity monitoring, subcriticality determination, and operational procedures for ADS similar to methods we adopted for thermal ADS. It should also demonstrate minor actinide transmutation by providing some dedicated positions in the core for special fuel assemblies. Finally, it should serve as a general multi-purpose neutron irradiation facility with a variety of core configuration.

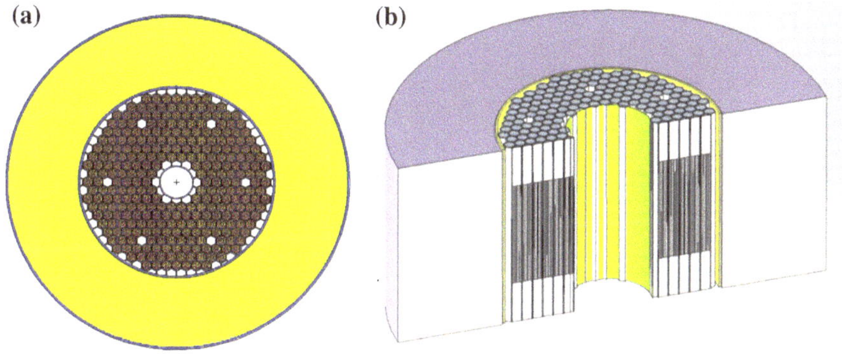

Fig. 4 **a** Cross section, **b** 3D cut-section view of proposed fast ADS assembly

There is an additional aim relevant from Indian perspective. India has proposed three-stage nuclear program to utilize abundant thorium resource within the country. Th–Pu MOX fuelled zero/low-power reactor driven in subcritical to critical mode can provide a very effective test bed for such a three-stage program. Currently, the available data on such a MOX fuel behavior is very preliminary.

3.1 Preliminary Design of Fast ADS

A preliminary design for fast ADS driven by D-T neutron source has been proposed. It consists of Th–Pu MOX fuel in a hexagonal matrix. Lead is being used both as diffuser of high-energy D-T neutron driving the assembly as well as neutron reflector. It is also being used to simulate lead coolant. The composition of MOX fuel is 64% ThO_2 and 36% PuO_2. The fuel pins have a diameter of 10 mm and length 500 mm. Each fuel assembly (FA) consists of seven fuel pins (arranged in two layers 1–6) in a lead matrix and has a pitch of 45 mm and length of 900 mm with two axial lead reflectors at the top and bottom of each fuel assembly. A total of 286 FAs constitute the core. The core is finally surrounded by radial lead reflector. Figure 4a, b shows schematic cross section and 3D cut view of the assembly. Control rods and safety rods have not yet been taken into account and their location and reactivity worth will be considered during detailed designing. A central cavity of 140 mm is provided for coupling to D-T neutron source.

3.1.1 Reflector-Controlled System

The system has been devised as a reflector-controlled system. In this system, the core and reflector block are decoupled and reflector can be made movable. During start-up, the reflector block move to align with the core and reflector is coupled to the core. In the event of any scram signal, the reflector will decouple from the core and drop down to bottom position. This fail-safe mechanism is similar to reflector control system employed earlier in PURNIMA-I [11]. The core of the system can be varied for different configuration depending on fast system using either Th–Pu MOX fuel or some part of it containing MA. Similarly, k_{eff} of the system can be varied depending on fuel composition.

Figure 5 shows schematic of reflector movement. The reactivity margin of the reflector is approximately 130 mk. Figure 6 shows the variation of effective multiplication factor with reflector height.

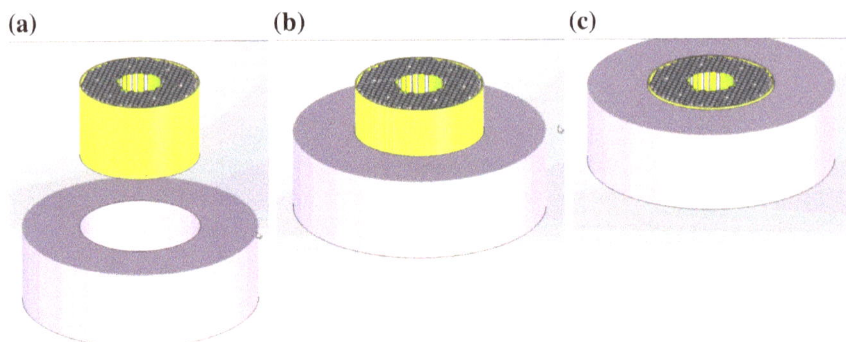

Fig. 5 Reflector position with respect to core, **a** fully OUT, **b** partial IN, **c** fully IN

Fig. 6 Variation of multiplication factor k_{eff} with reflector height

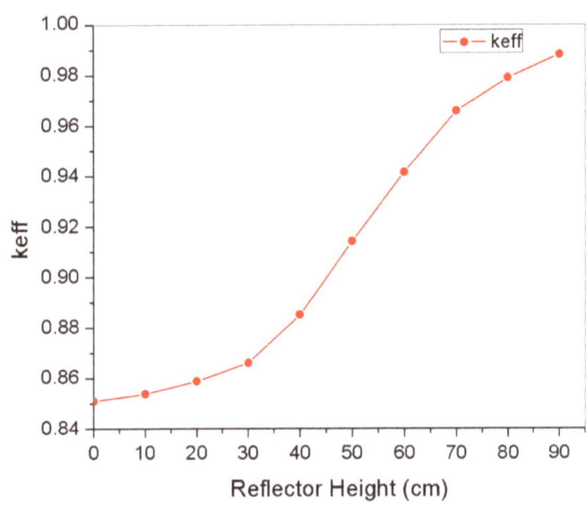

4 Conclusion

Following the successful implementation of thermal ADS BRAHMMA, we have proposed Th–Pu MOX fuel-based fast ADS and discussed some aspects of the system based on a preliminary design. This system shall be used to study the physics of fast systems for waste transmutation and physics of thorium–plutonium systems.

References

1. M. Gonzalez, Nuclear Waste Transmutation Enrique CIEMAT, European Physics Society: Nuclear Physics Board Valencia, 1st May 2004
2. A. Sinha, Tushar Roy, N. Yogesh Kashyap, M.Shukla Ray, T. Patel, S. Bajpai, P.S. Sarkar, S. Bishnoi, P.S. Adhikari, BRAHMMA: A Compact Experimental Accelerator Driven Subcritical Facility Using D-T/D-D Neutron Sources. Ann. Nucl. Energy **75**, 590–594 (2015)
3. A. Sinha, Tushar Roy, Yogesh Kashyap, Nirmal Ray, Mayank Shukla, Tarun Patel, P.S. Shefali Bajpai, Saroj Bishnoi Sarkar, Experimental subcritical facility driven by D-D/D-T neutron generator at BARC, India. Nucl. Instrum. Methods Phys. Res., Sect. B **350**, 66–70 (2015)
4. M. Shukla, S.S. Desai, Tushar Roy, Neutron spatial flux profile measurement in compact subcritical system using miniature neutron detectors. Nucl. Instrum. Methods Phys. Res., Sect. A **772**, 118 (2015)
5. C.M. Persson, A. Fokau, I. Serafimovich et al., Pulsed neutron source measurements in the subcritical ADS experiment YALINA-booster. Ann. Nucl. Energy **35**(12), 2357–2364 (2008)
6. N.G. Sjöstrand, Measurement on a subcritical reactor using a pulsed neutron source. Arkiv fysik **11**(13), 233–246 (1956)
7. G.I. Bell, S. Glasstone, *Nuclear Reactor Theory*, Van Nostrand Reinhold Company (1970)
8. A. Talamo, Z. Zhong, Y. Gohar, Monte Carlo and deterministic calculation of the bell and glasstone spatial correction factor. Comput. Phys. Commun. **183**, 1904 (2012)
9. S. Bajpai, T. Roy, N. Ray, Y. Kashyap, M. Shukla, T. Patel, A. Sinha, Evaluation of spatial correction factors for BRAHMMA subcritical assembly. Nucl. Sci. Eng. **181**, 361–367 (2015)
10. A. Billebaud, et al. The GUINEVERE project for accelerator driven system physics, in *Proceedings of Global 2009*, Paris, France
11. PURNIMA-1 Reactor. http://barc.gov.in/reactor/index.html

Radiological Impact Assessment for Near Surface Disposal of Thorium Waste

Faby Sunny and Manish Chopra

Abstract Thorium (^{232}Th) itself is not fissile and so is not directly usable in a thermal neutron reactor. However, it is fertile and upon absorbing a neutron will transmute to uranium-233 (^{233}U), which is a fissile fuel material. The thorium fuel fabrication may lead to a low-level waste comprising of ^{232}Th. This waste may be disposed of in the Near Surface Disposal Facility (NSDF). The very low probability event of leaching of the waste may lead to contamination of the groundwater system. This paper deals with the estimation of the radiological impact of thorium waste disposal in NSDF through groundwater drinking pathway using the Multiple Area Source Model (MASOM).

Keywords Radiological impact · Decay chain · Dose · Groundwater · NSDF

Nomenclature

B	is the width of the NSDF (m)
C	is the concentration of radionuclides in pore water (Bq m^{-3})
C_i	is the radionuclide concentration in the groundwater for instantaneous release of unit radioactivity (Bq m^{-3})
C_1	is the concentration of parent radionuclide in the groundwater (Bq m^{-3})
C_n	is the concentration of the nth progeny in the groundwater (Bq m^{-3})
D	is the annual effective dose through groundwater drinking pathway (Sv y^{-1})
DF_{ing}	is the ingestion dose coefficient of the radionuclide (Sv Bq^{-1})
D_w	is the drinking water consumption rate (L day^{-1})
D_x	is the dispersion coefficient in x-direction (m^2 s^{-1})

F. Sunny (✉) · M. Chopra
Radiation Safety Systems Division, Bhabha Atomic Research Centre,
Trombay, Mumbai 400085, India
e-mail: fabys@barc.gov.in

M. Chopra
e-mail: mchopra@barc.gov.in

A. K. Nayak and B. R. Sehgal (eds.), *Thorium—Energy for the Future*,
https://doi.org/10.1007/978-981-13-2658-5_14

D_y is the dispersion coefficient in y-direction ($m^2\ s^{-1}$)
H is the depth of the trench (m)
H_a is the thickness of the aquifer (m)
K_d is the distribution coefficient of the radionuclide for the aquifer material ($m^3\ kg^{-1}$)
K_{ds} is the distribution coefficient of the radionuclide for the waste material ($m^3\ kg^{-1}$)
K_L is the leach rate coefficient of the radionuclide from the NSDF (s^{-1})
L is the length of the NSDF (m)
λ is the radioactive decay constant of the radionuclide (s^{-1})
λ_1 is the radioactive decay constant of the parent radionuclide (s^{-1})
λ_n is the radioactive decay constant for nth progeny (s^{-1})
M is the inventory of the radionuclide (Bq)
v is the infiltration velocity ($m\ s^{-1}$)
Q is the disposal rate of the radionuclide into the NSDF ($Bq\ s^{-1}$)
ρ is the bulk density of the aquifer material ($kg\ m^{-3}$)
ρ_s is the bulk density of the waste material ($kg\ m^{-3}$)
R_d is the retardation factor in the aquifer
R_{ds} is the retardation factor in the waste material
$\psi(t)$ is the radioactivity release rate from the NSDF at any time t after disposal ($Bq\ s^{-1}$)
T is the period of dumping (s)
θ is the porosity of the aquifer material
θ_S is the porosity of the waste material in the trench
u is the groundwater velocity ($m\ s^{-1}$)
x is the down-flow distance (m)
y is the cross-flow distance (m)

1 Introduction

India has abundant reserves of thorium compared to uranium [1]. Considering the large thorium reserves in India, the future nuclear systems, in the third stage of Indian nuclear power program, will be based on ^{232}Th–^{233}U fuel cycle. The advanced heavy water reactor (AHWR) [2] is being designed in India to fulfill this need. The thorium fuel fabrication may lead to a low-level waste comprising of ^{232}Th. This waste may be disposed of in the Near Surface Disposal Facility (NSDF). The very low probability event of leaching of the waste may lead to contamination of the groundwater system. This paper deals with the radiological impact assessment (RIA) of thorium waste disposal in NSDF through groundwater drinking pathway using the Multiple Area Source Model (MASOM) [3].

2 Methodology

An analytical two-dimensional MASOM with decay chain transport is used for the RIA of the hypothetical NSDF.

2.1 Source Term Model

Thorium and its progeny may slowly leach into the groundwater depending on the leach rate coefficient. The resultant radioactivity release rate into the groundwater can be expressed using the equation [4]

$$\psi(t) = \frac{QK_L}{K_L + \lambda}\left[1 - \exp(-(K_L + \lambda)t)\right] \tag{1}$$

Leaching of radionuclides from the NSDF depends on the structural integrity of the concrete structures, distribution coefficient of radionuclides, and infiltration velocity. The leach rate coefficient, (K_L), can be estimated from the infiltration velocity as

$$K_L = \frac{\nu}{D\theta_S R_{ds}} \tag{2}$$

where

$$R_{ds} = 1 + \frac{\rho_s K_{ds}}{\theta_s} \tag{3}$$

2.2 Multiple Area Source Model (MASOM)

The major processes for dispersion and dilution of pollutants in the subsurface environment are advection, molecular and turbulent diffusion (hydrodynamic dispersion), interaction with solid phase, physical or radioactive decay, and chemical removal. Since the present study is applied to near surface disposal of radioactive waste, the environmental matrix of concern is saturated porous medium and the objective is to compute the concentrations of radionuclides in pore water of the aquifer as a function of space and time. The general mathematical model used in the study is given by Nair and Krishnamoorthy [5]

$$\frac{\partial C}{\partial t} = \frac{D_x}{R_d}\frac{\partial^2 C}{\partial x^2} + \frac{D_y}{R_d}\frac{\partial^2 C}{\partial y^2} - \frac{u}{R_d}\frac{\partial C}{\partial x} - \lambda C \tag{4}$$

The retardation factor of the radionuclide in the aquifer is defined as

$$R_d = 1 + \frac{\rho K_d}{\theta} \tag{5}$$

Equation (4) is solved by using the initial and boundary conditions as follows.
Initial conditions $C = 0$ at $t = 0$
Boundary conditions $C = 0$ at $x = \pm\infty$; $y = \pm\infty$ and

$$H_a \int_{-\infty}^{\infty} \int_{-\infty}^{\infty} C(x, y, t) \mathrm{d}x \mathrm{d}y = \frac{M \exp(-\lambda t)}{\theta R_d} \tag{6}$$

The solution of Eq. (4) for an instantaneous release of unit radioactivity is given as

$$C_i(x, y, t) = \frac{1}{\theta R_d} X(x, t)\, Y(y, t)\;\, Z(z, t) \exp(-\lambda t) \tag{7}$$

The equations for X, Y, and Z are given by

$$X(x, t) = \frac{1}{2L} \left\{ \mathrm{erf}\left[\frac{x + \frac{L}{2} - \frac{ut}{R_d}}{\sqrt{4\frac{D_x t}{R_d}}} \right] - \mathrm{erf}\left[\frac{x - \frac{L}{2} - \frac{ut}{R_d}}{\sqrt{4\frac{D_x t}{R_d}}} \right] \right\} \tag{8}$$

$$Y(y, t) = \frac{1}{2B} \left\{ \mathrm{erf}\left[\frac{y + \frac{B}{2}}{\sqrt{4\frac{D_y t}{R_d}}} \right] - \mathrm{erf}\left[\frac{y - \frac{B}{2}}{\sqrt{4\frac{D_y t}{R_d}}} \right] \right\} \tag{9}$$

$$Z(z, t) = \frac{1}{H_a} \tag{10}$$

The radionuclide concentration in the groundwater during dumping period can be evaluated using the convolution integral [3]

$$C(x, y, T) = \int_0^T \psi(T - \tau) C_i(x, y, \tau) \mathrm{d}\tau \tag{11}$$

The radionuclide concentration after termination of the dumping operation can be evaluated using the following convolution integral [4].

$$C(x, y, t) = \int_0^T \psi(T - \tau) C_i(x, y, T + \tau) \mathrm{d}\tau + \psi(T) \int_0^t \exp[-\lambda_e (t - \tau)] C_i(x, y, \tau) \mathrm{d}\tau \tag{12}$$

where $\lambda_e = \lambda + K_L$ and t is the post-dumping period whose origin is at the termination of the dumping period, T. The first integral in Eq. (12) represents the concentration of radionuclide in the groundwater during the post-dumping period due to the leaching of activity taking place during the dumping period. The second integral represents the additional contribution of radioactivity in the groundwater due to the leaching taking place after the termination of dumping.

2.3 *Ingrowths of Progeny*

^{232}Th has a long decay chain, terminating with stable ^{208}Pb isotope. The ingrowths of progeny in the NSDF as well as during transport in the groundwater are incorporated using the modified Bateman equation [6]. The calculation is complicated by the fact that the progeny will have chemical properties different from the parent radionuclide and will be adsorbed to greater or lesser extent in the solid phase. The concentration of the progeny in the liquid phase must, therefore, be corrected for its K_d.

The concentration of the nth progeny in the groundwater can be calculated as

$$C_n = \frac{\lambda_n}{\lambda_1} C_1 \left[\prod_{m=1}^{n-1} \lambda_m \right] \sum_{j=1}^{n} \left(\frac{e^{-\lambda j.t}}{\prod_{k \neq j}^{n} (\lambda_k - \lambda_j)} \right) \tag{13}$$

The concentration of the progeny is corrected for the difference in partitioning of parent and progeny between solid and liquid phases by multiplying C_n by ratio of the retardation factors of the parent and progeny [6, 7].

2.4 *Radiological Model*

The annual effective dose to members of the public due to consumption of groundwater can be evaluated by the equation [8]

$$D = C\, D_w\, DF_{\text{ing}} \tag{14}$$

3 Parameters

The aquifer characteristics, such as groundwater velocity, dispersivity, aquifer thickness, porosity, and the other parameters used in the model are given in Tables 1 and 2 [9–11]. The infiltration velocity used in the estimation is the design infiltration rate of a low-level waste disposal facility, which is 1×10^{-9} m s^{-1} [12]. A drinking water consumption rate of 2.2 L day^{-1} is used for the dose estimation. As ingestion dose

Table 1 Aquifer characteristics

Parameter	Value
Groundwater velocity	5.79×10^{-6} m s^{-1} (0.5 m day^{-1})
Longitudinal dispersivity	1 m
Transverse dispersivity	0.1 m
Aquifer thickness	12.5 m
Porosity	0.29
Bulk density	1560 kg m^{-3}
Depth of trench	3.0 m

Table 2 Other parameters used in the model

Radionuclide	K_d (waste material, m^3 kg^{-1})	K_d (aquifer material, m^3 kg^{-1})	Half-life (y)	DF_{ing} (Sv Bq^{-1})
^{232}Th	5.4×10^3	3×10^3	1.41×10^{10}	2.3×10^{-7}
^{228}Ra	9×10^3	4.9×10^2	6.75	5.9×10^{-7}
^{228}Ac	2.4×10^3	4.5×10^2	7.02×10^{-4}	4.3×10^{-10}
^{228}Th	5.4×10^3	3×10^3	1.913	7.2×10^{-8}
^{224}Ra	9×10^3	5×10^2	9.92×10^{-3}	6.5×10^{-8}
^{220}Rn	0	0	1.76×10^{-6}	–
^{212}Po	2.7×10^3	1.5×10^2	4.59×10^{-9}	–
^{212}Pb	5.4×10^2	2.7×10^2	1.21×10^{-3}	6.0×10^{-9}
^{212}Bi	4×10^2	1×10^2	1.15×10^{-4}	2.6×10^{-10}
^{208}Tl	2.7×10^2	1.5×10^1	5.8×10^{-6}	–

coefficients are not available for ^{220}Rn, ^{212}Po, and ^{208}Tl due to very short half-lives, their contributions to dose are not calculated.

Many countries over the globe use 10,000 y as the cutoff year for radiological impact assessment as it is not reasonable to assume same environmental conditions beyond this period as the major global changes like ice age occur during cycles of 10,000 y [13].

4 Results and Discussion

A hypothetical scenario of dumping of 1 Ci (3.7×10^{10} Bq) annual activity of ^{232}Th into a NSDF for a period of 50 years (i.e., a total of 50 Ci spread over 50 years) was envisaged. Figure 1 depicts the concentrations of ^{232}Th and its progeny in groundwater at extremely conservative distance of 100 m from the NSDF. The concentrations

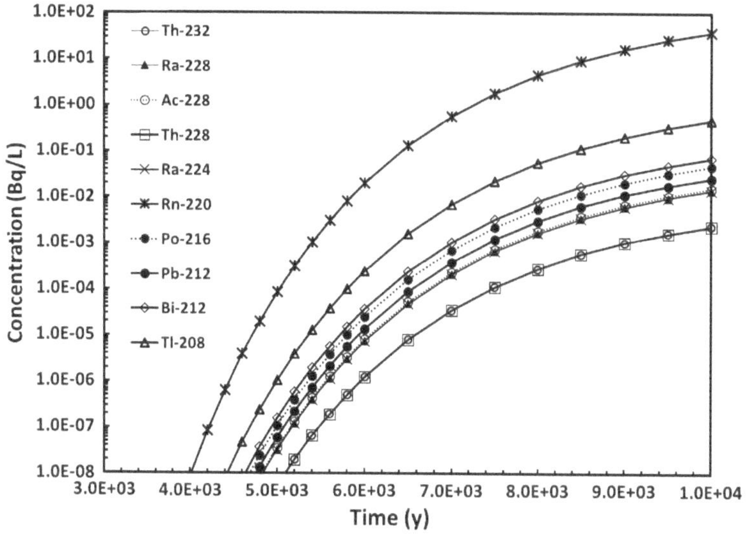

Fig. 1 Concentrations of ^{232}Th and its progeny in groundwater at 100 m from NSDF

are negligible up to 4000 years. After this period, the highest concentration is shown by ^{220}Rn followed by ^{208}Tl, ^{212}Bi, ^{216}Po, ^{212}Pb, ^{228}Ac and followed by ^{224}Ra and ^{228}Ra. The parent ^{232}Th has the least concentration in groundwater. The high concentration of ^{220}Rn is due to its zero K_d as it is soluble in water under confined conditions. The concentration of ^{220}Rn is about 38.4 Bq L^{-1} at 100 m distance after 10,000 y of disposal, whereas the concentration of ^{232}Th is about 0.0024 Bq L^{-1}. The radionuclide concentrations are translated into annual effective dose to the public through groundwater drinking pathway (Fig. 2). As the half-lives of ^{220}Rn, ^{216}Po, and ^{208}Tl are very short, their ingestion dose coefficients are not available at present and hence are considered to be zero.

The highest dose is due to ^{228}Ra (6.76×10^{-3} mSv y^{-1}), followed by ^{224}Ra (7.45×10^{-4} mSv y^{-1}) and then ^{232}Th (4.4×10^{-4} mSv y^{-1}). The total dose has a value of 8.23×10^{-3} mSv y^{-1} at 100 m which occurs after 10,000 years of disposal. In this scenario, it can be seen that the parent radionuclide contributes only 5.34% to the dose and the remaining 94.66% is contributed by the progeny; ^{228}Ra and ^{224}Ra being the main contributors with higher dose coefficient values and comparatively lower values of K_d as compared to the parent ^{232}Th.

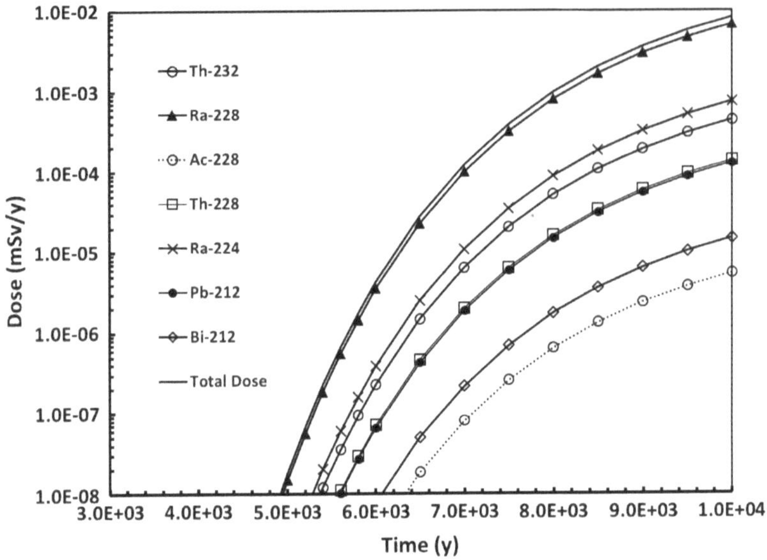

Fig. 2 Dose through the drinking water pathway due to ^{232}Th and its progeny at 100 m from NSDF

5 Conclusions

In the present scenario, with disposal of 50 Ci (1.85×10^{12} Bq) activity of ^{232}Th in NSDF over a period of 50 years, the total dose to the members of the public through groundwater drinking pathway after 10,000 y of disposal is 8.23×10^{-3} mSv y^{-1} at 100 m from the disposal facility. Also, it is concluded that the major contribution to the total dose is of the progeny rather than the parent itself. It signifies the importance of incorporation of in situ generation of progeny in model formulation and hence the dose due to these.

Acknowledgements The authors would like to thank Dr. K. S. Pradeepkumar and Dr. R. B. Oza of Bhabha Atomic Research Centre, Mumbai, India, for their help and support during the study.

References

1. IAEA, *Thorium fuel cycle—potential benefits and challenges* (Tecdoc 145, International Atomic Energy Agency, Vienna, 2005)
2. R.K. Sinha, A. Kakodkar, Design and development of AHWR—the Indian thorium fueled innovative reactor. Nucl. Eng. Des. **236**, 7–8, 683 (2006)
3. F. Sunny, S.T. Manikandan, R.N. Nair, V.D. Puranik, A multiple area source model to evaluate the groundwater quality at radioactive waste disposal sites. Environ. Geochem. **8**, 154–157 (2005)

4. R.N. Nair, T.M. Krishnamoorthy, SLBM-A *Fortran Code for Shallow Land Burial Low Level Radioactive Waste*. Rep. No. BARC/1997/E/030 (BARC, Mumbai, India, 1997)
5. R.N. Nair, T.M. Krishnamoorthy, Near field and far field radionuclides from shallow and land burial facility. Nucl. Technol. **114**, 235 (1996)
6. R.B. Codell, K.T. Key, G.A. Whelan, *Collection of mathematical models for dispersion in surface water and groundwater* (U.S. Nuclear Regulatory Commission, USA, 1982)
7. D.H. Lester, G. Jansen, H.C. Burkholder, *Migration of radionuclide chains through adsorbing medium* (Battelle, Pacific Northwest Laboratories, Richland, Washington, 1974)
8. IAEA, *Generic Models for Use in Assessing the Impact of Discharges of Radioactive Substances to the Environment*. Safety Reports Series No. 19 (International Atomic Energy Agency, Vienna, 2001)
9. M.I. Sheppard, D.H. Thibault, Default soil solid/liquid partition coefficients, KdS, for four major soil types: a compendium. Health Phys. **59**, 471–482 (1990)
10. IAEA, *Derivation of Activity Limits for the Disposal of Radioactive Waste in Near Surface Disposal Facilities* (IAEA-TECDOC-1380, International Atomic Energy Agency, Vienna, 2003)
11. IAEA, *International Basic Safety Standards for Protection Against Ionizing Radiation and for Safety of Radiation Sources*. Safety Series No. 115 (International Atomic Energy Agency, Vienna, 1996)
12. B.K. Tewary, et al., *Groundwater Modeling for Banduhurang Opencast Uranium Mine and Proposed Tailing Pond Area of UCIL*. BRNS Project Report GAP/006/EMG/DAE/BRNS/05-06 (CIMFR, Dhanbad, India, 2008)
13. R.N. Nair, V.N. Sastry, Safety assessment methodologies for near surface radioactive waste disposal facilities. Rad. Prot. Env. **24**(3), 544–563 (2001)

Part III
Thorium Based Fuel

$(Th\text{-}U)O_2$ MOX Fuel Fabrication and Dry Recycling of the Sintered Rejects

Sudhir Mishra, Kaushik Ghoshal, Joydipta Banerjee, Amrit Prakash, K. B. Khan and Arun Kumar

Abstract Thorium is an important nuclear material. Due to its abundance in large quantity in India, our preceptors have envisaged three-stage nuclear power program. Thoria-based fuel is planned to be used in the third stage of nuclear power program. $(Th\text{-}LEU)O_2$ mixed oxide (MOX) is the proposed fuel for AHWR-LEU-300 with UO_2 content varying from 13 to 30 w%. Fabrication of ThO_2-NU (natural uranium) fuel of similar composition with UO_2 varying from 13 to 30% has been carried out in kg scale at Radiometallurgy Division. Conventional powder metallurgical processes like mixing, cold compaction, and sintering were employed for fabrication of ThO_2-UO_2 fuel pellets. Sintered density up to 93–94% TD could be achieved. XRD result on sintered pellet showed single-phase formation. Fuel fabrication process is always associated with generation of green/sintered rejects. Recycling of the sintered reject is important to judiciously utilize the feed material and avoid the accumulation of the waste. As thorium dioxide is a stable compound, recycling of process rejects of thoria-based fuels either by dry or wet route is difficult unlike UO_2-based fuels. The stable oxidation state of thorium oxide limits oxidative-reductive processing of sinter rejects. Dissolution of thoria fuels in nitric acid is also difficult. The presence of UO_2 (13–30 w%) in the ThO_2-UO_2 fuel pellets makes it amenable to oxidative-reductive processing by modifying the fabrication parameters. A process for recycling of the sintered reject has been successfully worked out. This paper discusses the fabrication aspects of Th-NU MOX fuel and dry recycling of process rejects.

Keywords Fuel fabrication · Powder metallurgy · Recycling

S. Mishra (✉) · K. Ghoshal · J. Banerjee · A. Prakash · K. B. Khan · A. Kumar
Radiometallurgy Division, Bhabha Atomic Research Centre,
Trombay, Mumbai 400085, India
e-mail: sudhir@barc.gov.in

A. K. Nayak and B. R. Sehgal (eds.), *Thorium—Energy for the Future*,
https://doi.org/10.1007/978-981-13-2658-5_15

1 Introduction

Nuclear energy has a significant role to play in the contribution of the ever-growing power requirement of our country. India is pursuing the three-stage nuclear power program based on its scantily available uranium and abundantly present thorium [1–5]. The first stage of the program has been realized by operating pressurized heavy-water reactors (PHWRs). As a logical extension to this program, BARC is working on development of a thorium fuel cycle-based advanced heavy-water reactor (AHWR) [6]. A 300 MWe advanced heavy-water reactor having several advanced safety features is planned to be constructed as a technology demonstrator for large-scale utilization of thorium. The absence of fissile material constituent in natural thorium requires that fertile thorium has to be converted to fissile ^{233}U for its use. Thorium can be used along with either enriched uranium or plutonium for producing energy in a nuclear reactor. Fuel cycle strategies have been worked out for AHWR with both Pu and LEU as external fissile feed.

The fabrication of thoria-based fuel pellets by powder-pellet route involves conditioning of the powder to obtain the desired characteristics, mixing of the two different powder materials for obtaining the desired MOX fuel composition, pre-compaction and granulation to improve the flowability of powder, final compaction, and sintering at high temperatures to get high-density fuel pellets. The morphology of thorium oxide powder which is obtained by calcination of thorium oxalate is different from that of the natural uranium oxide powder. The calcined thoria powder has a "platelet" morphology (shown in Fig. 1a), whereas, the UO_2 powder, derived from ADU route has more "spherical" morphology (Fig. 1c). This requires premilling of the as-received thorium oxide powder in a planetary ball mill/attritor to improve its compactibility and sinterability. The morphology of ThO_2 powder obtained after milling is shown in Fig. 1b.

(a) As received ThO_2 powder (b) ThO_2 powder after milling in attritor for 3hrs (c) As received UO_2 powder

Fig. 1 Comparison of the morphologies of the powders

Table 1 Green density of (Th-U) MOX pellets at different compaction pressures

Composition	Compaction pressure (MPa)	Green density (%T.D)
ThO_2-15%UO_2	450	65.21
ThO_2-20%UO_2	300 375 450	61.06 63.63 65.55
ThO_2-22.5%UO_2	300 375 450	63.2 63.25 65.16
ThO_2-30%UO_2	375	66.43

Table 2 (Th-U) MOX pellet density sintered in different atmospheres

Composition	Sintering atmosphere	Pellet density (%T.D)
ThO_2-15%UO_2	Ar+8%H_2	92.06
	High purity N_2	92.23
ThO_2-20%UO_2	Ar+8%H_2	93.82
ThO2-22.5%UO_2	Ar+8%H2	92.28
ThO_2-30%UO_2	Ar+8%H2	92.52
	High purity N_2	94.83

2 Fabrication

Final compaction was carried out in a double acting hydraulic press for a range of pressures. Die wall lubrication of 1% stearic acid in acetone was employed during compaction. Table 1 shows the relationship between compaction pressure and green density.

2.1 *Sintering of Green Pellets*

Thoria has a higher melting point (3300 °C) in comparison to that of Urania (2800 °C). It is also perfectly stoichiometric and has a single oxidative state unlike that of Urania. The sintering characteristics therefore differ from that of the conventional UO_2 fuel. Experimental trials were carried out to study the effect of sintering atmosphere, viz. reducing (Ar+8% H_2) and inert (high purity N_2) at about 1700 °C. The sintered pellet density obtained for different sintering atmospheres is shown in Table 2.

The green pellets were loaded in alumina trays and sintered in the temperature range of 1680–1700 °C for 4–6 h with heating rate of 3 °C/min. Intermediate soaks were given at 150, 300, and 600 °C for moisture and binder/lubricant removal (zinc behenate), respectively. The effect of sintering atmosphere on the pellet density is shown in Table 2.

Table 3 O/M ratio of the pellets sintered in different atmosphere

Composition	Ar+H_2	High purity N_2
ThO_2-15%UO_2	2.006	2.000
ThO_2-20%UO_2	2.027	
ThO2-22.5%UO2	2.011	

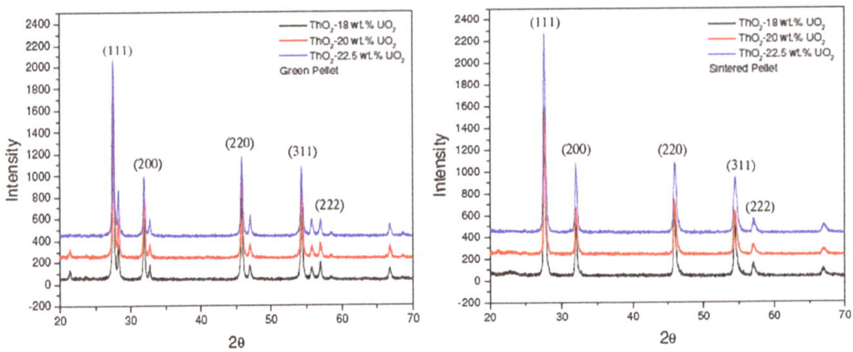

Fig. 2 XRD patterns of green and sintered ThO_2-UO_2 pellets

3 Characterization

The (Th-U) MOX pellets sintered in different atmospheres were characterized in terms of their density, oxygen-to-metal ratio (O/M), and phase contents. The O/M ratio was measured thermogravimetrically, and solid solution formation was estimated using X-ray diffractometry and metallography. Alpha autoradiography was carried out to see distribution. Table 3 gives the typical values of O/M ratio obtained for different atmospheres.

The X-ray diffraction patterns of the green and sintered pellets, using CuKα radiation and graphite monochromator, are shown in Fig. 2. It indicates the formation of complete sold solution between ThO_2 and UO_2.

For metallography, the sintered pellet was mounted in bakelite and ground using successive grades of emery paper. The final polishing was done using diamond paste, and the sample was etched thermally at 1600 °C in Ar+8% H_2 for 6 h. Figures 3 and 4 show the as-polished and etched microstructure of the pellets, respectively.

Alpha autoradiography technique has been employed to see the homogeneous distribution of uranium and thorium in mixed sintered $(Th,U)O_2$ pellets. Figure 5 shows the uniform distribution of U in the fuel matrix.

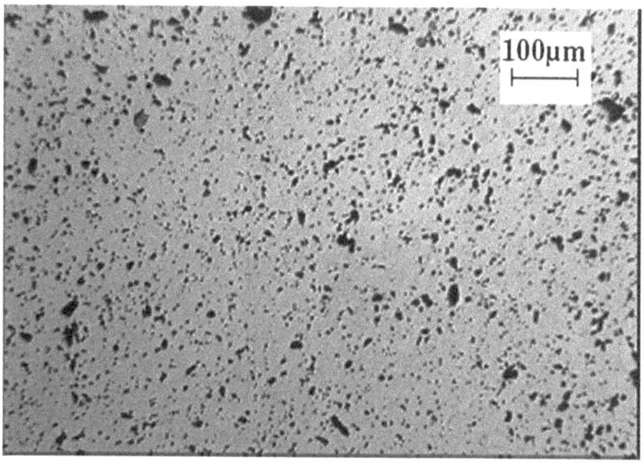

Fig. 3 As-polished microstructure of ThO_2-22.5%UO_2 showing uniform distribution of porosity

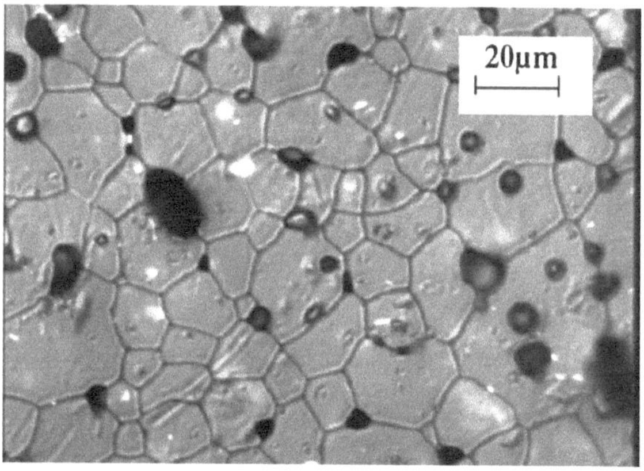

Fig. 4 Microstructure of ThO_2-18%UO_2 after thermal etching

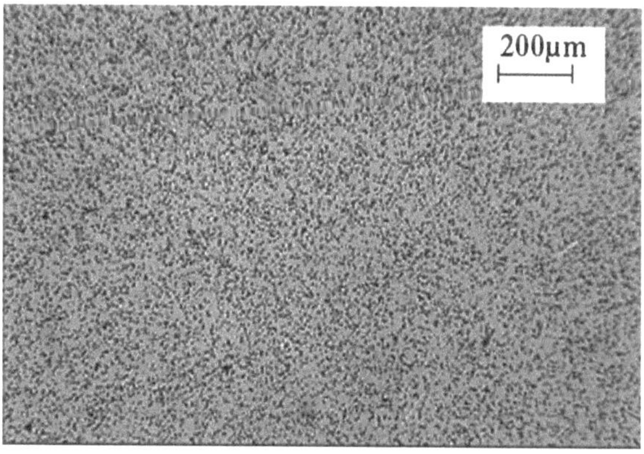

Fig. 5 Alpha autoradiograph of ThO_2-22.5%UO_2 pellet showing uniform distribution of U in the matrix

Table 4 Sintered pellet density measurements

Composition	Mass (g)	Diameter (mm)	Height (mm)	Density, g/cc/%T.D	Remarks
ThO_2+30% UO_2 (with 10 wt% reject)	(1) 8.42 (2) 8.36 (3) 8.61	10.42 10.40 10.39	10.39 10.33 10.64	9.51/92.40 9.53/92.63 9.55/92.80	Crack-free pellets
ThO_2+30% UO_2 (with 15 wt% reject)	(1) 8.62 (2) 8.58 (3) 8.52	10.40 10.40 10.41	10.64 10.59 10.61	9.54/92.71 9.54/92.71 9.44/91.73	Crack-free pellets
ThO_2+30% UO_2 (with 20 wt% reject)	8.39	10.42	10.37	9.47/92.07	Pellets observed to have cracks

4 Dry Recycling Process From Sinter Rejects of (Th,U)O_2 Fuel Pellets

4.1 Recycling of Sinter Rejects Mixed with Freshly Prepared ThO2-UO_2 Powder

Pellets were fabricated from sintered reject ThO_2-UO_2 (crushed and attritor milled) mixed with freshly prepared ThO_2-UO_2 powder. The fraction of sinter rejects in the powder mixture varied from 10 to 20 w%. Pellets of acceptable quality could be obtained by limiting the powder from rejected pellets to $\leq$15 wt% (Table 4).

4.2 Oxidative–Reductive Processing of Sinter Rejects

The sinter rejects (1 kg batch size) were crushed in a jaw crusher (<0.5 mm size) and subjected to calcination at 600 °C for 4 h in air. The calcined powder was further milled in an attritor to reduce the particle size (Sp.SA 4.5 m^2/g). Varying amounts of Nb_2O_5 powder were added as sintering aid to the milled powder. Sintering of the green pellets was done at 1700 °C for 6 h in reducing atmosphere (Ar+8% H_2). Sintered densities achieved from the pellets prepared by the above process are shown in Table 5.

4.3 Characterization

O/M (oxygen-to-metal ratio) of the sintered pellets was observed to be between 2.002 and 2.005. The effect of Nb_2O_5 on densification process was studied using a high-temperature dilatometer. The sample pellets having dimensions 7 mm diameter

Table 5 Sintered pellet density measurements prepared by oxidative–reductive processing of sinter rejects

Composition	Mass (g)	Dia. (mm)	Height (mm)	Density, g/cc/%T.D
ThO_2+13 w% UO_2 (with 0.5 w% Nb_2O_5)	(1) 10.37 (2) 10.24 (3) 10.33	10.60 10.60 10.59	12.16 12.09 12.10	9.66/95.45 9.60/94.86 9.70/95.84
ThO_2+30 w% UO_2 (with 0.5 w% Nb_2O_5)	(1) 11.30 (2) 11.20 (3) 11.19	10.68 10.68 10.69	13.24 13.10 13.16	9.53/92.61 9.54/92.79 9.47/92.11

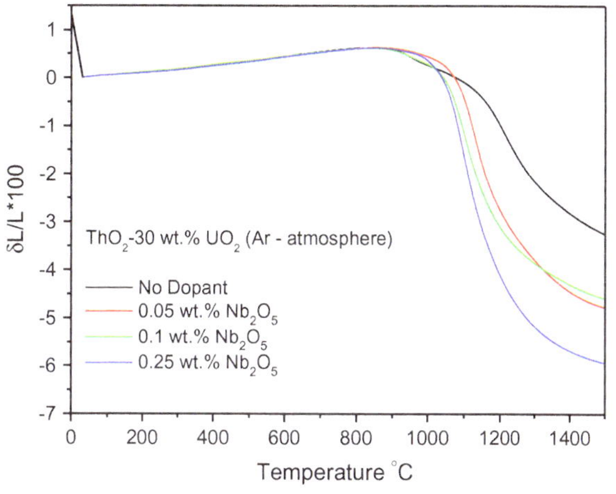

Fig. 6 Shrinkage versus temperature plot of ThO_2-30%UO_2 pellets with varying amount of dopant

and 8 mm height were heated at a predetermined heating rate, and the change in length was monitored. Figure 6 shows the shrinkage behavior (in the axial direction) of the pellets in high purity argon atmosphere. The percentage shrinkage ($\Delta L/L_0 \times 100$) values are plotted against temperature at heating rate of 3 °C per minute. It is observed that shrinkage progressively increases as the amount of dopant varied from 0.05 to 0.25 w%.

5 Conclusions

- Ex-oxalate-derived ThO_2 powder has platelet morphology resulting in poor interlocking during compaction. Hence, premilling of the as-received powder is required to break the platelet structure and improve compactibility.
- ThO_2 and ThO_2-based fuels have poor sinterability. ThO_2, being a perfectly stoichiometric compound with very high melting point, cannot be sintered to very high density (>96%T.D.) even at reasonably high temperature (1700 °C). Additives and sintering atmosphere (reductive and oxidative) have considerable effect

in the sintering behavior of ThO_2 and ThO_2-based fuels and help in bringing down the sintering temperature for obtaining high-density fuel pellets.

- Pellets of acceptable quality could be obtained by limiting the powder from rejected pellets to $\leq$15 wt% by recycling of sinter rejects mixed with freshly prepared ThO2-UO_2 powder.
- 100% recycling could be achieved by Oxidative–reductive processing of sintered rejects.

Acknowledgements Authors would like to acknowledge their sincere thanks to all the staff members of Radiometallurgy Division involved in this work for providing their support during the course of the work.

References

1. A. Kakodkar, Shaping the 3rd stage of Indian nuclear programme. Paper presented at annual conference of the Indian nuclear society, INSAC-2001, Indore, 2001
2. K. Balakrishnan, S. Majumdar, A. Ramanujam, A. Kakodkar, *The Indian perspective on thorium fuel cycles, IAEA TECDOC-1319* (Options and Trends, Thorium Fuel Cycle, 2002), p. 257
3. Department of Atomic Energy in India, Atomic Energy in India. A Perspective, Government of India, Sept 2003. (http://www.dae.gov.in/publ/persp/persp03/persp03.pdf)
4. K. Anantharaman, V. Shivakumar, D. Saha, Utilisation of thorium in reactors. J. Nucl. Mater. **383**, 119–121 (2008)
5. A. Kakodkar, Nu - Power. **23**, 22 (2009)
6. R.K. Sinha, A. Kakodkar, Design and development of the AHWR—the Indian thorium fuelled innovative nuclear reactor. Nucl. Eng. Des. **236**, 683–700 (2006)

Fluorination of Thorium Oxide by Ammonium Bifluoride and Its Reduction to Metal

Abhishek Mukherjee and Alok Awasthi

Abstract The fluorination of thoria (ThO_2) with ammonium bifluoride NH_4HF_2 to obtain oxygen-free thorium fluoride is reported. It is interesting to note that the fluorination starts right from room temperature to form ammonium thorium fluoride intermediate. On heating, the intermediate decomposes to form pure ThF_4. It was also seen that any unreacted ThO_2 reacts with the ThF_4 to form $ThOF_2$. The oxygen-free ThF_4 produced was then reduced by calcium to form metallic thorium. The advantage in the process is that it can make thorium in bulk form which greatly reduces the issue of handling powders. However, as the process involves the melting of thorium, selection of suitable crucible becomes a key to its success. Yttria crucible appears to be a good choice.

Keywords Thorium · Yttria crucible · Fluorination · Ammonium thorium fluoride

1 Introduction

Thorium is a source of fuel for nuclear power. Though thorium is fertile, it can be made fissionable via neutron absorption. Th^{232} converts to U^{233} in the process. The major advantage of the thorium-232—uranium 233 cycle over the presently used uranium-238—plutonium-239 cycle is the high net neutron yield of uranium-233. This translates to a smaller fuel and fissile inventory for a given power output. In practice, a thermal reactor, as is proposed in the Indian Advanced Heavy Water Reactor (AHWR) designs, is based entirely on the thorium-232—uranium-233 cycle [1]. Thorium metal, its alloys, the oxide, and the various mixtures containing thorium have all been suggested as usable forms for the fertile material. While ThO_2 is currently used as a fuel component [1] and as a blanket in fast breeder reactor [2], the use of thorium metal in a nuclear reactor would offer advantages in ease of

A. Mukherjee (✉) · A. Awasthi
Materials Processing and Corrosion Engineering Division, Bhabha Atomic Research Centre, Trombay, Mumbai 400085, India
e-mail: abhim@barc.gov.in

A. K. Nayak and B. R. Sehgal (eds.), *Thorium—Energy for the Future*,
https://doi.org/10.1007/978-981-13-2658-5_16

fabrication, high density, and good thermal conductivity. Additionally, metallic fuel provides high burnup capability, lower doubling time, and inherently passive safety characteristics [3].

Preparation of thorium metal in bulk quantities in an economical way is thus assuming importance, although the metal has been produced in laboratory scale since early twentieth century. The metal has been produced by a variety of routes. The most widely used route has been calciothermic reduction of thorium oxide [4]. This method leads to the production of thorium powders, which if not handled carefully can often lead to oxygen contamination in the final metal [4]. Oxygen containing thorium is brittle and difficult to shape by mechanical means [5]. The brittle thorium needs to be made ductile by electron beam melting purification/iodide refining, both of which are expensive. In addition, electron beam melting results in substantial metal loss due to the formation of vapor suboxide(s). Also, besides being radioactive, thorium powder is pyrophoric in nature and needs special precautions during handling [6].

Calcium reduction of thorium fluoride using $ZnCl_2$ as a heat booster and $CaCl_2$ as a slag fluidizer (popularly known as "Ames" process) has been used for the production of thorium ingot. The process leads to the formation of Th-Zn alloy (with around 8 wt% zinc), which requires extensive dezincification using vacuum treatment to produce a metal ingot with acceptable purity [4]. Thorium metal of reasonable quality in kilogram levels has also been produced by magnesium reduction of thorium chloride. However, several stages of purification steps are required in the above process [7, 8]. Fluoride being lesser hygroscopic and lesser volatile is generally preferred over the chloride.

In this paper, we report an alternative possibility of producing bulk thorium metal in a single step by direct calciothermic reduction of its fluoride, without the addition of $ZnCl_2$. As the dezincification step would not be required in such a case, it is expected that the process would be cost effective. With good electronic control systems being available now for control even at temperatures exceeding the melting point of thorium, the process should not be difficult to carry out. A similar method has been demonstrated by the authors for the reduction yttrium fluoride by calcium [9]. Thorium fluoride being lesser stable than yttrium fluoride [10], the process should be thermodynamically feasible. As liquid thorium is very reactive, a proper choice of crucible material becomes important. In this paper, we describe our experiences on the interaction of liquid thorium with the crucible materials as well.

Thorium occurs in nature as monazite mineral, which is a complex phosphate of rare earth, $REPO_4$ containing around 4–12% of ThO_2 and rare earth (RE) containing Ce, La, etc. [11]. After processing, the purest compound of thorium is usually thorium nitrate or oxalate, from which ThO_2 is obtained. This ThO_2 is the usual source of Th in the preparation of metal. Thorium fluoride can be prepared from ThO_2 using many fluorinating agents. Ammonium hydrogen fluoride (NH_4HF_2) was used in the present work. The advantages of using NH_4HF_2 as the fluorinating agent are discussed in our earlier work [12]. The aim was to produce pure ThF_4 without any formation of $ThOF_2$ phase. If the oxyfluoride phase is present in the starting material, its oxygen would reach the metal, belittling the advantages of using a fluoride route. The fluorination step is also studied in detail in the present work.

2 Experimental

2.1 *Materials*

The starting material ThO_2 (99% pure) was procured from Indian Rare Earths Ltd. ThO_2 powders was then mixed with commercial ammonium bifluoride solid (97% purity) in 20 and 100% excess over the stoichiometric requirements according to reaction (1).

$$ThO_2 + 4\,NH_4HF_2 = ThF_4 + 4\,NH_4F + 2\,H_2O \quad (1)$$

2.2 *DTA Studies*

The mixture was heated in a homemade differential thermal analyzer (DTA) to assess the temperature required for ThF_4 formation. The DTA was carried out in static air at 10 °C/min with alumina as the reference, and the end products were characterized. Cups made from vitreous silica were used.

2.3 *Product Characterization*

The X-ray diffraction of the freshly prepared samples was done at various times. XRD used was an Inel-make unit (model MPD) with Cu-$K_{\alpha 1}$ radiation at 30 mA 40 kV using a curved position-sensitive detector.

2.4 *Reduction*

Two types of crucibles were used for the reduction runs: (a) molybdenum crucible sourced from local vendors and (b) specially made yttria crucibles (25 mm dia and 85 mm length) of density of more than 90%. ThF_4 and 20% excess (over stoichiometry) calcium were loaded in these crucibles for preparation of thorium metal. The mixing of the reactants was done under argon atmosphere to prevent the oxidation of calcium. The crucible containing the charge was also loaded inside the furnace under argon cover. High pure argon (99.995%) was used in the present work. Charge was heated up to 1800 °C in an induction furnace [9] beyond the melting point of thorium (m.p 1750 °C) for slag/metal separation. The temperature was monitored through an optical pyrometer, through which it was also possible to observe the changes occurring in the charge. The end product was analyzed by XRF.

3 Results and Discussion

3.1 Fluorination of Thorium Oxide

The fluorination of oxides through NH_4HF_2 is known to occur through intermediate ammono-metallofluorides [12–14], and it begins right from the room temperature. According to authors' experience, the best way to prevent the formation of oxyfluorides or the retention of unreacted oxide in the product is to ensure the complete conversion of the starting oxide to an ammono-metallofluoride at room temperature itself, since on heating substantial amount of NH_4HF_2 evaporates even without reacting with the oxide.

The XRD of ThO_2 and NH_4HF_2 mixture with NH_4HF_2 being in 20% excess according to reaction (1), recorded immediately after mixing, is shown in Fig. 1b. This shows additional peaks besides the peaks of ThO_2. The additional peaks match with $(NH_4)_3ThF_7$ [JCPDS no: 711954] with slight peak shift. The mixture converts to $(NH_4)_4ThF_8$ [JCPDS no: 711190] and some unknown phase after 28 h (Fig. 1c). This is similar to the observations by Silva [13], although JCPDS data have slight variations in XRD peak locations. Besides, the time required for the formation of these phases is substantially more in Silva's [13] case, apparently as they used closed (although not-leak-tight) containers against the use of open container by us. There could be a few intermediate gas–solid reactions and, if so, the time required for completion of reaction would depend on the removal of the product gases from the reaction zone. However, Silva could observe complete conversion to $(NH_4)_4ThF_8$ phase after 28 days with 10% excess NH_4HF_2. Peaks of the unknown phase (Fig. 1c) match with some of those of the unknown phases of Silva [13].

Figure 2 shows the DTA of the freshly mixed charge with 20% excess NH_4HF_2 which indicates many peaks up to 500 °C. The product after this heating was $ThOF_2$ (Fig. 1d). It may be noted that NH_4F formed as a product in reaction (1) is in many cases known to further react with unreacted oxides, for example, possibly

$$ThO_2 + 8\,NH_4F = (NH_4)_4ThF_8 + 2\,H_2O + 4\,NH_3 \tag{2}$$

As the product of heating was oxyfluoride and many events are indicated by DTA, it is clear that the fluorination of ThO_2 using NH_4HF_2 is more complex than that for Y_2O_3 [12]. It occurs through many intermediate steps and requires further detailed studies. The unknown phase is likely to have oxygen present as ammono-thorium oxyfluoride ($NH_4F.Th_xO_yF_z$) or as hydrated ammono-thorium fluoride ($NH_4F.xThF_4.yH_2O$). ThO_2 had to be mixed with 100% excess NH_4HF_2 and heated to 500 °C for obtaining ThF_4 (Fig. 1e).

Therefore, it is clear that to obtain oxygen-free ThF_4 from ThO_2 using NH_4HF_2, substantially excess amount of NH_4HF_2 has to be used, although this may require further optimization.

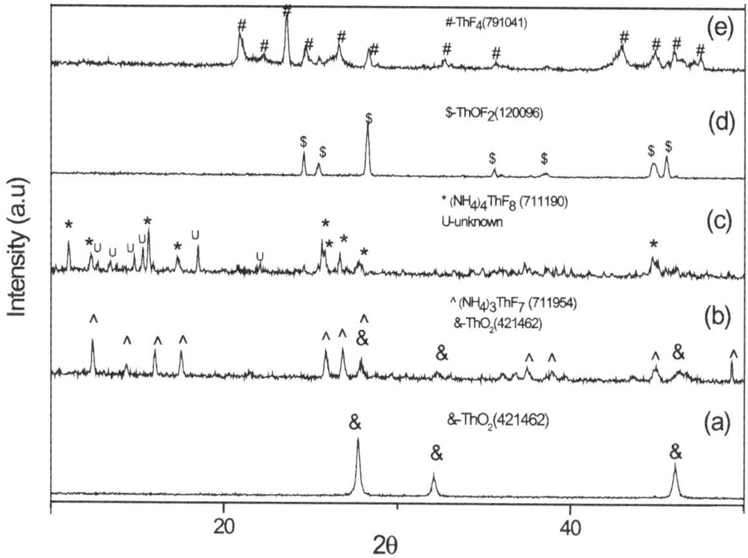

Fig. 1 XRD of **a** pure thoria (ThO_2), **b** mixture of thoria (ThO_2) with 20% excess ammonium bifluoride immediately after mixing, **c** the same mixture after 28 h holding at room temperature, **d** the mixture (with 20% excess NH_4HF_2) heated to 500 °C immediately after mixing, **e** product of freshly prepared mixture of thoria (ThO_2) with 100% excess ammonium bifluoride heated up to 500 °C immediately after mixing

3.2 *Preparation of Thorium Metal*

A controlled atmosphere induction heated reduction reactor, as described in [9], was used for the preparation of thorium metal. The chamber was initially tested for integrity by heating at 1900 °C for 2 h. ThF_4 was mixed with calcium and heated up to 1850 °C. Based on our experience with yttrium preparation [9], the reduction of ThF_4 was first carried out in molybdenum crucible. For taking out the thorium metal, the top part of molybdenum crucible was broken. A pinkish slag layer has formed on top of a 2-mm-thick thorium metal. The metal found to be conducting was showing thorium peaks by gamma spectroscopy. It was found sticking to molybdenum (Fig. 3), a behavior shown by yttrium reported earlier [9].

Yttria is apparently the only oxide which is more stable than ThO_2 and has higher melting point than thorium, and therefore, it looks attractive as the crucible material as it might result in pure thorium metal (instead of Th-Mo alloy). However, it is not known whether the two oxides dissolve in each other or form some binary oxides. Besides, in the case of reduction of fluoride, it is not known how the fluorides CaF_2 and ThF_4 would interact with yttria. In yttria crucible, the solidified thorium metal formed was button-shaped (sphere-like) (Fig. 3b) which indicated to the non-wetting characteristic of liquid thorium with yttria. No problems due to interactions with the fluorides were observed. The metal/slag was analyzed with XRF (Fig. 4). The metal

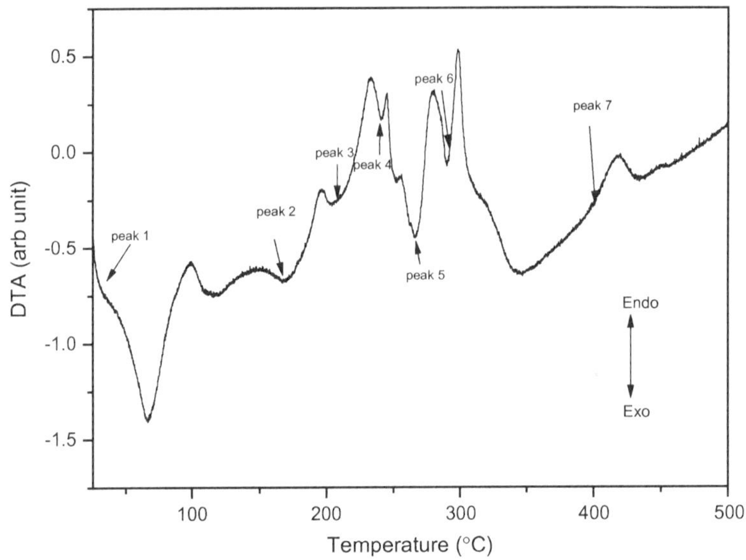

Fig. 2 DTA of freshly prepared charge of ThO_2 and 20% excess NH_4HF_2 with some apparent peaks/heat change events marked

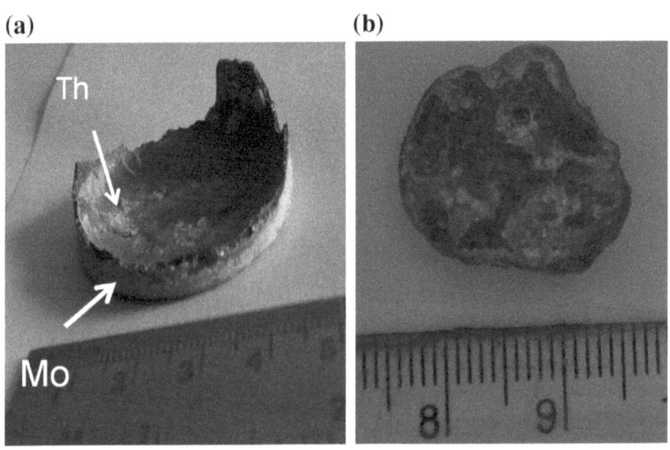

Fig. 3 Thorium metal prepared in **a** Molybdenum crucible and **b** specially fabricated yttria crucible

was found to be pure thorium. Peaks of Th L_α and Th L_β could be seen in the spectra. The yttrium peaks come from the yttria crucible. The actinium peaks were apparently from the radioactive decay of thorium. The slag contained yttria (from crucible) and CaF_2.

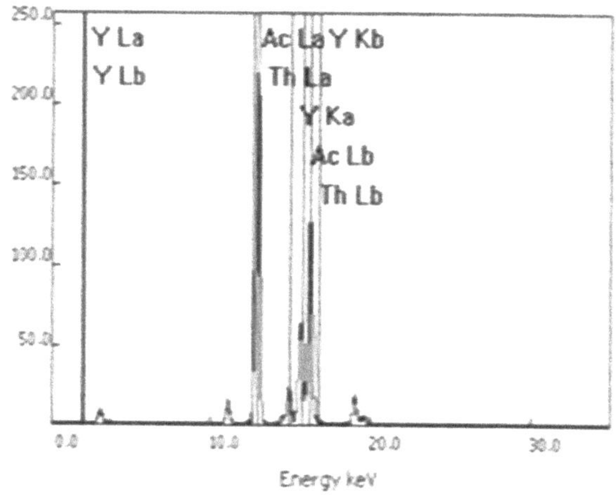

Fig. 4 XRF spectra of thorium metal prepared in yttria crucible

4 Conclusions

Fluorination of thoria with ammonium bifluoride begins right at room temperature. The final room temperature phase is $(NH_4)_4ThF_8$. However, DTA and XRD results point out to the existence of few unknown intermediate phases. Apparently, due to these phases, more NH_4HF_2 than stoichiometry is required to ensure complete conversion of ThO_2 to ammono-thorium fluoride and finally resulting in oxygen-free ThF_4. The mixture of ThO_2 with 100% excess NH_4HF_2 when heated to around 500 °C completely converts to ThF_4.

Bulk metallic thorium was obtained in a single step by reducing pure ThF_4 with calcium and heating the charge beyond the melting point of thorium. Contamination from crucible could be avoided by using densified yttria crucible.

Acknowledgements The authors gratefully acknowledge Dr. N. Krishnamurthy, on whose guidance the new route of thorium making could be attempted. Densified yttria crucible was custom-made by Dr Abhijit Ghosh for this work. We also thank Dr. R. K. Singhal for helping us in the XRF analysis of the thorium sample.

References

1. R.K. Sinha, A. Kakodkar, Nucl. Eng. Des. **236**, 683–700 (2006)
2. S.C. Chetal, V. Balasubramaniyan, P. Chellapandi, P. Mohanakrishnan, P. Puthiyavinayagam, C.P. Pillai, S. Raghupathy, T.K. Shanmugham, C.S. Pillai, Nucl. Eng. Des. **236**, 852–860 (2006)
3. Y.-I. Chang, Nucl. Eng. Technol. **39**, 161–170 (2007)
4. Interim report on metallurgy of thorium and thorium alloys (ORNL 1090. eng 26.), contract no.w740 s, ORNL, 1949–1951
5. B.G. William, Fabrication of metallic thorium, Google Patents, 1928
6. J.M. Dukert, Thorium and the third fuel, U.S. Atomic Energy Commission, Division of Technical Information, [Washington, D.C.], 1970

7. D.T. Peterson, W.E. Krupp, F.A. Schmidt, J. Less Common Met. **7**, 288–295 (1964)
8. P.L. Vijay, J.C. Sehra, C.V. Sundaram, K.R. Gurumurthy, R.V. Raghavan, BARC report **969**, 1–24 (1978)
9. A. Mukherjee, A. Awasthi, N. Krishnamurthy, Mineral Process. Extr. Metall. **125**, 26–31 (2016)
10. M. Binnewies, E. Milke, *Thermochemical data of elements and compounds*, 2nd edn. (Wiley-Vch Verlag Gmbh, Weinheim, 2002)
11. C.K. Gupta, N. Krishnamurthy, *Extractive metallurgy of rare earths* (CRC Press, Boca Raton, 2004)
12. A. Mukherjee, A. Awasthi, S. Mishra, N. Krishnamurthy, Thermochim. Acta **520**, 145–152 (2011)
13. G.W.C. Silva, C.B. Yeamans, G.S. Cerefice, A.P. Sattelberger, K.R. Czerwinski, Inorg. Chem. **48**, 5736–5746 (2009)
14. B.N. Wani, S.J. Patwe, U.R.K. Rao, K.S. Venkateswarlu, J. Fluor. Chem. **44**, 177–185 (1989)

Estimation of Plutonium Heterogeneity in Thoria–Plutonia Fuels

K. V. Vrinda Devi, J. N. Dubey, P. S. Somayajulu, I. H. Shaikh, Jyoti Gupta, S. D. Raut and K. B. Khan

Abstract Plutonium bearing mixed oxide (MOX) fuels requires to be characterized for plutonium homogeneity as it is a crucial characteristic for fuel performance. Conventional chemical analytical methods for characterization of homogeneity are difficult to be carried out on MOX fuels with thoria matrix due to poor solubility. Radiation imaging is an effective technique to monitor the distribution of plutonium in thoria matrix. Alpha autoradiography is an imaging technique which involves registration of alpha tracks using alpha-sensitive films known as solid-state nuclear track detectors (SSNTD). Manual evaluation of microscopic images from alpha autoradiographs is difficult to carry out where highly radioactive materials such as plutonium are involved due to large number of registered tracks. A method based on image analysis of alpha autoradiographs was explored. The method was applied on alpha autoradiographs of fuel samples with the same composition but having different levels of Pu homogeneity. Samples fabricated through two different routes, viz. the conventional powder–pellet (POP) route and coated agglomerate particle (CAP) route, were studied using this method. Studies on feasibility of this method for quantitative estimation of plutonium heterogeneity in MOX fuels are discussed in this paper.

Keywords MOX fuel · Thoria · Plutonium · Heterogeneity
Alpha autoradiography

1 Introduction

Advanced Heavy Water Reactor (AHWR) is proposed in view of utilization of India's vast thorium resources. The (Th, Pu)O_2 mixed oxide (MOX) being the initial fuel for AHWR, few fuel pins of $(Th_{0.99}, Pu_{0.01})O_2$ were fabricated at Advanced Fuel Fabrication Facility (AFFF), BARC, Tarapur, for preliminary experimental irradi-

K. V. Vrinda Devi (✉) · J. N. Dubey · P. S. Somayajulu · I. H. Shaikh · J. Gupta · S. D. Raut
K. B. Khan
Radiometallurgy Division, Bhabha Atomic Research Centre, Mumbai, Maharashtra, India
e-mail: kvvdevi@barc.gov.in

A. K. Nayak and B. R. Sehgal (eds.), *Thorium—Energy for the Future*,
https://doi.org/10.1007/978-981-13-2658-5_17

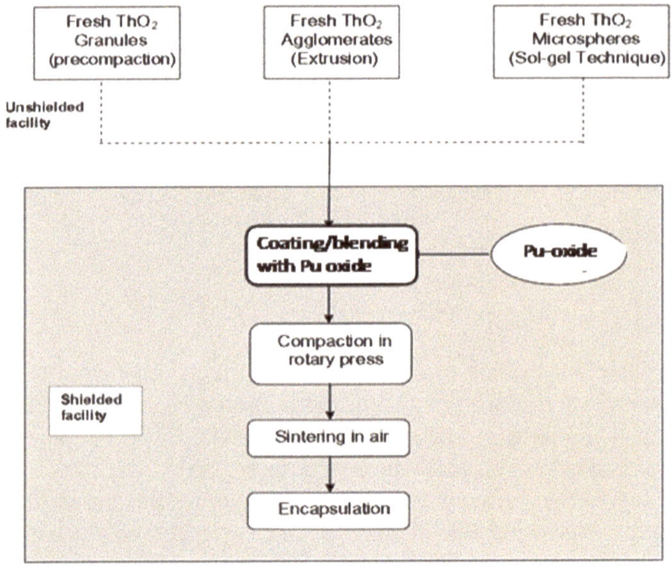

Fig. 1 Flow chart of CAP route of MOX fuel fabrication

ation. Different fuel fabrication techniques are being developed for thorium fuel fabrication, and standardization of these processes require conformance of the end product with the stringent specifications. Coated agglomerate particle (CAP) route is a newly developed route of fuel fabrication and highly relevant while involving highly radioactive nuclides.

The fuel for experimental irradiation in AHWR conditions was fabricated through two different routes, viz. conventional powder–pellet route (POP) and coated agglomerate particle (CAP) route. A comparative study of plutonium distribution in the fuel pellets fabricated through both these routes was carried out using alpha autoradiography combined with image analysis, and the details are described in this paper.

Powder metallurgical route is involved in powder mixing, milling, granulation, cold compaction and sintering [1]. The sintered pellets after inspection were encapsulated in Zircaloy-2 clad tubes filled with helium and welded with end plugs. This route has been referred as POP route in this paper.

CAP route of fabrication involved the following steps [2]. Thoria powder was mixed with appropriate amount of suitable binders (sodium stearate, polyethylene glycol, etc.) and extruded to get Thoria extrudes. These extrudes were converted into agglomerates/spheroids (equivalent to granules in POP route) using a spherodiser. The agglomerates were then coated with plutonium dioxide powder to achieve the required composition. These coated agglomerates were subjected to compaction, pellet sintering and pellet encapsulation. The flow chart of fabrication through this route is shown in Fig. 1.

Alpha autoradiography is a radiation imaging technique used to record distribution of a radionuclide with higher specific activity or higher energy of emitted radiation with respect to the matrix [3–5]. The technique involves exposure of α-sensitive solid-state nuclear track detectors (SSNTD) to the sample and evaluating the α-image with the help of a microscope after enhancing it by chemical etching. Since the specific activity of plutonium (0.06 Ci/g) is significantly higher than that of thorium (1.1×10^{-7} Ci/g), α-image of the MOX sample on the detector could be attributed to alpha particles emitted by plutonium and hence could be used to monitor Pu distribution in the fuel.

2 Materials and Methods

Alpha autoradiography of $(Th_{0.99}, Pu_{0.01})O_2$ pellets fabricated through conventional (POP) and CAP routes was carried out in Radiometallurgy division, BARC, for homogeneity studies. Pellet samples of CAP route which were identified to be having lower levels of Pu homogeneity compared to those from POP route were taken up for this study. One POP sample containing a Pu-rich agglomerate was also selected for comparative analysis.

The samples were prepared by cutting, grinding and polishing to achieve good surface contact with the α-sensitive film. Aluminium mylar of thickness 11 μm was placed on top of the detector to ensure minimum contribution from sub-surface alphas by elimination of non-perpendicular incidences. The sample was placed on the film for a pre-decided time period (30 s in this case) for exposure. The film was immediately bagged out and decontaminated to avoid any background irradiation. The decontaminated film was chemically etched to enhance the radiation effect caused on it and then viewed using a microscope (LEICA-DM ILM) with digital camera. Multiple frames from images of both the samples were grabbed for further analysis. The images were analysed using an image analysis software (metal power image analyser version 3.0.0.9 by Metal Power India (Pvt.) Ltd.) for estimating the alpha track density. Image analysis of alpha autoradiographs of Thoria-based fuels for compositional analysis has been previously reported [6, 7].

Cellulose nitrate and poly(allyl diglycol carbonate) (PADC) are two alpha-sensitive polymers which are used as solid-state nuclear track detectors (SSNTD) for alpha imaging. CR39 detector which is a commercially available form of PADC polymer was used in this study. Alpha exposure causes structural changes in the polymer by scission or cross-linking of the chemical bonds present in the molecular structure of the polymer. These changes make it prone to chemical reactions with alkalis such as NaOH or KOH which are used for etching of the film. When the film is treated with NaOH/KOH, the irradiated portion preferentially reacts with the alkali compared to the substrate and (some products of the reaction) partially gets removed from the substrate. This causes creation of craters on the irradiated portion of the film after etching and washing. These craters could be observed when viewed with a microscope as dark spots which are termed as alpha tracks. CR39 with 600 μm thickness was used in this study, and etching was carried out with 6N KOH solution at 72 °C for 90 min.

3 Results and Discussion

The images of $(Th_{0.99}, Pu_{0.01})O_2$ pellets from CAP route of fabrication revealed alternate regions with higher as well as lower densities of alpha tracks indicating Pu-rich and depleted zones, respectively. The microstructure of these samples also indicated a similar trend [8]. In this referred work, Somayajulu et al. discussed that a uniform microstructure was observed in POP pellet due to homogeneity in the powder mixture caused by mixing and milling operations. In the case of CAP pellets, which revealed a duplex microstructure of large as well as small grains since the constituents were not in the form of homogenized powders, the coated regions of PuO_2 on reduction to Pu_2O_3 generated higher defect concentration and diffused into the adjacent thoria lattice. The variation in concentration of Pu over the diffusion length is responsible for varied grain growth resulting in duplex microstructure in the pellet. Therefore, it can be concluded that the larger grains present in the microstructure indicate regions of higher defect (plutonium) concentrations leading to Pu heterogeneity.

Multiple frames each from rarer and denser regions of alpha tracks were acquired and analysed. Images of those regions which showed the lowest (in rarer region) and highest track densities (in denser region) are shown in Fig. 2a, b, respectively. Though alpha autoradiographs of POP pellets showed uniform track density (Fig. 3a), it was observed that the Pu-rich agglomerate in the POP sample was revealed as shown in Fig. 3b. The results were obtained by studying multiple frames of each of the autoradiographs for statistical consistency.

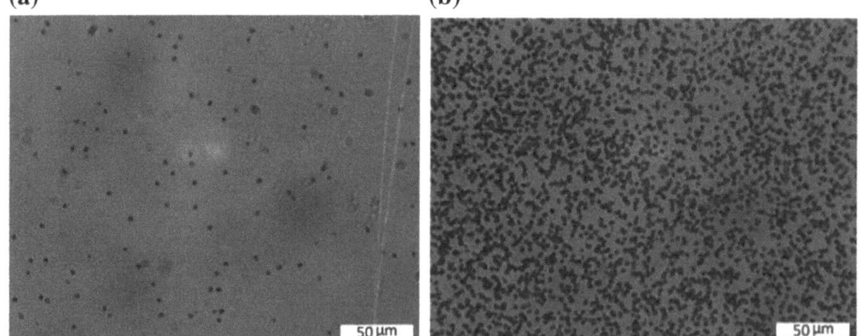

Fig. 2 Image frames from alpha autoradiographic images of $(Th_{0.99}, Pu_{0.01})O_2$ pellets fabricated through CAP, **a** region of less dense alpha tracks, **b** region of more dense alpha tracks

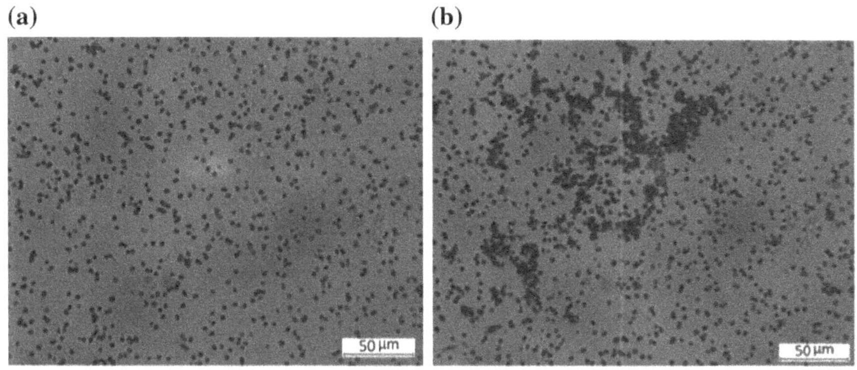

Fig. 3 **a** Alpha autoradiographic images of $(Th_{0.99}, Pu_{0.01})O_2$ pellets fabricated through POP, **b** region showing Pu rich agglomerate

Table 1 Results of alpha autoradiography image analysis

PuO_2 %	Type	Average track density ($\bar{x}$) (mm^2)
1	CAP dense	2259
1	CAP less dense	214
1	POP	1176

Estimation of correct track density was difficult in those frames where overlapping of alpha tracks was occurring. The clustered tracks had to be virtually segregated using image analysis features before counting the tracks. The average aspect ratio of individual tracks was estimated using the images with non-overlapped tracks. Number of superimposed tracks could be estimated with reasonable accuracy by estimating the aspect ratio of each individual feature (track) in the image. If a particular track had aspect ratio which is double that of the average aspect ratio of an individual track, it could be considered as superimposition of two tracks and so on. Every individual feature (superimposed or non-overlapped tracks) of the images was analysed by this method to virtually segregate the tracks before calculating the track density corresponding to each image. This method was applied for the first time to estimate the track density of alpha images registered by MOX fuel samples. Average value of all the frames corresponding to the same sample (homogeneous region or agglomerate region) was considered corresponding to that sample in case of POP pellet. The minimum and maximum values among the frames of the rarer and denser track density regions, respectively, were considered in case of CAP pellet sample. The results of image analysis are shown in Table 1.

4 Conclusion

Alpha autoradiography technique combined with image analysis was found to be effective for homogeneity studies of Thoria–Plutonia MOX pellets with 1% PuO_2 content which was fabricated for experimental irradiation in AHWR conditions. Pellets fabricated through powder–pellet (POP) route and coated agglomerate route (CAP) with different levels of plutonium homogeneity were compared, and it was found that alpha autoradiographic images of CAP pellets indicated alternate regions which were richer and depleted in Plutonium, respectively. The average track density registered by the homogenous POP pellet was 1176, and the statistical variation of only $\pm 3\%$ was observed among multiple frames of the image. The variation in alpha track density among different frames of image registered by CAP pellet was found to be from 214 to 2259 owing to heterogeneous distribution of plutonium. The average track density in different frames of the image of a CAP pellet was up to 81% less (rarer region) and 92% more (denser region) than the average track density of a homogenous POP pellet. Compared to chemical techniques, this method of characterization offers various advantages such as faster sample preparation due to elimination of sample dissolution, safety due to elimination of any high-level radioactive liquid waste and provides information on homogeneity on microscopic scale.

Acknowledgements The authors are grateful to all their colleagues at AFFF, BARC, Tarapur, for their cooperation during this work.

References

1. F. Glodeanu, J. Nucl. Mater. **126**, 181–183 (1984)
2. T.R.G. Kutty, K.B. Khan, P.S. Somayajulu, A.K. Sengupta, J.P. Panakkal, A. Kumar, H.S. Kamath, J. Nucl. Mater. **373**, 299–308 (2008)
3. T.M. Kegley, Metallography **5**, 113 (1972)
4. J.K. Ghosh, V.D. Pandey, P.R. Roy, PMAI Newsl. **6** (1&2) (1979)
5. W.J. Gruber, in *Proceedings 17th Conference on Remote System Techniques*, (American Nuclear Society 1969), p. 29
6. T.S. Rao, B.B. Shrivastv, J.N. Dubey, B.P. Patil, C.N. Chandrasekharan, V.D. Pandey, S. Majumdar, Radiat. Meas. **36**, 747–750 (2003)
7. I.H. Shaikh, J.N. Dubey, J. Gupta, B.B. Shriwastwa, K.V. Vrinda Devi, P.S. Somayajulu, K.B. Khan, A. Kumar, in *Proceedings of National Conference on Power from Thorium: Present Status and Future Directions* (Mumbai, 22–24 Dec 2014), pp. 103–104
8. P.S. Somayajulu, P.S. Ghosh, A. Arya, K.V. Vrinda Devi, D.B. Sathe, J. Banerjee, K.B. Khan, G.K. Dey, B.K. Dutta, J. Alloy. Compd. **664**, 291–303 (2016)

Evaluation of Thermal Properties of Thoria–Urania Fuel

Joydipta Banerjee, Santu Kaity, Koushik Bhandari, Arun Kumar and Srikumar Banerjee

Abstract Thoria–urania (ThO_2–UO_2)-based fuel system is the proposed fuel for the Advanced Heavy Water Reactor (AHWR) in India. As a part of thoria-based fuel development for India's nuclear power programme, thoria–urania (ThO_2–UO_2) solid solutions were prepared by Coated Agglomerate Pelletization (CAP) process which has many advantages over conventional powder pellet route for handling ^{233}U powders. Five compositions, namely ThO_2, ThO_2–4%UO_2, ThO_2–10%UO_2, ThO_2–20%UO_2 and UO_2 (all compositions are in wt%) were prepared and used in this study. Characterization of thoria–urania fuel was carried out in terms of both thermal expansion and hardness at elevated temperature (hot hardness). The data obtained on thermal expansion were compared with those available in the literature data and presented in the temperature range of 300–1800 K. It was found to be in good agreement with the reported data of other authors. The data on hot hardness of thoria–urania fuel are sparse and are presented here as a function of temperature from 300 to 1573 K. The experimental data on both thermal expansion and hot hardness for all the compositions were least squares fitted and analytical expressions obtained are presented as recommended data. The trend obtained for the above data in terms of both as a function of temperature and composition are critically evaluated and discussed.

J. Banerjee (✉) · S. Kaity · K. Bhandari
Radiometallurgy Division, Bhabha Atomic Research Centre, Trombay, Mumbai 400085, India
e-mail: joydipta@barc.gov.in

S. Kaity
e-mail: santuk@barc.gov.in

K. Bhandari
e-mail: koushik@barc.gov.in

A. Kumar · S. Banerjee
Bhabha Atomic Research Centre, Trombay, Mumbai 400085, India
e-mail: arunk@barc.gov.in

S. Banerjee
e-mail: sbanerjee@barc.gov.in

A. K. Nayak and B. R. Sehgal (eds.), *Thorium—Energy for the Future*,
https://doi.org/10.1007/978-981-13-2658-5_18

Keywords Thoria–urania · Fuel · Solid solutions · Thermal expansion
Hot hardness

1 Introduction

The long-term sustainability of indigenous nuclear power programme in India depends to a great extent on ^{232}Th–^{233}U 'closed' fuel cycle due to its vast thorium resources, which can be utilized for breeding of ^{233}U in thermal breeder reactors [1–5]. Beside this, thorium cycle also tends to reduce the proliferation of spent fuel and plutonium production; it reduces the long-lived minor actinides production compared with the uranium and plutonium cycle which is certainly an ecological advantage. The three-stage Indian nuclear power programme [6] was conceptualized accordingly to utilize its thorium reserves. A schematic diagram of three-stage Indian nuclear power programme is given in Fig. 1. To use thorium fuel in a reactor, it must be mixed with a fissile material such as enriched uranium or plutonium until it produces enough ^{233}U to sustain itself. Thoria (ThO_2) has a number of advantages over urania (UO_2), like higher melting point, better thermal transport and radiation stability ensuring better in-pile performance, more resistance to chemical interactions, low vapour pressure. Because of its very low solubility, the direct disposal of spent fuel in deep geological repository is also considered these days. Advanced Heavy Water Reactor (AHWR) is being developed for the related technology demonstration based on thorium. $(Th,^{233}U)O_2$ along with $(Th,Pu)O_2$ in the proposed fuel for the forthcoming Indian AHWR [7]. The fuel is designed to maximize generation of energy from thorium, to maintain self-sufficiency in ^{233}U.

Thermophysical and thermomechanical properties of nuclear fuel are important for predicting fuel performance under normal and off-normal conditions and also for developing a computer code for analysis of routine operation/ accidental conditions. In other words, evaluation of these thermal properties is important for judging in-pile performance of fuels in advance before its actual usage inside nuclear reactors. The available data on properties help in providing feed to the computer programmes which are available and/or are being developed for safety and performance analysis of nuclear fuels. These out-of-pile properties are the only benchmarks for deciding whether the fuel meets the specific design (specification) criteria. Studies on thoria–urania solid solutions, e.g. densification behaviour [8], sintering kinetics using master sintering curve approach [9, 10] and specific heats [11] have been carried out by the authors earlier. The present paper summarizes the microstructural observation and phase analysis carried out on those thoria–urania pellets and presents the experimental data generated on thermal expansion and hot hardness.

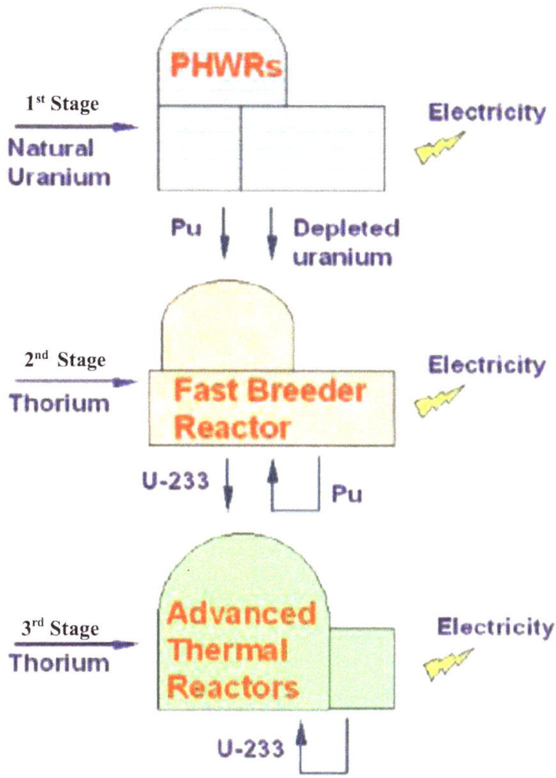

Fig. 1 Schematic of the three stage nuclear power programme of India [1]

2 Experimental

2.1 Starting Powders

ThO_2 powders used in this study are derived from oxalate route. The powders have platelet-like morphology. Typical size of powders varies from 1 to 2 μm. The SEM picture of starting ThO_2 powders is shown in Fig. 2. The present work was simulated using natural ^{235}U instead of ^{233}U. The ADU-derived UO_2 powders were used for the preparation of U_3O_8 powders. These UO_2 powders were heated in air at 400–600 °C for 4–6 h to convert to U_3O_8 powders. The SEM pictures of U_3O_8 powders, such obtained, are shown in Fig. 3.

Fig. 2 SEM pictures of ThO_2 powders at different magnifications

Fig. 3 SEM pictures of U_3O_8 powders at different magnifications

2.2 *Pellet Preparation*

The green pellets of ThO_2–4%UO_2, ThO_2–10%UO_2 and ThO_2–20%UO_2 were prepared by the Coated Agglomerate Pelletization (CAP) process [12, 13] using ThO_2 granules (−40 mesh) and U_3O_8 powder as the starting material. The flow sheet of CAP process is shown in Fig. 4. The green pellets of ThO_2 and UO_2 were fabricated by powder pellet route without addition of any dopants using the conventional steps, e.g. milling, pre-compaction, granulation and final compaction [14]. The green pellets of all five compositions, as mentioned above, were sintering at 1923 K for 4 h in Ar–8%H_2 atmosphere.

The details of pellet preparation for the present study are given elsewhere [11] which include (a) the characteristics of starting ThO_2, UO_2 and U_3O_8 powders, (b) impurity analysis of sintering atmosphere used, (c) 'Th' and 'U' content of

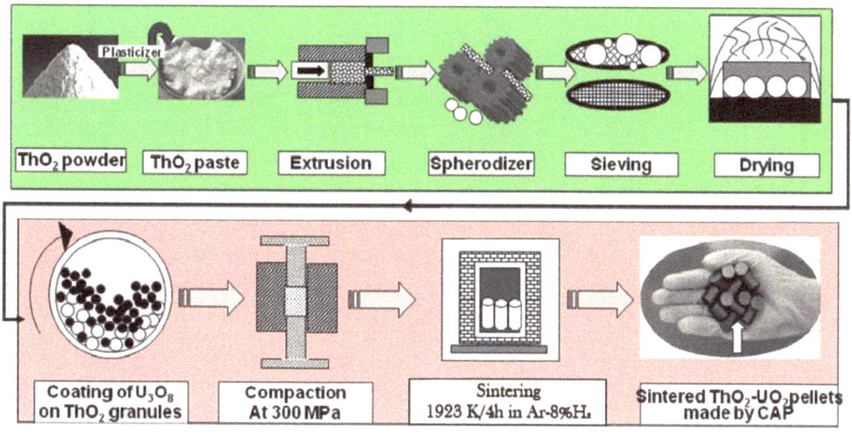

Fig. 4 Flow sheet adopted for fabrication of $(Th,U)O_2$ pellets by CAP process [12]

ThO_2–UO_2 sintered pellets, (d) impurity analysis of sintered pellets by ICP-AES, (e) density, grain size and O/M ratio of sintered pellets, (f) calculated lattice parameters of ThO_2, UO_2 and ThO_2–UO_2 solid solutions prepared from X-ray diffraction data etc.

2.3 XRD and Microstructure

X-ray diffraction studies of the sintered pellets were carried out using Cu Kα radiation and graphite monochromator. The XRD patterns were recorded at a 2θ scan rate of 1°/min where $20° \leq 2\theta \leq 70°$. Slow scan XRD at higher angles ($50° \leq 2\theta \leq 80°$) was carried out by step scan (step size: 0.004°; counting time: ~1 s) to ascertain homogeneity of sintered pellet.

For microstructural analysis, standard metallographic procedures were followed. Sintered pellets were cut in cut-off machine using diamond wheel and then mounted with Araldite. Mounted samples were ground in successive emery paper and were subsequently polished on polishing clothes using diamond paste. Polished samples were examined under optical microscope (model: Axiovet 40 Mat; make: Zeiss) for ensuring the removal of scratches. The samples were then etched thermally at 1873 K for 6 h. The etched samples were examined in SEM using tungsten filament. The grain size of the sintered pellets was measured by linear intercept method. The average grain size of thoria–urania pellets was observed to vary from 10 to 20 µm.

2.4 *Thermal Expansion*

The thermal expansion of sintered pellets was measured out in a vertical dilatometer (make: SETARAM, France). The pellets were heated up to 1773 K in high purity argon atmosphere (flow rate: 20 cm^3/min) with heating rate of 2 K/min. The pellets were cylindrical in shape with a diameter to length ratio of approximately one. The dilatometer was calibrated with Tungsten, NBS Standard Reference Material (SRM) 737. Temperature calibration of the dilatometer was checked using uranium metal sample, whose phase transformation temperatures are well known.

2.5 *Hot Hardness*

The high-temperature hardness (hot hardness) was measured in a hardness tester (make; Nikon, Model QM) using a diamond Vickers pyramid indenter. Samples were cut from sintered (cylindrical) pellet to a cube-shaped piece using a cut-off machine. These were metallographically polished on the top surface, and indentations were taken on this surface from room temp to 1573 K. The indentation was carried out in vacuum (<0.1 Pa) at an interval of 100 K at a load of 200 g and dwell time of 15 s. The load applied for making indentation over the material at any temperature was chosen in such a manner so as to avoid cracks at the indentation corners. After the indenter is withdrawn from the material surface, both the diagonals of the indentation were measured with an accuracy of ± 2 μm. Average of the diagonals was used to calculate the hardness. Four to five such indentations were made at each temperature, and the hardness was calculated for each of those. Average of these hardness data at each temperature was then plotted as a function of temperature.

3 Results and Discussions

3.1 *Microstructure and Phase Analysis*

The X-ray diffraction patterns for the sintered pellet are shown in Fig. 5. For comparison, the patterns obtained for pure ThO_2 and UO_2 are also shown in the same figure. All the sintered pellets were found to be of single-phase face-centred cubic (fcc) type with no traces of U_3O_8 in the matrix by XRD. It was observed that the lattice parameter decreases linearly with increase in UO_2 content and it obeys the Vegard's law [11]. However, there was some peak broadening observed for compositions having high UO_2 content at higher angles ($2\theta > 50°$) suggesting minor heterogeneity. The XRD pattern by slow speed step scan at higher angles ($50° \leq 2\theta \leq 80°$) obtained for compositions with higher UO_2 content (>4%) showed the presence of sub-peaks signifying the presence of variation of compositions within the matrix.

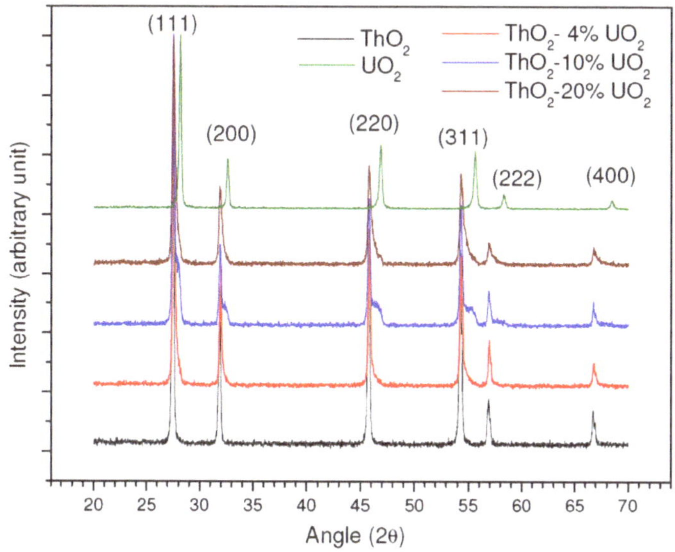

Fig. 5 XRD patterns of sintered ThO_2–UO_2 pellets

Table 1 summarizes the solid solution compositions corresponding to each sub-peak observed for all three thoria–urania compositions used in the present study assuming linear variation of lattice parameter with composition. The fractional amount of each of these solid solution compositions expected to be present in the matrix of the pellets as compositional variation in terms of micro-inhomogeneity is also included in the same table. For example, for the pellet having composition ThO_2–10%UO_2, there will be some areas rich in UO_2 and some areas rich in ThO_2. Therefore, the distribution of UO_2 in the matrix is not expected to be uniform in the entire matrix and localized variation in U/Th ratio other than the average composition of the material can be inferred from the presence of sub-peaks in slow-speed high-angle XRD pattern.

SEM pictures of three compositions of thoria–urania pellets obtained are shown in Fig. 6. The microstructure of the pellets sintered in Ar–8%H_2 shows a non-uniform and inhomogeneous grain structure especially with higher UO_2 content (>4%). This inhomogeneous microstructure for ThO_2–UO_2 fabricated by CAP process and sintered in Ar–8%H_2 has also been reported earlier [12, 13]. This could be attributed to sluggish reduction of U_3O_8 in-situ during sintering. The small grains observed in the microstructure (Fig. 6b, c) enveloping larger grains could be having different Th–U ratio than larger grains or the matrix. These small and circular grains are clearly visible to coalesce and form separate grains along the grain boundaries of larger grains. The heterogeneous microstructure observed by SEM in terms of small rounded grains are expected to be partially reduced U_3O_8 particles which got reduced to UO_2 later on during sintering and eventually all formed solid solution with ThO_2 in the matrix.

Table 1 Measured lattice parameters of different compositions found in the micro-inhomogeneity inside the thoria–urania matrix

S. No.	Composition	Average lattice parameter values calculated from five reflections ($50° \leq 2\theta \leq 80°$), nm	Corresponding approximate composition of solid solution (derived from the lattice parameter calculated)	Fraction of each solid solution calculated from the intensity of the peaks	Composition of major phase
1	ThO_2–4%UO_2	0.5582	$(Th_{0.9}U_{0.1})O_2$	0.40	
		0.5593	ThO_2	0.60	ThO_2
2	ThO_2–10%UO_2	0.5574	$(Th_{0.8}U_{0.2})O_2$	0.20	
		0.5584	$(Th_{0.9}U_{0.1})O_2$	0.60	$(Th_{0.9}U_{0.1})O_2$
		0.5604	ThO_2	0.20	
3	ThO_2–20%UO_2	0.5561	$(Th_{0.7}U_{0.3})O_2$	0.35	
		0.5573	$(Th_{0.8}U_{0.2})O_2$	0.50	$(Th_{0.8}U_{0.2})O_2$
		0.5598	ThO_2	0.15	

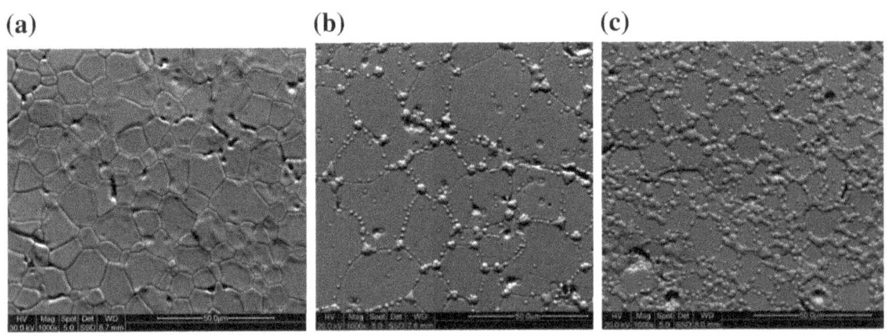

Fig. 6 Microstructure of **a** ThO_2–4%UO_2 **b** ThO_2–10%UO_2 and **c** ThO_2–20%UO_2

3.2 *Thermal Expansion*

Restrained thermal expansion of the fuel pellet surrounded by cladding generates thermal stresses inside the fuel. During the ‘power ramp’ of the reactors, the expansion of the fuel pellet and cladding takes place very fast. Too much of differential expansion between these two can lead to enormous stresses on the cladding material which can lead to rupture and failure of the fuel element. Therefore, the data on thermal expansion of nuclear fuels are very useful and important to fuel designers to ascertain this stress and predict the life of fuel elements.

Thermal expansion value of any material depends on temperature (T) as follows:

$$(\Delta L/L_0)_T = A + BT + CT^2 + DT^3, \tag{1}$$

Table 2 List of the authors worked on thermal expansion measurement on $(Th_{1-y}U_y)O_2$ ($0 \le y \le 1$) system and year of publications

S. No	Authors	Year	Remark
1	Fink [17]	2000	Review paper
2	Anthonysamy et al. [28]	2000	Measurement by XRD, (298–1973 K)
3	Tyagi and Mathews [27]	2000	Measurement by XRD, (298–1473 K)
4	Bakker et al. [18]	1997	Review paper
5	Momin et al. [20]	1991	Measurement by XRD, (298–1600 K)
6	Martin [16]	1988	Review paper
7	Rodriguez and Sundaram [25]	1981	Review paper
8	Touloukian et al. [15]	1970	Review paper
9	Springer et al. [21]	1967	Measurement (293–2273 K)
10	Turner and Smith [22]	1967	Measurement
11	Lynch and Beals [24]	1962	Measurement (upto 1173)
12	Kemper and Elliot [23]	1959	Measurement (293–1173 K)
13	Powers and Shapiro [26]	1959	Measurement, (Up to 1073 K)

Or,

$$(\Delta L/L_0)_T \times 100 = A + BT + CT^2 + DT^3 \quad (2)$$

where $(\Delta L/L_0)_T$ and $(\Delta L/L_0)_T$. 100 are fractional and percentage thermal expansion value respectively and A, B, C and D are constants. When the temperature changes by ΔT, the change in length ΔL is proportional to ΔT and the original length L_0

$$\Delta L = \alpha \times L_0 \times \Delta T, \quad (3)$$

where this proportionality constant, α is called the 'coefficient of linear thermal expansion' (CLTE) or simply 'coefficient of thermal expansion' (CTE).

There are several literatures available on ThO_2–UO_2 system. A list of the authors and the year of publications are given in Table 2.

For pure ThO_2, percentage thermal expansion was reported in the temperature range 150–2000 K by Touloukian et al. [15] as,

$$\left(\frac{\Delta L}{L_0}\right) \times 100(\%) = -0.179 + 5.097 \times 10^{-4}\,(T/\mathrm{K}) + 3.732 \times 10^{-7}\,(T/\mathrm{K})^2 - 7.594 \times 10^{-11}\,(T/\mathrm{K})^3 \quad (4)$$

where the percentage linear thermal expansion $[(\Delta L/L_0) \cdot 100]$ (%) is zero at 293 K.

Similarly, for pure UO_2, percentage thermal expansion was reported by Martin [16]. Fink in his review paper [17] on thermophysical properties of uranium dioxide has also recommended the data put forward by Martin. Two different equations for two different temperature ranges were given by Martin:

for $273\ \text{K} \leq T \leq 923\ \text{K}$

$$\left(\frac{\Delta L}{L_{273}}\right) \times 100(\%) = -0.27 + 9.082 \times 10^{-4}\,(T/\text{K}) - 2.705 \times 10^{-8}\,(T/\text{K})^2 + 4.391 \times 10^{-11}\,(T/\text{K})^3 \tag{5}$$

and for $923\ \text{K} \leq T \leq 3120\ \text{K}$

$$\left(\frac{\Delta L}{L_{273}}\right) \times 100(\%) = -0.328 + 1.179 \times 10^{-3}\,(T/\text{K}) - 2.429 \times 10^{-7}\,(T/\text{K})^2 + 1.219 \times 10^{-10}\,(T/\text{K})^3 \tag{6}$$

Using the result of these two extreme studies, Bakker et al. [18] have recommended the percentage linear thermal expansion data of $(Th_{1-y}U_y)O_2$ $(0 < y < 1)$ by linear interpolation method. He suggested two different relations in two different set of temperature ranges which are given below:

for $273\ \text{K} \leq T \leq 923\ \text{K}$

$$\begin{aligned}\left(\frac{\Delta L}{L_{273}}\right) \times 100(\%) = {} & -0.179 - y \times 0.087 + \left(5.097 \times 10^{-4} + y \times 4.705 \times 10^{-4}\right)(T/\text{K}) \\ & + \left(3.732 \times 10^{-7} - y \times 4.002 \times 10^{-7}\right)(T/\text{K})^2 \\ & - \left(7.594 \times 10^{-11} - y \times 11.98 \times 10^{-11}\right)(T/\text{K})^3\end{aligned} \tag{7}$$

and for $923\ \text{K} \leq T \leq 2000\ \text{K}$

$$\begin{aligned}\left(\frac{\Delta L}{L_{273}}\right) \times 100(\%) = {} & -0.179 - y \times 0.149 + \left(5.097 \times 10^{-4} + y \times 6.693 \times 10^{-4}\right)(T/\text{K}) \\ & + \left(3.732 \times 10^{-7} - y \times 6.161 \times 10^{-7}\right)(T/\text{K})^2 \\ & - \left(7.594 \times 10^{-11} - y \times 19.784 \times 10^{-11}\right)(T/\text{K})^3\end{aligned} \tag{8}$$

ThO_2 and UO_2 form an ideal solid solution between them. In the entire solid solubility range, the lattice parameter is observed to increase linearly following Vegard's law. The literature data on high-temperature vapour pressure measurement of the system also support their ideal solution behaviour [19]. From this observation, it can be reasonably assumed here that the linear change in lattice parameter could also occur at high temperature, thereby ensuring complete solid solubility between ThO_2 and UO_2 even at high temperatures. Extending this analogy, it was reported [18] that thermal expansion of their solid solution can be reasonably approximated as a function of temperature by taking account of weighted fraction of the individual constituents. Hence, thermal expansion data for compositions which are not covered in this study can be generated theoretically by linear interpolation method from Eqs. (7) and (8) as per respective weight fraction of individual constituents.

Anomalies are not uncommon in the reported thermal expansion data for $(Th,U)O_2$ system. The thermal expansion via lattice measurement of $(Th,U)O_2$ system was reported [20] using X-ray diffraction method. The values of CTE for both pure ThO_2 and $(Th_{0.8}U_{0.2})O_2$ in the temperature range of validity between 298 and 1600 K were reported which are found to be 9.5×10^{-6} K^{-1} and 7.1×10^{-6} K^{-1}, respectively. The CTE for $(Th_{0.8}U_{0.2})O_2$ is observed to be lower than either that of ThO_2 and UO_2. The authors did not provide any explanation for this, and hence, this data cannot be accepted. In the present study, CTE in the temperature range of 300–1573 K for ThO_2, ThO_2–20%UO_2 and UO_2 was found to be 9.8925×10^{-6}, 10.2194×10^{-6} and 11.0775×10^{-6} K^{-1}, respectively. A significant variation of the CTE data of $(ThO_2–UO_2)$ has been observed in the literature [20–24]. In a review article [25], an average CTE of 9.67×10^{-6} K^{-1} for ThO_2 (293–2273 K) and 12.5×10^{-6} K^{-1} for $(Th_{0.8}U_{0.2})O_2$ (1100–2400 K) were reported. In another report [26], the same average CTE value of 9×10^{-6} K^{-1} for both pure UO_2 and $(U_{0.064}Th_{0.936})O_2$ (up to 1073 K) was found. The authors obtained lower CTE value (8×10^{-6} K^{-1} up to 1073 K) for $(Th_{0.8}U_{0.2})O_2$. However, no explanation was given for this unusual behaviour.

Tyagi et al. [27] found CTE values for ThO_2 and ThO_2–2%UO_2 to be 9.58×10^{-6} and 9.74×10^{-6} K^{-1}, respectively, in the temperature range of 298–1473 K. The CTE value obtained in the present study for ThO_2 in the temperature range 300–1473 K is 9.73217×10^{-6} K^{-1}, and for ThO_2–4%UO_2, it is 9.85×10^{-6} K^{-1}; both these values were found to be in close agreement with those reported by Tyagi et al. The CTE value 10.33×10^{-6} K^{-1} for composition $(Th0_{.87}U0_{.13})O_2$ in the temperature range 298–1973 K as reported by Anthonysamy et al. [28] matches well with the value obtained in the present study for ThO_2–10%UO_2 which was found to be 10.21×10^{-6} K^{-1} in the temperature range 300–1773 K.

In the present study, the experimental data from dilatometer for each individual composition are least squares fitted to give a third-degree polynomial equation relating thermal expansion with temperature in the temperature range from 300 to 1800 K. The fitting errors for all these compositions are within ±1%.

For ThO_2: 300 K $\leq T \leq$ 1800 K

$$\left(\frac{\Delta L}{L_0}\right) \times 100(\%) = -0.19730 + 5.60599 \times 10^{-4}(T/\mathrm{K}) + 3.45902 \times 10^{-7}(T/\mathrm{K})^2 - 7.42760 \times 10^{-11}(T/\mathrm{K})^3 \quad (9)$$

For ThO_2–4%UO_2: 300 K $\leq T \leq$ 1800 K

$$\left(\frac{\Delta L}{L_0}\right) \times 100(\%) = -0.22017 + 6.62124 \times 10^{-4}(T/\mathrm{K}) + 2.53256 \times 10^{-7}(T/\mathrm{K})^2 - 4.65633 \times 10^{-11}(T/\mathrm{K})^3 \quad (10)$$

For ThO_2–10%UO_2: 300 K $\leq T \leq$ 1800 K

$$\left(\frac{\Delta L}{L_0}\right) \times 100(\%) = -0.22799 + 6.86287 \times 10^{-4}\,(T/\mathrm{K}) + 2.62207 \times 10^{-7}\,(T/\mathrm{K})^2 - 5.53501 \times 10^{-11}\,(T/\mathrm{K})^3 \quad (11)$$

For ThO_2–20%UO_2: 300 K $\leq T \leq$ 1800 K

$$\left(\frac{\Delta L}{L_0}\right) \times 100(\%) = -0.22587 + 6.72379 \times 10^{-4}.(T/\mathrm{K}) + 2.86975 \times 10^{-7}.(T/\mathrm{K})^2 - 6.19009 \times 10^{-11}.(T/\mathrm{K})^3 \quad (12)$$

For UO_2: 300 K $\leq T \leq$ 1800 K

$$\left(\frac{\Delta L}{L_0}\right) \times 100(\%) = -0.24915 + 7.54313 \times 10^{-4}\,(T/\mathrm{K}) + 2.68877 \times 10^{-7}\,(T/\mathrm{K})^2 - 4.94604 \times 10^{-11}\,(T/\mathrm{K})^3 \quad (13)$$

The polynomial fitted data for all the compositions, Eqs. (9)–(13), along with data from Touloukian [15], Martin [16] and Bakker [18] are shown together in Fig. 7. The data obtained in this study for ThO_2 and UO_2 were found to be in good agreement (within ±5%) with those reported by other authors in the literature in the entire temperature range from 300 to 1800 K. As can be seen from this figure, thermal expansion for each composition increases with increase in temperature. It can be correlated to the asymmetric relation between the potential energy verses inter-atomic distance of a material and is explained in the subsequent section.

In general, material expands with increase in temperature. Isotropic material expands in all three directions equally. For a long thin metal rod, this expansion would be observed as predominantly linear. For solids, thermal expansion has its origins in the nature of the bonding between atoms. At atomistic level, expansion of any material is observed due to the asymmetric nature between potential energy and inter-atomic distance curve. Here, asymmetry plays a more vital role than vibrational amplitude. For materials having a strong inter-atomic bonding, the trough of the curve will be very deep indicating smaller atomic separation. These materials have low coefficient of thermal expansion. With this logic, it is imperative that high melting materials having strong inter-atomic bonding would exhibit low thermal expansion coefficient.

The melting point of a material is inversely proportional to the coefficient of linear thermal expansion. All properties of solids can be correlated to the underlying atomic arrangements and inter-atomic forces acting from within the material. Coefficient of thermal expansion of a material is directly related to the attractive forces or bond energy between atoms. Materials having strong bonding show lower expansion coefficient than the materials having a weak inter-atomic bonding. So, materials having a high melting temperature—also due to strong atomic bonds—have low thermal

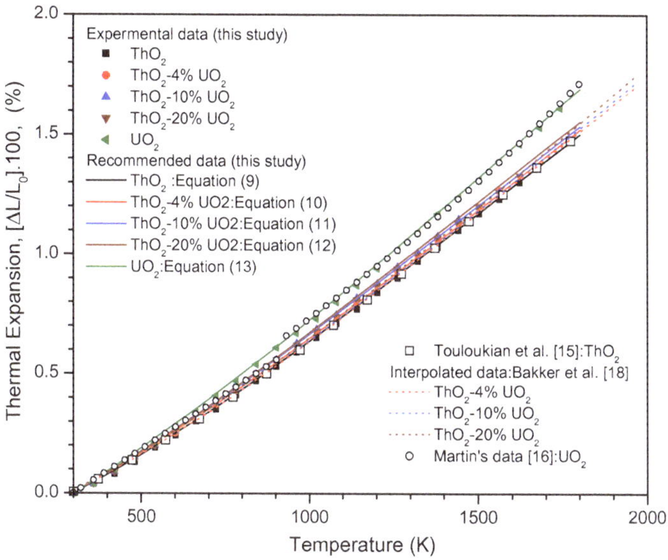

Fig. 7 Thermal expansion curve of ThO_2–UO_2

expansion coefficients and vice versa. Thermal expansion and coefficient of thermal expansion are proportional enteritis. Hence, it can be inferred from the foregoing discussion that melting point and thermal expansion are inversely proportional for a material. It is also evident from Fig. 7 that the thermal expansion at a particular temperature increases with increase in UO_2 content. This is as expected because UO_2 is having lower melting point than ThO_2 and thereby having lower inter-atomic bond energy.

3.3 *Hot Hardness*

Thermal toughness (hot hardness) data are useful to the fuel designers for the study of pellet-clad mechanical interaction (PCMI) where softening behaviour of the fuel with temperature can be studied. The hardness of a material is a measure of its resistance to plastic deformation and is directly related to the bond strength. With increase in temperature, the inter-atomic distances increase and the cohesive bond strength between the atoms decreases gradually resulting in the reduction of hardness with temperature.

Normally, hardness of any material varies exponentially with temperature [19, 29] which can be expressed, in general, in the following form:

$$H = A \times \exp(-B \times T) \tag{14}$$

where H is the hardness, T is the temperature, A is the intrinsic hardness, i.e. the value of hardness at absolute zero and B is the softening coefficient.

The values of A and B are dependent on deformation mechanisms of the material occurring at various temperatures. According to Westbrook [30], these constants signify some basic material characteristics as can be co-related to some fundamental physical parameter. He illustrated that parameter A can be related to inter-atomic bond strength of material. Hence, as discussed in Sect. 3.2, this parameter can be related to the melting point of the material. He established that the parameter A follows a linear relationship with melting point. However, the slope changes with the crystal structure of the material [30]. On the other hand, parameter B can be related to the increase in the inter-atomic spacing. Therefore, as discussed in Sect. 3.2, this represents the coefficient of linear thermal expansion. It was found that materials tend to group together as per their crystal structure when parameter B is plotted against thermal expansion [31].

The values of parameter A and B in the above equation change as we move from low temperature to high temperature. This infers to a change in deformation mechanism of the material with increase in temperature which is plausible. In this context, two different mechanisms have been suggested in the literature [30] which are slip at low temperature and dislocation climb and boundary sliding at high temperature ($>0.5T_m$). Metals and alloys are generally found to have a sharp transition in their hardness versus temperature profile at $\sim 0.5T_m$ where T_m is the melting point of the material in K. Ceramics are not generally found to have a sharp transition of this sort like their metals and alloys counterparts but are generally observed to have a smoother transition at ~ 0.3–$0.4T_m$. No such transition was observed for the any of the compositions in ThO_2–UO_2 solid solution undertaken in the present study. Hardness was observed to decrease with increase in temperature as is discussed in the following section.

Hardness values for ThO_2–UO_2 solid solutions were observed to decrease exponentially with increase in temperature. The experimental data from the hot hardness tester for each individual composition in the present study are least squares fitted to give an exponential decay curve as a function of temperature in the range from 300 to 1573 K. The fitting errors for all these compositions except ThO_2–20%UO_2 are within ±6%. A wide scatter of experimental data points was however observed for the composition ThO_2–20%UO_2. This can be attributed to the low density of this composition. As the density of the sample pellet was very low, the indentation was observed to overlap some area of the porosity on the surface of the sample. This led to the widespread scatter in hardness data. The temperature range of validity for these recommended equations is 300–1573 K. The least squares fitted equations for all the compounds are given below.

For ThO_2: $300\ \text{K} \le T \le 1573\ \text{K}$

$$H(\text{kg/mm}^2) = (1205.10718) \times \exp[-(0.00162) \times (T/\text{K})] \tag{15}$$

For ThO_2–4%UO_2: $300\ \text{K} \le T \le 1573\ \text{K}$

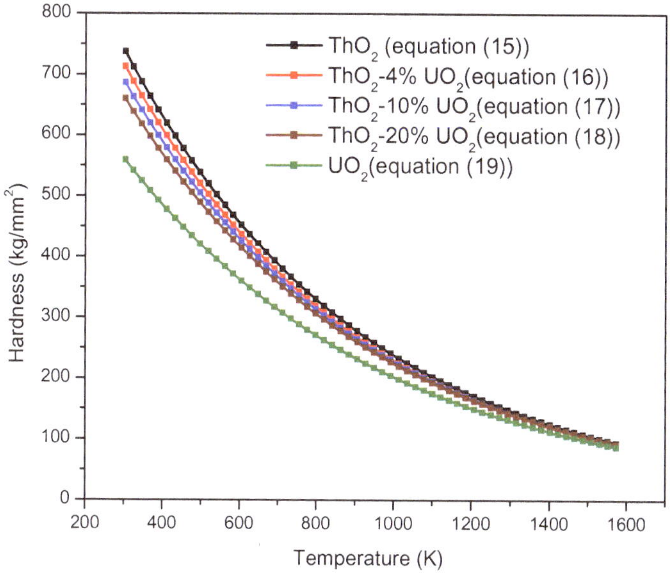

Fig. 8 Comparison of hot hardness data as a function of temperature for ThO_2, UO_2 and ThO_2–UO_2 solid solutions; the recommended data for the respective compositions were obtained by least squares fitting of the experimental data

$$H(\text{kg/mm}^2) = (1165.83167) \times \exp[-(0.00162) \times (T/\text{K})] \tag{16}$$

For ThO_2–10%UO_2: 300 K $\leq T \leq$ 1573 K

$$H(\text{kg/mm}^2) = (1108.87591) \times \exp[-(0.00158). \times (T/\text{K})] \tag{17}$$

For ThO_2–20%UO_2: 300 K $\leq T \leq$ 1573 K

$$H(\text{kg/mm}^2) = (1056.74448) \times \exp[-(0.00155) \times (T/\text{K})] \tag{18}$$

For UO_2: 300 K $\leq T \leq$ 1573 K

$$H(\text{kg/mm}^2) = (870.38832) \times \exp[-(0.00146) \times (T/\text{K})] \tag{19}$$

The least squares fitted curves of the experimental data for all the compositions are shown in Fig. 8. It is evident from the figure that irrespective of the composition, the hardness value decreases with increase in temperature. The rate of decrease of hardness is observed to be higher at low temperatures. The same rate is lower at high temperature. Hardness values are plotted with composition (wt%UO_2) at various constant temperatures, and this is shown in Fig. 9.

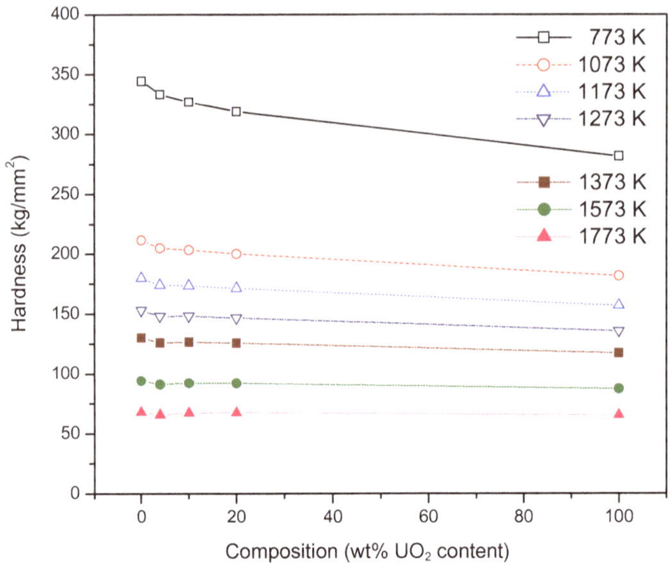

Fig. 9 Hardness as a function of UO_2 content

The hardness of a material is directly proportional to its melting point. ThO_2 has higher melting point (~3573 K) than UO_2 (~3033 K). Therefore, the hardness of ThO_2 is expected to decrease with addition of UO_2. The same trend was observed at all temperatures up to 1573 K for all the compositions as is evident from Fig. 8. The decrease in hardness with addition of UO_2 was not observed to be proportional with the UO_2 content. Figure 9 shows a small addition of UO_2 (from 4 to 10 wt%) decreases the hardness more than that with higher additions (from 10 to 20 wt%) especially at lower temperatures (773 K). However, the hardness data tends to merge at higher temperature (1573 K) for all compositions (Fig. 8). This could be attributed to softening behaviour of the material under the effect of temperature. Hardness data for ThO_2–10%UO_2 and ThO_2–20%UO_2 were observed to be much lower than that for ThO_2–4%UO_2, especially at lower temperature. This could be attributed to the larger grain size (~22 μm) for the composition ThO_2–10%UO_2 and lower density (~85% T.D) for the composition ThO_2–20%UO_2.

4 Summary

Thermal expansion of ThO_2–UO_2 solid solutions was found to increase with increase in temperature for a particular composition. At a particular temperature, thermal expansion was observed to increase with increase in UO_2 content.

The recommended equations generated from the experimental data are presented. The data obtained in this study for ThO_2–UO_2 solid solutions were found to be in good agreement (within ±5%) with those reported by other authors in the literature in the entire temperature range from 300 to 1800 K.

The hot hardness data for ThO_2–UO_2 solid solutions are sparse in the literature. The experimental hardness data at elevated temperatures for all the compositions were found to decrease exponentially with temperature up to 1573 K. The hardness data were observed to have some scatter due to porosity effect; the least squares fitted curves, however, shows a definite pattern with composition. Hardness of ThO_2–UO_2 solid solution decreases with increase in UO_2 content, and the decrease is more for lower UO_2 content and at low temperature. The decrease of hardness with temperature is not linear with the UO_2 content.

Acknowledgements We are thankful to Advanced Fuel Fabrication Facility (AFFF), Bhabha Atomic Research Centre (BARC), Tarapur, India, for providing the thoria, urania and thoria–urania pellets used in this study.

References

1. S. Banerjee, R.K. Sinha, S. Kailas, J. Physics: Conference Series 312 (2011) 062002
2. A. Kakodkar, Nu-Power 23 (2009) 22
3. A. Kakodkar, Shaping the 3rd stage of Indian Nuclear Programme, Presented at Annual Conference of the Indian Nuclear Society, INSAC-2001, Indore, 2001
4. K. Balakrishnan, S. Majumdar, A. Ramanujam, A. Kakodkar, in The Indian Perspective on Thorium Fuel Cycles, IAEA TECDOC-1319, Thorium Fuel Cycle: Options and Trends, Nov 2002, p. 257
5. K. Anantharaman, V. Shivakumar, D. Saha, Utilisation of thorium in reactors. J. Nucl. Mater. **383**, 119–121 (2008)
6. Department of Atomic Energy in India, Atomic Energy in India. A Perspective, Government of India, Sept 2003. (http://www.dae.gov.in/publ/persp/persp03/persp03.pdf)
7. R.K. Sinha, A. Kakodkar, Design and development of the AHWR —the Indian thorium fuelled innovative nuclear reactor. Nucl. Eng. Des **236**, 683–700 (2006)
8. Joydipta Banerjee, T.R.G. Kutty, Arun Kumar, H.S. Kamath, Srikumar Banerjee, Densification behaviour and sintering kinetics of ThO_2–4% UO_2 pellet. J. Nucl. Mater. **408**, 224–230 (2011)
9. A. Ray, J. Banerjee, T.R.G. Kutty, Arun Kumar, S. Banerjee, Construction of master sintering curve of ThO2 pellets using optimization technique. Sci. Sinter. **44**, 147–160 (2012)
10. Joydipta Banerjee, Aditi Ray, Arun Kumar, Srikumar Banerjee, Studies on sintering kinetics of ThO_2–UO_2 pellets using master sintering curve approach. J. Nucl. Mater. **443**, 467–478 (2013)
11. Joydipta Banerjee, S.C. Parida, T.R.G. Kutty, Arun Kumar, Srikumar Banerjee, Specific heats of thoria–urania solid solutions. J. Nucl. Mater. **427**, 69–78 (2012)
12. T.R.G. Kutty, K.B. Khan, P.S. Somayajulu, A.K. Sengupta, J.P. Panakkal, A. Kumar, H.S. Kamath, Development of CAP process for fabrication of ThO_2–UO_2 fuels Part I: Fabrication and densification behaviour. J. Nucl. Mater. **373**, 299 (2008)
13. T.R.G. Kutty, R.V. Kulkarni, P. Sengupta, K.B. Khan, K. Bhanumurthy, A.K. Sengupta, J.P. Panakkal, A. Kumar, H.S. Kamath, Development of CAP process for fabrication of ThO_2–UO_2 fuels part II: characterization and property evaluation. J. Nucl. Mater. **373**, 303 (2008)

14. T.R.G. Kutty, P.V. Hegde, J. Banerjee, K.B. Khan, A.K. Sengupta, G.C. Jain, S. Majumder, H.S. Kamath, Densification behaviour of ThO_2–PuO_2 pellets with varying PuO_2 content using dilatometry. J. Nucl. Mater. **312**, 224–235 (2003)
15. Y.S. Touloukian, R.K. Kirby, R.E. Taylor, T.Y.R. Lee, Thermal Expansion. Nonmetallic Solids (IFI/Plenum, New York, 1970)
16. D.G. Martin, The thermal expansion of solid UO_2 and (U, Pu) mixed oxides—a review and recommendations. J. Nucl. Mater. **152**, 94 (1988)
17. J.K. Fink, Thermophysical properties of uranium dioxide. J. Nucl. Mater. **279**, 1–18 (2000)
18. K. Bakker, E.H.P. Cordfunke, R.J.M. Konings, R.P.C. Schram, Critical evaluation of the thermal properties of $Th0_2$ and Th_1-yUy0_2 and a survey of the literature data on $Th_1-yPuy0_2$. J. Nucl. Mater. **250**, 1–12 (1997)
19. C.A. Alexander, J.S. Ogden, G.W. Cunningham, Battelle Memorial Institute Report BMI- 1789, 1967
20. A.C. Momin, E.B. Mirza, M.D. Mathews, High temperature X-ray diffractometric studies on the lattice thermal expansion behaviour of UO_2, ThO_2 and ($UO_2Th_{0.8}$) O_2 doped with fission product oxides. J. Nucl. Mater. **185**, 308 (1991)
21. J.R. Springer, E.A Eldrige, M.U. Goodyear, T.R. Wright, J.F. Langedrost, Battelle Memorial Institute Report BMI-X-10210, 1967
22. D.N. Turner, P.D. Smith, Australian Atomic Energy Commission report AAEC E183, 1967
23. C.P. Kemper, R.O. Elliot, J. Chem. Phys. **30**, 1524 (1959)
24. E.D. Lynch, R.J. Beals, Argonne National Laboratory Annual report for 1962. ANL-6677, 1962, p. 101
25. P. Rodriguez, C.V. Sundaram, Nuclear and materials aspects of the thorium fuel cycle. J. Nucl. Mater. **100**, 227 (1981)
26. R.M. Powers, H. Shapiro, Quarterly Technical Progress Report, Sylvania Corning Nuclear Corp. SCNC-301, 1959
27. A.K. Tyagi, M.D. Mathews, Thermal expansion of ThO_2–2 wt% UO_2 by HT-XRD. J. Nucl. Mater. **278**, 123–125 (2000)
28. S. Anthonysamy, G. Panneerselvam, S. Bera, S.V. Narasimhan, P.R. Vasudeva Rao, Studies on thermal expansion and XPS of urania–thoria solid solutions. J. Nucl. Mater. **281**, 15–21 (2000)
29. H.D. Merchant, G.S. Murthy, S.N. Bhadur, L.T. Dwivedi, Y. Mehrotra, Hardness-temperature relationships in metals. J. Mater. Sci. **8**, 437 (1973)
30. J.H. Westbrook, Temperature dependence of the hardness of pure metals. ASM Trans. **45**, 221 (1953)
31. E.R. Petty, Hardness and other physical properties of metals in relation to temperature. Metallurgica **56**, 231 (1957)

Thorium-Based Fuels for Advanced Nuclear Reactors: Thermophysical, Thermochemical, and Thermodynamic Properties

Dheeraj Jain, D. Das and B. N. Jagatap

Abstract India is developing thorium-based nuclear technologies with the aim to introduce them in its energy mix that is sustainable in the long term. Various reactor systems based on thorium fuel cycle are being developed. These include (i) thorium utilization in existing reactors (PHWRs), (ii) advanced heavy water reactor (AHWR), (iii) metallic-fueled reactors (fast/thermal), and (iv) molten salt reactor (MSR). Different fuel forms being studied include oxides/carbides, metallic alloys, and fluoride salts. Comprehensive information on the physicochemical properties of fuel is a primary input as required by reactor engineers and safety analysts to predict its performance under normal/off-normal reactor operation scenarios. The variable parameters essentially include temperature, composition, microstructure, and radiation field. A reliable database on thermophysical, thermochemical, and thermodynamic properties of fuel and related materials is therefore essential. This chapter presents an overview of R&D efforts in this direction carried out at Chemistry Group of Bhabha Atomic Research Centre (BARC). Important results and their implications in terms of fuel's performance potential are highlighted. While understanding of thorium-based oxide fuels (Th-U MOX, Th–Ce MOX, and SIMFUELS) has attained reasonable maturity, exciting potential of thorium-based in metallic fuel and fluoride salt fuel is being unearthed with recent experimental work. Few results on thorium-based metallic alloys are also presented. To conclude, glimpses of research efforts on thorium-based fluoride salts for Indian molten salt breeder reactor (IMSBR) are outlined.

D. Jain · D. Das · B. N. Jagatap (✉)
Chemistry Division, Bhabha Atomic Research Centre, Trombay, Mumbai, India
e-mail: bnj@barc.gov.in; jagatap@phy.iitb.ac.in

D. Jain
e-mail: jaind@barc.gov.in

D. Das
e-mail: dasd@barc.gov.in

B. N. Jagatap
Physics Department, Indian Institute of Technology Bombay,
Powai, Mumbai, India

A. K. Nayak and B. R. Sehgal (eds.), *Thorium—Energy for the Future*,
https://doi.org/10.1007/978-981-13-2658-5_19

Keywords Thorium-based fuels · AHWR · Thermophysical properties
Thermochemical properties

1 Introduction

Mastering the energy, economics, and environment dilemma has been the major focus of research and development in the energy sector today. Nuclear energy has by far proven itself as one of the safest, clean, environment-friendly, economically viable and sustainable, high energy density option. Development of safe and efficient nuclear energy technologies will continue to play an important role, particularly in countries that are large and growing rapidly. From the beginning of development of nuclear technology, thorium's potential as an inexhaustible and sustainable energy source has been widely acknowledged. This stems out of the facts that (i) thorium is highly abundant and widely available in economically beneficial extractable forms, (ii) nuclear attributes of thorium fuel cycle are superior to those of presently industrialized uranium fuel cycle, and (iii) physicochemical properties of thorium-based fuels are at par or better than uranium-based fuels. It would not be an exaggeration to say that since the initial thrust in nuclear science was focused on defense-related strategic applications, the industrial progress in uranium enrichment and plutonium production from natural uranium, essentially drove the technology toward light water reactors (BWRs/PWRs) and heavy water reactors (CANDUs/PHWRs/ACRs), respectively. Research on thorium fuel cycle, however, continued in major economic powers (USA, Europe, UK, and Japan) as well as in developing countries like India. These efforts have progressively led to understanding of thorium fuel cycle and recognition of its case as a promising option for long-term, safe, and sustainable source of energy. Indian nuclear fraternity in particular has been working on large-scale utilization of thorium for energy production in accomplishing India's three-stage nuclear power program, as envisaged by late. Dr. Homi Jahangir Bhabha. Sustained research efforts carried out over five decades have led to notable advancements, which include (i) mastering thorium production and fuel fabrication technology, (ii) irradiation experience on thorium-based fuels in research/power reactors, (iii) experience on reprocessing of thorium-based fuels, (iv) development and continued operation of world's only ^{233}U-based research reactor (KAMINI), (v) implementation of advanced heavy water reactor (AHWR) system, which is one of the safest reactor system with passive safety features designed to produce majority of power from thorium, (vi) experience on reactor physics aspects of AHWR through engineering-scale experiments in AHWR critical facility, etc. Development of thorium-based metallic fuels and molten salt fuels is also being pursued. Currently, India is poised to enter into the second stage of nuclear power program by commissioning the first prototype fast breeder reactor (PFBR) and that provides a testimony to the increasing maturity of thorium fuel cycle. India's vision to harness thorium energy would become reality when sufficient stockpile of plutonium, bred from its fast reactors, will be ready for blending with thorium in conventional reactors, and AHWR is constructed and commissioned, thereby introducing power production from thorium earlier than the envisaged program.

All these developments in the Indian context are made possible with multidisciplinary R&D efforts in basic and applied sciences and engineering. Innovations in chemical sciences hold key to several areas of front-end and back-end thorium fuel cycle, particularly in achieving reduced temperature of thoria sintering, production of pure thorium metal, spent fuel dissolution, etc. Nuclear fuel inside a reactor is a classical example of material evolving under thermal and radiation extremities, compositional complexity, and phase distribution. Detailed understanding of the physicochemical properties of fuel is of primary requirement to fuel designers and fuel performance analysts. This article presents highlights of applied research carried out in this direction on a variety of thorium-based fuel options, namely oxides, metallic alloys, and molten salts.

2 Thermophysical, Thermochemical, and Thermodynamic Characterization of Nuclear Fuels

Inside a fission-based nuclear reactor, complexity originates within the fuel at atomic/microscopic scale as neutron irradiation induces fission/activation resulting into formation of fission fragments, fission neutrons, activation products along with release of an enormous quantity of energy. The released energy, primarily coming out as kinetic energy of fission products, converts into heat after their interaction with fuel matrix. Along with the decay heat of radioactive products, this energy is eventually extracted by coolant flowing over the surface of cladded fuel pin. A nuclear fuel therefore experiences a host of simultaneously occurring processes, which include: (i) dynamic change in chemical composition due to accumulation of fission/activation products, (ii) changes in chemical state of fuel due to phases evolving from fuel-fission product (fp) interaction, fp-fp interaction and their temperature-dependent thermodynamic stabilities, (iii) constituent redistribution, driven by thermal and compositional gradients across the fuel pin, (iv) structural defect formation due to radiation damage and simultaneous defect annealing at high temperatures, (v) thermomechanical variations due to thermal expansion and fission-induced swelling, fission gas accumulation and partial release, (vi) thermal conductivity variations, and (vii) net energy storage due to heat capacity variations. Understanding the fuel performance under these conditions requires reliable physicochemical assessment, mainly as a function of temperature and burnup. Thermophysical, thermochemical, and thermodynamic characterization of fuel and related materials is therefore essential. Ideally, it requires detailed post-irradiation analysis (PIA) of fuel irradiated up to different burnups. Conducting PIA experiments on a large number of fuel pins is technologically challenging, expensive (as it requires hot cell facilities), time consuming as well as associated with men rem issues. PIA experiments are therefore carried out in unavoidable situations such as fuel pin failure analysis, fuel qualification and burnup assessment, etc. On the other hand, while not all but a great deal of information can be obtained from virgin fuels, SIMFUELS (synthesized in lab by mixing virgin fuel with stable isotopes of fission products for a given burnup) and individual fission product phases. Studies on these materials provide an indispensable dataset for new fuel types. To fuel designers, for example, it helps in finalizing fuel pin dimensions, fuel-clad gap for the estimation of linear power rating. For reactor physicists, temperature- and burnup-dependent thermophysical properties data help in simulating fuel's performance under a given neutron spectrum. For reactor engineers, such database is primary input for development and evaluation of fuel performance codes. Since fuel behavior under transient/off-normal reactor conditions would be governed by temperature-dependent variation in physicochemical properites, these data are also essential for safety analysis of a fuel/reactor system. To sum up, thermophysical, thermochemical, and thermodynamic characterization of fuels, SIMFUELS, and chemical phases formed during reactor operation is essential for overall evolution of reactor technology.

3 Thorium-Based Oxide Fuels

Majority of nuclear power reactors operating worldwide (PWRs/BWRs/PHWRs) use oxide fuels. Extensive research has been carried out on these fuels. Oxide fuel forms therefore become an obvious choice for any new reactor system. Reasons for dominance of oxide fuels in present nuclear technology include availability of industrially matured and economically viable fuel fabrication/reprocessing technologies and cumulative operational experience gathered over several decades. Oxides being inert matrix can be handled in large quantities (tonnes) under near ambient conditions. Thoria-based fuels are further attractive due to their redox inertness under chemical/thermal/radiation extremities. Among initial actinides, thorium is uniquely stable in its tetravalent oxidation state in bulk ceramic forms (oxides, carbides, nitrides, etc.). As a result, thoria-based fuel options have been widely studied. In India, fuel bundles exclusively containing high-density thoria pellets have long been irradiated in (i) thermal research reactors (J-rods in CIRUS and Dhruva reactors), (ii) fast breeder test reactor (FBTR, in reflector region), and (iii) Indian PHWRs (blanket region) to obtain irradiation experience. Thoria-based mixed oxides (MOX), namely Th-U and Th–Pu MOX, are the driver fuels for AHWR. Extensive research has been carried out to characterize these fuels and associated phases, and the following is a summary of progress made in this direction.

3.1 Thermophysical Properties of Thoria-Based Fuels/SIMFUELS

Thermal conductivity and thermal expansion are two important thermophysical properties for which reliable database are needed. Melting behavior (solidus/liquidus temperatures) and density information also help in predicting safe operational temperatures and attainable burnups. Thermal conductivity and thermal expansion data are required to be generated as a function of temperature, composition (burnup), oxygen potentials, porosity, and fuel microstructure. Care must be taken toward complete sample characterization, higher accuracy, and reproducibility of measurements while evaluating these properties, which in turn leads to higher confidence margins in computer codes/safety analysis codes and define precisely the safety margins for reactor operation.

Thermal conductivity of fuel affects various aspects of fuel behavior such as linear power rating, central line temperature, swelling, fission gas/volatile release, discharge burnup, etc. Efficiency of heat extraction from a nuclear fuel depends on thermal conductivity of fuel, clad, plenum gas, and other structural materials. Heat flow, particularly in solids, is governed by available heat carriers and their fractional contribution to net heat flow. These carriers are electrons, phonons, excitons, polarons, etc. Factors that influence their contribution are temperature, defects, composition, porosity, and microstructure. Thermal conductivity of a solid (λ_{total}) therefore can be represented as:

$$\lambda_{\text{total}} = \lambda_{\text{electron}} + \lambda_{\text{phonon}} + \lambda_{\text{exciton}} + \lambda_{\text{polaron}} + \lambda_{\text{radiation}} + \cdots \quad (1)$$

Factors influencing thermal conductivity of crystalline ceramics are recently summarized by Kutty et al. [1]. Experimentally, thermal conductivity can be evaluated by two different classes of methods, namely (i) steady-state methods and (ii) transient methods. In steady-state methods, specimen is brought to equilibrium heat flow under a predefined thermal gradient. Temperature measurements at various points inside the specimen are made, and thermal conductivity is directly evaluated using the general equation of unidirectional heat flow.

$$\mathrm{d}Q/\mathrm{d}t = \lambda.(-\mathrm{d}T/\mathrm{d}x).A \quad (2)$$

where $\mathrm{d}Q/\mathrm{d}t$ is heat flow rate; $\mathrm{d}T/\mathrm{d}x$ is temperature gradient along the direction of heat flow ($\mathrm{d}T/\mathrm{d}x < 0$); A is area cross section of conductor; and λ is thermal conductivity. Steady-state methods give highly accurate results but are time consuming, require large sample size, and are limited in applicable temperature range (RT-1200 K). An indigenously developed system operating over RT to 1250 K with ±3% accuracy has been extensively used for thermal conductivity studies on thoria and urania-based nuclear fuels at BARC (Fig. 1) [2]. In case of transient methods, a thermally equilibrated specimen is given a small thermal transient, and the heat flow response due to applied transient signal is measured in radial/axial direction. Using this response,

Fig. 1 Indigenously developed high-temperature thermal conductivity measurement setup

thermal diffusivity (α) is evaluated. Thermal conductivity (λ) and diffusivity (α) are related to each other by following relation:

$$\lambda(T) = \alpha(T) \times \rho(T) \times C_p(T) \quad (3)$$

where $\rho(T)$ and $C_p(T)$ are temperature-dependent density and heat capacity of the specimen, respectively.

Temperature dependence of density is derived from linear thermal expansion while specific heat can either be measured (Differential Scanning Calorimetry (DSC)/drop calorimetry) or estimated from Neumann-Kopp rule using the reported specific heat of constituents, wherever applicable. One of the most popular techniques based on transient method is flash technique [3]. It is fast and reliable, applicable at very high temperatures (up to 2800 K) and requires very small sample size.

Thermal expansion data is equally important and required to be evaluated for fuel, fission products phases, fuel-fission product interaction compounds, and SIM-FUELS as a function of temperature. Fission products may (i) dissolve in fuel lattice (rare earths, Zr, etc.), (ii) remain isolated in elemental/alloyed/intermetallic form (noble metals, Mo, Te, Cd, etc.), and/or (iii) make stoichiometric compounds/phases (alkalis, alkaline earths, etc.). Information on both bulk as well as lattice expansion behavior governs the optimization of fuel pellet-clad gap in such a way that the gap is not narrowed extensively to cause pellet-clad physical contact leading to pellet-clad interaction (PCI) under differential stress. Also, gap should not be kept unnecessarily wider, which further increases at high temperature, leading to poor heat transfer from fuel to clad due to deteriorating gap conductance. Thermal expansion data with porosity also gives useful information on fuel's thermomechanical behavior as a function of fission gas buildup. While lattice thermal expansion is evaluated by temperature-dependent diffraction techniques (HT-XRD, neutron diffraction, etc.), bulk thermal expansion is measured using high-temperature dilatometry. An indigenous dilatometer has been developed at Chemistry Division, BARC, for thermal expansion characterization of thoria and urania-based fuels [4].

Extensive work has been carried out to characterize thermal conductivity and thermal expansion of thoria-based fuels, and related materials as a part of task force activity on AHWR development program. All studies have been carried out on laboratory-synthesized oxide pellet samples that mimic the design basis of AHWR fuel. Thermal conductivity of UO_2 [5], ThO_2, $Th_{0.98}U_{0.02}O_2$ solid solutions [6], $(Th_{1-x}RE_x)O_{2-x/2}$ [7], and AHWR SIMFUEL have been evaluated. Suitable heat conduction models have been applied to delineate heat transfer mechanism in these ceramics as a function of temperature and composition. Thoria-based fuels have superior heat transport behavior compared to urania-based fuels, and it remains so with increasing temperature. Thermal conductivity decreases with temperature, and heat flow is dominated by phonon conduction. In general, thermal conductivity decreases with dissolution of fission products while it increases with precipitating metallic fission products. At very high temperatures (>2000 K for UO_2; >2700 K for ThO_2), thermal conductivity further increases, which has been attributed to increasing ambipolar contribution. This is useful for simulating heat extraction from fuel under severe accident scenario, where temperatures close to fuel melting are likely. Large database has been generated on virgin fuels, fuel-fission product interaction phases, and SIMFUEL compositions. More work is still required to (i) determine thermal conductivity of thoria-based SIMFUELS corresponding to extended burnups (50–150 GWd/T), (ii) ThO_2 doped with metallic fission product phases, and (iii) computational modeling for simulation of heat transfer efficiency in temperature range where experiments are challenging or not possible. Experimental data on thermal conductivity of liquid ThO_2 does not exist. Large uncertainty exists on thermal conductivity of Th–Pu MOX and needs to be addressed by further work on Th–Pu and Th–Ce MOX fuels (taking Ce^{4+} as surrogate of Pu^{4+} in fluorite lattice). Thermal conductivity also varies with microstructure, and therefore, it will be important to study microstructural variations in SIMFUELS corresponding to different burnups and correlate with thermal conductivity variation.

Thermal expansion studies have been carried out on ThO_2, UO_2 [8], Th-U MOX [9], Th–Ce MOX [10], Th-RE MOX [11, 12], Th-fp solid solutions [13], AHWR SIMFUEL, stoichiometric fuel-fp phases [14], etc. Results of these studies impart following attributes to thoria-based fuels: Thoria-based MOX fuels indicate better performance with relatively lower thermal expansion characteristics. With larger lattice parameter (a_{THO_2} = 560 pm; a_{UO_2} = 540 pm; a_{PuO_2} = 535 pm) and complete redox inertness of Th^{4+} in comparison to U^{4+} (UO_2 gradually converts to UO_{2+x}), thoria provides more space for fission product retention without excess swelling. It has also been identified that alkali thorates ($MThO_3$; M = Ba, Sr) and rare earth compounds show higher expansion behavior and are possible source of structural stress and swelling. Table 1 summarizes the results of thermal expansion studies on some representative fuel and fuel-fission product interaction phases.

Table 1 Average coefficient of thermal expansion for few thoria-based ceramics

S. No.	Oxide system	Range of 'x'	Average CTE (ppm/K) RT-1273 K
1	$Th_{1-x}U_xO_2$	$0.0 \le x \le 1.0$	9.10 (x = 0.0) to 10.81 (x = 1.0)
2	$Th_{1-x}Ce_xO_2$	$0.0 \le x \le 1.0$	9.10 (x = 0.0) to 11.58 (x = 1.0)
3	$Th_{1-x}Y_xO_{2-x/2}$	$0.0 \le x \le 0.125$	9.10 (x = 0.0) to 8.27 (x = 0.125)
4	$Th_{1-x}Nd_xO_{2-x/2}$	$0.0 \le x \le 1.0$	9.10 (x = 0.0) to 11.26 (x = 1.0)
5	$Th_{1-x}Ba_xO_{2-\delta}$	$0.0 \le x \le 0.05$	9.10 (x = 0.0) to 9.34 (x = 0.05)
6	$Th_{1-x}Sr_xO_{2-\delta}$	$0.0 \le x \le 0.10$	9.10 (x = 0.0) to 9.43 (x = 0.10)
7	$BaThO_3$	–	11.09
8	$SrThO_3$	–	10.49

In conclusion, superior thermophysical properties of thoria-based fuels would impart safer and superior operation of these fuels with relatively lower centerline temperatures, as compared to urania-based fuels. Further, they would have lesser fission gas release and lesser stored energy during severe transients. Lesser expansivity and higher solidus temperatures can allow for a narrower pellet-clad gap and high-temperature operation, respectively.

3.2 *Thermochemical and Thermodynamic Characterization of Thoria-Based Fuels/SIMFUELS*

Inside a nuclear reactor, when fuel undergoes fission/transmutation, the generated species (fission fragments and heavier nuclides) exhibit nuclear (with fission neutrons) and chemical reactions (solid-state reaction between fuel, fission products, and activation products) under prevailing thermochemical and radiochemical conditions. Chemical state of fuel is essentially governed by thermochemical stability of different phases. One fission event gives two fission fragments and thus, results into creation of additional nuclei, while transmutation and radioactive decay results into substitution and/or de-excitation of existing nuclei. In metallic fuel, chemical state of fuel evolves as heterogeneous mixture of alloys, intermetallics, and solid solutions along with fission gases/volatile products. Contrary to this, chemical state of fission products in ceramic fuels (oxide/carbide/nitride/composites) includes both ceramic as well as metallic phases along with gaseous/volatile products. This essentially depends upon nonmetal/metal potential (oxygen potential in oxides, carbon potential in carbides, etc.) and thermochemical stability of a given phase. Further, fuel's chemical state continues to evolve with burnup and temperature and has following consequences: (i) variation in thermophysical and thermochemical properties, (ii) constituent redistribution, i.e., material transport under the influence of concentration gradient, thermal gradient, and radiation damage, (iii) fuel swelling due to accumulated fission gases leading to varying fuel-clad gap, (iv) varying gap conductance due to fission gas

Fig. 2 Indigenously developed high-temperature transpiration system

release in plenum volume, and (v) modified clad properties due to radiation damage, corrosion by volatile fission products, pellet-clad interaction, etc. In oxide fuels, excess oxygen is liberated due to formation of higher fraction of metallic/lower oxidation state phases, compared to phases stable in higher oxidation states. Oxygen potential of fuel therefore increases with burnup and governs the extent to which the fuel can be irradiated. From above discussion, it is clear that detailed understanding of fuel's thermochemistry is essential to understand its performance potential in a reactor under actual operating conditions and predictable accidental situations. Such studies are carried out by (i) PIA on irradiated fuels to characterize chemical state at a given burnup and (ii) thermodynamic and thermochemical assessment of individual phases and SIMFUELS corresponding to various burnups, which include evaluation of (i) vaporization behavior, (ii) oxygen potential, (iii) oxygen transport studies, (iv) thermodynamic properties of fuel-fission product interaction phases, etc.

In case of thoria-based fuels for AHWR, extensive work has been carried out to evaluate thermodynamic and thermochemical properties of virgin fuel, SIMFUELs, and individual fuel-fission product phases. The subject has recently been reviewed by Bharadwaj et al. in a book chapter [14]. While detailing the results obtained on thoria fuels is out of the scope of present article, specific behavior of these fuels is summarized in what follows. It is noteworthy to mention here that several experimental systems (Knudsen effusion technique, transpiration technique, vapor pressure technique, galvanic EMF cell technique, manometric technique, etc.) have been indigenously developed for such studies. Figure 2 shows one such system, i.e., the high-temperature transpiration unit.

Similar to urania-based fuels, overall distribution of fission products in irradiated thoria-based fuels includes (i) fission products dissolved in oxide fuel matrix (Sr, Zr, Nb, Y, La, Ce, Pr, Nd, Pm, Sm), (ii) metallic phases (Mo, Tc, Ru, Rh, Pd, Ag, Cd, In, Sb, Te), (iii) precipitated oxide phases involving Ba, Sr, Zr, Nb, Mo, Rb, Cs, Te, etc., and (iv) fission gases/volatiles (Kr, Xe, Br, Te, I, etc.). Comparison of typical fission yields of urania- and thoria-based fuels indicate relatively higher yields of gaseous and lower yields of metallic fission products in thoria-based fuels under similar irradiation conditions. Investigations on phase stabilities, vapor pressure studies, and transport properties of these phases indicate that thermochemistry of thoria-based fuels is quite different, essentially due to redox inertness of Th^{4+} in all its nonmetallic condensed phases. This is in sharp contrast to urania-based fuels where uranium can stabilize as U^{4+}, U^{5+}, and U^{6+} under varying oxygen potential conditions. As a result, oxygen potential rises quite rapidly in thoria-based fuels, particularly during the onset of oxygen hyperstoichiometry. In comparison, oxygen potential buildup is slower in urania-based fuels since UO_2 can retain extra oxygen within the lattice (as UO_{2+x}). This results into slower oxygen transport controlled by self-diffusion in thoria-based fuels as compared to chemical affinity-controlled rapid oxygen transport in urania-based counterparts. The overall fuel performance and management of excess oxygen pressure in thoria-based fuels would therefore depend on (i) oxidation kinetics of molybdenum, which can make stable oxides and remain in metal/metal oxide equilibrium, (ii) formation kinetics and thermodynamic stability of binary/ternary/higher oxides of thorium with various fission products, and (iii) oxygen transport from fuel to clad as well as clad oxidation kinetics. Molybdenum oxidation is predicted at relatively lower burnups than usually seen in urania-based fuels. Apart from oxygen, transport behavior of reactive volatile fission products such as iodine, tellurium, etc., from fuel to clad would also determine clad corrosion. Analysis of thermodynamic stability of oxides phases with alkalis (M_2ThO_3) and alkali earths ($M'ThO_3$), which will be formed in thoria-based fuels (gray phases) indicates safer and superior performance of these fuels. In conclusion, thermochemistry of thoria-based fuels, which is mainly governed by buildup of oxygen potential, extent and mechanism of oxygen transport, and chemical transport behavior of volatile fission products, indicates good potential for safe irradiation up to high discharge burnups.

3.3 Transport Properties of Fission Gases and Volatile Fission Products in Thoria-Based Fuels

Fission gases (Kr, Xe) and volatile fission products (I, Te, Cs, Rb, etc.), which form in sizable proportion of total fission products, also influence fuel performance, especially at higher burnups (>60 GWd/T). Their transport and release behavior through the fuel matrix govern the extent of fuel swelling, which in turn help in estimating the onset of fuel-clad mechanical interaction (FCMI). Due to fission, these products are generated in monoatomic form inside the fuel matrix, and before escaping from fuel pellet in gaseous/vapor form, undergo significant redistribution. During this redistribution, they grow to form microbubbles (in grain interior), spherical/lenticular bubble (over grain surfaces), and interconnected bubbles/pores (along grain boundary region). With increasing burnup, higher release of fission gases leads to deterioration of gap conductance of pre-filled helium gas. This result in poor heat transfer and thereby increasing the fuel centerline temperature and further leading to increase in fission gas release. As a result, fission gas release behavior acts as one of the factors that limits the average discharge burnup for a fuel. Greater release of fission gases/volatiles is observed during transients, power ramps (startup) as compared to steady-state reactor operation. In particular, at extended burnups (>150 GWd/T), where fuel microstructure evolves as fine submicron-sized grains along with uniformly distributed closed porosity in the rim region of fuel pellet, relatively larger fraction of fission gases is retained within the fuel in the form of over pressurized pores. Volatile fission products, although less in concentration (other than Cs) as compared to fission gases, also affect the fuel performance. They have higher vapor pressure in free/bound state, and their escape from fuel pellet and deposition over clad surface can cause clad corrosion or hydride precipitation. Redistribution of both fission gases and volatile fission products occurs predominantly by thermally assisted transport under the influence of thermal and concentration gradients with a marginal contribution (1–3%) from athermal processes (recoil/knockout/irradiation induced). Dominant mechanism of diffusion varies with burnup. Fission gas/volatile release behavior is also influenced by fuel pellet microstructure, bulk defects (microcracks developed during thermal cycling), etc. As we go to higher burnups, transport occurs through interconnected pores, intergrain channels, etc. A review of fission gas/volatiles release behavior is presented by Das et al. [15].

Fission gas/volatiles release behavior is experimentally studied by PIA of trace irradiated fuel samples that have been fully characterized prior to irradiation. Irradiated samples are handled after sufficient delay to ensure decay of short-lived radioisotopes till safe handling limits. It is assumed that fission gases/volatiles trapped inside the matrix are uniformly distributed within the sample. Isothermal annealing of these samples under controlled atmosphere leads to release of certain fraction of trapped species, and release behavior can be studied by (i) immobilizing the released species in a suitable form and analyze it radiometrically (release fraction of individual species versus temperature) or (ii) evaluate the residual activity of irradiated sample after prolonged isothermal annealing at different temperatures (total release fraction vs temperature). The former approach is preferred as release behavior of individual species can be studied. Using the measured release fraction against temperature, intrinsic diffusivity and activation energy of diffusion for individual species can be obtained. Ion implantation of a given fission product in a specimen followed by vacuum annealing and detection by mass spectrometry is another approach to study fission gas/volatile release behavior. However, it is not preferred over trace irradiation method, which mimics the initial reactor irradiation conditions. It is important to note that these methods are reliable at lower burnups. Results obtained on high burnup fuels deviate from those predicted by atomic diffusion mechanism, as other modes of mass transport start contributing to overall release behavior. Other experimental methods include in-pile rod pressure measurement, gamma-ray spectrometry, PIA studies by ceramography or rod puncturing, etc.

Compared to UO_2, ThO_2-based fuels with similar density and microstructure in general exhibit slower fission gas release behavior. This is due to (i) lower diffusivity of Th^{4+} compared to U^{4+}, (ii) higher activation energies for vacancy and anion inter-

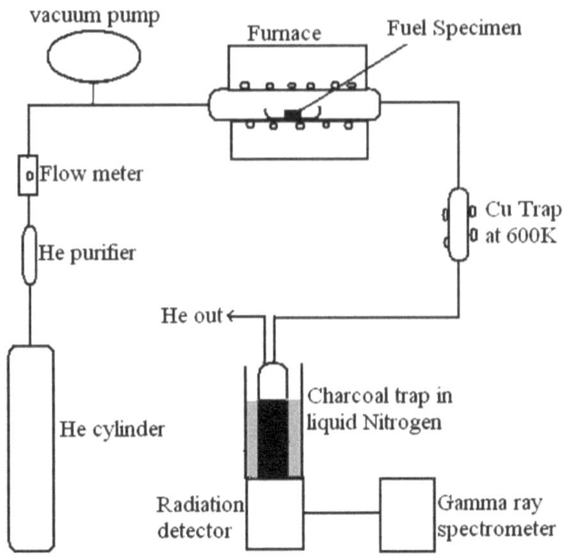

Fig. 3 Schematic of PIA setup used for Xe release studies in thoria-based fuels

stitial migration in ThO_2, (iii) redox inertness of Th^{4+} leading to absence of oxygen hyperstoichiometry-induced vacancy migration, (iv) slower diffusion due to lower temperature gradients in higher thermal conductivity thoria, etc. To study the transport behavior of fission gases and volatile fission products in AHWR fuel, extensive work has been carried out on both virgin fuel and high burnup SIMFUELS [16, 17]. PIA facilities have been indigenously set up to carry out these studies on trace irradiated fuels. Figure 3 shows schematic of PIA setup developed for Xe release studies in thoria-based nuclear fuels.

Experimental results are evaluated using statistical models. One example is Booth model [18], which assumes intragrain atomic diffusion to be rate-governing step in atomic transport of fission gases/volatiles. Experimental results, however, show lower release as compared to model's prediction. Using the experimental results, both lattice diffusion and grain boundary diffusion coefficients can be evaluated as a function of temperature, burnup, pellet microstructure, etc. To support experimental results, various theoretical approaches (first principle methods and molecular dynamics (MD) simulations) can be used to model atomic diffusion through fuel lattice.

Results obtained on fission gas/volatile fission product release behavior in trace irradiated (~5×10^{20} fission m^{-3}) AHWR fuels are summarized as follows: It is observed that high-density ThO_2 fuel with a nongranular microstructure would retain more fission gas as compared to urania-based fuels and would therefore be suitable for safe irradiation up to extended burnups. Due to lower release behavior of fission gases/volatiles, in combination with superior thermophysical properties, these fuels would be operable at higher power ratings. It has also been noted that increasing defect density (vacancies) would accelerate fission gas release. In case of AHWR SIMFUELS (thoria–urania solid solutions doped with fission products for 2 atom%

burnup), lowering of activation energies of diffusion for Xe, I, and Te, in comparison to virgin AHWR fuel ($Th_{0.98}U_{0.02}O_2$), indicates the role of defects/scattering centers on increasing ionic/atomic diffusivities. As compared to virgin fuel, Xe release is found to increase in SIMFUELS. Volatile fission products migration would be significantly lower in thoria-based fuels. Bulk diffusivities of iodine and tellurium are quite low compared to that in pure UO_2. Based on activation energies and intrinsic diffusivities, it has been identified that mechanism of atomic transport for fission gas (Xe) is quite different than that for volatile fission products (I, Te), indicating higher chemical affinity of later with the fuel matrix. In conclusion, assessment of fission gas/volatile fp transport behavior in thoria-based fuels reaffirms the potential of these oxides as safe and high burnup fuel options.

4 Thorium-Based Metallic Fuels

Fast reactors make the start of second stage of India's nuclear energy program. After the commissioning of U-Pu MOX-fueled PFBR, it is planned to construct a couple of oxide-fueled fast breeder reactors (FBR) in coming decade followed by another fleet of metallic-fueled fast breeder reactors with U-Pu-Zr alloys as driver fuels [19]. Metallic fuels are chosen as they offer a well-studied, safe, efficient, and proliferation-resistant fuel options for rapid expansion of nuclear energy capacity. Major advantages, inherent of metallic fuels over ceramic fuels, include faster breeding with shorter doubling time, convenient fuel reprocessing by pyrochemical electrorefining, inherent safety during transients, nonproliferative nature, extensive irradiation experience up to high burnups (10–20 atom%), etc. Excellent reviews on experience with metallic-fueled fast reactor technology are available [20]. Different aspects of U-Pu-Zr alloy fuel cycle including fuel fabrication, physicochemical assessment, test irradiation in fast breeder test reactor (FBTR), laboratory-scale reprocessing, etc., are already being studied at Indira Gandhi Centre for Atomic Research (IGCAR). In line with future program on metallic-fueled reactors and India's central strategy of utilizing its vast thorium resources, it is essential to explore the potential of thorium-based metallic fuels and develop them along with U-based metallic fuels. Table 2 presents general properties of thorium, uranium, and plutonium metals. It is evident that thorium has the best set of structural, physical, and thermophysical properties, looked for a nuclear fuel. For example, thorium has melting point (~2025 K) higher than uranium (~1405 K), which imparts better chemical stability, higher solidus temperatures of thorium-rich alloys, and amenability for fuel operation at higher centerline temperatures. High thermal conductivity and lower expansivity of thorium enable efficient heat extraction and lower fuel swelling, respectively. Isotropic cubic structure of thorium up to melting transition is significant in terms of isotropic swelling and less dimensional distortions in fuel element. Lower density of thorium allows for higher retention of fission products without much swelling. Higher solidus temperatures of thorium-based alloys tend to retard the diffusion of volatile/gaseous fission products, etc. Fuel can be made in the form of Th-U/Th–Pu binary alloys, Th-

Table 2 General properties of thorium, uranium, and plutonium metals

Property	Thorium	Uranium	Plutonium
Structure	FCC (RT-1673 K) BCC (1673-T_m)	Orthorhombic (RT-935 K) Tetragonal (935–1045 K) FCC (1045 K-T_m)	Six allotropes (Monoclinic at RT)
Lattice parameter (Å) *(for RT phase)*	5.07	$a = 2.85$ $b = 5.87$ $c = 4.95$	$a = 6.18$ $b = 4.82$ $c = 10.97$, $\beta = 101.80^o$
Melting point (K)	2025	1405	913
Boiling point (K)	5060	4405	3508
Theoretical density (g/cc at 298 K)	11.68	19.05	19.86
Specific heat at *RT* (J/mol K)	26.2	27.7	35.5
Thermal conductivity (W/m K) at 773 K	43.1	30	30
Coefficient of thermal expansion (ppm/K)	11.9 (30–600 K)	14.2 (30–600 K)	56 (mean value α-phase)

U-Zr/Th–Pu-Zr/Th-U-Pu ternary alloys, etc. In a recent article, Jain et al. [21] have reviewed the subject of thorium-based metallic alloys as advanced nuclear fuels in terms of their basic properties, inherent advantages/challenges, available literature information, irradiation experience, and future prospects. It comes out that along with inherently advantageous neutronic aspects of thorium fuel cycle, these alloys fuels have superior physicochemical properties compared to uranium-based alloys and are promising nuclear fuel candidates for future reactor technologies.

Compared to uranium-based metallic alloy fuels, limited literature is available on fabrication, irradiation experience, and reprocessing of thorium-based metallic alloys. Majority of work has been reported from USA and Europe citing the developments made during 1960–80. In view of India's program of thorium utilization and attractive features of metallic alloy fuels, early development of these fuels has been taken up at Bhabha Atomic Research Centre (BARC) with basic studies on preparation and characterization of thorium metal and alloys [22, 23] and their thermal and thermophysical characterization [24]. Following section summarizes the progress made on thermophysical characterization of binary and ternary alloys of thorium with uranium and zirconium.

4.1 Thermophysical Characterization of Thorium-Based Metallic Alloys

Reliable database on thermophysical properties of well-characterized thorium alloys, constituent metals, and related materials is essential for prediction of fuel's perfor-

Table 3 Average linear thermal expansion coefficients of Th-U binary alloys [24]

Sample → / Average CTE (*RT*-1100 K) ↓	Th metal	Th-10U alloy	Th-20U alloy
Measured values [24]	11.7×10^{-6}/K	11.9×10^{-6}/K	13.4×10^{-6}/K
Literature values	$12.2–12.6 \times 10^{-6}$/K	Not available	Not available
Oxidation during measurement	0.31 wt.%	0.45 wt.%	1.30 wt.%

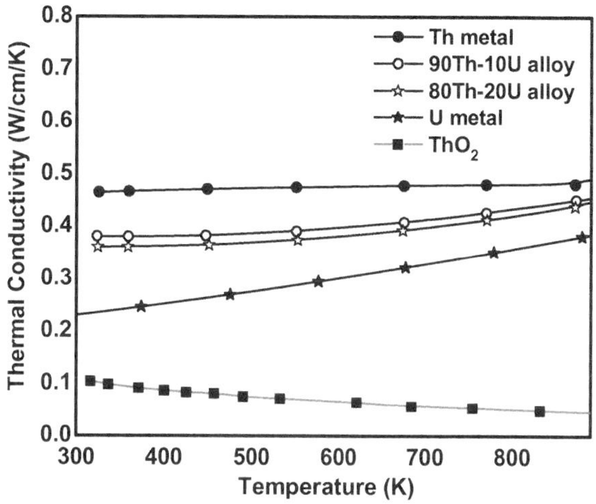

Fig. 4 Thermal conductivity of Th–U binary alloys and constituent metals as a function of temperature [24]

mance potential and safety analysis. Several alloys including 90Th-10U, 80Th-20U, 90Th-3U-7Zr, 90Th-7U-3Zr, 90U-10Zr, etc. (compositions in weight%) have been investigated. Alloys prepared by DC arc melting of constituent metals under inert atmosphere and characterized for their structure, phase morphology, and composition have been studied for their bulk thermal expansion, specific heat capacity, thermal diffusivity, and thermal conductivity as a function of temperature over RT—1200 K range. Phase distribution behavior of these alloys has also been studied to identify various phases present in (i) as-prepared and (ii) vacuum annealed alloy samples. These alloys have been characterized as multiphase alloys. Table 3 shows the variation of thermal expansion coefficient for these alloy systems. Figure 4 shows variation of thermal conductivity of Th-U binary alloys with temperature [24]. It is evident that thorium-based metallic alloys have excellent heat transfer properties, and furthermore, thermal conductivity of these alloys increases with temperature.

One essential attribute of these alloys is their higher reactivity toward air/moisture or contact materials at elevated temperatures. Extent of such reactions is seldom mentioned while reporting their thermophysical and thermodynamic properties and therefore can cause ambiguity in correct assessment of their performance potential. Care must be taken to minimize the extent of such surface reaction during high-temperature measurements and quantify them while reporting their physicochemical properties. Further work in this direction is focused on ternary alloys, SIMFUELS, and fuel-fission product alloys/intermetallic compounds.

5 Thorium-Based Molten Salt Reactors

Molten salt breeder reactor technology, based on thorium fuel cycle, is being considered as one of the options for third stage of Indian nuclear power program. Perhaps this is the most attractive candidate for thorium utilization with potential to achieve self-sustaining $^{232}Th/^{233}U$ fuel cycle [25]. In this direction, a conceptual design of Indian molten salt breeder reactor (IMSBR) without use of moderator/reflector has been developed and the essential material issues, which need to be addressed from basic scientific understanding, have been brought out [26]. This has given impetus to augment in-house basic and applied research on thermophysical, thermodynamic, and thermal studies on candidate salts systems and their interaction with IMSBR structural materials. A brief summary of the activities on candidate salt systems being pursued is presented below.

5.1 Thermophysical and Thermal Characterization of Thorium-Based Fluoride Salts for IMSBR

In case of molten salt reactor materials, the foremost requirement is development of various salt systems (fuel salt, blanket salt, coolant salt, etc.) and their thermal characterization. This includes preparation and purification of these salts with minimum moisture ingress and optimization of near perfect eutectic composition in multicomponent salt systems. Thermal analysis of these salts is therefore essential. Modification in experimental setups is also a prerequisite for handling these moisture sensitive salts. Selection of thermochemically inert accessories such as sample pans, crucibles, tubing, etc., is important in this regard. Thermal analysis on fuel (LiF-ThF_4-UF_4) and blanket salt systems (LiF-ThF_4, LiF-CaF_2-ThF_4, LiF-NaF-ThF_4) has resulted in identification of following salt compositions with true eutectic melting behavior: (i) fuel salt: 78LiF-20ThF$_4$-2UF$_4$ ($T_{\mathrm{eutectic}} = 823$ K; $\Delta H_{\mathrm{fusion}} = 106.6$ kJ/mol; composition in mol%), (ii) blanket salts: 70LiF-8CaF$_2$-22ThF$_4$ ($T_{\mathrm{eutectic}} = 773$ K) and 43.5LiF-32.5NaF-24ThF$_4$ ($T_{\mathrm{eutectic}} = 779$ K [27]. Heat capacity and enthalpy increment studies on these eutectic compositions have also been performed by drop calorimetric technique over the temperature range from 523 to 1073 K. Various other aspects such as temperature-dependent vapor pressure studies over these salts are being investigated.

Thermophysical properties that are required for performance evaluation of molten salt fuel systems are thermal conductivity and viscosity. Both need to be measured accurately over a wide temperature range as they influence heat transfer characteristics in molten salt systems. Indigenous development of experimental setups for carrying out these measurements on chemically harsh salt systems is required since commercial equipment for similar studies is rarely available and exorbitantly expensive. A couple of such systems are presently being developed [28] to characterize fluoride molten salts at moderate flow rates, where their behavior can be approximated to Newtonian fluids.

Another important aspect of fluoride salt systems is to understand their interaction with structural materials that are likely to be used for fabrication of MSBR flow tubes, reactor vessel, heat exchanger tubes, etc. Nickel-rich alloys such as Inconel, Hastelloy, etc., which have superior chemical resistant to fluoride salts at elevated temperatures, are potential candidates for this purpose. Fluoride-induced corrosion on various such materials is being investigated under prolonged exposure to various salt systems (fuel/blanket/coolant) at different temperatures.

To summarize, the current studies on thermal, thermodynamic, and thermophysical characterization of potential salt systems for IMSBR program would serve as essential input for performance assessment and simulation of these salt systems under reactor operating conditions.

6 Conclusions

Thorium-based nuclear fuels will play a dominating role in India's long-term nuclear energy program. Potential fuel options available with thorium cycle include ceramic oxides, metallic alloys, and molten salts, with distinct advantages of each fuel type for a specific reactor. Thermophysical, thermochemical, and thermodynamic characterization of such fuels is one of the most essential requirements for reactor engineers and safety analysts to assess the fuel performance in normal as well as transient reactor operating conditions. Significant developments have been made on thorium-based oxide fuels for Indian AHWR.

Detailed review of these studies as presented in the article indicates that thoria-based fuels would behave differently than urania-based systems with altogether distinct thermochemistry. These fuels would perform safely to higher burnups (>50 GWd/T) in thermal reactors and would be superior to conventional urania-based fuels. It is also indicated that future efforts should be dedicated toward characterization of extended burnup SIMFUELS and fuel-fission product compounds. Metallic alloys and fluoride salt systems based on thorium are also being investigated for their thermal, thermophysical, and thermodynamic properties. Early results indicate great potential of thorium-based metallic alloys as advanced nuclear fuel option for future reactors. Work on thorium-based molten salts in referenced to their thermal and thermodynamic assessment is also presented. In conclusion, these studies make essential platform for overall fuel development and performance assessment and contribute in development of thorium-based nuclear energy technologies.

Acknowledgements Authors are thankful to International Thorium Energy Alliance and organizers of International Thorium Energy Conference 2015 for providing the opportunity to present this work during the conference.

References

1. T.R.G. Kutty, J. Banerjee, A. Kumar, in *Thoria-Based Nuclear Fuels: Thermophysical and Thermodynamic Properties, Fabrication, Reprocessing and Waste Management*, ed. by D. Das, S.R. Bharadwaj, (Springer-Verlag, London, UK, 2013), pp. 11–71
2. C.G.S. Pillai, A.M. George, An improved comparative thermal conductivity apparatus for measurements at high temperatures. Int. J. Thermophys. **12**(3), 563–576 (1991)
3. W.J. Parker, R.J. Jenkins, C.P. Butler, G.L. Abbott, Flash method of determining thermal diffusivity, heat capacity and thermal conductivity. J. Appl. Phys. **32**, 1679 (1961)
4. N. Manoj, D. Jain, J.K. Gautam, K.C. Thomas, V. Sudarsan, C.G.S. Pillai, R.K. Vatsa, A.K. Tyagi, A simple, reliable, cost effective, and high temperature dilatometer for bulk thermal expansion studies on solids. Measurement **92**, 318–325 (2016)
5. C.G.S. Pillai, A.M. George, Thermal conductivity of uranium dioxide. J. Nucl. Mater. **200**, 78–81 (1993)
6. C.G.S. Pillai, P. Raj, Thermal conductivity of ThO_2 and $Th_{0.98}U_{0.02}O_2$. J. Nucl. Mater. **277**, 116–119 (2000)
7. D. Jain, C.G.S. Pillai, B.S. Rao, R.V. Kulkarni, E. Ramdasan, K.C. Sahoo, Thermal diffusivity and thermal conductivity of thoria-lanthana solid solutions up to 10 mol% LaO1.5. J. Nucl. Mater. **353**, 35–41 (2006)
8. A.K. Tyagi, M.D. Mathews, B.R. Ambekar, R. Ramachandran, Thermal expansion of ThO_2-2, 4 and 6 wt.% UO_2 by HT-XRD. Thermochim. Acta **421**(1–2), 69–71 (2004)
9. M.D. Mathews, B.R. Ambekar, A.K. Tyagi, Bulk and lattice thermal expansion of $Th_{1-x}Ce_xO_2$. J. Nucl. Mater. **280**(2), 246–249 (2000)
10. M.D. Mathews, B.R. Ambekar, A.K. Tyagi, Phase relation and bulk thermal expansion studies in thoria-neodia system. J. Alloy. Compd. **386**(1–2), 234–237 (2005)
11. M.D. Mathews, B.R. Ambekar, A.K. Tyagi, Phase relations and linear thermal expansion of cubic solid solutions in the $Th_{1-x}M_xO_{2-x/2}$ (M = Eu, Gd, Dy) systems. J. Nucl. Mater. **341**(1), 19–24 (2005)
12. A.K. Tyagi, M.D. Mathews, R. Ramachandran, Solubility limits and bulk thermal expansion of ThO_2:M^{n+} (M = Y^{3+}, Sr^{2+} and Ba^{2+}). J. Nucl. Mater. **294**(1–2), 198–201 (2001)
13. R.D. Purohit, A.K. Tyagi, M.D. Mathews, S. Saha, Combustion synthesis and bulk thermal expansion studies of Ba and Sr thorates. J. Nucl. Mater. **280**, 51–55 (2000)
14. S.R. Bharadwaj, R. Mishra, M. Basu, D. Das, in *Thoria-Based Nuclear Fuels: Thermophysical and Thermodynamic Properties, Fabrication, Reprocessing and Waste Management*, ed. by D. Das, S.R. Bharadwaj (Springer-Verlag, London, UK, 2013), pp. 107–156
15. D. Das, M. Basu, S. Kolay, A.N. Shirsat, *Thoria-Based Nuclear Fuels: Thermophysical and Thermodynamic Properties, Fabrication, Reprocessing and Waste Management*, ed. by D. Das, S.R. Bharadwaj, (Springer-Verlag, London, UK, 2013), pp. 157–204
16. M. (Ali) Basu, R. Mishra, S.R. Bharadwaj, D. Das, Thermodynamic and transport properties of thoria–urania fuel of advanced heavy water reactor. J. Nucl. Mater. **403**, 204–215 (2010)
17. A.N. Shirsat, M. (Ali) Basu, S. Kolay, A. Datta, D. Das, Transport properties of I, Te and Xe in thoria-urania SIMFUEL. J. Nucl. Mater. **392**, 16–21 (2009)
18. A.H. Booth, G.T. Rymer, *Determination of the Diffusion Constant of Fission Xenon in UO_2 Crystals and Sintered Compacts*. CRDC Report (CRDC-720) (1958)
19. S.C. Chetal, P. Chellapandi, P. Puthiyavinayagam, S. Raghupathy, V. Balasubramaniyan, P. Selvaraj, P. Mohankrishnan, B. Raj, Current status of fast reactors and future plans in India. Energy Proceedia **7**, 64–73 (2011)

20. W.J. Carmak, D.L. Porter, Y.I. Chang, S.L. Hayes, M.K. Meyer, D.E. Burkes, C.B. Lee, T. Mizuno, F. Delage, J. Somers, Metallic fuels for advanced reactors. J. Nucl. Mater. **392**, 139–150 (2009)
21. D. Jain, V. Sudarsan, A.K. Tyagi, Thorium-based metallic alloys as nuclear fuels: present status, potential advantages and challenges. SMC Bull. Soc. Mater. Chem. **4**(1), 27–40 (2013)
22. S. Das, R. Kumar, S.B. Roy, A.K. Suri, Developmental study on metallic thorium and uranium-thorium alloy. *BARC Newsletter*, Founder's Day Special Issue, pp. 7 2–76 (2011)
23. S. Das, S. Kaity, R. Kumar, J. Banerjee, S.B. Roy, G.P. Chaudhari, B.S.S. Daniel, Characterization of microstructural, mechanical and thermophysical properties of Th-52U alloy. J. Nucl. Mater. **480**, 223–234 (2016)
24. D. Jain, C.G.S. Pillai, Thermophysical characterization of thorium and uranium based metallic fuels and alloys. ITAS Bull. Indian Therm. Anal. Soc. **4**(2), 10–22 (2011)
25. R.C. Robartson, Conceptual Design Study of A Single-fluid Molten-salt Breeder Reactor. ORNL Report (ORNL-4541) (1971)
26. A. Basak, I.V. Dulera, *Fluoride Salts Requirements for Indian Molten Salt Breeder Reactor*. Tech. Rep. MSBR/RED/HTRS/2014/02, pp. 1–13 (2014)
27. R. Mishra et al., Unpublished Results
28. P.A. Hassan et al., Unpublished Results

Determination of Room Temperature Thermal Conductivity of Thorium—Uranium Alloys

Santanu Das, Santu Kaity, R. Kumar, Joydipta Banerjee, S. B. Roy, G. P. Chaudhari and B. S. S. Daniel

Abstract Room temperature thermal conductivity values of selected composition of Th-U alloys were experimentally measured employing Transient Plane Source (TPS) technique using slab sample geometry. Experimental values were fitted in an empirical model relating room temperature thermal conductivity with alloy composition. The trend has been explained based on actual microstructural features. The room temperature thermal conductivity values of Th-U alloys were found to be superior to that of other prevailing metallic and ceramic fuel candidates.

Keywords Thermal conductivity · Thorium–uranium · Alloy · Microstructures TPS

S. Das (✉)
Integrated Fuel Fabrication Facility, Bhabha Atomic Research Centre, Mumbai, India
e-mail: sadas@barc.gov.in; reachoutsantanu@gmail.com

S. Kaity · J. Banerjee
Radiometallurgy Division, Bhabha Atomic Research Centre, Trombay, Mumbai 400085, India
e-mail: santuk@barc.gov.in

J. Banerjee
e-mail: joydipta@barc.gov.in

R. Kumar · S. B. Roy
Chemical Engineering Group, Bhabha Atomic Research Centre, Mumbai 400085, India
e-mail: rajk@barc.gov.in

S. B. Roy
e-mail: mrsued@barc.gov.in

G. P. Chaudhari · B. S. S. Daniel
Indian Institute of Technology Roorkee, Roorkee, India
e-mail: chaudfmt@iitr.ernet.in

B. S. S. Daniel
e-mail: s4danfmt@iitr.ernet.in

A. K. Nayak and B. R. Sehgal (eds.), *Thorium—Energy for the Future*,
https://doi.org/10.1007/978-981-13-2658-5_20

1 Introduction

Energy is essential for mankind to survive and cater to further development. To have a reasonable human development index (HDI), we would require nearly ten times more number of nuclear reactors even if half of the total required power is generated by nuclear fission [1, 2]. It would certainly require more nuclear fuel to run so many reactors. U^{235} is the only naturally occurring fissile isotope that undergoes fission to produce energy in a nuclear reactor. Other fissile isotopes can be produced in a reactor, e.g. Pu^{239} from U^{238} and U^{233} from Th^{232}. It has become clearer now that mining of uranium for its fissile content is at best can be a temporary solution. This has been indicated that although vast quantities of high-cost fissile uranium can conceivably be considered a source of energy, the cost of utilizing it would be prohibitive [3]. Rather, a more economical solution to the long-term energy supply problem is to produce all future fissile material by irradiating fertile materials in the fuel elements of a nuclear reactor. It was realized early in the history of nuclear energy that U^{235} alone would not be able to cater to the global power requirement. Both U^{238} and Th^{232} are more abundant in nature than U^{235}. More plutonium can be produced by conversion of U^{238} that is consumed. The *breeding ratio* for plutonium in case of fast reactors is higher than that in case of thermal reactors. U^{233} production by conversion from Th^{232}, on the other hand, is possible both in thermal as well as in fast reactors. Thorium (Th^{232}) has an edge over uranium (U^{238}) as far as utilization of a fertile material is concerned. Nevertheless, thorium has received much less research attention compared to that of uranium. Research endeavors on thorium metal and its alloys are even lesser. Metallic uranium and plutonium have certain disadvantages, e.g. anisotropic crystal structure, irradiation growth, swelling, unwanted low melting eutectic formation. On the other hand, thorium has isotropic crystal structure, high melting point and negligible fission gas swelling, etc. Moreover, utilization of thorium offers added advantage of inherent proliferation resistance [4].

At present, nearly half of the yearly requirement of uranium is met by uranium produced (primary source) worldwide while the other half comes from civilian and nuclear stockpiles (secondary source). By 2025, this secondary source is expected to cater only 2–6% of the global requirement [4]. It is imperative that we look around for another source of nuclear fuel to sustain the growth and the importance of thorium is realized here. However, being a fertile metal thorium alone cannot act as a fuel for nuclear reactor. It can be used in nuclear reactor in combination with other fissile nuclides. Thorium metal can be alloyed with a fissile metal, and the alloy can then act as a fuel having fertile and fissile components together. *Natural or enriched uranium metal, if alloyed with thorium, the resulting alloy can possibly be a candidate fuel.* Thorium–uranium alloy fuel will have (i) *at least* some of the advantages of thorium metal (e.g. higher thermal conductivity, isotropic crystal structure, high-temperature use) over uranium (ii) benefits of metallic fuel and (iii) lesser concern for proliferation. *After all, the biggest advantage of thorium–uranium alloy fuel will be that it will be serving as fuel as well as source for further fuel generation.*

Research efforts on metallic thorium and its alloys were initiated during 1950–1970. Interest in thorium waned away because of discoveries of new uranium resources and improvement in its availability. Thorium–uranium alloy fuel has not been explored much except in few cases [5–8]. Lack of a continuing systematic effort to define the limitations of these fuels, efforts to define its microstructures that can result from fabrication and composition variables are prime reasons for this system not having witnessed much development [9]. Recent work on microstructural and mechanical characterization of various compositions of Th-U alloy systems attempts to bridge this gap in research [10–12]. Since our ultimate objective is to find suitable composition(s) of Th-U alloy for a possible reactor application, thermal properties need to be assessed alongside the study of microstructural and mechanical behavior. One of the key thermal properties required for nuclear fuel application is thermal conductivity. Thermal conductivity indicates the level of *power density* (thermal power per unit core volume) and *specific power* (thermal power per unit fuel mass) that can be achieved in a fuel material. Thermal conductivity also influences power level, restructuring and transport processes inside the fuel [13]. A prospective candidate fuel ideally should have higher thermal conductivity compared to that of prevailing fuel materials. The rise in center-line temperature and an excessive temperature gradient from center to surface in fuel material can be avoided in fuel material with higher thermal conductivity.

Focused research pertaining to the aspect of thermal conductivity of Th-U alloys is extremely less. In the present study, room temperature thermal conductivity (λ_{RT}) values of selected Th-U alloys were experimentally determined. Th-U alloys have been so selected from Th-U phase diagram that they cover the complete composition range of the phase diagram. Hence, one can select a composition and can reasonably predict thermal conductivity value for it. Also, an explanation, based on microstructural characteristics, has been attempted for the observed trend of room temperature thermal conductivity of Th-U alloys. An empirical correlation between room temperature thermal conductivity and alloy composition has also been proposed.

2 Experimental

Thorium metal used in this study was produced by calciothermic reduction of thoria (ThO_2). However, this reduction reaction produces thorium powder which, in the present case, was pressed into small pellets and melted in non-consumable arc in inert atmosphere. The arc-melted fingers were used as stock thorium metal. Uranium was used in the form of slugs (12 mm diameter, ~10 mm height). Required amounts of thorium and uranium were arc melted to produce thorium–uranium alloys of desired compositions. Thorium and thorium–uranium alloy fingers were jacketed inside copper tube in inert atmosphere and hot rolled at 850 °C using a small hand operated rolling set-up. Rolled metal and alloys were then de-jacketed and polished on all sides. Two identical square slabs of length ~25 mm and thickness ~1.5 mm were required for each composition. Specimens of required dimension were prepared using low-speed cut-off machine fitted with SiC cutting wheel. Details of thorium powder, its processing, purity, alloy compositions, characterizations of metal/alloy, etc., have been dealt elsewhere [11].

Thermal conductivity of metal and alloys at room temperature was determined by Transient Plane Source (TPS) technique using slab sample geometry [14–16]. A sheet of material with a definite thickness is referred to as slab. The samples were polished to ensure good thermal contact with the TPS sensor and to minimize contact resistance at its surface. Nickel spiral sensor (radius 3.189 mm) with Kapton insulation was placed between two slab samples. The TPS unit was calibrated using Inconel 600 standard sample. TPS technique works on a specially designed Wheatstone bridge with TPS sensor as one of the arms. A Keithley 2400 source meter supplies a constant voltage across the bridge. The bridge is balanced automatically before the measurement. It becomes increasingly unbalanced with the increase in resistance of the sensor. A Keithley 2000 digital voltmeter fitted with a scanner or multiplex card, records the unbalanced voltage. Using these recorded voltages, it is possible to determine the increase in temperature of the sensor and consequently thermal conductivity of the sample. Similar experiments were also carried out with pure uranium and thorium metal slabs to compare the thermal conductivity data.

3 Results and Discussion

In the past, Th-U alloy system has been explored only to a limited extent. As a consequence, published information on microstructural features of Th-U alloys and its influence on various physical properties are also scarce. Systematic study on microstructural features of various Th-U alloy compositions has been made recently [10–12]. Thermal conductivity of a binary alloy of two actinides is a complex subject. Here, we have also attempted to establish a simple correlation of room temperature thermal conductivity with the gross microstructural features of Th-U alloys. Hence, it would be prudent to discuss microstructural features of various Th-U alloy compositions preceding the discussion on its thermal conductivity.

3.1 Microstructural Features

Th-U phase diagram [17] indicates that these two metals have very limited mutual solid solubility. Hence, resultant microstructures are mostly biphasic. Details of investigation on as-cast microstructures have been dealt elsewhere [11]. For the present purpose, only basic features of such microstructures will be discussed. Hereafter, Th-x wt% U alloy will be described as Th-xU throughout the paper. Th-U alloys are divided into three categories for ease of discussion, viz. (i) high thorium, (ii) intermediate and (iii) high uranium Th-U alloys. High thorium represents Th-3U and Th-7U while high uranium represents Th-80U and Th-96U alloys. The interme-

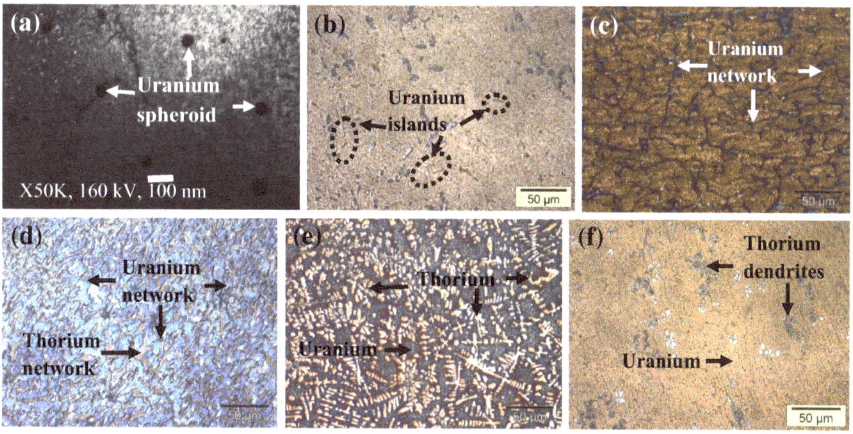

Fig. 1 **a** Transmission electron micrograph of microstructure of Th-3U alloy. Optical micrographs of various Th-U alloys, hot rolled at 850 °C; **b** Th-7U, **c** Th-30U, **d** Th-52U, **e** Th-80U and **f** Th-96U

diate compositions correspond to Th-30U and Th-52U and represent purely biphasic microstructures. Representative microstructures are presented in Fig. 1.

High thorium alloys are basically continuous matrix of thorium having some dispersion like presence of uranium in it. In Th-3U, optical microscopy cannot reveal the presence of uranium. Transmission electron microscopy was used to capture spheroid like shapes (Fig. 1a) of uranium in thorium-rich matrix. These spheroids are typically 50–70 nm in size. The overall microstructure resembles natural dispersion of uranium spheroids in thorium-rich matrix. Alloy microstructure of Th-3U and its various other aspects have been dealt elsewhere elaborately [10]. The microstructure of Th-7U (Fig. 1b) has two types of uranium shapes in thorium-rich matrix, viz. (i) irregular-shaped islands (2–5 μm) and (ii) spheroids/ellipsoids of uranium.

The intermediate compositions (Th-30U, Th-52U) produce quite different microstructures. Well-developed and obvious grain structures with uranium present in the grain boundaries typify the microstructures of these alloys. Essentially, these microstructures consist of interpenetrating, continuous and fully developed networks of uranium and thorium-rich phases. It is expected that heat flow will face alternate networks of two different phases having different thermal conductivity.

The microstructures (Fig. 1e, f) of high uranium Th-U alloy composition are composed of (majority) uranium and small amount of thorium. Thorium-rich dendrites are distributed in uranium-rich matrix in Th-80U alloy (Fig. 1e). These dendrites typically look like herringbone. Th-96U is eutectic composition, and hence, eutectic reaction product occupies major portion of the microstructure. However, fine eutectic product can be observed in higher magnification [11]. Thorium is present here as un-connected dendrites which are dispersed throughout the uranium matrix. Higher thermal conductivity of thorium, hence, does not help in obtaining higher thermal conductivity of the alloy. The thermal conductivity of the high uranium alloys, hence,

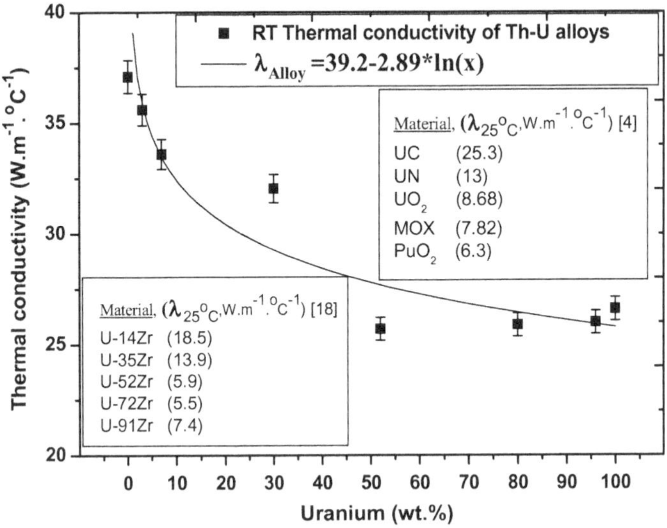

Fig. 2 Room temperature thermal conductivity values of Th-U alloys are shown against uranium content. Model fitted to the experimental data is also shown in the figure. Room temperature thermal conductivity of few oxide, non-oxide ceramic and alloy fuel candidates are also provided for comparison

is dominated by that of uranium. It can be viewed as if the thermal conductivity of uranium is decreasing, albeit very gently, with increasing addition of thorium.

3.2 Thermal Conductivity

Experimentally determined thermal conductivity values are plotted against the uranium content (wt%) in Fig. 2. Room temperature thermal conductivity values of many fuel materials are also shown in the graph for ease of comparison. It can be seen that thermal conductivity values are decreasing with increase in uranium content up to ~50 wt% U. There is a steep decrease in the thermal conductivity of thorium on small addition of uranium (<7 wt% U). Least among the experimentally determined thermal conductivity values of the selected alloy compositions were observed in Th-52U. With further addition of uranium, thermal conductivity increases with increase in uranium content. However, the increment is not very sharp and rather gentle.

The room temperature thermal conductivity (λ_{RT}) of high thorium alloys is expected to be close to that of unalloyed thorium as presence of uranium and its contribution are insignificant. Similarly, λ_{RT} of high uranium Th-U alloys will be more influenced by the presence of large amount of uranium-rich phase in the microstrcture. The overall trend finds similarity with the results in other alloys and compounds used in nuclear context. In U-Zr system, lowest thermal conductivity is observed at

~70 at.% Zr addition, and it is attributed to δ-phase formation [18]. However, in Th-U system existence of a similar phase formation is not reported in the literature. In as-cast as well as in hot-rolled microstructures of Th-52U, two fully developed networks of thorium and uranium are seen. Volume occupancy of thorium and uranium in Th-52U alloy composition is ~0.4 and ~0.6 for uranium ($\rho = $ ~19.1 g cc^{-1}) and thorium ($\rho = $ ~11.6 g cc^{-1}), respectively. A rule-of-mixture estimate predicts the thermal conductivity to be ~32 W m^{-1} $^{o}C^{-1}$ against experimentally determined value of ~26 W m^{-1} $^{o}C^{-1}$. At the cross section of this alloy specimen, area fractions occupied by two phases will be numerically same to their respective volume fractions. However, heat will not be conducted between two opposite faces, by any of these two phases, in a straightforward manner. The networks are woven and interpenetrating type. Heat flow will be *interfered* at every interface it comes across. The flow of any physical quantity, if hindered at interfaces inside the material, suffers some kind of scatter and losses. Uranium-rich and thorium-rich networks have different thermal conductivities. It is expected that heat flow gets substantially interrupted at the interfaces of two materials having alternately high and low thermal conductivities. The material having lower value will control the heat flow to the next interface. A biphasic alloy having such a configuration is expected to have lower values of thermal conductivity. This can possibly explain the observation at macroscale.

3.3 Schematic Presentation of Microstructure and Empirical Correlation

A schematic presented below (Fig. 3) can help us visualize the above conditions in the microstructures. Thorium metal in unalloyed condition serves as continuous matrix (Fig. 3a). If heat is applied to one surface, it transmits through the continuous path to reach the opposite surface in a unidirectional flow. When uranium is added to thorium and the former is present as a second phase with lower thermal conductivity (Figs. 1a, b, 3b), situation is no more the same as before. The heat flow has to encounter the second phase having lower thermal conductivity. Overall, thermal conductivity will be reduced because of unavailability of the continuous thorium matrix. Every interface created because of the presence of uranium particle or islands will hinder and reduce the flow of heat which in turn will reduce the overall thermal conductivity. Spheroids of uranium will create an interfacial region which will be equal to their surface area. Hence, spheroids will reduce the heat flow in proportion to their total surface area. Islands too will influence similarly. With increasing addition of uranium, formation of continuous network of uranium is favoured (Figs. 1c, d, 3c). The flow of heat is hindered maximum in the alloy having maximum interfaces inside. Toward high uranium side of the phase diagram, the microstructures are occupied more by uranium-rich phase, and hence, the overall thermal conductivity of alloy is more influenced by that of uranium (Figs. 1e, f, 3d). However, the interface created by the presence of thorium still hinders the heat flow in a similar manner till the

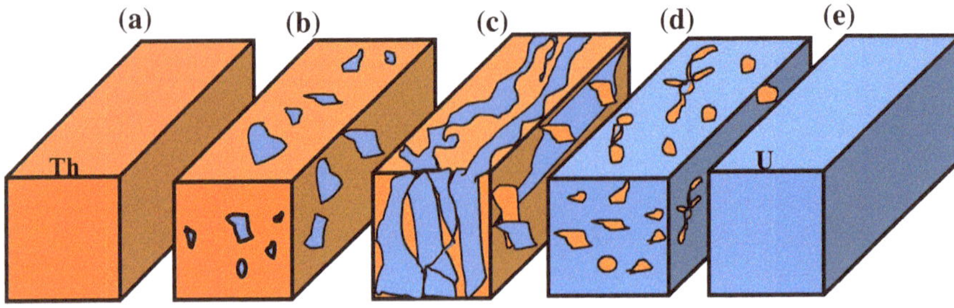

Fig. 3 A simple schematic is presented to explain the relation of thermal conductivity of Th-U alloys and its microstructural features. Microstructural features are represented as follows: **a** unalloyed thorium, **b** high thorium Th-U alloys, **c** Th-52U alloy, **d** high uranium Th-U alloys and **e** unalloyed uranium

microstructure is fully occupied by uranium for unalloyed uranium (Fig. 3e). The nature of decrease in the experimental values with weight percentage of uranium (x) is that of a natural logarithm.

Research works related to assessment of thermal conductivity in two component heterogeneous systems [19], steels [20], uranium dioxide using molecular dynamics [21], dispersion fuel [22] are available. Not much information, except that reported by Bentle [6], on variation in thermal conductivity with composition of Th-U alloys is available. Prediction of effective thermal conductivity in metallic alloys and composites is a complex subject and involves many microscopic as well as macroscopic phenomena [23–25]. A large number of empirical relations are available based on simplified models to deal with presence of second phase, impurities, grain size and size distribution, grain orientation, shape and size of second phase, dispersions, etc. Effective medium theory was developed to deal with such approximations of physical properties of heterogeneous bodies through combination of homogeneous isotropic phases [26]. Rigorous analysis and prediction is not in the scope of this paper. However, in the absence of complete lack of information on this alloy system, an effort has been made to relate thermal conductivity with its composition and microstructural feature.

Nechtelberger [27] has related change in thermal conductivity of ferrite ($\boldsymbol{\lambda}$) in cast iron with that of pure iron ($\boldsymbol{\lambda_o}$) and concentration of solutes present in the material using the expression shown below.

$$\lambda = \lambda_o - \ln \sum x$$

where x is solute concentration expressed in wt%.

We draw a similarity from the above expression in the present proposed model to corelate experimentally determined room temperature thermal conductivity values of Th-U alloys with its microstructural features. In line with the preceding discussion based on the schematic, the proposed model to simulate the physical phenomena inside the alloy is proposed to have a form as below.

$$\lambda_{\text{Alloy}} = \lambda_{\text{Th}} - A_{*} \ln(x)$$

where λ_{Alloy} is overall thermal conductivity of alloy, the term A is associated with interfacial area, λ_{Th} is thermal conductivity of thorium and x is weight percent of uranium in the alloy. The model fitted in the experimental data represents the trend seen and has the data scatter on both the sides of the fitted curve (Fig. 2).

Thermal conductivity values of Th-U alloys containing 3, 5, 8, 10, 12, 18 and 20 wt% U, in the temperature range 20–1000 °C, are calculated [5]. In this, it was shown that addition of uranium has resulted in decreases in thermal conductivity of thorium. For all the alloys up to Th-20 wt% U, thermal conductivity values have increased with temperature. However, no information is available for Th-U alloys containing more than 20 wt% U.

4 Summary and Conclusions

- Room temperature thermal conductivity of thorium decreases with addition of uranium up to ~50 wt% U.
- Least value was observed in Th-52U having interpenetrating, interwoven, continuous networks of thorium and uranium-rich phases.
- Room temperature thermal conductivity $\left(\lambda_{RT,Th-Ualloy}\right)$ remains nearly similar to that of unalloyed uranium beyond ~50 wt% uranium in Th-U alloys.
- The room temperature thermal conductivity of thorium-rich Th-U alloys remains higher than most of the metallic fuels.

High thorium Th-U alloys, with its dispersion like microstructure and fairly high overall room temperature thermal conductivity, can be considered for specific applications requiring dispersion type fuel material. The thorium-rich continuous phase will serve as a conducting matrix to transfer the heat fast away from the fission occurring in the uranium spheroids. However, final application depends on many other criteria including processing, fabrication, in-pile performance, reprocessing aspects.

Acknowledgements Authors sincerely acknowledge the help received from Shri Suman Neogy, MSD, BARC for TEM; Shri Dulal Mandal, Shri G. D. Vaidya, Shri N. K. Vernekar, Shri D. B. S. Bhandari, UED, BARC for their assistance in conducting experiments.

References

1. World Nuclear Association, [Online] http://www.world-nuclear.org/info/Facts-and-Figures/World-Nuclear-Power-Reactors-and-Uranium-Requirements/
2. United Nations Development Programme, Human development report (1999)
3. A.J. Lane, Economic incentive for thorium reactor development, in *Thorium Utilization in Power Reactors* (Vienna, 1965)
4. IAEA, *Thorium Fuel Cycle-Potential Benefits and Challenges, TECDOC 1450* (IAEA, 2005)
5. G.L. Copeland, *Evaluation of Thorium-Uranium Alloys for the Unclad-Metal Breeder Reactor, ORNL-4557* (USAEC, 1970)
6. G.G. Bentle, A physical metallurgical study of thorium rich thorium-uranium alloys, NAA-SR-2069, in *Evaluation of Thorium Uranium-Alloys for Unclad-Metal Breeder Reactor*, ed. by G.L. Copeland (ORNL, 1970, USAEC, 1958)
7. G.H. Bannister, R.C. Burnett, J.R. Murray, Ageing and hot hardness characteristics of certain thorium alloys. J. Nucl. Mater. **2**(1), 51 (1960)
8. B.R. Hayward, P. Corzine, Thorium-uranium fuel elements for SRE, in *Second Int'l Conference on the Peaceful Uses of Atomic Energy* (Geneva, 1958) p. P/785
9. The use of thorium in nuclear power reactors, WASH 1097 (Department of Reactor Development Technology, USAEC, 1969)
10. S. Das, R. Kumar, S. Kaity, S. Neogy, K.N. Hareendran, S.B. Roy, B.S.S. Daniel, G.P. Chaudhari, Characterization of microstructural, mechanical and thermal properties and ageing study of Th–3 wt.% U alloy. Nucl. Eng. Des. **282**, 116–125 (2015)
11. S. Das, S.B. Roy, G.P. Chaudhari, B.S.S. Daniel, Microstructural evolution of as-cast Th-U alloys. Prog. Nucl. Energy **88**, 285–296 (2016)
12. S. Das, S. Kaity, R. Kumar, J. Banerjee, S.B. Roy, G.P. Chaudhari, B.S.S. Daniel, Characterization of microstructural, mechanical and thermophysical properties of Th-52U alloy. J. Nucl. Mater. **480**, 223–234 (2016)
13. B. Raj, An overview of R&D of fast reactor fuel cycle. Int. J. Nucl. Energy Sci. Technol. **1**(2/3), 164 (2005)
14. S.E. Gustafsson. Rev. Sci. Instrum **62**(3), 797 (1991)
15. S.E. Gustafsson, International Patent Application No. PCT/SE89/00137
16. Hot Disk Thermal Constants Analyser, Instruction Manual, Software version 5.9, Hot Disk (Sweden, 2007)
17. D.E. Peterson, The Th-U (thorium-uranium) system. Bull. Alloy Ph. Diagr. **6**, 443 (1985)
18. Y. Takahashi, M. Yamawaki, Y. Yamamoto, Thermophysical properties of uranium-zirconium alloys. J. Nucl. Mater. **154**, 141 (1988)
19. R.L. Hamilton, O.K. Crosser, Thermal conductivity of heterogeneous two-componenet systems. Ind. Eng. Chem. Fundam. **1**, 187 (1962)
20. M.J. Peet, H.S. Hasan, H.K.D.H. Bhadeshia, Prediction of thermal conductivity of steels. Int. J. Heat Mass Transf. **54**, 2602 (2011)
21. K. Yamada, K. Kurosaki, M. Uno, S. Yamanka, Evaluation of thermal properties of uranium dioxide by molecular dynamics. J. Alloys Compd. **307**, 10 (2000)
22. H.J. Ryu, Y.S. Kim, J.M. Park, H.T. Chae, C.K. Kim, Performance evaluation of U-Mo/Al dispersion fuel by considering a fuel-matrix interaction. Nucl. Eng. Technol. **40**, 409 (2008)
23. K. Almadhoni, S. Khan, Evaluation of the effective thermal properties of aluminum metal matrix composites reinforced by ceramic particles. Int. J. Curr. Eng. Technol. **5**(4) (2015)
24. Bin Wan, K. Yue, L. Zheng, X. Zhang, The effective thermal conductivity of composite materials with spherical. Adv. Mater. Res. **239–242**(1662–8985), 1870–1874 (2011)
25. J.K. Chen, S.F. Chen, On thermal conductivity of an in-situ, in *Metal, Ceramic and Polymeric Composites for Various Uses,* ch. 10, ed. by Dr. J. Cuppoletti (INTECH, 2011)
26. T.C. Choy, *Effective Medium Theory, Principles and Applications* (Oxford Science Publications, New York, USA, 1999)
27. E. Nechtelberger, *The Properties of Cast Irons up to 500 °C*. Technical report (Technicopy Ltd.,1980)

Thermo-physical Properties of ThC and ThN from First Principles

Vinayak Mishra and Shashank Chaturvedi

Abstract Thorium carbide and nitride are potential candidates for their use as fuel materials in fast breeder reactors. Therefore, knowledge of their thermo-physical properties at high temperatures is necessary. In this paper, we present results of the first-principle calculations of properties such as specific heat, volume and bulk modulus at high temperatures. The all-electron FPLAPW method has been used with LDA, GGA, LDA+U and GGA+U exchange–correlation potentials for performing these calculations. It has been found that GGA gives the most accurate ground state properties such as lattice constant, bulk modulus and pressure derivative of the bulk modulus. Therefore, first principles results obtained from GGA calculations have been combined with quasi-harmonic approximations to calculate the thermo-physical properties. Calculated thermo-physical properties are in reasonably good agreement with published experimental results.

Keywords Thorium carbide · Thorium nitride · Specific heat
High-temperature bulk modulus · FPLAPW

1 Introduction

The carbides and nitrides of thorium possess outstanding physical properties such as high melting point, high thermal conductivity, high density and good compatibility with the coolant (liquid Na) [1, 2]. Therefore, these materials are potential candidates for their utilization in fast breeder reactors. In order to use these materials as advanced fuels, it is very important to know their thermo-physical properties at

V. Mishra (✉)
Computational Analysis Division, Bhabha Atomic Research Centre,
Visakhapatnam 530012, AP, India
e-mail: vinayakm@barc.gov.in

S. Chaturvedi
Institute for Plasma Research, Bhat, Gandhinagar 382428, Gujarat, India
e-mail: shashank@ipr.res.in

A. K. Nayak and B. R. Sehgal (eds.), *Thorium—Energy for the Future*,
https://doi.org/10.1007/978-981-13-2658-5_21

high temperatures, which are required for modelling fuel pin behaviour. Only a few authors have attempted theoretical studies of thermo-physical properties of these materials at high temperature. Their studies of these properties are based either on pseudo-potentials or on molecular dynamics with empirical potentials. In the present work, we report results of a few useful properties of ThN and ThC, calculated from all-electron FPLAPW+QHA method.

2 Computational Details

The calculations have been performed within the framework of density functional theory employing full-potential linearized augmented plane wave (FPLAPW) method, as implemented in the WIEN 2k code [3]. Exchange–correlation potentials based on LDA and GGA both have been used. Perdew–Wang parameterization [4] of LDA and Perdew–Burke–Ernzerhof parameterization [5] of GGA have been used. The basis function is expanded up to $R_{MT}\ K_{max} = 9$, where R_{MT} is the muffin tin radius and K_{max} is the plane wave cut-off parameter. The R_{MT} values were selected as 2.2 for Th and 1.6 for C and N. The self-consistent cycles were run until the energy convergence criterion of 10^{-5} Ry was reached, and 5000 k-points were taken in the calculations.

Thermal effects have been included using quasi-harmonic approximation. The (p, T) equilibrium states are obtained by minimizing Gibbs free energy (G), with respect to V. where G can be expressed as

$$G(V; p, T) = E(\mathrm{V}) + p\mathrm{V} + A_{\mathrm{vib}}(\Theta(\mathrm{V}); T) \tag{1}$$

The contribution A_{vib} can be calculated by using quasi-harmonic Debye model.

$$A_{\mathrm{vib}}(\Theta(\mathrm{V}); T) = nkT\left[9\Theta/8T + 3\ln\left(1 - e^{(-\Theta/T)}\right) - D(\Theta/T)\right] \tag{2}$$

where n is the number of atoms per formula unit, k is the Boltzmann's constant, Θ is the Debye temperature, and $D(\Theta/T)$ is the Debye integral. Once the equilibrium state for a given (p, T) has been obtained, then the required thermodynamic properties can be determined by using the corresponding volume in the appropriate thermodynamic equations.

3 Results and Discussion

3.1 *Ground State Properties Using LDA, GGA, LDA+U and GGA+U*

We first performed calculations of the total energy as a function of volume using the FPLAPW method. Next, we fitted the calculated *E-V* data with Birch–Murnaghan equation of state, in order to obtain lattice constant (a), bulk modulus (B) and pressure derivative of the bulk modulus (B_P) at the ground state. The results of lattice constant a, bulk modulus B and pressure derivative of the bulk modulus B_P for ThN are shown in Figs. 1, 2. The experimental values of these ground state properties are also shown for the comparison. In Figs. 1 and 2, the results for U = 0.0 eV are the results corresponding to pure LDA/GGA calculations. The importance of the results for nonzero values of U is described in the subsequent paragraphs of this section. As can be seen in the figures (Figs. 1, 2), in the case of ThN, a, B and B_P results obtained from the BM fitting of our GGA (LDA) calculations are 5.178 (5.1055) A, 174.6 (200.2) GPa and 4.11 (3.32), respectively (for U = 0.0). The experimental results of these properties taken from the literature [6, 7] are: a = 5.16 A, B = 175 GPa and B_P = 4.0. In the case of ThC, our GGA (LDA) calculations predict the a, B and B_P as 5.355 (5.268) A, 131.9 (141.4) GPa and 3.0 (2.8), respectively. The published experimental results, a and B of ThC, are 5.344 A and 125 GPa, respectively [1]. Thus, LDA tends to over-bind; as a result, LDA lattice constants are too small and bulk moduli are too large. Though GGA over-corrects the lattice constant, the GGA results are in better agreement as compared to LDA.

In order to check whether the results could be further improved by adding the extra orbital-dependent onsite Coulomb repulsion, we have performed LDA+U and GGA+U calculations for the ground state properties of these materials. Here, U is the Hubbard parameter. It is well known that LDA+U/GGA+U calculations improve the results in the case of actinides and their compounds for which conventional first-principle exchange–correlation potentials LDA and GGA fail to predict accurate physical properties. However, LDA+U/GGA+U calculations are not fully first principles as one needs to tune the Hubbard parameter, U parameter. The tuned values of U for other actinides (e.g. uranium and plutonium) are reported to be approximately 4–5 eV in the literature. Therefore, for these thorium compounds also, we have varied the U value from 0.5 to 5.0 eV in the steps of 0.5 eV and have calculated the lattice constant, bulk modulus and pressure derivative of the bulk modulus. Although we have done these calculations for both materials, we have shown the results only for ThN as the trends of the results for ThC were similar.

We can see in Fig. 1 that similar to LDA/GGA calculations, LDA+U underestimates the lattice constant with respect to the experiment due to the over-binding of the compound and GGA+U overestimates due to the under-binding. Even after tuning the Hubbard U parameter 'ON', the same trend is maintained. In both cases LDA+U/GGA+U, the calculated lattice constant monotonically increases with increasing U parameter. The GGA lattice constant (without U) was already slightly

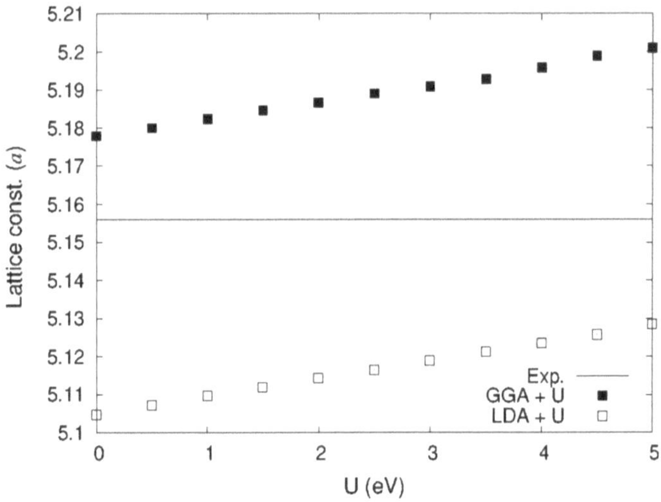

Fig. 1 Lattice constant of ThN as a function of Hubbard parameter U with LDA+U and GGA+U calculations. Unit of lattice constant is angstrom. Experimental result has been taken from Ref. [6]

larger than the experimental lattice constant, and the inclusion of extra Coulomb potential makes it even more large. Hence, GGA+U calculations do not improve the lattice constants. The equilibrium volumes per formula unit (V), calculated with LDA+U/GGA+U, should show similar trends; therefore, V versus U results are not shown.

Lattice constant calculated with LDA+U approaches nearer to the experimental lattice constant with increasing U. However, as can be seen in Fig. 2, the bulk modulus calculated using LDA+U increases with increasing U, which was already (with U = 0 eV or pure LDA calculations) much larger than the experimental bulk modulus. Previously mentioned over-binding effect can be seen in the results of the bulk modulus also in Fig. 2 as the LDA+U bulk modulus is always larger than the GGA+U results. Hence, though LDA+U improve the lattice constant, it fails to improve the bulk modulus. GGA+U bulk modulus does not show any definite trend with increasing U, and it always remains near to the experimental value. But we have already seen that GGA+U failed in the correct prediction of the lattice constant. Similarly, we can see in Fig. 3 that the LDA+U and GGA+U both fail to make any improvement in the results of the pressure derivative of the bulk modulus. Hence, LDA+U/GGA+U calculations do not make any improvement in the ground state properties of ThN. LDA+U and GGA+U are used to handle the strong correlation of f electrons. However in thorium, the $4f$ electrons belong to the core and $5f$ electrons do not contribute to chemical bonding. Due to these reasons, GGA is sufficient for handling the correlation effects and extra 'U' term is not required. Therefore, in our subsequent calculations we have used GGA for the exchange and correlation.

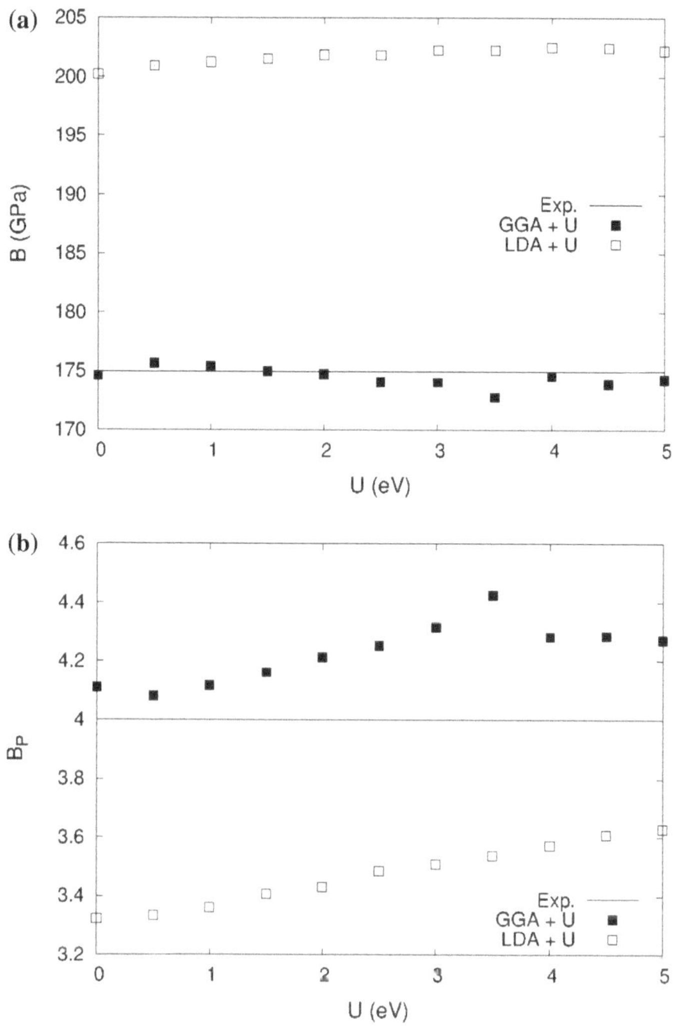

Fig. 2 **a** Bulk modulus of ThN as a function of Hubbard parameter U with LDA+U and GGA+U calculations. Experimental result has been taken from Ref. [7], and **b** derivative of the bulk modulus of ThN as a function of Hubbard parameter U with LDA+U and GGA+U calculations. Experimental result has been taken from Ref. [7]

3.2 *Pressure-Induced Phase Transformation*

The materials with B_1 phase can transform to B_2 phase on applying pressure. Such transformations may restrict the use of the materials as they lead to a sudden change in the material properties. In order to test the stability of the B_1 phase of ThC and ThN in the compression and expansion, we have calculated the volume-dependent

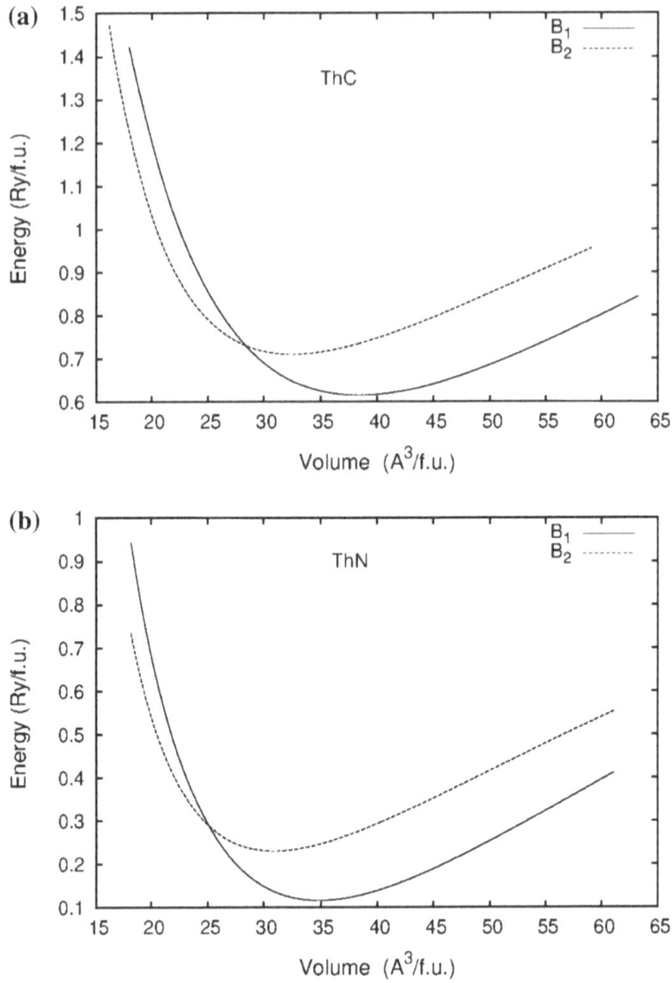

Fig. 3 **a** Comparison of total energies of B_1 and B_2 phases of ThC. B_1 phase is stable near normal density and in the expansion. B_2 phase becomes stable at compression and **b** comparison of total energies of B_1 and B_2 phases of ThC and ThN. B_1 phase is stable near normal density and in the expansion. B_2 phase becomes stable at compression

total energies of their B_2 phase also. The total energies of these two structures have been compared in Fig. 3. We can see that at equilibrium and expanded volumes the B_1 phase of both material is stable. However, at compressed volumes, 28.4 A^3 for ThC and 25.1 A^3 for ThN, the total energy of B_2 phase becomes smaller than the total energy of their B_1 phase. This indicates a phase transformation on compressions up to these volumes.

In order to find pressures at which this B_1–B_2 phase transition takes place, we have compared their Gibbs free energies at various pressures. The Gibbs free energy versus P curves of these materials shows that for ThC this transition takes place at 42 GPa and for ThN this transition takes place at 75.5 GPa. Hence, these materials can be used without any problem up to these pressures.

3.3 Volume, Bulk Modulus and Specific Heats at High Temperatures

As these materials are used at very high temperatures, it is necessary to understand the high-temperature behaviour of the bulk modulus of these materials. Equilibrium volumes are obtained for each $(p - T)$ by minimizing the Gibbs free energy as described in the section computational details. Using the $(p - \mathrm{V})$ points for each T in the following formula, one can calculate the bulk modulus at high temperatures.

$$B_T = -\mathrm{V}\left(\frac{\partial p}{\partial \mathrm{V}}\right)_T \qquad (3)$$

The evolution of the volume per formula unit of ThC and ThN with temperature is reported in Fig. 4 for zero pressure. We can observe in Fig. 4 that the volume per formula unit increases with temperature, as expected.

The temperature evolution of the bulk modulus has also been shown in the bottom part of Fig. 5. Bulk modulus decreases with the temperature. We have fitted the temperature evolution of the bulk modulus with the Wachtmann formula [8]:

$$B_{\mathrm{T}} = B_0 - b_1 T\, e^{(-T_0/T)} \qquad (4)$$

where B_0 is the bulk modulus at absolute zero, T is the temperature, and b_1 and T_0 are fitting parameters. At very high temperature, the temperature dependence becomes linear according to this formula which is consistent with the obtained bulk modulus versus temperature data. We can see that the fitting is excellent for ThC. For ThN, the fitting is slightly deviated. The fitting parameters are b_1 =0.0154 GPa/K, T_0 =118.1 K for ThC and b_1 =0.0363 Gpa/K, T_0 =46.5 K for ThN. Temperature-dependent bulk modulus is not available in the literature; therefore, our results can serve as a prediction.

Heat capacity C_{p} is important as it is one of the thermodynamic functions. Also, it is essential for the calculation of thermal conductivity from thermal diffusivity. The heat capacity (C_{p}) can be calculated by using the following equation

$$C_{\mathrm{p}} = C_{\mathrm{V}}(1 + \alpha\gamma T) \qquad (5)$$

Here, α is the thermal expansion coefficient, γ is the Gruneisen parameter, and C_{V} is the specific heat at constant volume which can be calculated by using the following formula

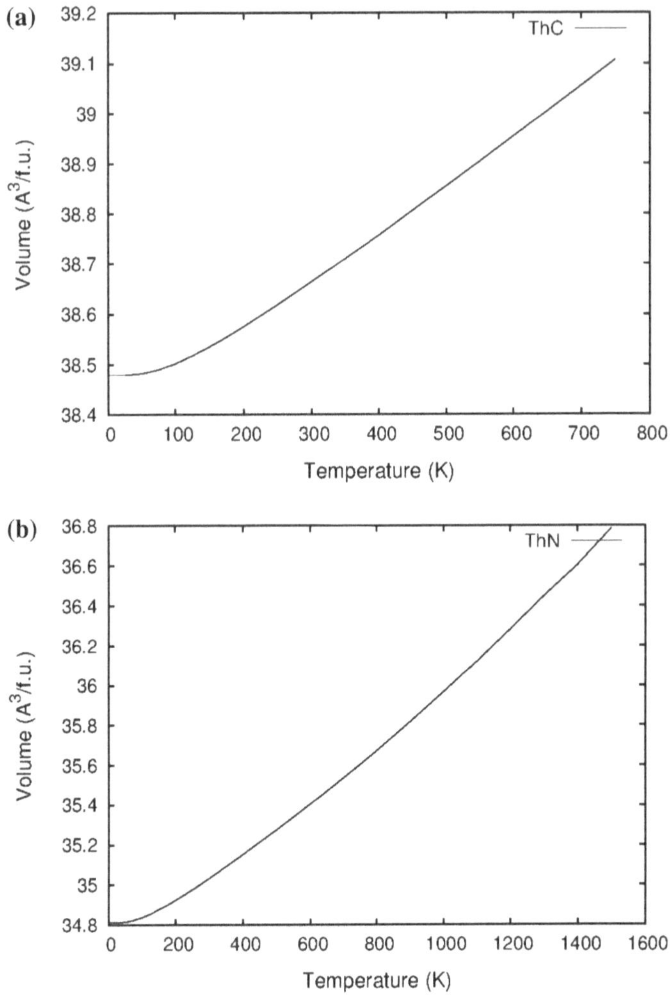

Fig. 4 **a** Variation of volume per formula unit of ThC with temperature and **b** variation of volume per formula unit of ThN with temperature

$$C_V = 3nk\left[4D(\Theta/T) - \frac{3(\Theta/T)}{e^{(\Theta/T)} - 1}\right] \tag{6}$$

Figure 6 shows the calculated specific heats at constant pressure and volume, respectively, for zero pressure. The temperature dependence of C_V and C_p follows a T^3 law at low temperatures. As is usual, C_V tends to the $3R$ limit for higher values of

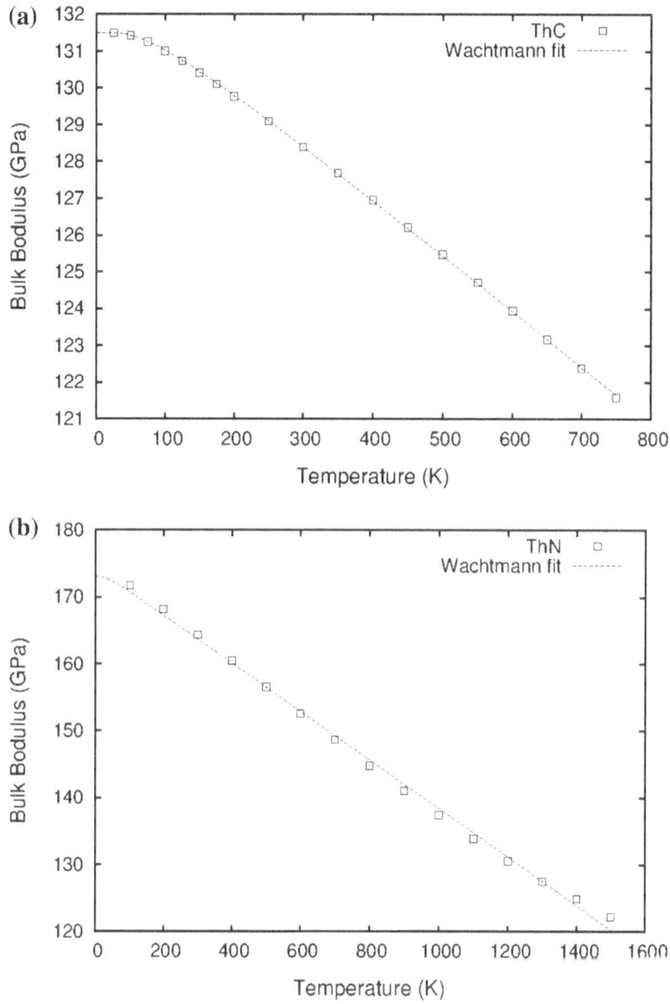

Fig. 5 **a** Variation of the bulk modulus of ThC with temperature and **b** variation of the bulk modulus of ThN with temperature

T, where *R* is the gas constant. For the temperatures larger than 300 K, C_p shows a nearly linear dependence with temperature. Our calculated values are compared with the experimental data from [9–11]. We can see reasonably good agreement between the calculated and experimental results.

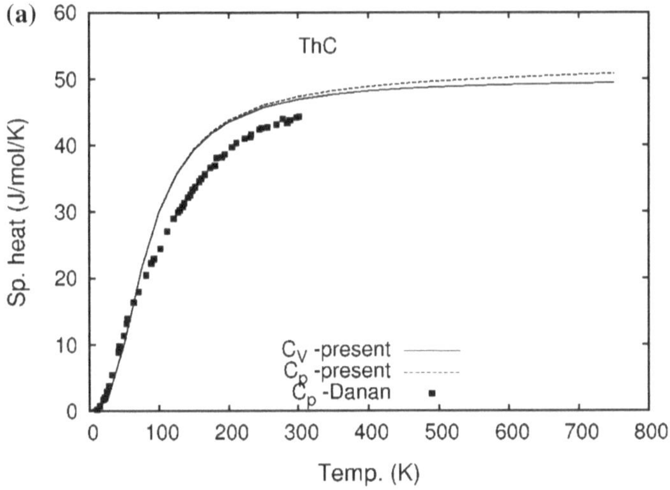

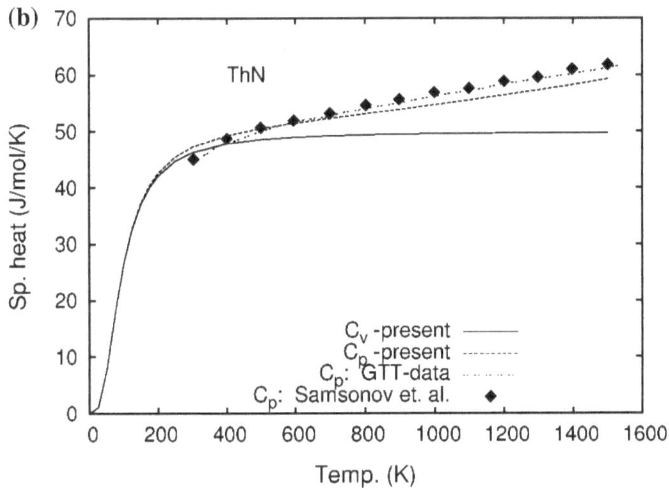

Fig. 6 **a** C_p, and C_V results of ThC are from present calculations, and experimental results are from Danan [9]; **b** C_p, and C_V results of ThN are from present calculations, and experimental results are from Samsonov and Vanitskii [10], GTT-data [11]

4 Conclusion

In summary, we have performed first-principle calculations of the ground state properties such as a, B and B_P, of ThC and ThN, using LDA, GGA, LDA+U and GGA+U. We have found that GGA gives more accurate ground state properties. LDA+U and GGA+U do not make any improvements in the results. Therefore, for calculations of other properties we have used only GGA. We have also tested the energetic stability of the B_1 phase, which is the ground state phase of both compounds, as compared to their B_2 phase, and have found that at high pressure B_2 phase becomes stable. The phase transition pressures are 42 and 75.5 GPa for ThC and ThN, respectively. We have also calculated the temperature-dependent thermo-physical properties such as volume, bulk modulus and specific heats. Our results of specific heat C_p are in reasonably good agreement with the available experimental data.

References

1. H. Kleykamp, Thorium carbides, in *Gmelin Handbook of Inorganic and Organometallic Chemistry*, vol. C3, Eighth edn. (Thorium Supplement, Springer, Berlin, 1987)
2. R. Benz, A. Naoumidis, Thorium compounds with nitrogen, in *Gmelin Handbook of Inorganic Chemistry*, vol. C6, Eighth edn. (Thorium Supplement, Springer, Berlin, 1992)
3. P. Blaha, K. Schwarz, G.K.H. Madsen, D. Kvasnika, J. Luitz. *WIEN 2k-Userguide: An Augmented Plane Wave Plus Local Orbital Program for Calculating Crystal Properties* (University of Technology, Vienna, Austria, 2014). Available from http://www.wien2k.at/reguser/textbooks/usersguide.pdf/
4. J.P. Perdew, Y. Wang, Accurate and simple analytic representation of the electron-gas correlation energy. Phys. Rev. B **45**, 13244 (1992)
5. J.P. Perdew, S. Burke, M. Ernzerhof, Generalized gradient approximation made simple. Phys. Rev. Lett. **77**, 3865 (1996)
6. F.A. Wedgwood, Actinide chalcogenides and pnictides. III. Optical-phonon frequency determination in UX and ThX compounds by neutron scattering. J. Phys. C: Solid State Phys. **7**, 3203 (1974)
7. L. Gerward, J.S. Olsen, U. Benedict, J.P. Itié, J.C. Spirlet, The crystal structure and the equation of state of thorium nitride for pressures up to 47 GPa. J. Appl. Cryst. **18**, 339 (1985)
8. J.B. Wachtman Jr., W.E. Tefft, D.G. Lam Jr., C.S. Apstein, Exponential temperature dependence of young's modulus for several oxides. Phys. Rev. **122**, 1754 (1961)
9. J. Danan, Chaleur specifique de 2 a 300 K du monocarbure de thorium (French), specific heat of thorium monocarbide from 2 to 300 K (English translation). J. Nucl. Mater. **57**, 280 (1975)
10. G.V. Samsonov, I.M. Vanitskii, *Handbook of Refractory Compounds* (Plenum publishing corporation, New York, 1982)
11. The SGTE Pure Substance and Solution Database, GTT-Data Services (1996), (http://gtt.mch.rwth-aachen.de/gtt-web/data/sgte-databases)

Irradiation Behaviour of Thoria-Based Fuels

Prerna Mishra, B. N. Rath, V. P. Jathar, H. N. Singh, P. M. Satheesh, R. S. Shriwastaw, K. M. Pandit, Anil Bhandekar, G. K. Mallik, Sunil Kumar and J. L. Singh

Abstract Different aspects of thorium fuel cycle have been studied for the development of thorium fuel-based Advanced Heavy Water Reactor (AHWR) in India. In order to study the irradiation performance of thoria-based mixed oxide (MOX) fuels, short-length fuel pins assembled in two types of fuel clusters were irradiated in the pressurized water loop (PWL) of CIRUS reactor. Post irradiation examination (PIE) of the fuel pins from the two fuel clusters was carried out at the hot cells facility of BARC. Non-destructive examinations of the fuel pins from the cluster included visual examination, leak testing, fuel pin diameter measurement, ultrasonic testing, eddy current testing, gamma scanning and gamma spectrometry. Destructive examinations included fission gas analysis and microstructural examination. Microstructural characterization on the fuel samples was carried out using optical microscopy, scanning electron microscopy, β-γ autoradiography and α-autoradiography techniques. Results of PIE show good performance of the fuel up to 18.5 GWd/t(HM) burn-up under irradiation. This paper presents the salient observations of the examinations carried out on the pins of the fuel clusters.

Keywords Thoria · Mixed oxide fuel · Post irradiation examination
Microstructure · Burn-up

1 Introduction

India has about four times more thorium resources than uranium. Utilization of thorium for large-scale energy production is a major goal in our three-stage nuclear power programme [1, 2]. In order to study the irradiation performance of thoria-based MOX fuels, short-length fuel pins assembled in two types of fuel clusters were

P. Mishra (✉) · B. N. Rath · V. P. Jathar · H. N. Singh · P. M. Satheesh · R. S. Shriwastaw
K. M. Pandit · A. Bhandekar · G. K. Mallik · S. Kumar · J. L. Singh
Post Irradiation Examination Division, Bhabha Atomic Research Centre,
Trombay, Mumbai 400085, India
e-mail: prernam@barc.gov.in

A. K. Nayak and B. R. Sehgal (eds.), *Thorium—Energy for the Future*,
https://doi.org/10.1007/978-981-13-2658-5_22

irradiated in the PWL of CIRUS reactor. The two fuel clusters were designated AC-6 and BC-8. Five fuel pins containing (Th-4%Pu)O_2-sintered fuel pellets, encapsulated in a free-standing Zircaloy-2 cladding tube, were arranged in the AC-6 cluster and were irradiated to a peak fuel burn-up of 18.5 GWd/t(HM). The BC-8 cluster comprised of fuel pins containing UO_2, ThO_2, (Th-6.75%Pu)O_2 and (U-3%Pu)O_2 pellets encapsulated in collapsible, graphite-coated Zircaloy-2 cladding. The average burn-up of this fuel cluster was 4.5 GWd/t(HM), and the peak burn-up of the highest rated fuel pin of the cluster was 10.8 GWd/t(HM). Post irradiation examination of the thoria-based fuel clusters was carried at the PIED hot cells using various non-destructive and destructive techniques to assess the performance of the fuel. The results of these examinations have been reported earlier in detail by Prerna et al. [3, 4], and result of similar work carried out on thoria-based MOX fuels in other countries has been reported by different authors [5–7].

2 Fuel Cluster Details

2.1 AC-6 Fuel Cluster

The AC-6 fuel cluster consisted of five short-length fuel pins containing (Th-4%Pu)O_2 fuel pellets and one helium filled pin assembled in a circular geometry for experimental irradiation. The fuel pellets were encapsulated in free-standing Zircaloy-2 clad tubes sealed at either ends with end plugs using tungsten inert gas welding (TIG). Helium was used as the filler gas in all the fuel pins. The schematic diagram of a single fuel pin is shown in Fig. 1. Table 1 provides further details of the fuel pins of the cluster.

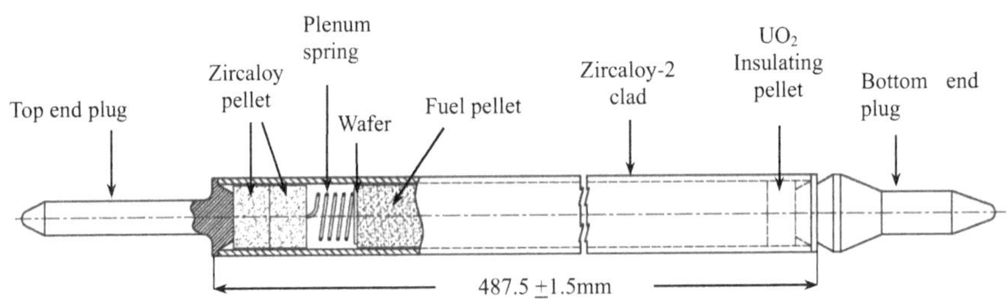

Fig. 1 Schematic illustration of a typical experimental MOX fuel pin of AC-6 cluster

Table 1 Details of the thoria-based MOX fuel pins in the clusters

Cluster	AC-6	BC-8
Clad type	Free standing	Collapsible
Number of pins	6 (5 nos. Th–Pu MOX)	12 (2 nos. Th–Pu MOX)
PuO_2 content	4%	6.75%
Pellet diameter, mm	12.22	14.3
Pellet length, mm	12.0	13.9
Pellet density, % TD	92–94%	94.6
Cladding outer wall diameter, mm	14.2	15.26
Cladding wall thickness, mm	0.8	0.38
Cold plenum length, mm	20	nil

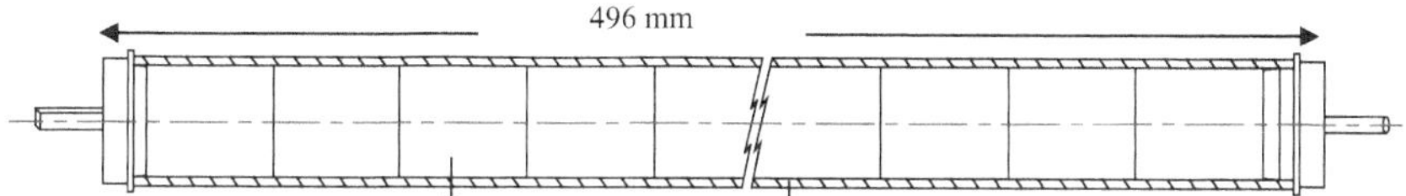

Fig. 2 Schematic diagram of a typical fuel pin of the BC-8 cluster

2.2 *BC-8 Fuel Cluster*

Twelve fuel pins with collapsible cladding and fuel pellets of different chemical compositions were assembled in a two-tier cluster. The tier 1 had three pairs of fuel pins, with one pair each of natural UO_2 pellets, $(U\text{-}3\%Pu)O_2$ and $(Th\text{-}6.75\%Pu)O_2$. Fuel pins of similar composition were placed on diametrically opposite ends of the cluster. The tier 2 consisted of six fuel pins containing ThO_2 pellets. The fuel pellets were encapsulated in graphite-coated Zircaloy-2 cladding, which is hermetically sealed with end plugs using resistance welding. Helium was used as the filler gas in all the fuel pins. Table 1 provides few more details of the fuel pins of the cluster. The schematic of a typical thoria MOX fuel pin from BC-8 cluster is shown in Fig. 2.

3 Irradiation Details

The experimental fuel pin clusters were irradiated in the PWL of CIRUS research reactor. The thermal neutron flux, the temperature and pressure of the coolant in the loop were 5×10^{13} n/cm^2/s, 240 °C and 105 kg/cm^2, respectively. The linear heat rating of the fuel pins was in the range 35–42 kW/m. The AC-6 fuel cluster was irradiated up to a calculated fuel burn-up of 18.5 GWd/t(HM). The BC-8 fuel cluster was irradiated up to a calculated average fuel burn-up of 4.5 GWd/t(HM). The burn-up of the peak-rated fuel pin was calculated to be 10.8 GWd/t(HM).

Fig. 3 As-received AC-6 fuel cluster inside the hot cell

Fig. 4 Periscopic view of the two tiers of fuel pins in the BC-8 fuel cluster

4 Post Irradiation Examination

Post irradiation examination of the fuel pins from the two fuel clusters was carried out at hot cells facility of BARC. After preliminary examination of the fuel cluster, dismantling of the fuel pins was carried out. Various non-destructive and destructive techniques were used for examination of the fuel pins inside the hot cells using master–slave manipulators and remotized equipments. The non-destructive techniques included visual examination, leak testing, fuel pin diameter measurement, ultrasonic testing, eddy current testing, gamma scanning and gamma spectrometry. Destructive examination included fission gas analysis and microstructural examination. Microstructural characterization on the fuel samples was carried out using optical microscopy, scanning electron microscopy, β-γ autoradiography and α-autoradiography techniques.

5 Pie Results

5.1 Visual Examination

Visual examination was carried out on the fuel clusters as well as individual pins of the cluster using a wall-mounted periscope. Figures 3 and 4 show the photograph of the fuel clusters inside the hot cell. No abnormality, discoloration of the cladding surface or defect of any type was visible on the surface of the cladding of the fuel pins.

5.2 Diameter Measurement

Diameter of the fuel pins of both the fuel clusters was measured using a remotely operated dial gauge. The diameter of the fuel pins from AC-6 cluster was found to be within the manufacturing tolerances, indicating negligible dimensional change during irradiation. Diameter measurements on the fuel pins of BC-8 cluster showed a reduction in the diameter as compared to their as-fabricated diameter. Reduction in diameter of ~0.1 and ~0.15 mm was observed in ThO_2 and (Th-6.75% Pu)O_2 fuel pins, respectively.

5.3 Leak Testing

Leak testing was carried out on the fuel pins of the two clusters in the hot cell using liquid nitrogen and alcohol leak test method. Each fuel pin was immersed in liquid nitrogen for 5–7 min; subsequently, the fuel pin was transferred into a tank containing alcohol. No leaks could be detected, indicating that all the fuel pins were intact.

5.4 Ultrasonic Testing

Ultrasonic testing (UT) was carried out using two 10 MHz line-focused immersion probes fitted at 27° in the probe carriage for detection of axial and circumferential defects. Multichannel ultrasonic flaw detector was used for slow helical scan combining axial probe translation and rotation of the fuel pin. The cladding of the fuel pins from both the clusters did not show presence of defect in the cladding.

5.5 Eddy Current Testing

Eddy current testing (ECT) was carried out on the fuel pins of the AC-6 cluster. ECT of the fuel pins revealed the presence of a defect in the cladding of the fuel pin, TH-2. The eddy current signals obtained from the defect location of the fuel pin TH-2 is shown in Fig. 5. This tight cladding defect probably got resealed, subsequently; hence, it was not detected during the leak testing of the fuel pin.

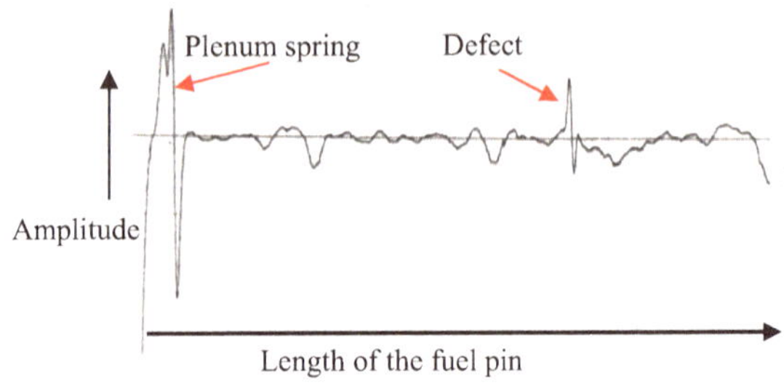

Fig. 5 ECT signal from defect location of fuel pin TH-2

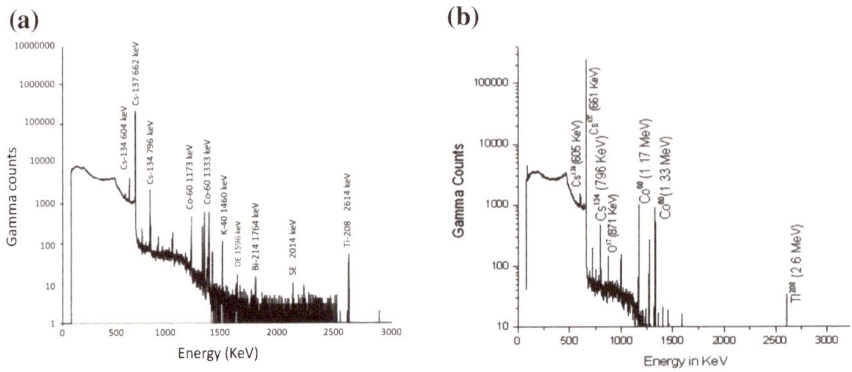

Fig. 6 Typical gamma-ray spectrum of (Th–Pu)O_2 fuel pin of **a** AC-6 cluster and **b** BC-8 cluster

5.6 Gamma Spectrometry and Gamma Scanning

A liquid nitrogen-cooled HP Ge detector and a PC-based multichannel analyser (MCA) were used for gamma spectroscopy and isotopic gamma scanning. The gamma-ray spectrum obtained from the thoria-based fuel pins of the two clusters showed peaks of Cs^{137}, Cs^{134}, Co^{60} and Tl^{208}. Figure 6a, b shows a typical gamma-ray spectrum from a fuel pin of AC-6 cluster and BC-8 cluster, respectively.

Gamma scanning over the length of the fuel pins of AC-6 cluster showed higher counts of Cs-137 near the top and bottom portion of the fuel pins probably due to migration of Cs to these locations which are relatively cooler. Figure 7a illustrates the scans for three fuel pins of the AC-6 cluster. Significant variation in the Cs^{137} activity was observed in the thoria MOX fuel pins of BC-8 cluster as a function of axial position in the reactor as shown in Fig. 7b. The Cs^{137} counts of these fuel pins were found to increase towards the reactor centre and follow the flux profile in the reactor.

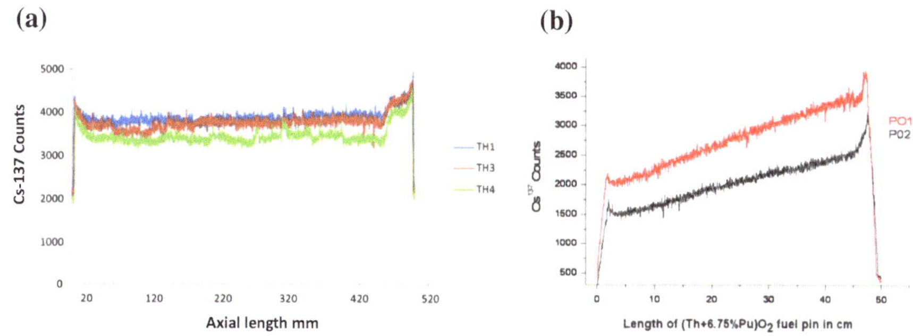

Fig. 7 **a** Gamma scan of fuel pins from AC-6 cluster and **b** gamma scan of (Th-6.75% Pu)O_2 fuel pins from BC-8 cluster

Table 2 Results of fission gas release measurements in fuel pins from AC-6 and BC-8 clusters

Pin Nos.	Fuel composition	Internal pressure at RT (atm)	% FGR
AC-6 cluster			
TH-4	(Th-4%Pu)O_2	4.4	0.6
TH-5	(Th-4%Pu)O_2	3	0.4
TH-2	(Th-4%Pu)O_2	No fission gas in the pin	
BC-8 cluster			
P-01	(Th-6.75% Pu)O_2	2.87	0.35
P-02	(Th-6.75% Pu)O_2	1.52	0.54

5.7 *Fuel Pin Puncturing and Fission Gas Analysis*

Fuel pins were punctured for the measurement of the amount of released fission gases. The volume of the released fission gases and the void volume in the fuel pin were measured to arrive at the pressure of the gas inside the fuel pin. A dual-column gas chromatograph and a quadrupole mass spectrometer were used to analyse the chemical composition and isotopic composition, respectively, of the collected gases.

Fission gas analysis was carried out on three fuel pins, TH-2, TH-4 and TH-5, of AC-6 cluster, and the results are given in Table 2. The fuel pin TH-2 did not show presence of fission gas inside the pin indicating defect in the cladding as indicated by the eddy current test. The fission gas release in other fuel pins was found to be less than 1%. The results of fission gas release measurement in the thoria MOX fuel pins of the BC-8 fuel cluster are given in Table 2. The results show very low fission gas release in the fuel pins.

(a) (b) (c)

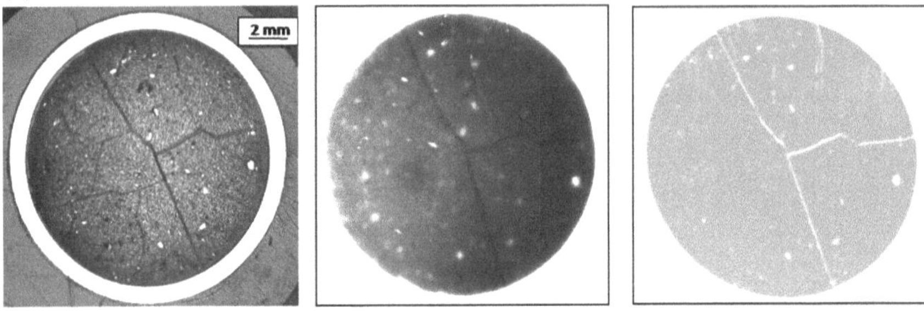

Fig. 8 **a** Photomacrograph, **b** β-γ autoradiograph and **c** α-autoradiograph of fuel section from fuel pin TH-5 of AC-6 cluster

5.8 Metallographic Examination

Metallographic samples from fuel pins of the two fuel clusters were prepared inside the hot cells and examined using a remotized metallograph. β-γ autoradiography of the metallographic samples was carried out to study the distribution of the fission products (mainly Cs) across the fuel cross section. Re-distribution of plutonium in the fuel cross section was studied by carrying out α-autoradiography.

5.8.1 AC-6 Fuel Cluster

Samples from fuel pins TH-2 and TH-5 were examined. Figure 8a–c shows the photomacrograph, β-γ autoradiograph and α-autoradiograph of the fuel section from fuel pin TH-5. Radial cracks were observed in the fuel. No columnar grain formation or grain growth was observed in the fuel. The β-γ autoradiographs revealed an asymmetric distribution of β-γ activity across the fuel cross section. Higher activity (increased darkness indicates higher β-γ activity) was observed in the cracks in the fuel. The α-autoradiograph of the fuel sections revealed a uniform Pu activity along the cross section.

Uniform oxide layer was observed on the outer surface of the cladding, whereas a discontinuous oxide layer was noticed on the inner surface. Average oxide layer thickness on the outer and inner surfaces of the cladding was 1.3 and 0.9 µm, respectively.

Examination of metallographic samples taken from the fuel pin, TH-2, at the defect location (as detected during eddy current testing) was carried out. Figure 9a, b shows the photomacrograph and β-γ autoradiograph of a section of the pin near the defect in the cladding. The cladding defect in the photomacrograph appeared as cracks on the inner surface of the cladding. A higher β-γ activity (as dark regions) was noticed in the fuel close to region of defect in the cladding as shown in Fig. 9b. Detailed examination of the cladding revealed a massive zirconium hydride blister

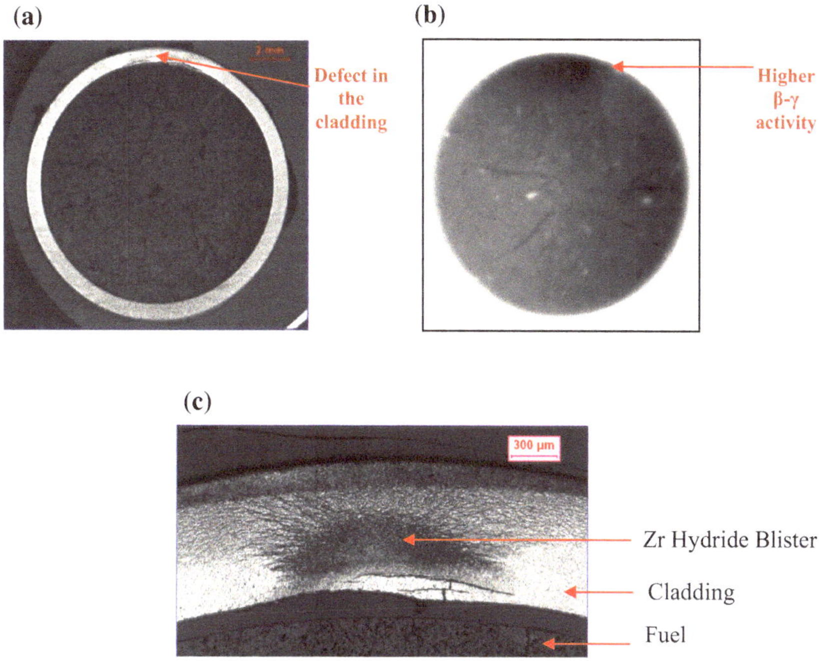

Fig. 9 **a** Photomacrograph of fuel section from TH-2, showing clad defect and **b** β-γ autoradiograph of fuel section from TH-2, **c** defect region in the cladding of the fuel section from fuel pin TH-2

on the inner surface of the cladding along with cracks in the cladding as shown in Fig. 9c. The blister was 2.2 mm in diameter and 0.64 mm in depth. A 150-μm-thick hydride layer was observed towards the outer surface of the cladding opposite the hydride blister.

Scanning electron microscopy of fuel

Due to the chemical inertness of thoria, conventional etching techniques to reveal the fuel microstructure were not suitable. Hence, replicas were prepared from the fractured pieces of the fuel, and the replicating foil having the impression of the grains was examined under a scanning electron microscope (SEM).

Figure 10a shows the SEM photomicrograph of the replica of fuel surface showing the size and morphology of the grains of fuel from fuel pin TH-5. The average grain size was found to be 14 μm. A bimodal grain size distribution was observed in few regions, with larger grains up to 30 μm in size and small grains of 2–5 μm size. The average grain size in the fuel and the size distribution are similar to that for the as-fabricated fuel, indicating no grain growth during irradiation. Examination of the grains sticking on the replica foil showed presence of submicron size fission gas bubbles on the faces of smaller grains as shown in Fig. 10b.

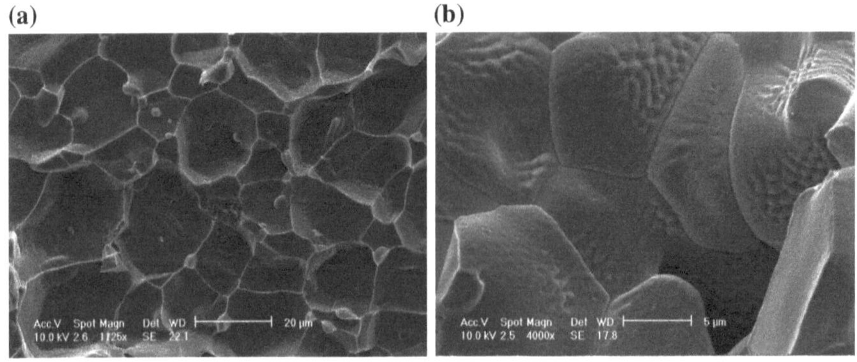

Fig. 10 **a** Impressions of fractured grains on replicating tape and **b** grain faces of fuel showing fine fission gas bubbles

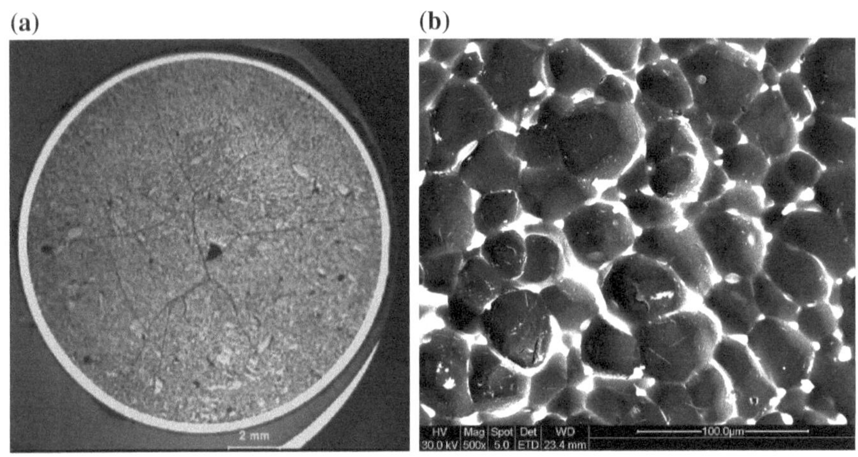

Fig. 11 **a** Photomacrograph of the fuel section from P-02 fuel pin and **b** replica of fractured piece of $(Th–Pu)O_2$ fuel

5.8.2 BC-8 Fuel Cluster

Figure 11a shows the photomicrograph of the fuel section taken from the fuel pin, P-02. Fine radial cracks and white particles were observed in the fuel cross sections taken from the MOX fuel pins. These particles did not show β-γ activity or α-activity. The porosity distribution observed in the fuels was comparable to that in the as-fabricated fuel.

Examination of the replica prepared from the fractured pieces of the $(Th–Pu)O_2$ fuel from pin P-02 was carried out, and Fig. 11b shows the impression of the grains of the fuel from the centre of the fuel section. The average grain size in the fuel was observed to be 30 µm, which was similar to the grain size in the as-fabricated fuel. No grain growth was observed in the fuel section.

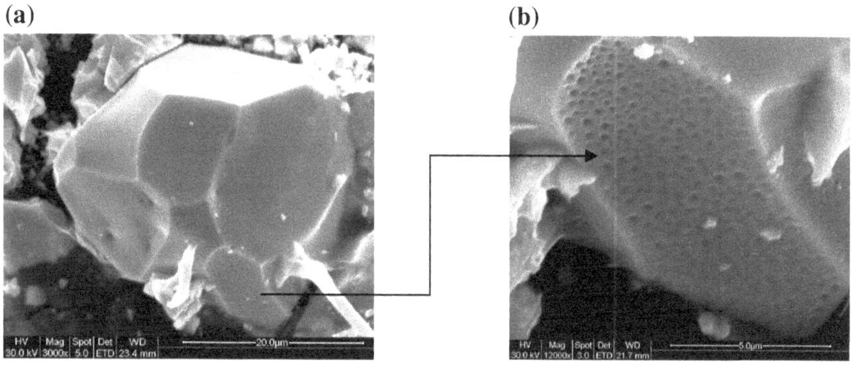

Fig. 12 **a** Single grain of the Th–Pu MOX fuel and **b** face of the grain covered with fission gas bubbles

Fractured pieces of grains of fuel were sticking to the replicating foil. Surface of most of the grains did not show any feature. Few of the grains revealed fission gas bubbles on one of the faces, as shown in Fig. 12a, and the magnified view of the grain face is shown in Fig. 12b.

Examination of fuel sections did not indicate any interaction between fuel and cladding in the fuel pins. Lesser and localized corrosion of the cladding on the inner surface was observed in the (Th–Pu) MOX fuel pins as compared to (U–Pu) MOX fuel pins of BC-8 cluster. Oxide layer could be seen at a very few locations with the average thickness 1.2 μm in (Th–Pu) MOX fuel pin, whereas a continuous layer was observed in the cladding of (U–Pu) MOX fuel pins with the average thickness being 3.3 μm. Continuous oxide layer was observed on the outer surface of the clad of all the fuel pins of the cluster as they were exposed to the same coolant during irradiation. The average oxide layer thickness on the outer surface of the cladding of the fuel pins was in the range of 1.6–2.5 μm.

6 Discussion

Performance of a fuel is evaluated on the basis of the following parameters:

(i) Dimensional stability,
(ii) Thermal performance and
(iii) Chemical stability during irradiation.

Irradiation performance of thoria-based MOX fuel pins from the two clusters has been evaluated from PIE results on the basis of the above fuel performance parameters.

(i) *Dimensional stability*

The cladding of the fuel pins of BC-8 cluster was collapsible type, which experienced creep of the cladding into the fuel clad gap under coolant pressure. Permanent plastic deformation of the cladding leads to slight reduction in diameter of the fuel pins. Whereas the cladding of the fuel pins in AC-6 cluster was free-standing type, no dimensional change in the fuel pins was observed.

Stehle et al. [8] have discussed on the dimensional behaviour of fuel pins, which is governed by in-reactor volume changes of UO_2 due to densification and swelling. According to Ioan [9], the main contribution to gaseous swelling comes from the intergranular bubbles. Fuel grains examined from AC-6 cluster showed submicron size fission gas bubbles on the faces of only a few small grains. This is due to lower mobility of fission gas atoms at the operating temperature as reported by Sah et al. [10]. Fuel grain examined from BC-8 cluster revealed fission gas bubbles on one of the faces. Presence of fission gas bubbles on a face of the grain of $(Th–Pu)O_2$ fuel and other faces devoid of the bubbles denotes the presence of a plutonium rich region at that face. High concentrations of plutonium termed as high fission areas in $(U–Pu)O_2$ fuel were also observed by Harrison et al. [11]. Absence of fission gas bubbles at grain boundaries in the fuel section and observation of fine fission gas bubbles only on few grains or a surface of a grain indicate absence of swelling in the fuel. Absence of dimensional changes in the fuel pins is supported by the microstructural observations of fission gas bubbles. The observations show dimensional stability of the fuel.

(ii) *Thermal performance*

Olander [12] has shown that the temperature distribution in a fuel pin influences fuel performance in two important ways, i.e. high fuel centre temperatures and steep temperature gradients. During PIE, fuel thermal performance is assessed by measuring the amount of fission gas release, fuel restructuring and fission product distribution in the fuel.

The main mechanism of fission gas release in oxide fuel is diffusion of gas atoms and grain boundary sweeping, which in turn depends on the fuel temperature has been discussed by Sah et al. [10, 13]. Percentage fission gas release in the pins of both the clusters was estimated to be less than 1%. The low fission gas release in the fuel pins indicates lower fuel temperature as compared to UO_2 leading to lower diffusion of fission gas atoms.

Extent of restructuring in fuel refers to the alterations in the microstructure of the fuel due to high fuel centre temperature and steep temperature gradient [12]. Pellet crack pattern, microstructure, fission product migration in the fuel, etc., depend on the fuel temperature. The presence of few fine radial cracks, absence of columnar or grain growth and average grain size in the fuel observed to be similar to that in the as-fabricated fuel indicates no restructuring in the fuel pins of the two clusters, which is due to the low fuel centre temperature during irradiation. Absence of migration of fission product (caesium) from centre to periphery also indicates a low fuel centre temperature.

(iii) *Chemical stability during irradiation*

Inherent chemical inertness and existence in single valency state of ThO_2 provide better chemical stability [14] during normal operation and defected conditions as compared to UO_2-based fuels.

Localized and discontinuous oxide layer observed on the inner surface of the cladding of (Th–Pu) MOX fuel pins as compared to the thicker and continuos oxide layer in (U–Pu) MOX fuel pins of BC-8 cluster indicates better chemical stability of the Th-based fuel pins as compared to U-based fuels.

7 Conclusion

Post irradiation examination of two fuel clusters irradiated up to a fuel burn-up of 18.5 GWd/t(HM) and 10.2 GWd/t(HM) has been carried out to assess the performance of thoria based fuels.

1. PIE observations showed that fission gas release in the fuel pins was negligible ($<1\%$), there was very less cracking in the fuel, and there was no evidence of fuel restructuring. No swelling was observed in the fuel, and there was no dimensional change in the fuel pins. Fuel–clad interaction and fission product migration were also absent in the fuel pins.
2. Out of the five fuel pins in AC-6 cluster, four fuel pins remained intact during irradiation. One fuel pin showed an internal hydride defect on the ID surface of the clad, which might have formed due to higher moisture content in the fuel pellets. Strict control of moisture in the pellets is necessary to avoid such defects in fuel pins.
3. Thoria-based MOX fuels have performed well during irradiation. Good performance of the fuel was attributable to lower operating fuel centre temperature because of high thermal conductivity of the fuel.
4. The performance of a nuclear fuel is judged by its dimensional stability, the extent of fission gas release and the nature and extent of fuel cladding interaction. In all these respects, the performance of the fuel pins is found excellent. The results provide confidence for use of this fuel in the forthcoming AHWR.

Acknowledgements The authors express their gratitude to all the colleagues in post irradiation examination division for their assistance in carrying out the hot cell activities. The authors would like to acknowledge the guidance provided by Shri D. N. Sah, Ex. Head, PIED, in preparation of the manuscript.

References

1. Atomic Energy in India: A Perspective. Government of India, Department of Atomic Energy, Sept 2003
2. Thorium Fuel Cycle—Potential Benefits and Challenges. *IAEA-TECDOC-1450*, May 2005
3. P. Mishra, H.N. Singh, V.P. Jathar, R.S. Shriwastaw, P.M. Satheesh, P.B. Kondejkar, G.K. Mallik, J.L. Singh, S. Anantharaman, Irradiation Behaviour of Th–Pu and U–Pu MOX Fuels, in *Proceedings of the 12th International Conference on CANDU Fuel*, Kingston, Ontario, Canada, 15–18 Sept 2013
4. P. Mishra, J.L. Singh, J.S. Dubey, K.M. Pandit, V.P. Jathar, B.N. Rath, P.M. Satheesh, R.S. Shriwastaw, P.K. Shah, A. Bhandekar, S. Kumar, P.B. Kondejkar, H.N. Singh, S. Anantharaman, Post Irradiation Examination of Thoria–Plutonia MOX Fuel from AC-6 Cluster, *BARC Report*, BARC/2014/E/003
5. M. Karam, F.C. Dimayuga, J. Montin, Post-Irradiation Examination of CANDU fuel bundles fuelled with (Th, Pu)O_2, in *Proceedings of the 11th International Conference on CANDU Fuel*, Niagara Falls, Ontario (2012)
6. P.G. Boczar, P.S.W. Chan, G.R. Dyck, R.T. Jones, J.D. Sullivan, P. Taylor, Thorium Fuel-Cycle Studies for CANDU Reactors, IAEA Meeting on Thorium Fuel Utilization: Options and Trends, *IAEA-TECDOC-1319* (Nov 2002)
7. G. Horhoianu, G. Olteanu, D.V. Ionescu, Irradiation Behaviour of PHWR Type Fuel Elements Containing UO_2 and (Th, U)O_2 Pellets, *IAEA Technical Meeting on "Fuel integrity during normal operation and accident conditions in pressurized heavy water reactors (PHWRs)*", Romania, Bucharest, 24–27 Sept 2012
8. H. Stehle, H. Assman, G. Wunderlich, Uranium Dioxide properties for LWR fuel rods. Nucl. Eng. Des. **33**, 230–260 (1975)
9. I. Arimescu, Fuel Swelling Importance in PCI Mechanistic Modelling, in *Proceedings of Pellet Clad Interaction in Water Reactor Fuels*, France, 9–11 Mar 2004
10. D.N. Sah, Basic mechanisms of fission gas release and high burn-up issues. Met. Mater. Process. **18**(1), 27–34 (2006)
11. N.F. Harrison, T. Janathasing, F.C. Dimayuga, Post-Irradiation Examination of MOX fuel with varying plutonium homogeneity, in *Proceedings of the 11th International Conference on CANDU Fuel,* Niagara Falls, Ontario, (2012)
12. D.R. Olander, *Fundamental Aspects of Nuclear Reactor Fuel Elements*, Chapter 10, Technical Information Centre, Virginia, USA
13. D.N. Sah, U.K. Viswanathan, C.S. Viswanadham, K. Unnikrishnan, B.N. Rath, Blind prediction exercise on modeling of PHWR fuel at extended burnup. J. Nucl. Mater. **383**, 144–149 (2008)
14. K. Anantharaman, V. Shivakumar, D. Saha, Utilisation of thorium in reactors. J. Nucl. Mater. **383**, 119–121 (2008)

PuO_2 Agglomerate Detectability in $(Th,1\%Pu)O_2$ Fuel—A Monte Carlo Simulation Study

K. V. Vrinda Devi, K. Biju and K. B. Khan

Abstract Advanced heavy water reactor (AHWR) is being developed for thorium utilisation and various technologies associated with this are being studied. Experimental irradiation of $(Th_{0.99},Pu_{0.01})O_2$ MOX fuel has been carried out in AHWR conditions. Fabrication and characterisation of this fuel were carried out through conventional powder metallurgical route similar to $(U,Pu)O_2$ MOX fuel. Homogeneity of fissile material distribution is a crucial characteristic of the fuel, and quality control procedure of the fuel incorporates inspections to ensure its conformance with specifications in this regard. Gamma autoradiography (GAR) is one of the techniques carried out for this wherein plutonium-rich agglomerates; if any, present in the matrix appear as dark spots in the gamma image of the fuel. However, it is difficult to characterise the agglomerate based on the GAR image as multiple factors such as plutonium richness and size of the agglomerate, location on the fuel pellet influence the formation of the image significantly. Hence, a theoretical simulation study was carried out to estimate the minimum detectable size of a 100% rich plutonium agglomerate located on the surface of $(Th_{0.99},Pu_{0.01})O_2$ MOX fuel by GAR. Monte Carlo methods using FLUKA2011.2C code was applied for this. Details of the simulations and the results obtained are discussed in this paper.

Keywords MOX · Plutonium homogeneity · GAR · Monte Carlo

1 Introduction

$(Th_{0.99},Pu_{0.01})O_2$ MOX was the fuel for experimental irradiation in AHWR conditions. MOX fuel pellets of this composition were fabricated in Advanced Fuel

K. V. Vrinda Devi (✉) · K. B. Khan
Radiometallurgy Division, Bhabha Atomic Research Centre, Mumbai, Maharashtra, India
e-mail: kvvdevi@barc.gov.in

K. Biju
Health Physics Division, Bhabha Atomic Research Centre, Trombay, Mumbai 400085, Maharashtra, India

A. K. Nayak and B. R. Sehgal (eds.), *Thorium—Energy for the Future*,
https://doi.org/10.1007/978-981-13-2658-5_23

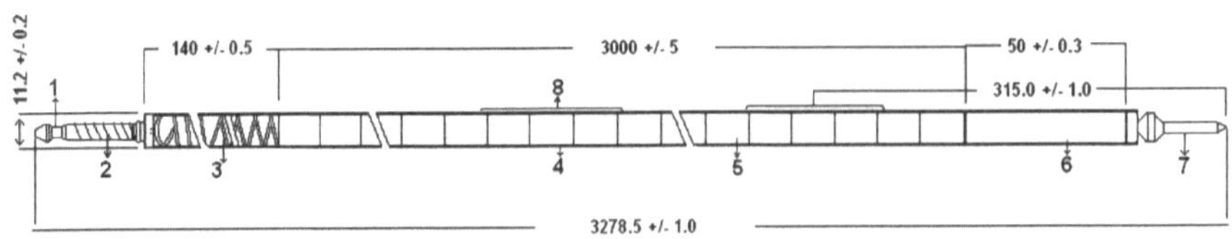

1. Upper end plug (Zr-2) 5. Fuel pellet (MOX)
2. Expansion spring (Inconel) 6. Plenum tube(Zr-2)
3. Plenum spring (SS302) 7. Lower end plug (Zr-2)
4. Clad tube (Zr-2) 8. Spacer pad (Zr-2)

Fig. 1 Schematic of $(Th_{0.99},Pu_{0.01})O_2$ MOX fuel pin [2]

Fabrication Facility (AFFF), BARC, Tarapur through conventional powder metallurgical route and encapsulated in Zircaloy-2 clad tubes filled with helium [1]. Schematic of the fuel pin is shown in Fig. 1.

Radiation assay and radiation imaging are two usually implemented non-destructive testing methods for Pu distribution in MOX fuels. Gamma autoradiography (GAR) and passive gamma scanning (PGS) were routinely employed for homogeneity inspection in U,Pu MOX fuels at AFFF [3]. Attempts were done to quantify the Pu-rich agglomerates detected using PGS technique in MOX fuel pins [4]. It was found that the minimum detectable limit (MDL) depends on the experimental conditions in both these non-destructive techniques. Hence, the need for estimation of MDL under the given experimental conditions was felt and a simulation study was carried out. Monte Carlo simulation method for radiation emission and transport was earlier used to simulate the actual gamma emission from and transport through $(U,Pu)O_2$ MOX and (U,Pu)C fuels for fast reactors [5, 6] and the same method was adopted for ThO_2 and 1% PuO_2 fuel with AHWR fuel configurations.

2 Methodology

Gamma autoradiography is a process in which the encapsulated fuel is imaged on an X/γ ray film by exposing the film with the fuel pin placed on it for sufficient period of time and later processing the film to obtain the radiographic image. Formation of the gamma image of MOX fuel is majorly due to the gamma rays from plutonium as the specific activities of plutonium isotopes are much higher than that of thorium. Hence, the distribution of plutonium could be monitored by analysing the GAR image. However, it will be difficult to detect Pu-rich agglomerates if they are present deep inside the fuel pellet due to the self-attenuation of plutonium gammas in the high-density fuel matrix. If a small agglomerate is present within the limited path length (~2 mm) but not on the outermost surface, the image will be magnified because of the diffusion of gamma rays while traversing the matrix before reaching the film. The darkness of the spot also could be reduced during this traverse. Hence, it is

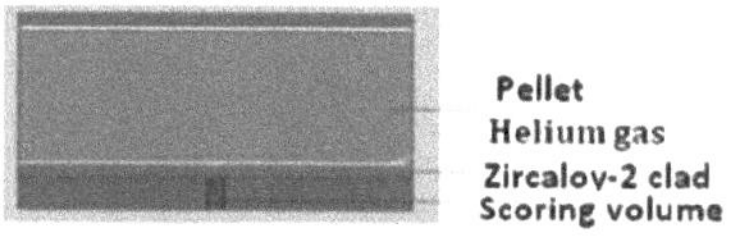

(a) Simulation for background gamma fluence homogeneous pellet

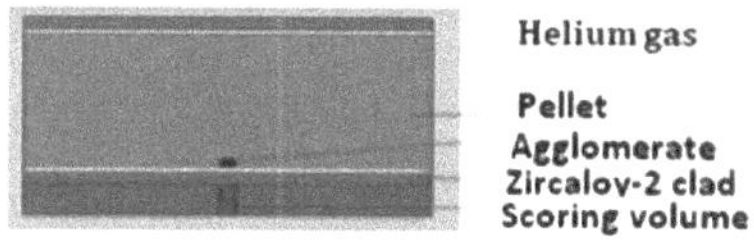

(b) Simulation for gamma fluence from a pellet containing a plutonium agglomerate

Fig. 2 Gamma assay/imaging geometry simulated

difficult to characterise the agglomerate directly based on the image it created during GAR. So, it was necessary to estimate the minimum detectable size of a plutonium agglomerate under the most favourable conditions of detection so that one could be sure that any detected agglomerate is having size more than that. This will help in ensuring the conformance of MOX fuel for AHWR with its stringent specification which does not allow any plutonium-rich agglomerate of more than 100 μm in size.

A validated Monte Carlo code FLUKA2011.2C [7, 8] was used to compute the photon fluence spectra outside the MOX fuel. FLUKA is a general-purpose Monte Carlo particle transport code for simulating interactions of ionizing radiations with matter. Monte Carlo method involves the actual mathematical simulation of transport of the particle through the medium using random sampling techniques [9]. FLUKA can simulate the propagation and interaction of photons of energies from 1 keV to thousands of TeV in very complex geometries modelled using the combinatorial geometry (CG) package.

The process of GAR of AHWR experimental fuel pin was simulated using Monte Carlo method of radiation transport. The actual values of fuel pellet composition and details such as dimensions and density were used for the simulation. Details of helium gas and Zircaloy-2 clad such as density and composition were also input to the code. The exact geometry of the encapsulated fuel pin was also simulated.

The concept of simulation was as follows. The photon fluence from a homogeneous pellet is estimated by simulating a fuel pin encapsulated with a homogeneous $(Th_{0.99},Pu_{0.01})O_2$ MOX fuel pellet with helium cover gas, etc., as the background. The background photon fluence is called X and the statistical fluctuation of the background is called $\pm\Delta X$. Further, photon fluence from a pellet of the same composition having a pure PuO_2 agglomerate located at the point of contact between the clad with the film was calculated along with its statistical fluctuation ($Y \pm \Delta Y$). Simulations were carried out with different sizes of agglomerates ranging from 50 to 150 μm. The background ($X \pm \Delta X$) was also averaged to the same scoring volume in each case so that the fluencies are directly comparable and checked if $X \pm \Delta X < Y \pm \Delta Y$ which is necessary for detection.

The photon fluence was scored outside the fuel pin in the scoring volume as shown in Fig. 2a, b.

The agglomerate was considered to be of 100% PuO_2 and in contact with the inner surface of the clad location whose outer surface made contact with the film. These conditions were assumed to simulate the most favourable conditions for detection.

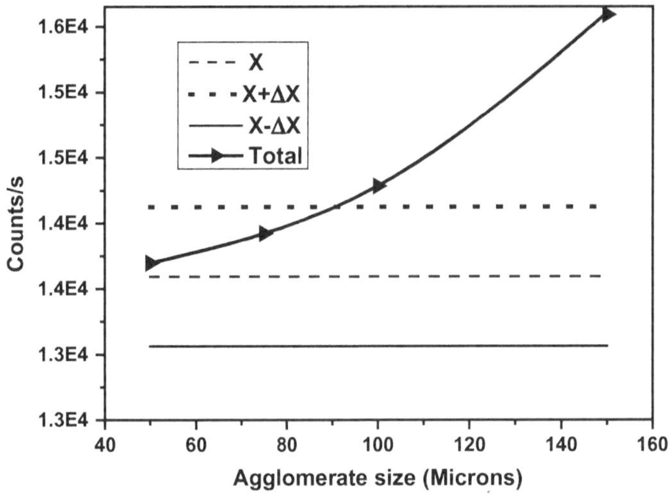

Fig. 3 Photon fluence as a function of agglomerate size

3 Results and Discussion

Results from the simulated radiation emission and transport are shown in Fig. 3.

The dotted line represents the average background gamma fluence and the lines parallel to that represent the statistical fluctuation in the background fluence. The line denoted as 'Total' represents the gamma fluence from the agglomerate along with the background. It could be observed that 'Total' exceeds the background when the agglomerate is at least 100 μm in size. This shows that it is not possible to detect a 100% PuO_2 agglomerate having size less than 100 μm using radiation assay/imaging in $(Th_{0.99},Pu_{0.01})O_2$ MOX with AHWR fuel configurations. So, any detected agglomerate could be inferred as having size of 100 μm or more. Since the specification for Pu micro-homogeneity is not to have any plutonium agglomeration of more than 100 μm size, this result shows that the technique of GAR could be used for screening the fuel for any allowable plutonium micro-heterogeneity during fabrication.

4 Conclusion

Thoria matrix mixed oxide fuels are being fabricated, characterised, and experimentally irradiated under thorium utilisation program. The conventional chemical dissolution tests for monitoring plutonium micro-homogeneity is not preferred in thoria matrix fuels due to poor solubility of thoria in HNO_3 without the assistance of HF. Radiation imaging and radiation assay techniques were earlier explored for

MOX fuels with urania matrix. Gamma autoradiography technique was being used for homogeneity screening of the fuel but quantification of the agglomerate size if detected, was difficult. In this study, a minimum detectable limit under the most favourable conditions of detection is obtained theoretically, so that any agglomerate; if detected in GAR could be inferred to be having size more than that. Monte Carlo simulations were used using FLUKA for theoretical estimation of the minimum detectable size of Pu-rich agglomerate in $(Th_{0.99},Pu_{0.01})O_2$ MOX fuel and were found to be 100 μm under the most favourable conditions for detection. Since the MDL was found to be equal to the maximum allowable size for a plutonium agglomerate, any detected agglomerate indicated non-conformance with the specification for AHWR experimental fuel. Hence, GAR could be an effective non-destructive technique to characterise MOX fuel of this composition for Pu homogeneity.

Acknowledgements The authors are grateful to all their colleagues at AFFF, BARC, Tarapur for their encouragement and support during this work.

References

1. F. Glodeanu, J. Nucl. Mater. **126**, 181–183 (1984)
2. N. Kumar, A.K. Hinge, N. Walinjkar, D.B. Sathe, A. Prakash, D. Mukherjee, T.R.G. Kutty, Mohd. Afzal, J.P. Panakkal, *Proceedings of the National Seminar & Exhibition on Non-Destructive Evaluation NDE 2011*, 8–10 December 2011, pp. 101–104
3. J.P. Panakkal, K.V. Vrinda Devi, D. Mukherjee, R.S. Wadhwani, H.S. Kamath, *WCNDT'96*, New Delhi, Dec 1996
4. K.V. Vrinda Devi, J.P. Panakkal, Nucl. Eng. Des. **255**, 132–137 (2013)
5. K.V. Vrinda Devi, K. Biju, K.B. Khan, A. Kumar, *5th ISMC-2014*
6. K.V. Vrinda Devi, K. Biju, K.B. Khan, A. Kumar, *59th DAE SSSP-2014*
7. A. Fasso', A. Ferrari, J. Ranft, P.R. Sala, *FLUKA: A Multi-Particle Transport Code*. CERN-2005-INFN/TC_05/11, SLAC-R-773 (2005)
8. G. Battistoni, S. Muraro, P.R. Sala, F. Cerutti, A. Ferrari, S. Roesler, A. Fasso', J. Ranft, The FLUKA code: description and benchmarking, in *Proceedings of the Hadronic Shower Simulation, Workshop 2006, Fermilab, 6–8 September 2006. AIP Conference Proceeding*, vol. 896, pp. 31–49 (2007)
9. E.D. Cashwell, C.J. Everett, *A Practical Manual on the Monte Carlo Method for Random Walk Problems*, LA-2120, Physics and Mathematics, Los Alamos Scientific Laboratory, New Mexico (1957)

Automated Fabrication of AHWR Fuel in Shielded Cells: Challenges and Initiatives

S. Panda, Saurabh Gupta, A. Saraswat, Vijay Ahire, Sanjeev Sharma, P. S. Somayajulu and K. Jayarajan

Abstract With the growing demand of various innovative reactor designs and use of thorium as fuel, importance of automation for fuel fabrication is realized. In order to handle the fuel remotely during its fabrication cycle, various stages of its fabrication are required to be automated. A first of its kind, full-scale mock-up facility is recently developed at BARC to demonstrate automated fabrication, inspection and handling of (Th-^{233}U)-based fuel pins. In order to optimize the operations and to avoid spread of contamination, equipments in the facility are arranged in multiple interconnected shielded cells. One of the major challenges in handling of such fuel pin is requirement of intra-cell and inter-cell autonomous transfer of long fuel pins. This paper describes a uniquely designed rail-guided vehicle, which automatically triggers opening and closing of heavily shielded inter-cell door during intra-cell material transfer. Further, in this paper, the concept of dual robotic arms for precise handling of pins during its fabrication is presented. All the quality inspection tests are performed remotely and autonomously. Results of the inspection are available online at control station, and operator is alerted when monitored parameters go beyond the acceptable tolerance band. In order to avoid radiation damage to electronic components, controllers and data acquisition system are located outside the cells. Novel methods presented in this paper produced consistent results with considerable reliability during mock-up trials, thus ensuring their suitability for the real system.

Keywords AHWR · U-233 · Thorium · Automated fuel fabrication

S. Panda (✉) · S. Gupta · A. Saraswat · S. Sharma · P. S. Somayajulu · K. Jayarajan
Bhabha Atomic Research Centre (BARC), Trombay, Mumbai 400085, Maharashtra, India
e-mail: panda@barc.gov.in

V. Ahire
Symec Engineers (I) Pvt. Ltd., Navi Mumbai 400705, India

A. K. Nayak and B. R. Sehgal (eds.), *Thorium—Energy for the Future*,
https://doi.org/10.1007/978-981-13-2658-5_24

1 Introduction

Advanced heavy water reactor (AHWR) [1] is a unique reactor concept, designed to utilise the vast thorium resources available in India. It will be using mixed oxides of (Th-Pu) and (Th-^{233}U) as fuels. Fertile thorium is converted to fissile ^{233}U in reactor through neutron capture, and the same will be used as fuel in AHWR. ^{233}U produced in reactor is always contaminated with small quantities of ^{232}U. As a few daughter products of ^{232}U in its decay chain are hard gamma emitters, the fuel containing ^{233}U can only be handled behind shielded cells. Mixed oxide (MOX) fuel for AHWR will be fabricated by powder-pellet/ceramic processes, starting from PuO_2, ThO_2 and $^{233}UO_2$. Although well-established processes and techniques are available for fuel fabrication for operating thermal reactors and fast reactors, using UO_2 and PuO_2 fuel, we have limited experience of fuel fabrication in shielded facility. Automation in fuel fabrication is also limited to equipment level and not to the plant level. Therefore, a first of its kind, mock-up facility [2, 3] is being set-up at BARC to demonstrate fully automated fuel fabrication, including pellet fabrication, pin fabrication and assembly of fuel cluster.

The reactor core of AHWR consists of 452 fuel clusters of 4300 mm length and 118 mm diameter. The cluster has 54 fuel pins, arranged in three rings around a central rod [4]. Each fuel pin consists of an 11.2 mm diameter Zircaloy clad tube, welded with two end plugs and filled with MOX pellets (Fig. 1). Eighteen spacer pads are welded to the clad tube.

The major operations to be performed on the clad tube or fuel pin are insertion of pellets into the tube, end-plug welding, buffing the welded area, various dimensional measurements, weight checking, visual inspection, X-ray radiography, He-leak detection and gamma scanning. In addition, it involves handling material among various equipments in various stations.

2 Mock-Up Facility

Large length, combined with small diameter, and presence of spacer pads makes the automated material handling difficult. Fuel pin fabrication will be performed in a series of shielded cells (hot cells). The physical constraints arising from hot cells are

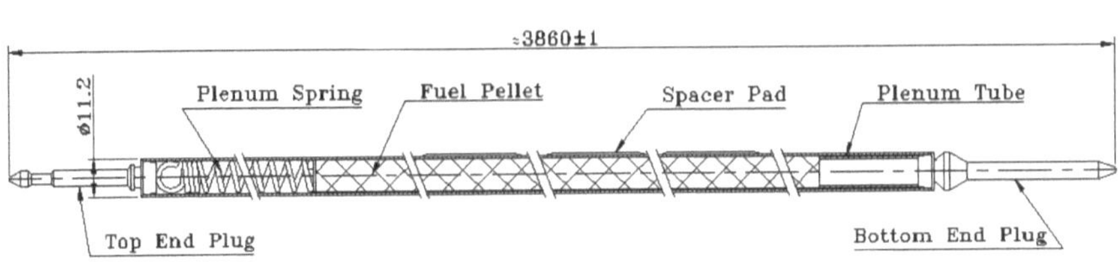

Fig. 1 Fuel pin

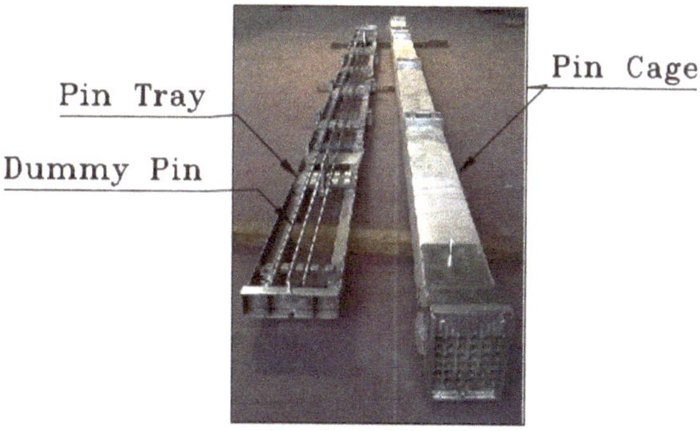

Fig. 2 Dummy pin, tray and cage

respected in the design of the mock-up system. Width of the hot cells is limited to the reach of master–slave manipulators (MSMs) for accessing the desired areas. In addition, no permanent structure/equipment is placed above the automated stations, in the layout for handling the fuel pins with MSM.

We have designed a fuel pin cage (Fig. 2) with a capacity of 36 pins for transferring the clad tubes to the pin handling plant and for transferring the fabricated pins to the cluster assembly plant. The cage doors can be opened only in the cell using automated tools. The cage containing pins will be transferred to the cluster assembly plant in a shielded flask. Within the plant, the pins or clad tubes will be transferred from one hot cell to another in a pin tray. The tray and cage have features for locating them accurately at the designated stations.

The mock-up plant has three simulated hot cells [5]. A cage containing 36 empty clad tubes are brought to the plant, and another cage containing 36 fabricated pins are taken out of the facility. Automated operations begin with pulling out clad tubes from the cage and end with inserting accepted pins into another cage. Photograph of the set-up is shown in Fig. 3.

Fig. 3 Mock-up facility for pin fabrication

3 Material Transfer Tools

In the facility, the long clad tubes and pins are transferred between cells and within a cell for feeding them to different processing or inspection systems. Major material transfer mechanisms developed are as listed below.

3.1 Pin and Tray Handling Robot

For vertical and transverse transfer of the pins and clad tubes, they have to be lifted from the support. As the long pin cannot be lifted by a single device, a pair of cooperative robot was developed for pin handling. The overhead mounted robot pair is capable of lifting and transferring the long clad tubes, pins and trays in coordination.

3.2 RGV and Inter-cell Door

Instead of transferring single pin or clad tube, six of them are loaded into a pin tray (Fig. 2) and the pin tray is transferred between cells. A pair of pin and tray handling robot loads the pins on the pin tray, and the same robot pair loads the tray on a rail-guided vehicle (RGV). Rails are laid in the cells and in the common walls for moving the RGV (Fig. 4). Contact lines laid along the rail and bus bar contactors on the RGV are used to supply power to RGV. Inter-cell doors are embedded in the common walls for the RGV to pass through. The normally closed door is opened only during the transfer of RGV.

3.3 Cage Transfer Tool

Pin cage weighing about 200 kg has to be transferred among different stations within a cell. A heavy-duty cage transfer tool, which is similar to pin and tray handling tool, was developed for this.

3.4 Tools for Longitudinal Transfer of Pin

In about ten stations, the pins and clad tubes are to be moved axially to feed them to various process and inspection systems. The axial movement ranges from few millimetres to the full length of the pin. Depending upon the axial travel, accuracy

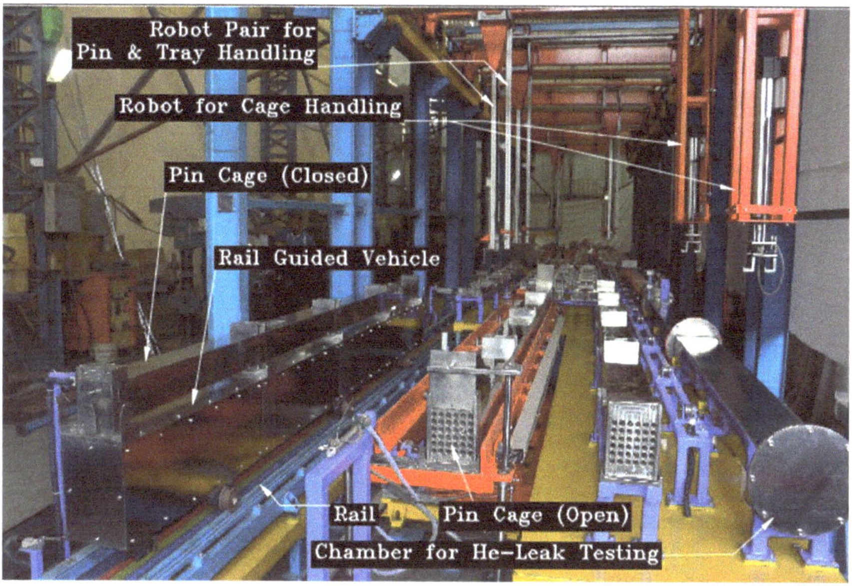

Fig. 4 View of the automation system from one end of facility

and space available, different mechanisms are used for pin transfer. The pushing force is limited to avoid damage of the pin during pushing.

3.5 NC Table for Cage Positioning

Pin cage has 36 pins or clad tubes arranged in a 6 × 6 array (Fig. 2). The cage is placed on a 2-axis NC table for aligning the required clad tube to the clad tube-unloading tool. The same set-up is used for loading the fabricated pin into the specified location in the cage. For pushing and pulling, the clad tube (or pin) is gripped using a pneumatic gripper and pulled (or pushed) in small increments.

3.6 Pin Clamping and Rotation Set-Up

The pin has to be rotated for welding, buffing, X-ray radiography, TIR, weld bead measurement and visual inspection. Wherever, positive rotation is required, the pins are clamped on a rotating collet; otherwise, the pins are supported on a set of rotating rollers. The rotation is either continuous or intermittent. Wherever required, the pins are clamped and guided to avoid vibration and wobbling of the pin.

4 Inspection Systems

Automated inspection techniques are developed for metrology and weight measurement. X-ray radiography, helium leak detection and gamma scanning are proven methods in fuel fabrication. Therefore, in place of actual measurement, only the necessary automated motions of the pins are demonstrated.

4.1 Weight Measurement Set-Up

Weight of the fuel pin is measured by placing the pin on a frame mounted on a load cell-based weighing balance. The pin weight is measured with ± 1 g accuracy.

4.2 TIR and Weld Bead Measurement Set-Up

A LVDT-based set-up is developed to measure concentricity of the top end plug with reference to the clad tube diameter and to measure weld bead size. Three LVDTs are placed in contact with clad tube, weld bead and top end plug. The pin is clamped and rotated to get the TIR and weld bead size.

4.3 Pin Length Measurement

After pushing the pin against a stopper, the length of pin is measured using an LVDT.

4.4 Bow Measurement

Bow of fuel pins is usually measured by placing the pin on a surface plate and visually examining the gap between the pin and the surface plate. If a gap is visible, feeler gauges are inserted in the gap to measure the gap. As this technique is not amenable for automation, we have developed a novel technique for bow measurement. The set-up consists of a surface plate, a set of active rollers and a set of LVDTs. The rollers can be lifted, lowered and rotated. In the lifted condition, the rollers can rotate the pin placed on it. In the lowered condition, the pin gets transferred on the surface plate. Readings of the LVDTs, which are mounted below the surface plate, provide the bow. The pins are rotated to measure the bow in other planes of the pin.

4.5 Visual Inspection

A high-resolution camera, which is mounted on a moving frame, is used for visual inspection of the pin. The pin is rotated and scanned by the moving camera across its length. Visual inspection is not automated; the operator takes a decision based on the image available to him at the control station.

4.6 X-ray Radiography

X-ray radiography is done to assess the weld quality. The pin is fed into the set-up, and the weld area is exposed to X-ray. The pin is rotated in steps to expose the weld in different angles.

4.7 He-Leak Detection

The set-up consists of a helium leak test (HLT) chamber, a tray conveyor, tray pusher and a tray. The HLT chamber has an automated leak-tight door to load or unload tray through it.

4.8 Gamma Scanning

The pin is fed through a gamma scanner to confirm the presence of fuel pellets at all locations inside the clad tube. The system includes a passive gripper and a motorized chain drive for full-length axial travel of the pin.

5 Process Equipments

Processes required for AHWR fuel pin fabrication are similar to those in the fabrication of fuel pins of other reactors. Therefore, we have not duplicated the process equipments in our system. However, we have demonstrated necessary pin handling and associated equipment movements in our system. The major tools used are listed below.

5.1 *Top End-Plug Welding*

Material handling tools required for end-plug welding are developed. The pin is fed to the welding chamber, stopped at specified location, clamped and rotated at the required speed.

5.2 *Buffing*

The buffing set-up consists of two buffing wheels: one motorized and the other an idler. The welded area of the pin is brought near the wheels; the buffing wheels are pressed against the pin; and the pin is rotated between the pressed wheels.

6 Control System

A remote control station was developed to monitor and to control the large number of processes. Apart from the centralized station, each cell also has separate touchscreen interface for monitoring and controlling the sub-systems. Centralized controller, referred as supervisor control and data acquisition (SCADA), monitors and controls the activities in the three cells and the movement of RGV. Synchronization of various processes is essential in the system. Therefore, real-time feedback of the status and non-completion of an action within the pre-defined time is reported. It helps to identify the problem, before they turn into significant failures. Centralized station periodically obtains data from remote input modules and controls various actuators through programmable logic controller (PLC). Win CC, an open architecture, SCADA, is installed on an industrial PC. It is configured for visual display of status, alarm handling, logging and report generation. Images from CCTV cameras are also available at the control station. Figure 5 shows overview of the control diagram for controlling this complex system. This system aids in visualizing real-time process data and facilitates configuration and monitoring of process data (Fig. 6).

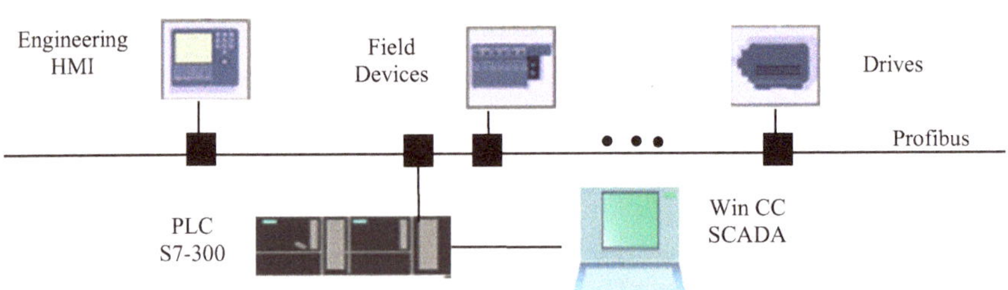

Fig. 5 Overview of the control system for AHWR fuel pin fabrication

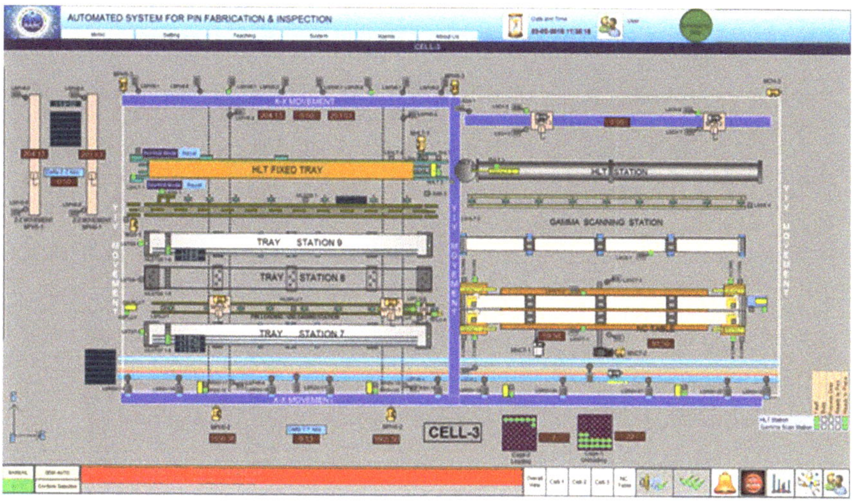

Fig. 6 SCADA display of one of the cells

7 Conclusions

An attempt has been made to develop a comprehensive full-scale mock-up system to demonstrate all functional and qualitative requirements of automated AHWR fuel pin fabrication. In this mock-up system, many novel tools and techniques for automated transfer of the pin and the cage have been developed. A few conventional techniques have been modified for automated inspection. Constraints arising from the hot cell, such as space limitations, maintainability, access and interference with remote handling tools, spread of contamination, etc., are considered in the design. Another outcome of the mock-up facility is optimized layout of hot cell and equipments, which reduces the number of material transfers, reduces the number of equipments, optimally utilise hot cell space and provides access for maintenance. A hybrid layout, combining the desired features of fixed-position layout, product-oriented layout and process-oriented layout, is adopted in this mock-up demonstration. This mock-up demonstration facility is a first step towards commercialization of automated fuel fabrication in a shielded facility.

Acknowledgements Implementation of such large and complex system would not have been possible, if we did not have a support of many individuals and organizations. Authors hereby duly acknowledge support and guidance from Shri Arun Kumar, Ex-Director, Nuclear Fuels Group and Dr D. N. Badodkar, Associate Director, Design, Manufacturing and Automation Group for development of the system. We are also thankful to Shri R C Sharma, Ex-Director, Reactor Group for giving us permission to install the system in Reactor Group premises. We thankfully acknowledge support from all staff members of Division of Remote Handling and Robotics, Nuclear Fuels Group and Reactor Group.

References

1. R.K. Sinha, A. Kakodkar, Design and development of the AHWR—the Indian thorium fuelled innovative nuclear reactor. Nucl. Eng. Des. **236**, 683–700 (2006)
2. S. Panda, S. Gupta, S. Sharma, K. Jayarajan, A. Saraswat, P.S. Somayajulu, Automation system for fuel fabrication in shielded facility, in *Indian Nuclear Society Annual Conference* (2014)
3. K. Jayarajan, R. Sahu, S. Panda, M.N. Rao, M. Singh, S. Gangotra, U. Basak, A. Kumar, P.B.S. Sengar, *Remotised Fabrication Scheme for AHWR Fuel Pin, NMD-ATM* (2007)
4. K. Anatharaman, V. Shivakumar, R.K. Sinha, Design and fabrication of AHWR fuel, in *Characterization and Quality Control of Nuclear Fuels* (2002)
5. S. Panda, S. Gupta, A. Saraswat, V. Ahire, S. Sharma, P.S. Somayajulu, K. Jayarajan, Mock-up facility for AHWR fuel pin fabrication: a step towards fully automated fuel fabrication, in *Thorium Energy Conference (ThEC15), Mumbai* (2015)

Online Fuel Failure Detection and Damage Severity Analysis for Thorium-Based AHWR Fuel Matrix—An Empirical Analysis

R. Rajalakshmi, Roshini Robin, K. Umashankari, A. Rama Rao and P. K. Vijayan

Abstract A clad failure results in the escape of fission products from the fuel to coolant. Continued operation of the reactor with the presence of failed fuel would cause excessive radioactive contamination of the main heat transport (MHT) system and its associated components. Therefore, online detection and precise location of failed fuel in the core are necessary for the safe and healthy operation of the reactor and to reduce the man-rem exposure. For the development of online system for iodine and gaseous fission product monitoring for AHWR, an empirical analysis was carried out to compute these fission product release rate data for thorium-based AHWR mix-oxide fuel consisting of $(Th–U^{233})O_2$ and $(Th–Pu)O_2$. The release rate and activity concentration rates in the coolant were calculated for various types of fuel failures, and a feasibility study was carried out for online gaseous fission product and iodine monitoring using HPGe detector and high-resolution gamma-ray spectrometer system. Further, this paper also discusses the different methodologies for identifying severity of fuel damage.

R. Rajalakshmi (✉) · R. Robin · A. Rama Rao · P. K. Vijayan
Bhabha Atomic Research Centre, Reactor Engineering Division,
Trombay, Mumbai 400085, India
e-mail: rajlaxmi@barc.gov.in

R. Robin
e-mail: roshini@barc.gov.in

A. Rama Rao
e-mail: arr@barc.gov.in

P. K. Vijayan
e-mail: vijayanp@barc.gov.in

K. Umashankari
Bhabha Atomic Research Center, Reactor Physics Design Division,
Trombay, Mumbai 400085, India
e-mail: umak@barc.gov.in

A. K. Nayak and B. R. Sehgal (eds.), *Thorium—Energy for the Future*,
https://doi.org/10.1007/978-981-13-2658-5_25

Keywords Fuel failures · Tramp fissile content · Counts · Escape time constant
Surface-exchange coefficient

Nomenclature

R_i^0 Release rate in atoms/s of fission products from fuel pellets into fuel-clad gap.
B_i Fission product birth rate in the fuel in atoms/s.
l Typical AHWR fuel length in meters i.e. 3.5 m.
y_c Cumulative yield of the nuclide i in atoms/fission.
P Linear power rating of the defective fuel pin in kW/m.
D' Empirical diffusion coefficient in s^{-1}.
X Number of defective fuel pins.
N_i Inventory in fuel gap in atoms.
υ_i Escape time constant, s^{-1}.
α Surface-exchange coefficient in (m s^{-1}) between the gap and coolant.
E Defect size in mm^2.
λ_i Decay constant of species i, s^{-1}.
W Reactor water mass, kg.
β_c Reactor water cleanup (RWCU) system removal time constant, s^{-1} defined as / assuming 100% efficient.
f RWCU flow rate, kg/s.
β_s Steam removal time constant, ss^{-1} which is defined as F/W.
ε Iodine carryover $= \frac{\text{the concentration of species } i \text{ in the condensate}}{\text{the concentration of the species } i \text{ in the reactor water}}$.
F Steam flow rate, kg/s.
υ_s Sample volume of the detector in cc.
η Detector efficiency for a particular gamma energy.
I_γ Yield of a particular energy gamma radiation from a radioisotope.
w_s Weight of the sample in mg.

1 Introduction

Fuel failures have always been an important operating concern for nuclear power plants and play a critical role in health, safety, and the economics of an operating reactor. A clad failure results in the escape of fission products from the fuel to coolant. Continued operation of the reactor with the presence of failed fuel would cause excessive radioactive contamination in the MHT system, difficulty in the maintenance of MHT equipment and piping, increase in man-rem, and activity in the reactor building, distortion in the defective fuel element due to its continued irradiation, and hence making its removal from the location difficult. Hence, online fuel failure detection is very essential which will reduce the man-rem exposure and enable early detection.

The thorium-based advanced heavy water reactor (AHWR), being developed by BARC, has a composite fuel cluster consisting of $(Th–U^{233})O_2$ and $(Th–Pu)O_2$. AHWR is pressure tube-type heavy water-moderated and boiling light-water-cooled natural circulation reactor. Hence, as in BWR type of reactors the gaseous fission products released as a result of fuel failures reaches the off-gas system. Thus for AHWR, online fuel failure detection by noble gas monitoring from the off-gas system and iodine isotopes monitoring from the purification system are envisaged. As there is not much data pertaining to fuel failure fission product release rates and its activity concentrations of volatile and non-volatile isotopes in the coolant are available for thorium-based fuel configuration, an empirical analysis was carried out. This analytical work comprises the fission product release rate calculations, activity concentrations in coolant and off-gas systems, online fuel failure detection mechanism, and damage severity analysis methodologies for AHWR fuel matrix.

2 AHWR Fuel Matrix

AHWR uses a composite fuel cluster containing both $(Th–U^{233})$ and (Th–Pu) MOX fuel pins in the equilibrium core configuration. The AHWR core has 452 number of fuel clusters each having 54 fuel pins. The fuel cluster lattice configuration is shown in Fig. 1. AHWR equilibrium core fuel cluster also has an axial enrichment of 4% Pu in lower half and 2.5% Pu in upper half. The equilibrium core inventories of the fertile and fissile isotopes of Th, U, and Pu of the inner and outer pins for the top and bottom halves of the cluster are separately given in Table 1 [4].

The % fissile content in the inner and the outer pins for an average top incore burnup of 15000 MWd/t and average bottom incore burnup of 24,000 MWd/t is as follows:

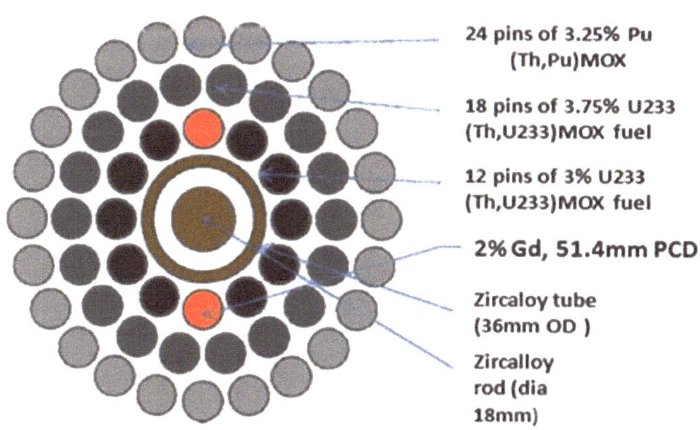

Fig. 1 AHWR fuel cluster lattice configuration

Table 1 Fuel assembly fertile and fissile isotopic inventories for the AHWR equilibrium core

Isotope	Isotope inventories of g/kg of HM				Fuel assembly inventory (kg)	Core inventory (kg)
	Top (2.5% Pu) avg. incore BU 15 GWd/T		Bottom (4% Pu) avg. incore BU 24 GWd/T			
	Inner pins (30)	Outer pins (24)	Inner pins (30)	Outer pins (24)		
Th233	956.63	962.68	951.53	941.94	111.03	50,189.04
U233	26.65	8.62	23.96	10.79	2.140	967.23
U234	1.67	0.45	2.35	0.8	0.162	73.41
U235	0.15	0.04	0.29	0.08	0.017	7.84
Total U	**28.48**	**9.11**	**26.63**	**11.69**	**2.321**	**1049.31**
Pu239	0	2.54	0	1.85	0.114	51.36
Pu240	0	5.77	0	7.68	0.348	157.36
Pu241	0	2.34	0	3.23	0.144	65.17
Pu242	0	1.18	0	2.27	0.089	40.36
Total Pu	**0**	**11.86**	**0**	**15.13**	**0.699**	**315.77**

Inner pins: Top half = 99.4% U233, 0.559% U235, 0% Pu239, 0% Pu241.
Bottom half = 89.40% U233, 10.6% U235, 0% Pu239, 0% Pu241.
Outer pins: Top half = 63.66% U233, 0.29% U235, 18.759% Pu239, 17.28% Pu241
Bottom half = 67.6% U233, 0.5% U235, 11.6% Pu239, 20.25% Pu241.

3 Fission Product Release Rate Calculations of Gaseous Fission Products and Iodine Isotopes for AHWR Fuel

Considering the fuel failure detection, among the fission products, a trustworthy calculation of source strengths can only be made using specific radio nuclides such as fission product noble gas isotopes and iodine isotopes. It has been generally accepted that diffusion is the primary release mechanism for volatile fission products from oxide fuel pellets. According to the equivalent sphere model proposed by Booth, the release rate of nuclide '*i*' into the fuel-clad gap is given by:

$$R_i^0 = 3B_i X\left(D'/\lambda\right)^{1/2} \tag{1}$$

where B_i is given as,

$$\begin{aligned} B_i = & \frac{1}{200}\left(\frac{\text{fission}}{\text{Mev}}\right) * l\ (\text{m}) * \frac{1}{1.602 \times 10^{-19}}\left(\frac{\text{Mev}}{\text{MJ}}\right) \\ & * \frac{1}{1000}\left(\frac{\text{MJ}}{\text{kW s}}\right) * y_c\left(B_u\right)\left(\frac{\text{atoms}}{\text{fission}}\right) * P\left(\frac{\text{kW}}{\text{m}}\right) \end{aligned} \tag{2}$$

$$B_i\left(\frac{\text{atoms}}{\text{s}}\right) = F_{\text{f}} * y_{\text{c}} * P\left(\frac{\text{atoms}}{\text{s}}\right) \tag{3}$$

and D' is given as,

$$D' = 9.0 \times 10^{-12} * \frac{10^{0.114P}}{P^2} \tag{4}$$

The release rate mechanism of fission products from defective duel into the reactor coolant can be theoretically modeled as a fission product mass conservation relation given as,

rate of change of the fission product inventory in the fuel clad gap
= release rate from the fuel pellets − decay − leakage into the coolant − burnup rate

$$\frac{\mathrm{d}N_i}{\mathrm{d}t} = R_i^0 - \lambda_i N_i - \upsilon_i N_i - \phi\sigma_i N_i \tag{5}$$

Upon integration,

$$N_i = \frac{R_i^0}{\lambda_i + \phi\sigma_i + \upsilon_i}\left(1 - \mathrm{e}^{-(\lambda_i+\phi\sigma_i+\upsilon_i)t}\right) \tag{6}$$

At steady state, when $t \to \infty$

$$N_i = \frac{R_i^0}{\lambda_i + \phi\sigma_i + \upsilon_i} \tag{7}$$

The release rate in atoms/s to the reactor coolant $= R_i = N_i\upsilon_i = \frac{R_i^0\upsilon_i}{\lambda_i+\phi\sigma_i+\upsilon_i}$
Where υ_i is the escape time constant in s^{-1} which is a function of the defect geometry and size, generally defined as,

$$\upsilon_i = \frac{2n\alpha}{l} \tag{8}$$

n describes the defect geometry. $n=1$ for a mid-rod defect, $n=0.5$ for an end element defect.

The surface-exchange coefficients for iodine and noble gases are given as follows:

$$\alpha_I = 10^{0.7\log_{10}E-7.5} \tag{9}$$

$$\alpha_N = 7.1 \times 10^{-6} \tag{10}$$

Substituting for R_i^0, $R_i = \frac{\upsilon_i}{\lambda_i+\phi\sigma_i+\upsilon_i} * \frac{A}{\sqrt{\lambda}}$ where $A = 3F_{\text{f}}y_{\text{c}}P\sqrt{D'}$

The burnup rate $\phi\sigma_i$ is generally small and can be neglected, thus the release rate is,

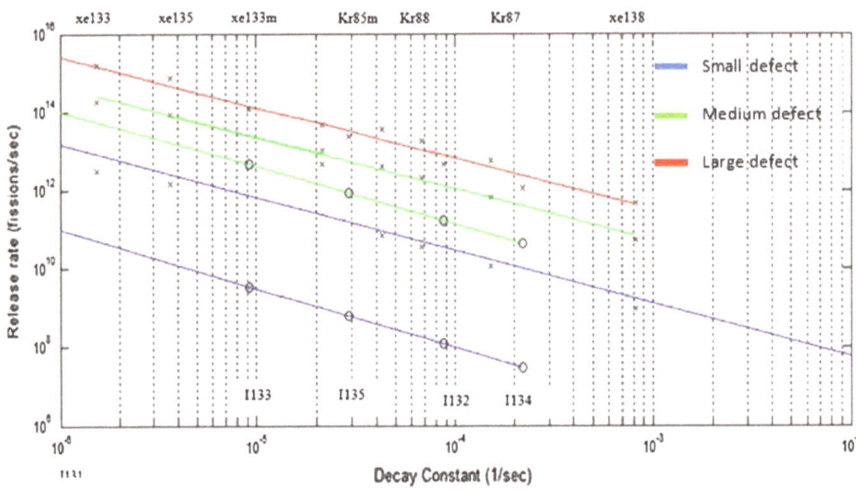

Fig. 2 Release rate versus decay constant for noble gases and iodine isotopes released from various types of defects in the AHWR fuel

$$R_i = \frac{v_i}{\lambda_i + v_i} * \frac{A}{\sqrt{\lambda}} \tag{11}$$

The release rate in fissions/s is

$$\frac{R_i}{y_c} = \frac{v_i}{\lambda_i + v_i} * \frac{3F_f P\sqrt{D'}}{\sqrt{\lambda}} \tag{12}$$

The activity release rate in Bq/s is

$$A_i = R_i\lambda_i = \frac{v_i\lambda_i}{\lambda_i + v_i} * \frac{A}{\sqrt{\lambda}} \tag{13}$$

The release rate calculations were done using MATLAB for the AHWR equilibrium core operating at 100% FP for the following three types of defects, whose results are shown in Fig. 2.

- A small pinhole defect of 1 mm^2 in a fuel pin
- A medium defect of 100 mm^2 in 23 fuel pins
- A large defect of 28 slits each 36 * 0.3 mm (=300 mm^2) from 100 pins.

Iodine exists in the form of insoluble CsI compound in fuel pins. Thus to permit iodine release, water or steam should enter the fuel pin to make iodine soluble. Thus as observed in the graph for small leaks, only noble gases escape and enter the coolant whereas release rate of iodine isotopes is very less. As the defect size increases, release rate of iodine increases. Thus from the graph, it can be seen that from small defect to large defect, the difference between the release rates of iodine and noble gases reduces and for large defects, the release rates of both will be almost of same magnitude. Furthermore, it is seen that as the defect size increases the slope of the fission product release rates increases due to the release of short-lived fission products.

4 Online Fuel Failure Detection System for AHWR

The fuel integrity analysis is carried out for the AHWR equilibrium core from the release rate concentrations of the noble gas isotopes of Xe-138, Kr-87, Kr-88, Kr-85m, Xe-135, and Xe-133 and iodine isotopes of I-131, I-132, I-133, I-134, and I-135. The proposed AHWR failed fuel detection system consists of online nuclide-specific high-resolution HPGe detector and gamma-ray spectrometer system for monitoring concentration of fission product activities released from the defective fuel. The sample for online noble gases monitoring is taken after the condenser before the delay tank. Iodine isotopes are monitored from the coolant samples taken from the purification circuits before the filters and ion exchangers. The fission product activity levels from the failed AHWR fuel for different types of defects and the corresponding detector counts calculations are discussed below.

4.1 Activity Level Calculations of Noble Gas and Iodine Isotopes from the Off-Gas and Purification Systems at the Detector Location for Different Type of Fuel Failures

The activity (Bq/kg) of the noble gases in the off-gas is given by:

$$C_i = \frac{A_i}{F_{\text{steam}}} \tag{14}$$

A_i is the activity in Bq/s of species *i* released from the fuel defect into the coolant.
F_{steam} is the steam flow rate in kg/s

Iodine activity in the reactor coolant taking into consideration iodine removal from the reactor water cleanup system and fraction carryover to steam is given by:

$$C_i = \frac{A_i}{W(\lambda_i + \beta_c + \beta_s)} \tag{15}$$

Measurement of the fission product activity by the HPGe detector system is given by

Table 2 Calculated activity values from fuel defects and from tramp fissile content at the detector location

Isotopes	Activity in μC/l for 0.1% fuel failure			
	Small defect	Medium defect	Large defect	From tramp fissile content
Xe133	0.219336	0.516091	2.269657	0.07374
Kr85m	0.032799	0.077381	0.339256	0.024138
Kr88	0.063944	0.151054	0.662644	0.057294
Xe135	0.187767	0.444048	1.945587	0.037135
Xe138	0.01108	0.026152	0.114811	0.026127
Kr87	0.031517	0.074349	0.326631	0.040849
I134	0.001919	0.114861	1.071855	0.065782
I131	0.055951	2.338468	16.14142	0.043236
I133	0.030149	1.722761	15.28965	0.072944
I132	0.004531	0.2699	2.511782	0.05809
I135	0.012872	0.760958	6.99466	0.058886

$$R_{\text{offgas}}\ (\text{atoms/ml}) * v\ (\text{ml}) * \mathrm{e}^{-\lambda t_d} * \left(1 - \mathrm{e}^{-\lambda t_c}\right) * \eta * I_\gamma \quad (16)$$

$$C_i\left(\frac{\text{Bq}}{\text{kg}}\right) * w\ (\text{kg}) * t_c(\text{s}) * \eta * I_\gamma \quad (17)$$

The activity in μC/l was calculated for 0.1% fuel failures of the three types of defects for the defect locations in inner pins, outer pins, top and bottom half of the fuel cluster separately. From the calculations, the activity levels of noble gases and iodine isotopes are less for the outer pins which is in the detectable range of the detector. The corresponding data for the outer pins for all the types of defects are given in Table 2. The total tramp fuel content (for 54 * 452 fuel pins) for the AHWR equilibrium core, i.e., after 6 years of irradiation was calculated using ORIGEN code and the corresponding activity is given in Table 2.

4.2 *Detector Output Counts Calculations for Noble Gas and Iodine Radioisotopes for Various Types of Fuel Failures*

For the calculations, commercially available coaxial HPGe detector of crystal length 5.2 cm, relative efficiency 19.3%, resolution 1.87 keV for the Co^{60} gamma ray at 1332.5 keV and a source-detector distance 3 cm is considered. The calculated detector counts for noble gas and iodine isotopes considering 0.1% fuel failures for small, medium, and large defects are shown in Figs. 3 and 4, respectively.

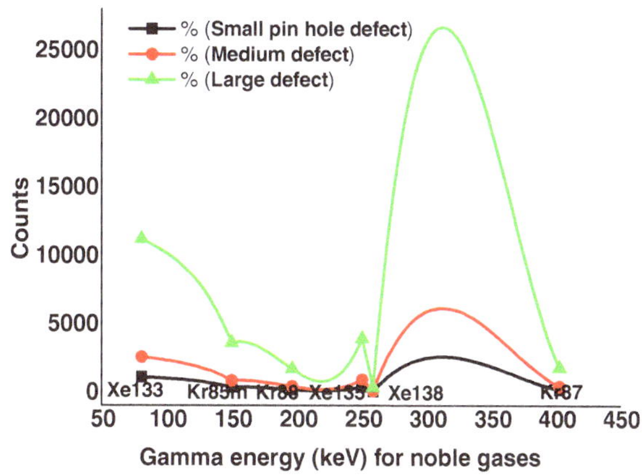

Fig. 3 Detector counts for noble gas isotopes

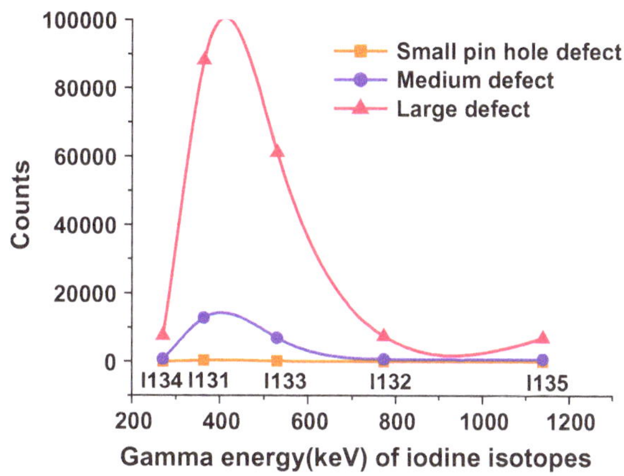

Fig. 4 Detector counts for iodine isotopes

5 Damage Severity Analysis and Discussion

5.1 Feasibility Analysis from the Release Rate into Coolant at the Defective Pin Location

The release rate of noble gases from the defective fuel for small defects (1 mm^2) is 100 times higher than iodine isotopes release rates. The release rate values for the longest-lived isotopes are I133 ($t1/2 = 20.8$ h) is of order of 10^{10} fissions/s and Xe133 ($t1/2 = 5.25$ days) is of order of 10^{12} fissions/s. For medium to large defects, the order of release rates of noble gas and iodine isotopes are almost same. Thus, for smaller defects detection of noble gas isotopes are preferred as compared to iodine isotopes.

5.2 Feasibility Analysis from the Detector Output Counts

Detector output counts were calculated for 0.1% fuel failures for the AHWR fuel matrix. For medium to large defects, I131 and I133 detector counts are in 10^4 range as compared to all other noble gas isotopes which come in 10^3 range. Among the iodine isotopes, the activity level of I131 is high as compared to other iodine isotopes. Activity levels at the detector location of I131 for medium and large defects are around 3 and 16 μCi/l. This level is far above the activity level from the tramp fissile inventory which comes around 0.043236 μCi/l. For small defects, detector counts for the noble gas isotopes are appreciable as compared to iodine isotopes. Monitoring of Xe133 and Xe135 isotopes are preferred for small defects. The activity levels of Xe133 are 0.2193 μCi/l with a background activity of 0.07374 μCi/l.

5.3 Feasibility Analysis from the Release Rate Ratios

An approach for the defect size identification is from the fission product release rate ratio calculations from short- and long-lived isotopes of different species and of same species. This approach relies on the variation in hold-up between noble gases and iodine. The table for the release rate ratios of same and different isotopic species calculated for all the three types of defects for the AHWR fuel matrix is shown in Tables 3 and 4, respectively

On comparing the release rate ratio values from Tables 3 and 4, the release rate ratio of different isotopic species can be considered as a good indicator of fuel failure detection. From the calculations, Xe133/I131 release rate ratio can be considered for

Table 3 Release rate ratio of same isotopic species

Ratio	Small defect	Medium defect	Large defect
I131/I133	27.771	20.312	15.798
Xe133/Xe135	0.702	16.759	16.821
Xe133/Kr88	85.828	85.377	85.591
Xe135/Kr88	122.192	5.094	5.0881

Table 4 Release rate ratio of different isotopic species

Ratio	Small defect	Medium defect	Large defect
Xe133/I131	3.3	1.86	1.18
I131/Xe135	0.02	9.03	14.2
I131/Kr88	2.6	46	72.4
Xe133/I133	917	37.7	18.7
Xe135/I133	1310	2.25	1.11
I133/Kr88	0.09	2.26	4.58

identifying different types of defects. The release rate ratio of Xe135/I133 can be considered for identifying the onset of the fuel failure by the identification of small pinhole defect. The release rate ratio of I131/Kr88 can also be considered for the identification of different types of fuel failure.

5.4 Estimation of the Burnup for Defective Fuel Rods

The measurements of Cs-134 and Cs-137 activities in the reactor coolant can be used as an indicator of the exposure of a failed fuel rod from which the fission products are released. Cs-137(30y) is produced directly from fission, and Cs-134 (2.1y) is produced by neutron activation on Cs-133, a stable fission product. The activity of Cs-134 increases proportionately to the square of the fuel exposure. Thus as the fuel exposure increases, ratio of Cs-134 to Cs-137 increases and can be used for the estimation of defective fuel rod burnup.

6 Conclusion

An empirical analysis was carried out for the online gaseous fission product and iodine monitoring for AHWR fuel for different types of fuel failures. Calculations for the release rate of fission products from the defective fuel, its activity concentrations in the reactor coolant and off-gas system, and the corresponding detector output counts were computed. Feasibility of detection of 0.1% fuel failures in AHWR was analyzed by comparing the calculated results with the background activity due to tramp fissile material in the AHWR equilibrium core. Methodologies for identifying the severity of fuel damage and burnup of the defective fuel are discussed.

References

1. C.C. Lin, *Radiochemistry in Nuclear Power Reactors*, Chap. 3, Nuclear Science Series, Pleasanton California (1996)
2. C.C. Lin, J.M. Skarpelos, Monitoring of fission product release in a boiling water reactor. J. Radio Anal. Nucl. Chem. **220**(2), 173–181 (1997)
3. Safety Analysis Report of AHWR, 30 Apr 2015
4. S. Samanta, A. Kumar, U. Kannan, Physics DBR of Spectrum Averaged One-Grouped Cross Sections for Gaseous Fission Products in Advanced Heavy Water Reactor Fuel Assembly
5. A. Gultekin, G. Kaynak, O. Gurler, Determination of full Energy peak efficiency of HPGe detector from 59.5 to 1332.5 keV. Int. J. Pure Appl. Phys. **44**, 281–286 (2006)

Part IV
Reactor Physics

Xenon Dynamics in AHWR-LEU with Coolant Density and Fuel Temperature Feedback

Arindam Chakraborty, Amit Thakur, Baltej Singh and P. D. Krishnani

Abstract The static core calculations assume average neutron flux; as a result, the xenon concentration, coolant density, and fuel temperature are also assumed to have core average values. In order to study xenon dynamics for different perturbation (like refueling of channels and withdrawal of regulating rods), explicit xenon calculations with coolant density and Doppler feedback were carried out for AHWR-LEU core. The calculations were carried out at small time steps (say 0.1 h intervals). Time-dependent quarter core power distribution was studied for different reactivity perturbation.

Keywords AHWR-LEU · Xenon dynamics · Coolant density feedback
Doppler feedback

1 Introduction

Xenon plays a very important role in nuclear reactor due to its high thermal neutron capture cross section. Due to flat core power distribution in AHWR-LEU [1], it is neutronically loosely coupled in nature. The effect of neutronic loose coupling is manifested in small difference in the eigenvalues between fundamental mode and higher modes. Small perturbation in reactor core can excite the higher modes resulting in spatial xenon oscillation. It leads to distortion in power distribution keeping total power unchanged. It has been observed that due to small perturbation in some part of the reactor, the power distribution starts oscillations such that the total power remains unchanged. A detailed study has been performed to see the effect on power distribution due to operational transient like movement of regulating rod (RR) and refueling of channels. The study also includes the reactivity feedback due to change in coolant density and fuel temperature.

A. Chakraborty (✉) · A. Thakur · B. Singh · P. D. Krishnani
Reactor Physics Design Division, Bhabha Atomic Research Centre,
Trombay, Mumbai 400085, India
e-mail: arindamc@barc.gov.in

A. K. Nayak and B. R. Sehgal (eds.), *Thorium—Energy for the Future*,
https://doi.org/10.1007/978-981-13-2658-5_26

2 Methodology

The core follow-up studies along with many other core calculations of AHWR-LEU have been performed by core follow-up code FEMFOL [2]. The studies have considered saturated xenon conditions during operation. The FEMFOL was modified [3] to incorporate the time-dependent xenon calculation to study the operational transient in AHWR-LEU. The scheme followed for the modification is shown in Fig. 1. The time-dependent xenon and iodine concentration is given in Eqs. (1) and (2). The solution [4] of the two coupled differential equation is given in Eqs. (3) and (4).

$$\frac{\mathrm{d}I_i^t}{\mathrm{d}t} = \gamma_I \Sigma_{fi} \phi_i^t - \lambda_I I_i^t \tag{1}$$

$$\frac{\mathrm{d}X_i^t}{\mathrm{d}t} = \gamma_X \Sigma_{fi} \phi_i^t + \lambda_I I_i^t - \left(\lambda_X + \sigma_X \phi_i^t\right) X_i^t \tag{2}$$

$$I_i^{t+1} = \frac{\gamma_I \Sigma_{fi} \phi_i^t}{\lambda_I}\left(1 - \mathrm{e}^{-\lambda_I \Delta t}\right) - I_i^t \mathrm{e}^{-\lambda_I \Delta t} \tag{3}$$

$$\begin{aligned} X_i^{t+1} = & \frac{(\gamma_I + \gamma_X)\Sigma_{fi}\phi_i^t}{\lambda_X + \sigma_X \phi_i^t}\left[1 - \mathrm{e}^{-(\lambda_X + \sigma_X \phi_i^t)\Delta t}\right] \\ & + \frac{\gamma_I \Sigma_{fi}\phi_i^t - \lambda_I I_i^t}{\lambda_X - \lambda_I + \sigma_X \phi_i^t}\left[\mathrm{e}^{-(\lambda_X + \sigma_X \phi_i^t)\Delta t} - \mathrm{e}^{-\lambda_I \Delta t}\right] + X_i^t \mathrm{e}^{-(\lambda_X + \sigma_X \phi_i^t)\Delta t} \end{aligned} \tag{4}$$

It is assumed that the neutron flux in each mesh remains constant for a given small interval of time (or time step). The absorption characteristics of each fuel mesh (lattice cell) shall be incremented corresponding to the estimated concentration of ^{135}I and ^{135}Xe, and new flux distribution is estimated at each time step. In the next time step, the updated flux distribution is used to estimate the corresponding concentration of ^{135}I and ^{135}Xe for updating the absorption characteristics of each mesh, and new flux distribution is obtained at each fuel mesh. This process is repeated for all the time steps for each fuel mesh. The neutron flux in each mesh, at each time step, is obtained through iterative process till the maximum relative difference in the neutron flux between the two successive steps is less than 10^{-4}.

Core calculations of AHWR-LEU equilibrium core have been performed considering the burn-up and flux distribution corresponding to 5876 FPD of AHWR-LEU core follow-up. Whole core of AHWR-LEU has been divided into four radial zones. Equilibrium core layout of AHWR-LEU is shown in Fig. 2.

3 Perturbation Due to Simultaneous Refueling of Four Channels

Xenon dynamics study with coolant density and Doppler feedback was carried out for a case where a mini batch of four channels (one each in four quadrants) was refueled

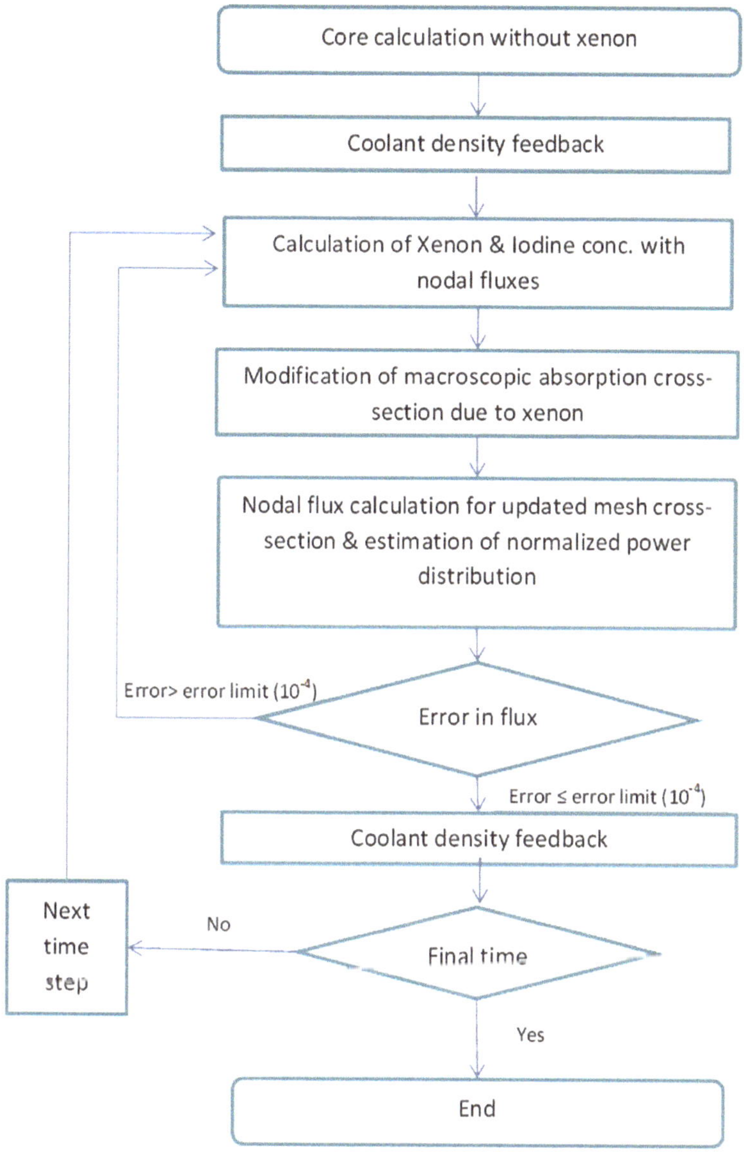

Fig. 1 Flowchart for explicit xenon calculation with coolant density and fuel temperature feedback

simultaneously in saturated xenon condition. Locations of four channels belonging to a mini batch are (13,20), (21,14), (20,21), and (14,13). The simultaneous refueling of four channels did not cause any significant variations in the spatial power distribution with respect to time because each of the quadrants was perturbed to the same level.

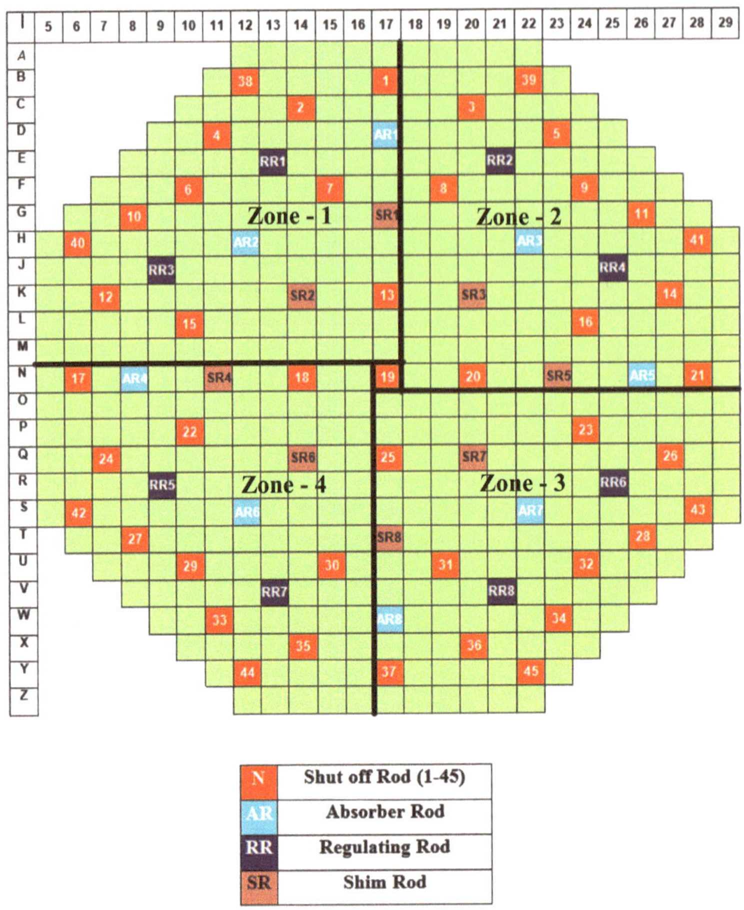

Fig. 2 Four radial zones in AHWR-LEU core

4 Perturbation Due to Sequential Refueling of Four Channels with Some Time Interval

The actual refueling operation is expected to have some time interval between two consecutive refueling operations. Therefore, refueling of the four channels of AHWR-LEU was simulated by refueling one channel after another with certain time intervals, and three cases with different time intervals of 3, 6, and 9 h were considered. For all the cases, same sequence of refueling of channels (located at (13,20), (21,14), (20,21), and (14,13)) was considered. Variation in the quadrant power of four radial zones with time for all the cases was studied with reactivity feedbacks and without reactivity feedbacks. The comparative results are given in Figs. 3a, b, 4a, b and 5a, b.

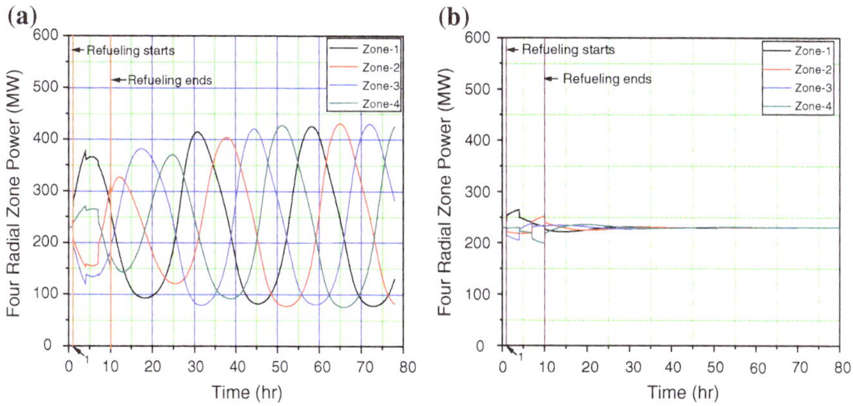

Fig. 3 **a** Variation of power of four radial zones with time at 100% power without any feedback (interval between two refueling is 3 h), **b** variation of power of four radial zones with time at 100% power with feedback (interval between two refueling is 3 h)

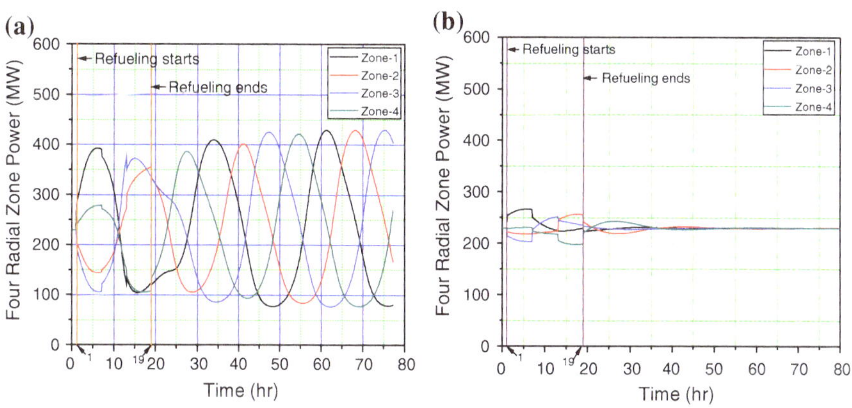

Fig. 4 **a** Variation of power of four radial zones with time at 100% power without any feedback (interval between two refueling is 6 h), **b** Variation of power of four radial zones with time at 100% power with feedback (interval between two refueling is 6 h)

5 Perturbation Due to Withdrawal of Single Regulating Rod

The time-dependent variation in the powers of four radial zones was also estimated after the full withdrawal of one regulating rod (RR) located at (9,13) with coolant density and Doppler feedback. Variation in the quadrant power of four radial zones with time was studied with reactivity feedbacks and without reactivity feedbacks. The comparative results are given in Fig. 6a, b.

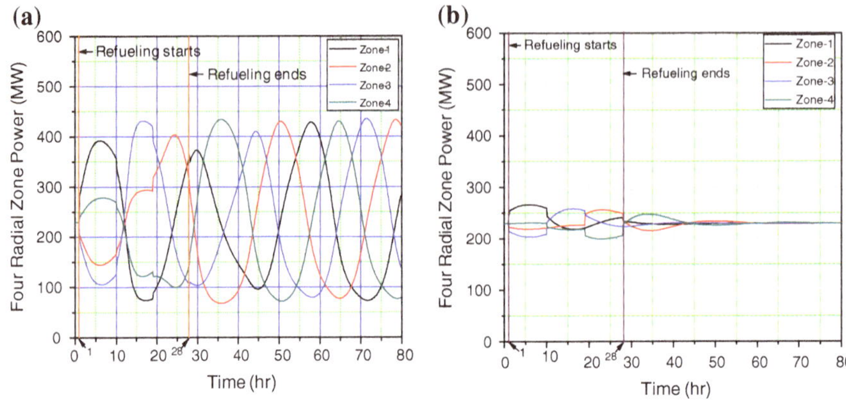

Fig. 5 **a** Variation of power of four radial zones with time at 100% power without any feedback (interval between two refueling is 9 h), **b** variation of power of four radial zones with time at 100% power with feedback (interval between two refueling is 9 h)

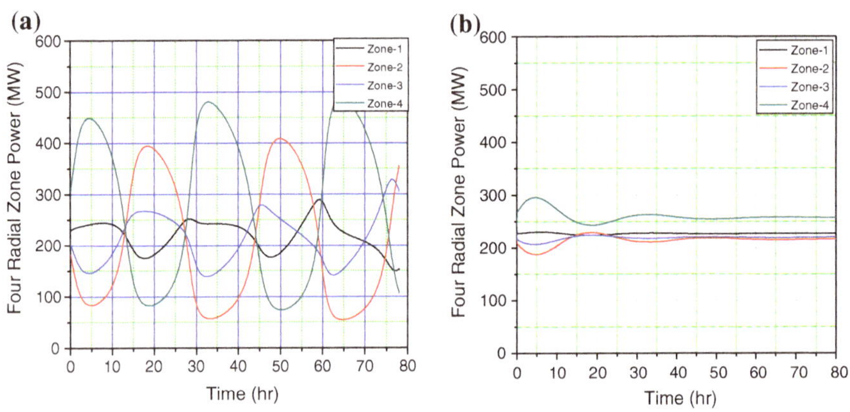

Fig. 6 **a** Variation of power of four radial zones with time at 100% power without any feedback, **b** variation of power of four radial zones with time at 100% power with feedback

6 Results and Discussions

The present study has provided very valuable information about xenon dynamics of the AHWR-LEU during operational transients like refueling of a fuel channel and the full movement of reactivity device (say RR/AR/SR).

The xenon dynamic calculations for the refueling of a mini batch of four channels showed that the simultaneous refueling of four channels of the mini is not expected to introduce any xenon instability. However, the successive refueling (one channel after another channel) of four channels after some interval of time is expected to introduce xenon instability in the radial zones when the feedbacks due to coolant density and Doppler were not considered; however, variations in radial power distribution diminish when the reactivity feedbacks are duly accounted.

The xenon dynamic calculations have also shown that the outward movement of single RR (fully or partially) is expected to introduce xenon instability. It was also observed that xenon instability due to outward movement of single rod is not expected.

Acknowledgements We are thankful to Sh. A. S. Pradhan, Head Reactor Physics and Shielding Section, RSA Group, NPCIL for providing the complete technical support, infrastructures, and guidance. We are highly thankful to Smt. S. Ray and Dr. S. Mishra for having a variety of many fruitful technical discussions. The authors of the report are also thankful to D. Chandrakar and Arnab Dasgupta, RED, BARC for providing channel power to coolant density correlation table and channel power to fuel temperature correlation table.

References

1. A. Chakraborty, et al., *Physics Chapter on Safety Analysis Report (Preliminary) on AHWR-LEU*, Report No. RPDD/HWRPS/AHWR/213/2012 dated 31 Dec 2012
2. A. Kumar, R. Srivenkatesan, *Nodal Expansion Method for Reactor Core Calculation*, Technical Report No. BARC-1249 (1984)
3. S. Ray, et al., *Design Manual on Reactor Physics*, TAPP-3&4/DM/01100/1
4. W.M. Stacey, *Nuclear Reactor Physics*, 2nd edn. (Wiley-VCH Verlag GmbH & Co. KGaA)

LORA Analysis Using Thorium in 220 MWe PHWR Fuel

Manish Raj, M. P. S. Fernando, Suma Nair and S. G. Ghadge

Abstract Thorium is not a fissile material and cannot either start or sustain a chain reaction. Therefore, a reactor using thorium would also need either enriched uranium or plutonium to initiate the chain reaction and sustain it until enough of the thorium has converted to fissile uranium (U^{233}). U^{233} can be produced on a regular basis by irradiating thoria pins in existing PHWRs. In order to retrieve and reprocess the irradiated fuel, the bundle is designed with few thoria pins and rest SEU pins. Another way is to uniformly mix the plutonium with thorium in all the pins. The reactivity load due to thoria pins is compensated by SEU/plutonium. The objective is to irradiate thorium for production of U^{233} and compensate the burnup loss by using SEU in separate pins of PHWR fuel bundle. In this paper, the loss of reactivity accident has been discussed for 220 MWe PHWR. The analysis has been carried out using computer code MRIF-FUEL (Jain, Yadav in Point kinetics code MRIF-FUEL, Theoretical Physics Division, BARC, 1989, [1]). The input data are generated using computer code CLUB (Krishnani in CLUB, a multigroup integral transport theory code for lattice calculations of PHWR cells, BARC/1992/E/017 [2]).

Keywords Thorium · SEU · Bundle-shift scheme · AR · RR · Regulation

1 Introduction

Uranium-233 is produced by the neutron irradiation of thorium-232. When Th^{232} absorbs a neutron, it becomes Th^{233}, which has a half-life of only 22 min. Th^{233} decays into protactinium-233 through beta decay. Pa^{233} has a half-life of 27 days, and beta decays into U^{233}. The design of reactor regulating system is such that the malfunction of any single component will leave the system in safe condition. However, simultaneous failures of either channel A/B together with transfer circuit may lead to loss of regulation accident (LORA). In the present analysis, LORA

M. Raj (✉) · M. P. S. Fernando · S. Nair · S. G. Ghadge
Nuclear Power Corporation of India Ltd., Mumbai, India
e-mail: rmanish@npcil.co.in

A. K. Nayak and B. R. Sehgal (eds.), *Thorium—Energy for the Future*,
https://doi.org/10.1007/978-981-13-2658-5_27

scenario of absorber rods (AR) and regulating rods (RR) moving together (half of total number) have been considered. The analysis is carried out for equilibrium core of different thorium fuel cycles for 220 MWe PHWR.

2 Cases Considered for LORA Analysis

The following six cases with different bundle-shift schemes (BSS) have been considered for the LORA analysis.

1a: 1 pin Th + 18 pin SEU (1.2 wt%)—2 BSS	1b: 1 pin Th + 18 pin SEU (1.2 wt%)—8 BSS
2a: 7 pin Th + 12 pin SEU (2.2 wt%)—2 BSS	2b: 7 pin Th + 12 pin SEU (2.2 wt%)—8 BSS
3: All pin 2.2% Pu-ThO_2—2 BSS	4: All pin 2.4% Pu-ThO_2—2 BSS

Figures 1 and 2 describe the position of thoria pins in fuel bundle considered for study along with SEU pins (Cases 1 and 2).

The point kinetics analysis considers six delayed neutron groups and reactivity feedback due to fuel and coolant temperatures. The contribution from delayed photoneutrons is not considered in the analysis. The different thorium fuel cycles considered for the analysis are listed below. The point kinetics and other parameters, such as effective delayed neutron fraction, prompt neutron generation time, LORA worth, and PSS worth, relative to the NU core are shown in Fig. 3.

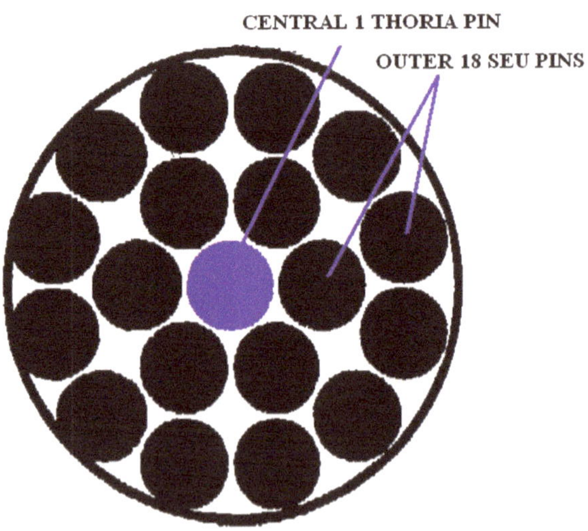

Fig. 1 Nineteen-element fuel cluster (1 pin thoria)

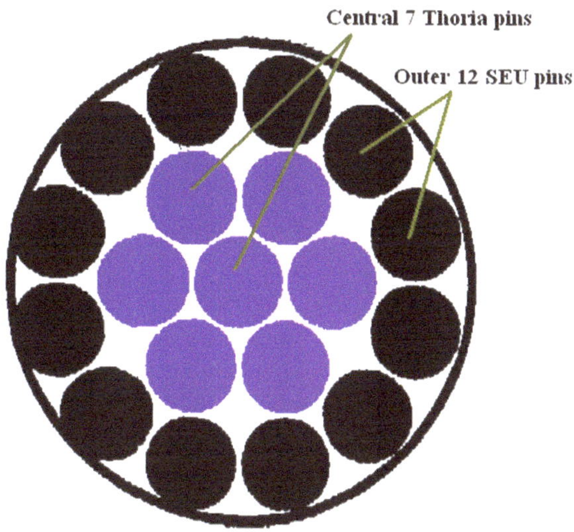

Fig. 2 Nineteen-element fuel cluster (7 pin thoria)

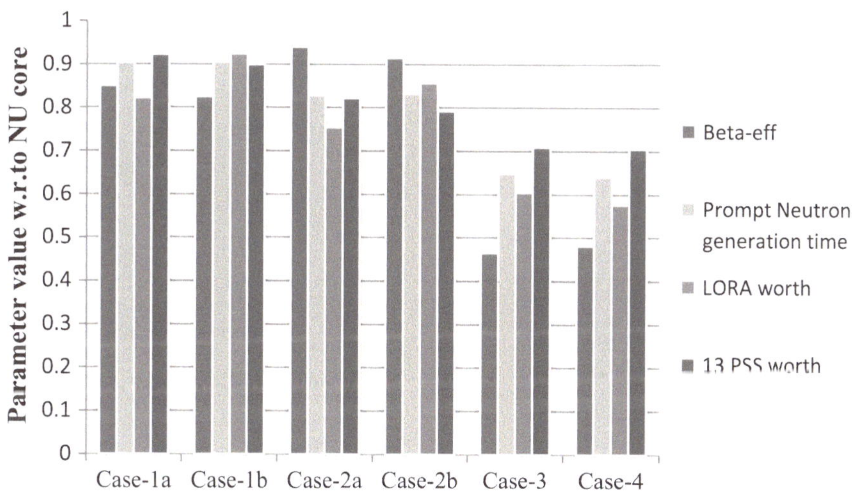

Fig. 3 Point kinetics and other parameters relative to NU core

The corresponding values of effective delayed neutron fraction, prompt neutron generation time, LORA worth, and PSS worth for NU core are 5.542 mk, 0.755 ms, 5.5 mk, and 29.7 mk, respectively. Figures 4 and 5 give the reactivity feedback coefficients due to fuel temperature and coolant temperatures, respectively.

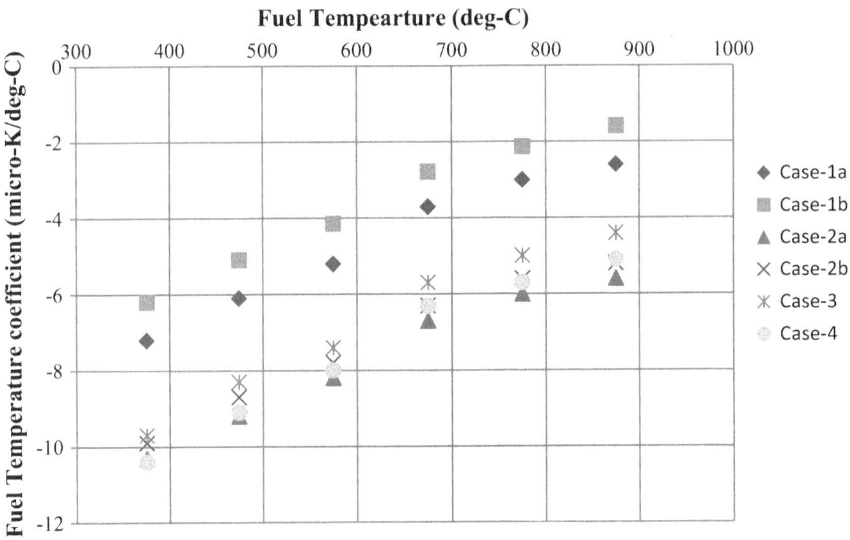

Fig. 4 Reactivity feedback due to fuel temperature

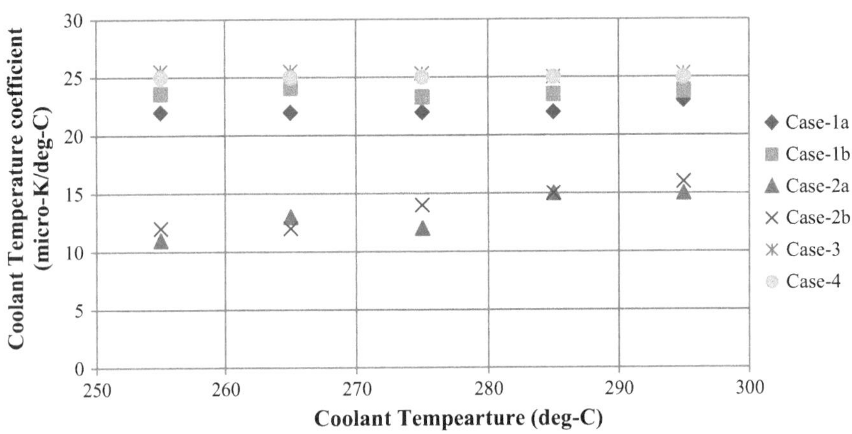

Fig. 5 Reactivity feedback due to coolant temperature

3 Results and Discussion

The results of LORA analysis for all six thorium fuel cycles are given in Table 1 along with full NU equilibrium core for comparison purposes.

It can be noted from these results that the cases are comparable to the NU equilibrium core, even though delayed neutron fractions for U^{233} and $Pu^{239/241}$ are lower and the worth of reactivity devices is lower. The LORA reactivity due to withdrawal of AR and RR has also reduced compared to NU core.

Table 1 Output of LORA analysis (AR and RR moving together)

Parameter	Case 1a	Case 1b	Case 2a	Case 2b	Case 3	Case 4	NU core
Trip sensed at (s)	8.9	7.8	11.64	10.11	8.88	10.08	6.49
Max. relative Power	111.7	112.1	111.2	111.4	111.3	111.2	112.31
Max. net reactivity (mk)	0.23	0.24	0.21	0.23	0.12	0.12	0.33
Final net reactivity (mk)	−20.89	−20.1	−16.77	−15.50	−14.9	−14.7	−21.85
T-fuel (center line) (°C)	1005	1002	1005	1002	1000	1003	994.5
T-coolant (average) (°C)	270.9	270.8	271	271	270.9	270.94	270.8

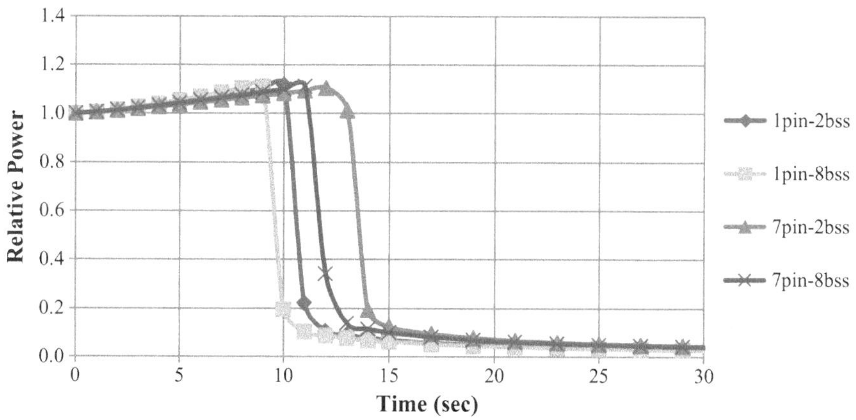

Fig. 6 Relative power variation with time for (Th + SEU) pin configurations

The final subcriticality margins for all cases are above 10 mk. The centerline fuel temperatures during the transient have reached about 1000 °C. Considering the various peaking factors such as axial, radial, and bundle peaking factors, the maximum fuel temperature would be about 2000 °C which is much below the melting point of fuel. The peaking factors are generated from lattice code CLUB, and the resultant peaking factor considering the effect of all the factors comes out to be about 1.92. Figures 6 and 7 give the relative power variation for all the cases with Th + SEU and with Th + PU, respectively.

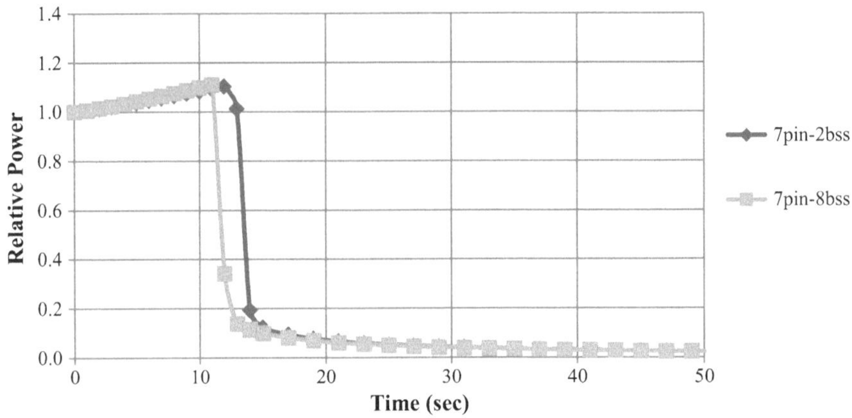

Fig. 7 Relative power variation with time for (Th + Pu) fuel

4 Conclusion

The cases considered for the analysis are comparable to the NU core from transient analysis point. This is due to the fact that in spite of reduction in effective delayed neutron fraction (due to higher discharge burnup as well as the presence of fissile isotopes of Pu in core), the LORA worth for all these cases has also been reduced. Furthermore, the feedbacks are stronger compared to the NU core. The final subcriticality margins for all cases are above 10 mk. The centerline fuel temperatures during the transient have reached about 1000 °C. Considering the various peaking factors such as axial, radial, and bundle peaking factors, the maximum fuel temperature would be about 2000 °C which is much below the melting point of fuel.

References

1. V.K. Jain, R.D.S. Yadav, *Point Kinetics Code MRIF-FUEL*, Theoretical Physics Division, BARC (1989)
2. P.D. Krishnani, CLUB, *A Multigroup Integral Transport Theory Code for Lattice Calculations of PHWR cells*, BARC/1992/E/017

Improvement in Estimation of Distribution Algorithm (EDA) for Fuel Loading Pattern Optimization in AHWR

Amit Thakur, Baltej Singh, Anurag Gupta, V. Duggal, K. Bhatt and P. D. Krishnani

Abstract The initial core of AHWR-LEU consists of 444 fuel channel locations, and it is proposed to load two differentially reactive clusters (namely, type-1 (more reactive) and type-2) in these locations for flux flattening. This loading pattern optimization problem (LPO) has been solved by applying genetic algorithm (GA) and estimation of distribution algorithm (EDA). It has been observed that genetic algorithm (GA) is more efficient for AHWR initial core LPO problem as compared to EDA. To improve the efficiency of EDA, different modifications have been suggested in this paper. As a first modification, the tournament selection method (similar as in GA) was applied in EDA for diversity. In the second modification, we have tried to improve the EDA by restricting the random numbers in a specified range without compromising randomness.

Keywords Loading pattern optimization (LPO) · Evolutionary algorithm AHWR · Genetic algorithm

1 Introduction

The initial core of AHWR-LEU [1] (Advanced Heavy Water Reactor with Thorium and Low-Enriched Uranium as fuel) consists of 444 fuel lattice locations and 69 locations for control and shutdown devices. For initial flux flattening, minimum two types of differentially reactive clusters have to be loaded in these 444 fuel lattice locations. The two types of cluster are named as type-1 and type-2. The initial core loading pattern optimization of AHWR-LEU requires simulation of 2^{62} ($\sim10^{18}$)

A. Thakur (✉) · B. Singh · A. Gupta · P. D. Krishnani
Reactor Physics Design Division, Bhabha Atomic Research Centre,
Trombay, Mumbai 400085, India
e-mail: thakur@barc.gov.in

V. Duggal · K. Bhatt
Computer Division, Bhabha Atomic Research Centre,
Trombay, Mumbai 400085, India

A. K. Nayak and B. R. Sehgal (eds.), *Thorium—Energy for the Future*,
https://doi.org/10.1007/978-981-13-2658-5_28

different combinations of arranging two types of fuel clusters in 444 locations. Simulation of one loading pattern requires 1–2 min of computation time; therefore, it is not practical to simulate all these combinations to choose the optimized loading pattern. These kinds of problems can be efficiently solved by population-based algorithms like simulated annealing (SA) [2], ant colony optimization (ACO) [3], genetic algorithm (GA) [4–6], and estimation of distribution algorithm (EDA) [7]. The loading pattern optimization problems in light water reactors (LWRs) and pressurized heavy water reactors (PHWRs) [8] have been solved by using SA, ACO, GA, and EDA. In this paper, EDA has been used to optimize the loading pattern initial core of AHWR-LEU [9]. During our earlier analysis [9], it was observed that the optimization results are very sensitive to internal parameters like weighting factor 'α' and for AHWR initial core LPO, adequate value of α is estimated to be 0.5. However, when we compared the analysis results with GA, it was observed that GA gave similar results with lesser population size in each generation. For GA, a population size of 240 is sufficient as compared to EDA where population size of 1200 is adequate. To improve the efficiency of EDA, different modifications have been suggested in this paper. It has been observed that [9] the higher value of α may lead to a trap in local minima for lesser population size in each generation. The best solutions of the previous generation are used to update the distribution function in EDA. Therefore, to avoid the chances of trap in local minima, tournament selection can be used to choose the candidates for updation of distribution function. The tournament selection method can be applied in similar way to GA. As another modification, the algorithm is made more focused by restricting the random numbers in a specified range without compromising randomness.

Neutron Transport Theory computer code ITRAN [10, 11] has been used to perform the lattice calculations for the AHWR-LEU initial core clusters. The calculations were performed by using 69-group library based on ENDF-B VI.8 nuclear data obtained from IAEA [12]. Two group cross sections for both the clusters were generated using ITRAN. Diffusion theory code FEMINA [13] based on nodal expansion has been used to do the core-level calculations with two group cross sections generated using ITRAN.

2 Loading Pattern Optimization Problem for AHWR-LEU Initial Core

As discussed in the introduction, there are 444 fuel lattice locations, 24 control rods (8 regulating rods (RRs), 8 shim rods (SRs), and 8 absorber rods (ARs)) and 45 shutoff rods for primary shutdown system (SDS-1). The LEU in (Th, LEU) MOX is considered to consist of 19.75% of U-235 and 80.25% U-238. There are two types of cluster being considered for flux flattening in initial core of AHWR-LEU. The two types of clusters have different initial excess reactivity. The description of type-1 and type-2 cluster has been given in Table 1.

Table 1 Description of AHWR-LEU initial core clusters

Cluster type	Fuel type	Gd content	LEU content (%) in various rings of cluster			K_∞ (0 MWd/Te Xe sat.)
			Inner ring (12 pins)	Middle ring (18 pins)	Outer ring (24 pins)	
Type-1	(LEU-Th) MOX	No Gd	13	13	13	1.084
Type-2		5% Gd in 2 pins of inner ring	13	13	13	0.9813

If we try to load two types of clusters in 444 lattice locations, then the total number of possible configurations is 2^{444} (~10^{133}) with which these two types of fuels can be loaded in the core. The use of core symmetries can reduce the problem size to 2^{62} (~10^{18}). These numbers of combinations are still very large to be evaluated in a practical timescale to choose the best loading pattern. A combinatorial optimization problem like this one can be solved by defining an objective function and then maximizing it.

The main objectives of initial core loading pattern optimization problem are to maximize K-effective while keeping safe and continuous operation. In short, the target parameters for initial core optimization are:

1. Maximization of K-effective;
2. SDS should meet design requirement of 63 mk;
3. Maximum channel power (MCP) should be less than design limit of 2.6 MW;
4. Maximum mesh power (MMP) should be less than design limit of 200 kW.

The objective function has been defined using penalty method in a similar way to reference [8] and is given below:

$$\begin{aligned} \text{OF} = {} & A_1 \cdot k\text{-eff} - A_2 \cdot (\text{MCP-2.6}) - A_3 \cdot (\text{MMP-200}) \\ & - A_4 \cdot (\text{63.0-worth of SDS-1}) \end{aligned}$$

A_1, A_2, A_3, and A_4 are constants. The design limits for *MCP* and *MMP* are 2.6 MW and 200 kW. For optimization studies, *SDS-1* (43 rods, two maximum worth rods failure criteria) worth requirements of 63.0 mk have been considered.

If MCP of an LP is <2.6, then $A_2 = 0$

Similarly, if MMP <200, $A_3 = 0$

And worth of SDS-1 >63.0, $A_4 = 0$

Hence, the penalty due to design parameters not meeting their design limit has been considered but any LP which is having very high margins in peaking or SDS-1 worth will not be preferred. The criteria are to maximize the K-effective without compromising safety of reactor and full power operation.

3 Estimation of Distribution Algorithm (EDA)

EDA belong to class of algorithms which are population-based, and optimized solution is achieved in an evolutionary way. It has similarities to GA and other swarm-based algorithms like ACO. The main part of EDA is to generate solution pool using a distribution function. The distribution function is then updated by taking feedback of current best candidates. In this algorithm, all candidates generated by initial distribution function are evaluated to estimate their objective function values. The candidates are arranged in descending order as per objective function values. The top few candidates are selected to update the distribution function. In other words, first the objective function value for all the candidates in the present candidate pool is evaluated and then the probability distribution is tailored by considering some weight to current best candidates. In the present analysis, probability distribution function has been updated by using univariate marginal distribution algorithm (UMDA [7]). The description of this EDA in steps is given below:

Step-1 Initial distribution function is used to generate pool of initial candidate solutions (different loading patterns). In this case, the population of $N = 1200$ is considered.

Step-2 All N candidates are simulated using 3D diffusion theory code FEMINA to estimate their objective function values.

Step-3 Best $M < N$ candidates are selected depending on their objective function values.

Step-4 The distribution function is updated

$$\mathrm{DF}(t+1) = \mathrm{DF}(t) \cdot (1-\alpha) + \alpha \cdot X$$

α is a constant and its value is between 0 and 1. The structure of X is similar to DF(t)
where

$$X = \frac{1}{M}\sum_{i=1}^{M} X_i(t)$$

$X_i(t)$ can be 0 or 1. The value of $X_i(t)$ is assigned depending on fuel type in a lattice location.

Step-5 The updated distribution function is used to generate pool of new candidate solution.

Step-6 Go to Step-2 and the cycle is repeated till convergence is achieved or insignificant improvement for few generations.

Computer code FEMINA was commissioned on distributed memory parallel computer system AGGRA at BARC for evaluation of N candidates (loading patterns) independently. In the present study, M is considered as 25% of N. EDA with $\alpha = 0.5$ requires a higher population size of 1200 in each generation to behave in a similar

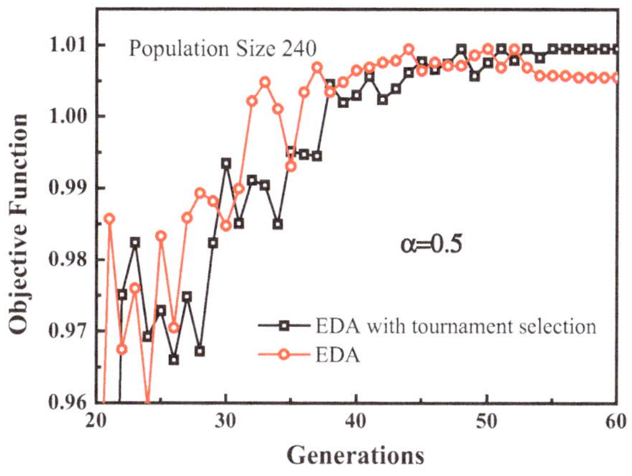

Fig. 1 Tournament selection applied in EDA

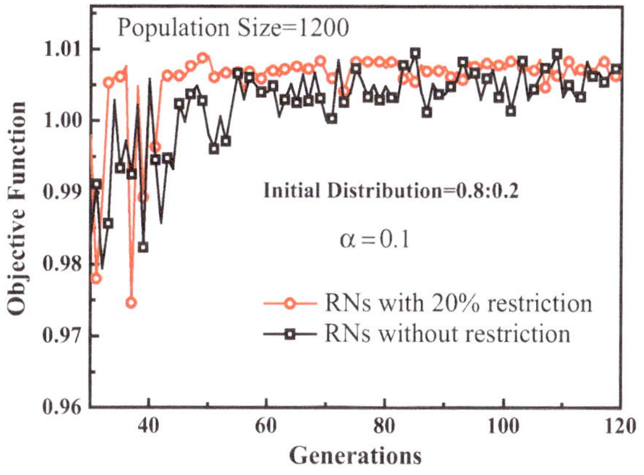

Fig. 2 Restriction criteria for random numbers in EDA

way as GA [9]. The convergence is achieved very fast in less than 60 generations. However, the search is influenced by the best solution in the previous pool. Therefore, to enhance the diversity, instead of choosing best candidates from the pool, tournament selection can be applied to choose the better candidates to modify the distribution function. In this way, solution which is at bottom of the pool can also be picked. A few simulations were carried out with $\alpha = 0.5$, population size of 240. It was observed that the optimized loading pattern is same to as observed by GA. Figure 1 gives comparison of best value of objective function in each generation for simple EDA and EDA with tournament selection. It is observed that EDA with tournament selection converges to higher value. However, the best value of objective function is 1.0095 which is same as observed in GA. Therefore, the efficiency of EDA has been improved. Now EDA gives optimized loading pattern as achieved by GA with same population size of 240 instead of earlier requirement of 1200.

In the next modification, we have put the restriction criteria for random numbers to generate the loading patterns within 20% of present DF. We considered a lower value of $\alpha = 0.1$. In this way, search direction will be less biased in direction of best solutions of the previous generation. To improve the efficiency of EDA, restriction criteria for random numbers to generate the loading patterns within 20% of present DF were considered.

For initial core LPO problem of AHWR-LEU, we have 444 channel locations and two types of fuels, namely type-1 and type-2, where type-1 is having higher k-infinity than type-2. When we use the quarter core symmetry to reduce the problem size, there are 111 locations. If we suppose that there are ~80% channels of type-1 and 20% channels of type-2 [14] and define our initial distribution function (DF) in this way, then while generating loading patterns, we have to generate very small set (111 numbers only) of random numbers. Such a small set is not truly random, and it is observed that 40% of loading patterns generated are having variation of more than ±20% from the average DF value. To avoid local minima, if we consider small value of α (=0.1), then the convergence of objective function is very slow. In order to improve the converged value of objective function as well as for faster approach to converged value and without falling in local minima, we have tried to restrict the loading patterns generated in a domain of ±20% of the averaged DF at each iteration.

For doing this, first we generate a set of 111 random numbers and analyze that the LP generated by this set is in ±20% of the averaged DF. If it is NOT, then we ignore this set and generate a new set and the process is repeated, till we get the desired random number set.

Figure 2 shows the comparison for $\alpha = 0.1$ with and without restriction criteria of 20%. It was observed that the approach to higher value of objective function is faster; however, as number of iterations is increased, same value of objective function and similar loading patterns are achieved. Figure 3 shows the best loading pattern achieved which is same in both the modifications. However, lesser population size is required after first modification and lesser number of generations is required after second modification. Therefore, the efficiency of EDA has improved after modifications.

	13	14	15	16	17	18	19	20	21	22	23	24	25
A	1	1	1	1	1	1							
B	S	1	1	1	1	S	1						
C	1	1	1	S	1	1	1	1					
D	AR	1	1	1	1	1	S	1	1				
E		1	1	1	RR	1	1	1	1	1			
F	1	1	S	1	1	1	1	S	1	1	1		
G	SR	2	1	1	1	1	1	1	1	S	1	1	
H	2	2	2	2	1	AR	1	1	1	1	1	S	1
J	1	2	2	2	2	1	1	1	RR	1	1	1	1
K	S	2	2	SR	2	2	1	1	1	1	S	1	1
L	1	2	1	2	2	2	1	S	1	1	1	1	1
M	1	1	2	2	2	2	2	1	1	1	1	1	1
N	S	1	1	S	1	2	SR	1	1	AR	1	S	1

1	Type-1 fuel
2	Type-2 fuel

S	Shut-off rod
SR	Shim Rod
RR	Regulating rod
AR	Absorber rod

Fig. 3 Optimized loading pattern in quarter core

4 Conclusions

In the present study, a computer code has been developed for initial core LPO problem for AHWR-LEU using EDA. The basic methodology of EDA has not been altered, but we have successfully improved the efficiency of EDA for best value of OF value. Presently, the initial core LPO problem (for AHWR) is considered, where we have considered only two types of clusters. Future work is planned on for considering these effects on batch-type refueling problems as well as other evolutionary optimization algorithms are also being explored. We have tried various modifications in EDA for improving the performance and search for a better-optimized LP. Restriction criteria on random numbers and use of tournament selection method are considered. The present study has lead to slight improvement in the performance.

A search for a better-optimized loading pattern has led us to a better understanding of the algorithm. In general, different results may be expected for same modifications when applying EDA on different problems.

References

1. A. Thakur, B. Singh, N.P. Pushpam, V. Bharti, U. Kannan, P.D. Krishnani, R.K. Sinha, Fuel cycle flexibility in AHWR with the use of Th-LEU fuel, in *Proceedings of International Conference on Future of HWRs*, Ontario, Canada, 2011
2. J.G. Stevens, K.S. Smith, K.R. Rempe, Optimization of pressurized water reactor shuffling by simulated annealing with heuristics. Nucl. Sci. Eng. **121**, 67–88 (1995)
3. L. Machado, R. Schirru, The ant-q algorithm applied to the nuclear reload problem. Ann. Nucl. Energy **29**, 1455–1470 (2002)
4. D.E. Goldberg, *Genetic Algorithms in Search Optimization and Machine Learning* (Addison-Wesley, Reading, New York, 1989)
5. G.T. Parks, Multiobjective pressurized water reactor reload core design by nondomindated genetic algorithm search. Nucl. Sci. and Eng. **124**, 178–187 (1996)
6. J.L.C. Chapot, F.C.D. Silva, R. Schirru, A new approach to the use of genetic algorithms to solve the pressurized water reactor's fuel management optimization problem. Ann. Nucl. Energy **26**, 641–655 (1999)
7. S. Jiang, A.K. Ziver, J.N. Carter, C.C. Pain, A.J.H. Goddard, S.J. Franklin, H. Phillips, Estimation of distribution algorithms for nuclear reactor fuel management optimization. Ann. Nucl. Energy **33**, 1039–1057 (2006)
8. S. Mishra, R.S. Modak, S. Ganesan, Optimization of thorium loading in fresh core of Indian PHWR by evolutionary algorithms. Ann. Nucl. Energy **36**, 948–955 (2009)
9. A. Thakur, B. Singh, A. Gupta, V. Duggal, K. Bhat, P.D. Krishnani, Performance of estimation of distribution algorithm for initial core loading pattern optimization of AHWR-LEU. Ann. Nucl. Energy **96**, 230–241 (2016)
10. P.D. Krishnani, CLUB-A multigroup integral transport theory code for analysis of cluster lattices. Ann. Nucl. Energy. **9**, 255–260 (1982)
11. P.D. Krishnani, Interface current Method for PHWR cluster geometry with anisotropy in the angular flux at interfaces. Ann. Nucl. Energy **9**, 287–296 (1982)
12. http://www-nds.iaea.org/wimsd/
13. A. Kumar, R. Srivenkatesan, Nodal expansion method for reactor core calculation. Technical report no. BARC-1249 (1984)
14. A. Thakur, B. Singh, U. Kannan, N.P. Pushpam, A. Kumar, P.D. Krishnani, R. Srivenkatesan, Initial core optimization of AHWR, in *Proceedings of National Symposium on Radiation Physics (NSRP-18)*, Udaipur, India, pp. 139–141, 2010

Energywise Contributions of Th, Pu and U Isotopes to the Reactivity Feedbacks of (Th-LEU) Fuelled AHWR

Anindita Sarkar, Umasankari Kannan and P. D. Krishnani

Abstract Advanced heavy water reactor (AHWR) has been designed with thorium-based fuel cycle and boiling light water as coolant. The coolant void reactivity (CVR) and fuel temperature reactivity (FTR) are the main reactivity feedback mechanisms. In the present work, we have attempted to understand the physics behind CVR and FTR by estimating the energywise isotopic contributions for AHWR fuelled with (Th, LEU) fuel. The normalised reaction rate approach was used to quantify the individual contributions from ^{232}Th, U, Pu, H and other isotopes. It has been shown that ^{232}Th capture plays a major role in making these coefficients negative. This led us to study the sensitivities of these isotopes to the reactivity coefficients to the basic nuclear cross-section set. The sensitivity due to nuclear data to CVR and FTR from ENDF/B-VI.8 and JENDL 3.3 has been calculated. It is shown that FTR is less sensitive to the dataset whereas the CVR is 16% lower in the JENDL3.3 dataset as compared to the ENDF/B-VI.8.

Keywords AHWR-LEU fuel · CVR · FTR · D5 cluster · Isotopic reaction rates

1 Introduction

AHWR is a boiling water reactor with light water as the coolant and heavy water as the moderator. The LEU design of AHWR has been considered for the current purpose [1]. The fuel is composed of thorium and LEU oxide with 19.75% of U^{235} and 80.25% of U^{238}. The LEU cluster having 30% LEU in inner 12 pins, 24% LEU in

A. Sarkar (✉) · U. Kannan · P. D. Krishnani
Reactor Physics Design Division, Bhabha Atomic Research Centre, Trombay, Mumbai 400085, India
e-mail: anindita@barc.gov.in

U. Kannan
e-mail: uma_k@barc.gov.in

P. D. Krishnani
e-mail: pdkrish@barc.gov.in

A. K. Nayak and B. R. Sehgal (eds.), *Thorium—Energy for the Future*,
https://doi.org/10.1007/978-981-13-2658-5_29

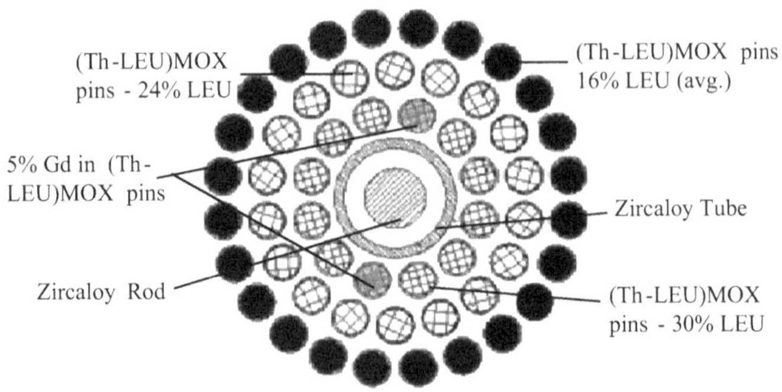

Fig. 1 Cross-section of AHWR-LEU fuel cluster

middle 18 pins and an average of 16% LEU in outer 24 pins is considered for these calculations [1]. Two of the inner 12 pins contain 5% gadolinium to suppress the initial excess reactivity. The cross-section of the AHWR-LEU fuel cluster is given in Fig. 1.

The spread in the safety reactivity coefficients like CVR and FTR due to the uncertainty in nuclear data has been already established [2, 3]. The effect of individual isotopes and their contributions to the safety coefficients can be calculated based on the approach detailed by Ganda and Greenspan [4]. In this method, the normalised reaction rate differences give a direct estimate of the components from each constituent energy and isotope. In an earlier work, the energywise isotopic components of CVR for an older fuel design of AHWR with (Th, ^{233}U) MOX and (Th, Pu) MOX fuel has been done, where the effect of ^{232}Th, ^{233}U and Pu isotopes on CVR had been quantified in the 69 group WIMS library structure [5].

In this paper, the current fuel design with LEU fuel has been considered. Here too, the energywise contributions of the individual isotopes to these reactivity feedbacks have been quantified through isotopic reaction rates. The break-up of the individual feedbacks have been done in the 172 WIMS energy group structure. This has been carried out in two parts; in the first part, the contribution of different isotopes to CVR and FTR has been calculated, and in the second part, the sensitivity in contribution due to difference in basic nuclear data has been studied.

2 Methodology

A number of techniques to quantify the reactivities either by integral parameters or by normalised reaction rates exist in literature. In our present study of AHWR lattice, normalised reaction rate approach by Ganda and Greenspan [4] has been used. This method of determining the constituent contribution to reactivity coefficients is subdivided into two classes—normalised per absorbed neutron and normalised per fission neutron. Components of reactivities are determined by normalisation to one fission neutron for problems in which the major contributors to the reactivity are fissile isotopes. In such problems, it is convenient to rank the isotopes by the number of fission neutrons they emit per neutron absorbed in the system. In normalising 'per absorbed neutron', the system k_∞ is expressed as

$$k_\infty = \eta.\mathrm{f} = \frac{\sum_j \nu_j \Sigma_{fj}\phi_j}{\sum_i \Sigma_{ai}\phi_i} = \frac{\nu_1\Sigma_{f1}\phi_1}{\Sigma_{a1}\phi_1}.\frac{\Sigma_{a1}\phi_1}{\sum_i \Sigma_{ai}\phi_i} + \frac{\nu_2\Sigma_{f2}\phi_2}{\Sigma_{a2}\phi_2}.\frac{\Sigma_{a2}\phi_2}{\sum_i \Sigma_{ai}\phi_i} + \cdots \tag{1}$$

$$= \eta_1.\check{f}_1 + \eta_2.\check{f}_2 + \cdots \tag{2}$$

where $\eta_i = \frac{\nu_i\Sigma_{fi}\phi_i}{\Sigma_{ai}\phi_i}$, $\check{f}_i = \frac{\Sigma_{ai}\phi_i}{\sum_j \Sigma_{aj}\phi_j}$.

Thus, the contribution to reactivity is computed as

$$\rho = \frac{k_\infty^p - k_\infty^n}{k_\infty^p.k_\infty^n} = \frac{\left(\eta_1.\check{f}_1|_p - \eta_1.\check{f}_1|_n\right) + \left(\eta_2.\check{f}_2|_p - \eta_2.\check{f}_2|_n\right) + \cdots}{k_\infty^p.k_\infty^n} \tag{3}$$

Here, n stands for normal operating power and p stands for perturbed state.

Normalisation to per fission neutron is used when absorbing nuclei like burnable poison is the major contributor of reactivities. In this approach, the system k_∞ is expressed as

$$k_\infty = \eta.f = \frac{\nu\Sigma_f^{\mathrm{fuel}}\phi_{\mathrm{fuel}}}{\Sigma_a^{\mathrm{tot}}\phi_{\mathrm{tot}}} \tag{4}$$

Thus,

$$\frac{1}{k_\infty} = \frac{\Sigma_{a1}\phi_1}{\nu\Sigma_f^{\mathrm{fuel}}\phi_{\mathrm{fuel}}} + \frac{\Sigma_{a2}\phi_2}{\nu\Sigma_f^{\mathrm{fuel}}\phi_{\mathrm{fuel}}} + \cdots = \bar{\Sigma}_{a1} + \bar{\Sigma}_{a2} + \cdots \tag{5}$$

where the absorption of system constituent i, per fission neutron is, $\bar{\Sigma}_{ai} = \frac{\Sigma_{ai}\phi_i}{\nu\Sigma_f^{\mathrm{fuel}}\phi_{\mathrm{fuel}}}$

Isotopic contribution to reactivity is calculated as

$$\rho = \frac{k_\infty^p - k_\infty^n}{k_\infty^p.k_\infty^n} = \left(\bar{\Sigma}_{a1}^n - \bar{\Sigma}_{a1}^p\right) + \left(\bar{\Sigma}_{a2}^n - \bar{\Sigma}_{a2}^p\right) + \cdots \tag{6}$$

The D5 cluster of AHWR has been simulated as circular arrayed cluster using WIMS-D lattice code [6] with 172 group nuclear data library. The Boltzmann's transport equation has been solved for this highly heterogeneous cluster using collision probability methods with white boundary conditions. The reaction rates from fissions and captures for each nuclide can be calculated from the effective cross-section and the neutron flux. The multi-group spectrum has been obtained from lattice simulations, and the spectrum corresponding to core average burnup is shown in Fig. 2. The neutron spectrum over the lattice has been studied to understand its effect on the capture or absorption from individual isotopes.

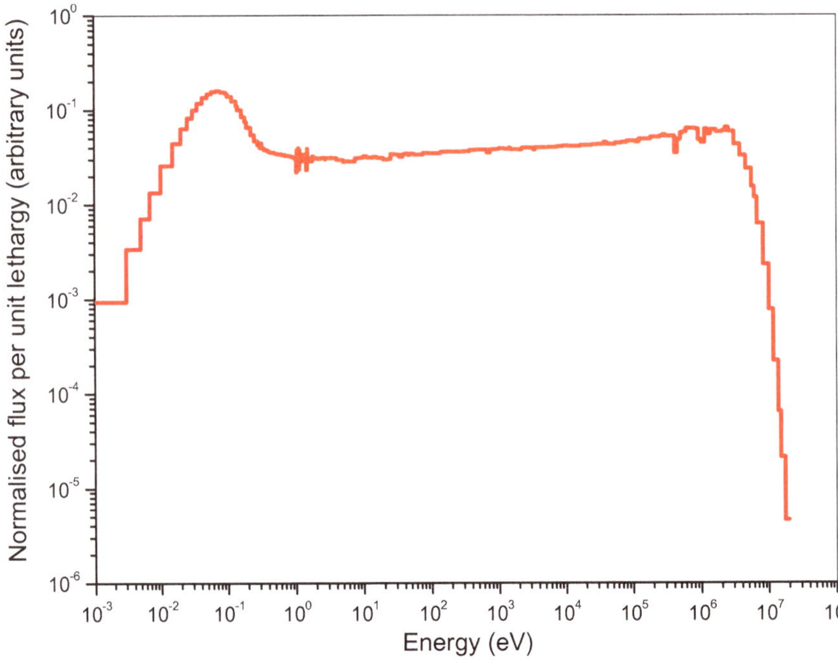

Fig. 2 AHWR spectrum in 172 neutron energy groups from lattice simulations at core average burnup

3 Isotopic Components of CVR

As a first step, the fission reaction rates and absorption reaction rates of each isotope have been estimated using the flux from the transport equation solution. Energy-wise isotopic contribution to CVR and FTR has been calculated with these reaction rates using 'normalisation to per fission neutron' approach of Ganda and Greenspan [4]. The major energy regions have been defined as fast region ranging from 20 to 9.118 keV, resonance from 9.118 to 0.625 eV and thermal below 0.625 eV. The components of CVR in fast, epithermal and thermal energy ranges are −9.8 mk, −52.9 mk and +55.0 mk, respectively. The total CVR at any instantaneous burnup step would be an integral sum over all the energies. The individual components have also been determined using Eq. 6. The isotopic contribution to CVR in 172 energy groups from few nuclides at core average burnup i.e., at 30 GWD/T has been shown in Fig. 3.

From the isotopic contribution, it is shown that H and Pu isotopes have a positive component whereas, ^{232}Th, ^{233}U and ^{238}U have components of negative magnitude. When the coolant voids, the absorption in hydrogen is reduced which increases the neutron thermalisation and hence a higher reactivity. Thus component of hydrogen in CVR is positive and of large magnitude. Due to voiding, the average neutron energy

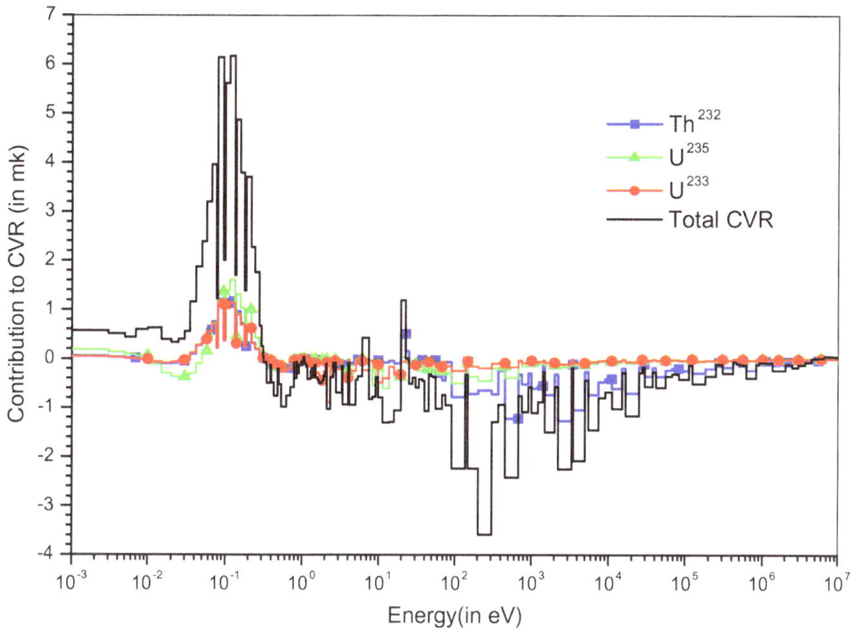

Fig. 3 Energywise isotopic contributions to CVR

increases from 0.397 to 0.69 eV, which implies spectral hardening, and this facilitates the absorption in the resonance region. This effect results in negative contribution from resonant absorbers like ^{232}Th, ^{233}U, ^{235}U and ^{238}U. The maximum negative magnitude is from epithermal captures in ^{232}Th. Thermal captures in ^{235}U and ^{239}Pu decrease with spectral hardening due to voiding, which gives a positive contribution to CVR in thermal region.

4 Isotopic Component of FTR

The isotopic contribution to FTR in 172 energy groups from all nuclides at core average burnup has been calculated and a few are shown in Fig. 4. It is very clear from the figure that the captures in resonance energies give a large negative contribution. Epithermal absorptions in Th^{232} and U^{238} are the major negative contributors of FTR, due to Doppler broadening of the capture resonances of these nuclides. The energywise component of FTR is −0.17 mk, −5.12 mk and 0.002 mk in fast, epithermal and thermal energies, respectively. The 0.3 eV fission resonance of ^{239}Pu gives a positive contribution to FTR.

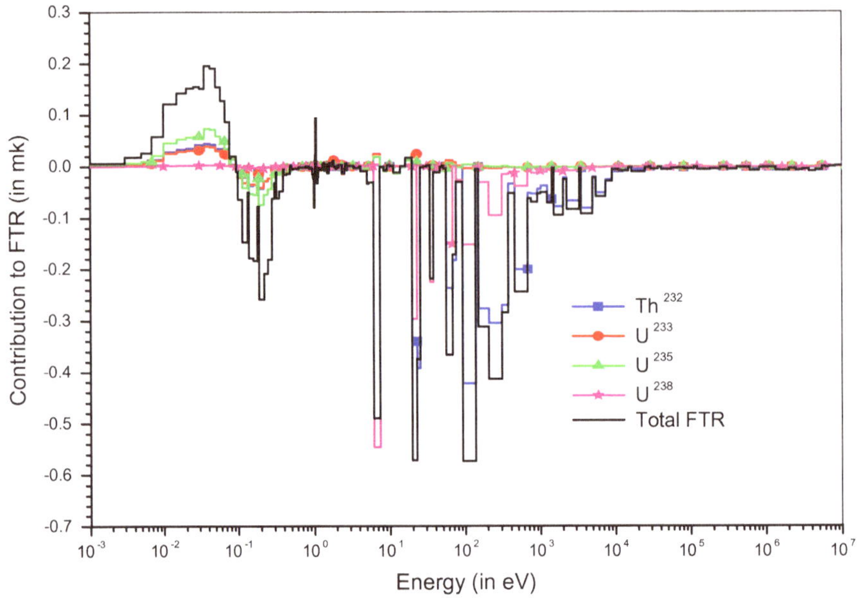

Fig. 4 Energywise isotopic contributions to FTR

5 Sensitivity Studies of the Isotopic Component of the Reactivities

JENDL 3.3 dataset has been recommended for Th-U fuel systems as the resolved energy range for ^{233}U has been increased to 150 eV as compared to 60 eV in ENDF/B-VI.8. ^{232}Th cross-sections are also significantly different in the two datasets [3]. From the sensitivity study done earlier on the total reactivity swings, it was shown that JENDL-3.3 dataset results in 16 and 5.6% difference in CVR and FTR, respectively as compared to the ENDF/B-VI.8 data [2].

In this paper, the deviations in the individual isotopic components were studied to quantify this sensitivity. The comparison between the individual isotopic components in the CVR and FTR between ENDF/B-VI.8 and JENDL 3.3 for AHWR-LEU fuel at core average burnup is given in Table 1. The CVR as calculated from JENDL 3.3 data is −6.46 mk and is lower than the value from ENDF/B-VI.8 which is −7.72 mk. It is observed that the deviation in CVR from JENDL 3.3 set is mainly due to the contributions from Th^{232} and U^{233} which is −18.58 and −30.56%, respectively, and that due to plutonium isotopes is only −2.33%. However, it is evident from the table that the isotopic contributions to FTR are not sensitive to the basic nuclear data.

Table 1 Isotopic contributions to CVR and FTR in AHWR lattice at core average burnup

Nuclides	Components of CVR (in mk)		Components of FTR (in mk)	
Dataset used	ENDF/B-VI.8	JENDL 3.3	ENDF/B-VI.8	JENDL 3.3
Th^{232}	−11.09	−9.03	−3.24	−3.03
Pa^{233}	−0.54	−0.57	2.4E−03	2.0E−03
U^{233}	−3.23	−3.57	8.6E−02	8.9E−02
U^{235}	−1.44	−1.00	0.196	0.196
U^{238}	−3.30	−3.40	−1.69	−1.69
Total U	−8.81	−8.90	−1.53	−1.54
Pu^{239}	3.07	3.07	−0.31	−0.31
Pu^{241}	0.61	0.61	−2.8E−02	−2.7E−02
Total Pu	2.57	2.51	−0.53	−0.52
H_2	17.32	17.53	−0.015	−0.015
Zr	−2.84	−3.57	−0.07	−0.07
Fission products	−3.13	−3.15	0.09	0.09
Other nuclides	−1.20	−1.28	0.02	0.007
Total	−7.72	−6.46	−5.27	−5.06

6 Conclusions

The energywise isotopic components of CVR and FTR have been estimated for AHWR-LEU fuel, and it is observed that Th^{232} plays a major component of CVR. The break-up of the components has given an insight into role of fission and captures to the reactivity feedback coefficients. Sensitivity to the different nuclear datasets also was studied for the isotopic components. It was observed that the CVR is highly sensitive to basic cross-sections, and the variation in CVR is mostly attributed to difference in nuclear data of Th^{232}. It is worth noting that the isotopes of the thorium cycle, namely ^{232}Th and ^{233}U received more attention and have been revised in the JENDL 3.3 dataset. The isotopic contributions to FTR for different nuclear data libraries are comparable.

References

1. A. Chakraborty et. al., AHWR-LEU SARP, BARC, March 2013
2. A. Sarkar, U. Kannan, P.D. Krishnani, Sensitivity studies for (Th, LEU) fueled AHWR with different 172 multi- group nuclear cross section sets, in *National Conference on Power from Thorium (NCPTh)*, Mumbai, 22–24 Dec 2014
3. U. Kannan, A. Kumar, A. Sarkar, S. Ganesan, P.D. Krishnani, Sensitivity studies on nuclear data for thorium fuelled advanced heavy water reactor (AHWR), in *Proceedings of 5th AASPP Workshop on Asian Nuclear Reaction Data Development, INDC(IND)0048*, pp. 133–139

4. F. Ganda, E. Greenspan, Analysis of reactivity coefficients of hydride-fueled PWR cores. Nucl. Sci. Eng. **164**, 1–32 (2010)
5. U. Kannan, S. Ganesan, Analysis of coolant void reactivity of advanced heavy water reactor (AHWR) through isotopic reaction rates. Nucl. Sci. Eng. **167**(12), 105–124 (2011)
6. J.R. Askew, F. Fayers, P. Kemshell, *A General Description of the Lattice Code WIMS* (British Nuclear Energy Society, 1966)

Power Flattening Study of Ultra-Long Cycle Fast Reactor Core

Taewoo Tak, Jiwon Choe, Yongjin Jeong, Jinsu Park, Deokjung Lee and T. K. Kim

Abstract A design study of ultra-long cycle fast reactor with a rated power of 1000 MWe (UCFR-1000) was performed to flatten its radial power distribution throughout the cycle depletion of the core. It is expected that the flattening of the radial power distribution lowers the power peak factor at the center and consequently the maximum peak temperature of the fuel and clad. Thorium fuel has lower reproduction factor than uranium; thus, thorium-loaded fuel zone has relatively lower power. A lot of test cases were introduced that have thorium-loaded fuel in the inner fuel region to lower the center power peak. Each test case had a different zoning strategy; different fuel forms were loaded for both the upper blanket region and the lower low enrichment uranium (LEU) region. The final evaluation models were designed by loading the thorium-loaded fuel only in the inner blanket for the 1-zone case, in both blanket and LEU regions for the 2-zone case, and in an additional zone for the other cases. It was demonstrated that the fuel zoning has a significant impact on the power flattening, and a gradual fuel zoning can achieve a flattened radial power distribution, and can consequently reduce the maximum power peak.

Keywords SFR · UCFR · Power peaking factor · Power flattening · Thorium fuel

T. Tak · J. Choe · Y. Jeong · J. Park · D. Lee (✉)
Ulsan National Institute of Science and Technology, Ulsan, South Korea
e-mail: deokjung@unist.ac.kr

T. Tak
e-mail: ttwispy@unist.ac.kr

J. Choe
e-mail: chi91023@unist.ac.kr

Y. Jeong
e-mail: yjjeong09@unist.ac.kr

J. Park
e-mail: ccpj0310@unist.ac.kr

T. K. Kim
Argonne National Laboratory, Illinois, USA
e-mail: tkkim@anl.gov

A. K. Nayak and B. R. Sehgal (eds.), *Thorium—Energy for the Future*,
https://doi.org/10.1007/978-981-13-2658-5_30

Nomenclature

BOC	Beginning of cycle
EOC	End of cycle
LEU	Low enriched uranium
MOC	Middle of cycle
NC	Narrow center
NU	Natural uranium
UCFR	Ultra-long cycle fast reactor
WC	Wide center

1 Introduction

The sodium-cooled fast reactor (SFR) technology became the most promising reactor concept for the future among the six Generation IV (Gen IV) reactor systems and the SFR research has been performed in many countries. Various SFR concepts have been proposed, and each design has several advantages and disadvantages at the same time by the SFR nature. Among the SFR concepts, some burnup strategies have been applied for the long cycle operation. The ultra-long cycle fast reactor (UCFR) has been developed with the target of a 60-year operation without the refueling, and it has a power rating of 2600 MWth [1]. The UCFR adopts a breed-and-burn strategy to achieve such a long cycle, and the strategy can be achieved by using low enrichment uranium (LEU) as an igniter, and natural uranium (NU) as a blanket material. The operational feasibility of the UCFR had been reported only from the neutronics point of view, and subsequently the optimization of UCFR was performed to consider thermal-hydraulic feedback analysis and to mitigate the power peaking issue. In addition, a study of the spent fuel utilization for the blanket region in UCFR has been performed [2]. It was noticed that the high reactor power rating combined with high peaking could cause a very high local power deposition and consequently violates the limits of temperature or fast neutron fluence in the fuel and clad regions [3]. A power flattening is one of the ways to mitigate the radial power peaking factor for which there are several design strategies. Thorium fuel was proposed as a blanket material at the center of the core to flatten the radial power distribution of UCFR-1000, which results in the reduction in the radial peaking factor at the core center at the middle of cycle (MOC) [4].

In this study, several designs of UCFR-1000 were developed to flatten the power distribution, not only at MOC but also throughout the whole cycle. They have an inner fuel region whose radius is 36.7% of the whole fuel region in contrast to the previous thorium-loaded UCFR-1000 with an inner fuel region of only one-fourth of the radius of the whole fuel region. In addition, the LEU zoning has been performed to flatten the radial power distribution at the beginning of cycle (BOC) too.

Table 1 Core design parameters

Parameters	Value
Thermal power (MWth/MWe)	2600/1000
Cycle length (effective full power years)	60 (once through)
Equivalent core diameter (m)	5.9
Fuel pin overall length (cm)	340
Active core height (cm)	240
Average linear power (W/cm)	158.7
Core volume (kL)	42.4
Average volumetric power density (W/cc)	61.3
LEU enrichment (wt%)	11.9

2 Core Design Strategy

UCFR-1000 is the reference core for this study, and the core design variations have been introduced to flatten the radial power shape. In this paper, thorium is used in the core center region to make the radial power shape of the core flattened with its relatively weak breeding performance. In addition, fuel zoning was performed by the fuel form variation of a driver or a blanket for each case.

2.1 Core Design Requirement

For the power flattening study, five core designs with variations in the inner core region and in the LEU zoning region are introduced along with their geometry and fuel compositions. Table 1 shows the core design requirements of UCFR-1000 that all five designs should have in common: the power rating, the cycle length, and the geometry. The base fuel form is U-10Zr with thorium fuel partially added. The enrichment of LEU was determined to satisfy the criticality through the cycle of the cores, except for in the last test case.

2.2 Core Layout

Figure 1 shows the top view core layout of UCFR-1000 in the *x*-*y* cross section. The equivalent core diameter is 5.9 m and that of the outer and inner cores are 5.0 m and 1.8 m, respectively. There are a total of 19 control assemblies: 13 primary and 6 secondary. The primary system uses natural boron as an absorber material, and the secondary system uses 90% enriched boron. The reflector, cladding, and structure material are HT-9, and sodium is used for the coolant material.

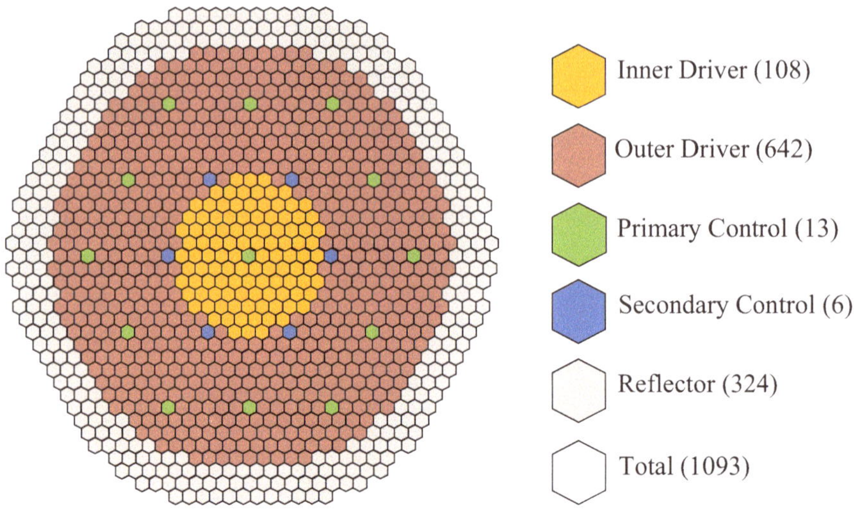

Fig. 1 Radial core layout of UCFR-1000

Fuel zoning was performed from the reference core which has only one fuel form of U-10Zr. It was performed only for the blanket region in 1-zone case (Case 1), and for the LEU region as well as the blanket region in 2-zone case (Case 2), and the other complex zoned cases. Figure 2 shows the core layout in the *x*-*z* cross section, which shows the fuel zoning state of each case. Case 1 core has only one fuel form of U-10Zr in the bottom driver region, while Case 2 has various forms of fuels for power flattening even at BOC. The other cases have different case names according to "LEU and blanket" zoning condition, with wide center (WC) and narrow center (NC) regions. They have one more LEU zone at the bottom of their LEU region, so that the power shape is more flattened than that of Case 2. Each case has a different inner blanket region size. The inner blanket has thorium-loaded fuel to flatten the power shape at the MOC.

3 Results and Discussion

The analysis of the UCFR core designs and performance evaluation were done using McCARD code, which solves a continuous energy neutron transport equation using Monte Carlo method [5]. Figure 3 shows the multiplication factor behavior of each core model. The reference case has a lifetime of 60 years, and the multiplication factors are smaller than 1.03. The initial *k*-effective values of thorium-loaded cores are smaller than those of the reference core, which is caused by the thorium fuel loading. Case WC & WC and Case NC & WC have many points that do not achieve criticality in their operation time, which is due to the fact that they have a relatively

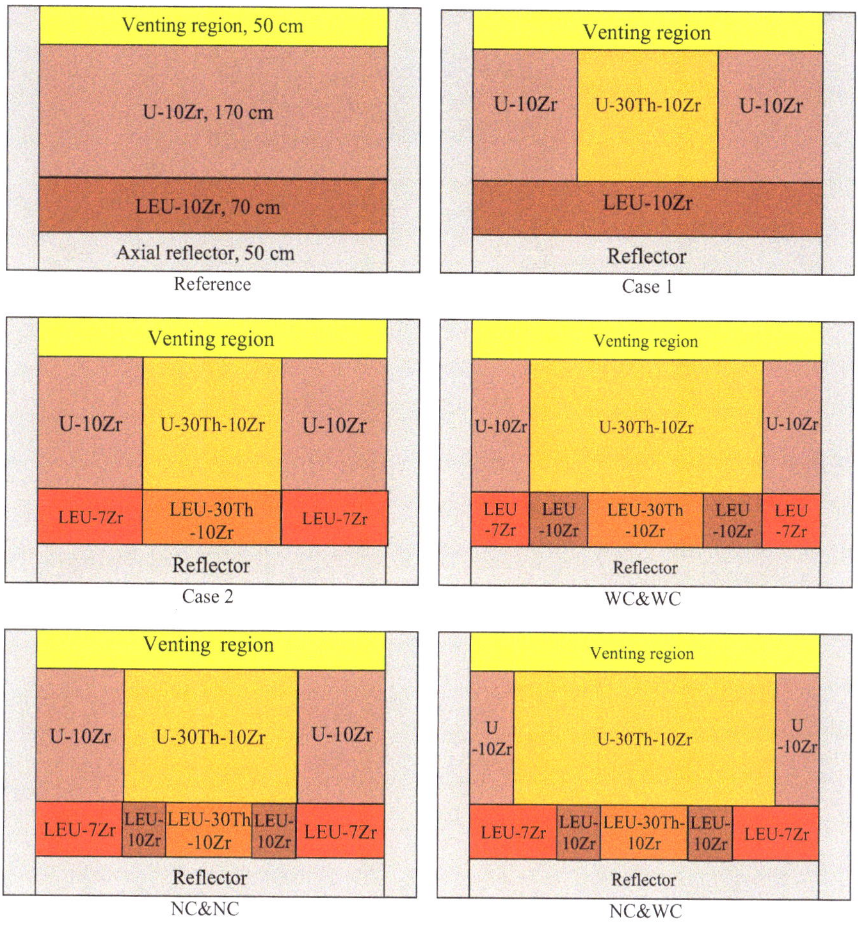

Fig. 2 Core layouts of UCFR-1000 in x-z plane

large central thorium fuel region that is insufficient to make criticality. In the other cases, it can be found that larger inner core regions have lower multiplication factors.

The power flattening effect has been analyzed in the following figures. Figure 4 shows the normalized radial power distribution every 20 year. The values have been normalized with the axially integrated power of each assembly. There are control assemblies not only in the center, but also at 99 and 198 cm from the center; so the center spot in each graph is extrapolated, and the two other spots in each graph are interpolated by averaging the two values of both sides. The figures show that the test cases achieve radial power flattening in comparison with the reference core. The peak of the reference core appears in the center fuel region at BOC and moves to the peripheral region as the core burns. The peak of the Case 1 core also moves from the center to the peripheral region, but the peak factor at the center has decreased,

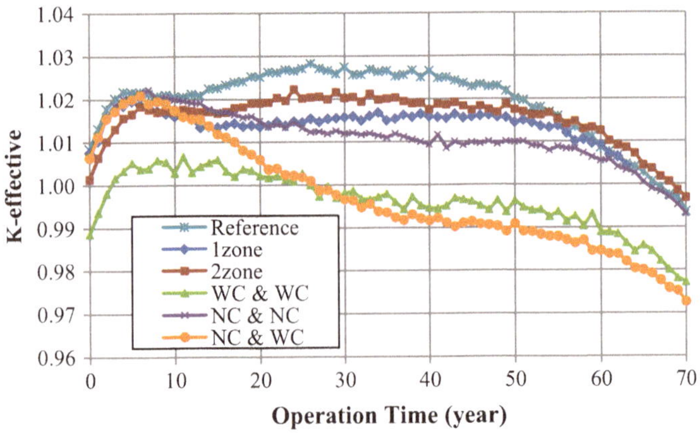

Fig. 3 *k*-effective versus operation time

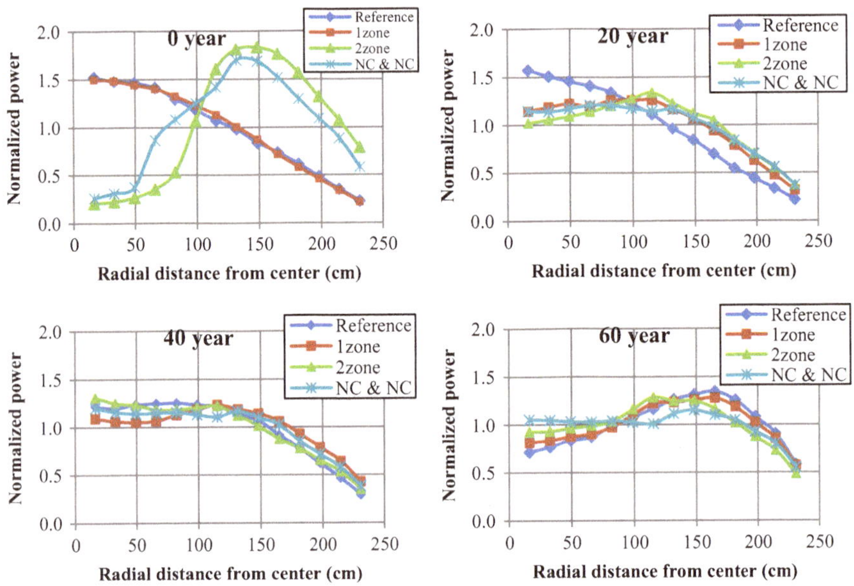

Fig. 4 Normalized radial power distribution

especially at year 20, from 1.57 to 1.14 due to the effect of thorium fuel in the inner core region. On the other hand, the peaks of the Case 2 and Case NC & NC cores appear in the peripheral region at BOC due to the LEU zoning and then move to the center. These two cases show more flattened radial power distribution than Case 1 throughout the operation time, except for the initial state. At the end of cycle (EOC), at year 60, it is noticeable that Case 2 and Case NC & NC have more flattened shapes while the reference and Case 1 have peripheral peaks.

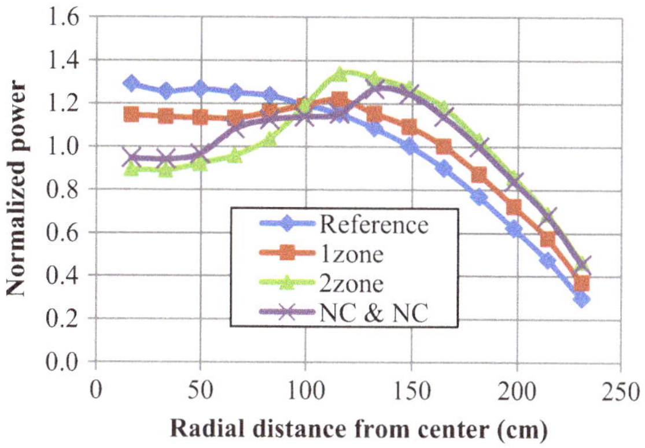

Fig. 5 Average normalized radial power distribution

Table 2 Maximum normalized power

	BOC	MOC	EOC
Reference	3.70	2.03	2.46
1-zone	3.75	1.98	2.13
2-zone	3.66	1.93	1.97
NC & NC	3.49	1.91	2.05

There are differences between Case 2 and Case NC & NC in the radial power distribution especially at year 0 and year 60. Case NC & NC has one more zone in its LEU region while Case 2 has only two zones. This is noted in the radial power distribution figure for year 0. The peripheral peak of Case NC & NC is lower than that of Case 2, and the distribution of Case NC & NC at year 0 and year 60 is more flattened than Case 2. At year 60, Case NC & NC has a maximum peak of 1.15 while the reference case has a peripheral peak of 1.34.

Figure 5 shows the average value distribution of the normalized radial power distribution every 10 years. The standard deviation from their average is 0.30, 0.24, 0.23, and 0.21 for the reference case, Case 1, Case 2, and Case NC & NC, respectively, which means that Case NC & NC has the most flattened shape for the radial power distribution throughout the operation time.

Figure 6 shows the normalized axial power distribution of the center fuel assembly at BOC, MOC, and EOC that shows the active core movement. At BOC, the active core of each case stays within the bottom 70 cm of the core where the peaking factor is greater than other states, and this needs to be lowered. The active core moves to the top of the core as the core burns and breeds. The speed of the active core movement is fastest for the reference case and is slower for Case 1 due to the thorium blanket loading and flattened radial power shape. Case 2 and Case NC & NC have slower movements of the active cores than Case 1 because they have more flattened radial power shapes, which make the speed of the active core movements slower. The maximum normalized power is summarized in Table 2.

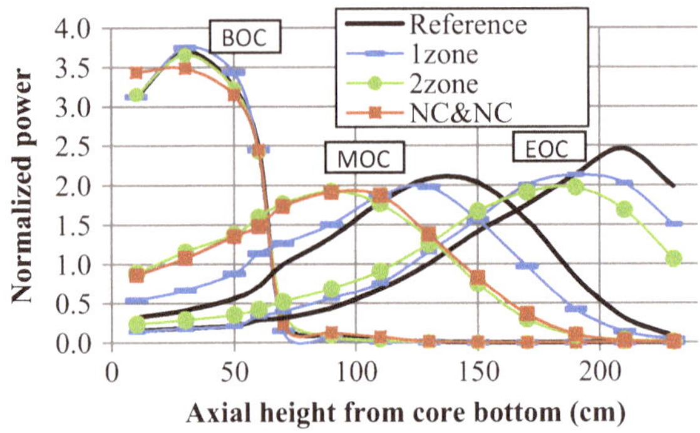

Fig. 6 Normalized axial power distribution

4 Conclusions

The flattening study of the radial power distribution of UCFR-1000 has been performed and compared for the several test models. The test models were designed by loading the thorium fuel only in the inner blanket for the 1-zone case, in both blanket and LEU regions for the 2-zone case, and in an additional zone for the other cases. It has been confirmed that more gradual fuel zoning can achieve more flattened radial power distribution and consequently decreases the maximum power peak. Case 1 has a center peak at BOC like the reference core, but the peaking factor decreases due to the thorium fuel in the inner fuel region, which is shown by the fact that at year 20, it has decreased from 1.57 to 1.14. In contrast to the reference and Case 1 cores, the peaks of Case 2 and Case NC & NC cores occur in the peripheral regions at BOC, which cause more flattened power shapes at MOC and EOC than in Case 1. At EOC, Case NC & NC has a maximum peak of 1.15 while the reference case has a maximum peak of 1.34. As a next step for this study, reactivity coefficient analyses from thermal feedback would be required to demonstrate inherent safety features.

Acknowledgements This work was supported by the National Research Foundation of Korea (NRF), grant funded by the Korean government (MSIP).

References

1. T. Tak, D. Lee, T.K. Kim, Design of ultra-long cycle fast reactor employing breed-and-burn strategy. Nucl. Technol. **183**, 427–435 (2013)
2. T. Tak, D. Lee, T.K. Kim, Ser Gi Hong, Optimization study of ultra-long cycle fast reactor core concept. Ann. Nucl. Energy **73**, 145–161 (2014)
3. H. Seo, J. Kim, I. Bang, Subchannel analysis of a small ultra-long cycle fast reactor core. Nucl. Eng. Des. **270**, 389–395 (2014)
4. T. Tak, H. Lee, S. Choi, D. Lee, Power flattening study for ultra-long cycle fast reactor UCFR-1000, in *Transactions of America Nuclear Society Summer Meeting*, 16–20 June 2013
5. H. Shim et al., McCARD: Monte Carlo code for advanced reactor design and analysis. Nucl. Eng. Technol. **44**, 161–176 (2012)

3D Space-Time Analysis of Anticipated Transient Without Scram in CHTR with Fuel Temperature Feedback

D. K. Dwivedi, Anurag Gupta and P. D. Krishnani

Abstract The compact high temperature reactor (CHTR) is being developed as technology demonstrator and critical facility for Indian high temperature reactor program. Physics design of envisaged core of Th-U^{233}-based TRISO-coated particle-fueled CHTR is in advance stage, and various core configurations have been proposed. Reactor core operation at high temperature necessitates sophisticated safety and anticipated transient analyses including postulated loss of regulation accident, LOCA as well as intentional power setback transient in CHTR. An efficient IQS module in 3D space-time analysis code ARCH with adiabatic Doppler feedback capability has been developed for reactor transient simulation. For temperature feedbacks in reactivity transients in CHTR, 1D radial heat conduction in multichannel-based TH module has also been incorporated with IQS module in code ARCH. The coolant hydraulics in incorporated TH module of ARCH-IQS is limited to given mass flow rate and inlet temperature in every channel. The capability of fuel temperature feedback in the code has been checked with AER benchmark analysis. Some of the results of ARCH-IQS code for benchmark (AER-DYN-002) with Doppler feedback are presented in this paper. 3D space-time analyses of anticipated transient without scram (ATWS) cases of CHTR have been carried out with only fuel temperature feedback, and results are also discussed in this paper.

Keywords Thorium · High temperature reactor · ATWS
AER benchmark analysis · Integrated neutronics–TH

1 Introduction

India is pursuing a comprehensive high temperature reactor program to fulfill its needs for several crucial applications of nuclear energy other than those for grid-based

D. K. Dwivedi (✉) · A. Gupta · P. D. Krishnani
Reactor Physics Design Division, Bhabha Atomic Research Centre,
Trombay, Mumbai 400085, India
e-mail: ddwivedi@barc.gov.in

A. K. Nayak and B. R. Sehgal (eds.), *Thorium—Energy for the Future*,
https://doi.org/10.1007/978-981-13-2658-5_31

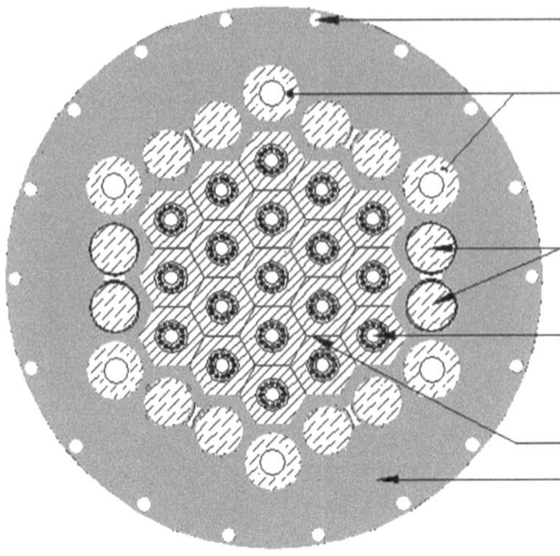

Fig. 1 Cross-sectional view of CHTR core

electricity generation, such as hydrogen production from thermochemical splitting of water and unattended nuclear power packs [1–3]. As a technology demonstrator and critical facility, 100 kWth compact high temperature reactor (CHTR) is being designed based on the guidelines such as use of thorium fuel cycle, passive core heat removal by natural circulation of coolant, compact design to minimize weight, and longer core life [3, 4]. As CHTR is a new conceptual design with high core temperature distribution during normal operation, it requires sophisticated safety and anticipated transient analyses including postulated control rod ejection accident such as loss of reactivity devices accident (LORA), loss of coolant accident (LOCA), as well as power setback transients. Several reactor transient and safety simulation tools have been developed and reported in the literature in the way of coupled neutronics (N)–thermal-hydraulics (TH) as well as integrated N–TH code system under larger multiphysics multiscale framework [5]. In the view of that 1D radial heat conduction in multichannel-based module for temperature feedbacks have been incorporated in IQS module of 3D space-time analysis code ARCH [5, 6]. The IQS module of ARCH has been benchmarked by AER benchmark problems in case of without any temperature feedback [7] as well as with Doppler feedback in simulation.

In case of model based on 1D radial heat conduction [8] in multichannel for TH feedback in ARCH-IQS, the radial heat conduction equations are simultaneously being solved for every axial mesh in each fuel assembly/channel for computation of heat transfer from fuel to coolant in the core [5, 9, 10]. For simplifying the problem in the model, adiabatic boundary conditions have been considered at outer radial boundary of moderator region and also at axial boundaries of conducting meshes in the CHTR fuel assembly, without any significant error in the simulations. The outlet coolant temperature of every axial mesh of the channel/fuel assembly is considered to be the inlet for the next axial mesh in the channel. For computation of heat transfer coefficients at conducting–convective interface of the meshes, Cheng and Tak correlation [11] is being considered for LBE-cooled CHTR core. For temperature feedback of reactivity, the linear interpolation scheme for the instantaneous value of multigroup cross-sectional data is being adopted with help of generated temperature-dependent cross-section dataset [5].

2 CHTR Core Description

In the current envisaged design, 100 kWth power CHTR core consists of 19 BeO moderator blocks in the hexagonal form with centrally located graphite fuel tubes (Fig. 1). The central portion of graphite tube acts as coolant channel with liquid metal (Pb–Bi eutectic) flowing between the bottom to top plenum upward and returning through downcomer tubes [3]. The coolant temperature at core inlet is 900 °C. The average outlet temperature of coolant is envisaged to be at 1000 °C. There are 12 equi-spaced longitudinal fuel bores of 10 mm diameter in each graphite tube. These bores are to be filled with approximately 35 mm long fuel compacts in the form of graphite matrix embedded with TRISO-coated particles. The core average fuel temperature is about 1000 °C. The TRISO particles are like microspheres of (U+Th) carbide kernel (500 μm diameter) coated with three types of layers of soft pyrolytic carbon (90 μm thickness), SiC (30 μm thickness), and hard carbon layer (inner/outer 30/50 μm thickness) [3]. In the present design of CHTR, core contains 2.7 kg of U^{233} in 8 kg of (U+Th) fuel for effective 15 full power years of operation. Gadolinium as burnable poison is mixed with fuel in kernel of the TRISO particles of central fuel assembly [3].

In conceptual design, the reactor has 12 control rods made of Ta and W in outer coolant channels for power regulation and setback. It has two independent shutdown systems, i.e., primary and secondary. Primary shutdown system contains six shutoff rods similar to CRs and will be inserted in six inner coolant channels. Whereas secondary having 12 axially moving cylindrical BeO reflector blocks being taken out of the core for activation. It has six burnup compensation rods axially inserted into fixed BeO reflector blocks for coarse control of initial excess reactivity [5]. It is envisaged that CHTR core temperature is to be kept at 200 °C at hot zero power during startup. The simulation of postulated ATWS case in CHTR has been carried out for the core configuration during startup [12] with adiabatic Doppler feedback. Another case of postulated ATWS under hot operating CHTR core configuration during BOC has also been studied with only Doppler feedback computed with incorporated 1D radial heat conduction-based TH model in code ARCH-IQS, and results are discussed in detail in the next section.

3 Results and Discussions

The current international trend is to perform integrated neutronics–thermal-hydraulics studies under larger multiphysics multiscale framework. In-house multi-physics capability is being developed for safety and transient studies of CHTR, and efforts have been made to incorporate more efficient improved quasi-static (IQS) model [7] in the 3D space-time analysis code ARCH [6] with adiabatic fuel temperature feedback model as well as with 1D radial heat conduction model for temperature feedbacks [5, 10].

3.1 AER Benchmark Analysis

The viability of 3D space-time kinetics code ARCH-IQS without [7] as well as with Doppler feedback has been examined with AER benchmark (Dyn002) [12], and the predictions of transient power, integral power, core reactivity, and peak fuel temperature have been compared with KIKO3D [13]. The predictions with ARCH-IQS for parameters like nuclear power and maximum fuel temperature in the core during transient have been found in good agreement with benchmarked code with fuel temperature feedback in simulation, and graphs are shown in Fig. 2a, b.

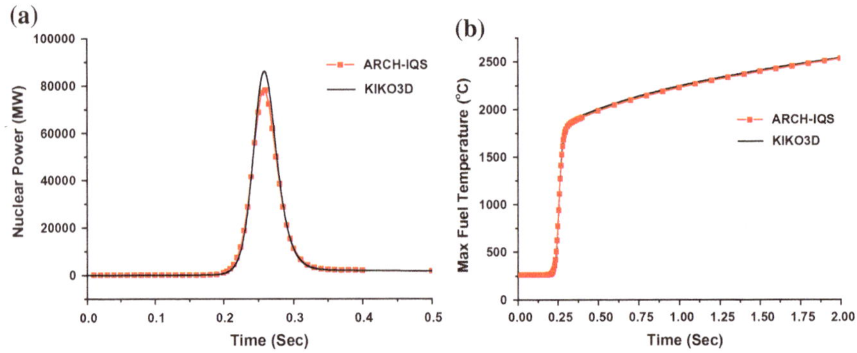

Fig. 2 **a** AER benchmark of ARCH-IQS: nuclear power variation with time. **b** AER benchmark of ARCH-IQS: fuel temperature versus time

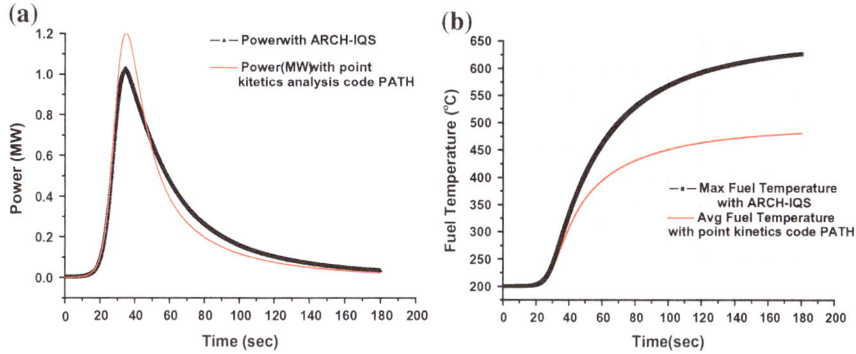

Fig. 3 **a** (Th-U^{233}) fueled CHTR transient with ARCH-IQS and PATH: power versus time. **b** (Th-U^{233}) fueled CHTR transient with ARCH-IQS and PATH: fuel temperature versus time

3.2 Analysis of ATWS in CHTR with Adiabatic Fuel Temperature Feedback During Startup

The ATWS case in (Th-U^{233}) fueled CHTR has been simulated with consideration of adiabatic fuel heating for Doppler feedback and also compared with point kinetics code PATH [12]. In this case, 2.47 mk reactivity has been inserted in the core in 2.8 s, and transient rise in power is arrested only by Doppler feedback from adiabatic heating of fuel. The 5-group condensed homogenized cross section of CHTR lattice at various core temperature have been generated with collision probability based code ITRAN [14] for the computation of temperature dependence of cross-sectional parameters in Doppler feedback. The estimation of reactivity due to withdrawal of control rod and fuel temperature coefficient (FTC) have been computed with code ARCH. The value of FTC is found to be $-1.7 \times 10^{-5}/^{\circ}\mathrm{C}$ (around 200 °C), and in case of LORA, the reactivity addition is found to be 2.47 mk (with $\beta_{\mathrm{eff}} = 4.66$ mk) during approach to criticality. The power variation and maximum fuel temperature in the core during ATWS in (Th-U^{233}) fueled CHTR with ARCH-IQS have been compared with point kinetics analysis and presented in Fig. 3a, b. The differences between the maximum fuel temperature predicted by ARCH-IQS and average fuel temperature predicted with PATH are due to effect of power peaking in the core.

3.3 Analysis of ATWS in Hot Operating Core of CHTR with Heat Conduction Model of TH Feedback

The case of ATWS in (Th-U^{233}) fueled CHTR has also been studied with incorporated model of 1D radial heat conduction for TH feedback [5, 10] in ARCH-IQS. In the present TH module, the solution of time-dependent heat conduction equation of each hexagon axial meshes in the fuel zone is based on corner-mesh finite difference method. In the model, heat conductions in axial and azimuthal direction as well as inter-hexagonal meshes are ignored for simplicity. It is considered that heat generated in each hexagon mesh of CHTR fuel assembly is only being ultimately transferred to coolant flowing in its coolant channel mesh. The coolant equations have been taken as lumped model [9, 10] with given inlet temperature and mass flow rate in each channel in the core. The heat transfer coefficients have been computed using Cheng and Tak correlation for liquid metal as coolant in CHTR [8, 11]. The temperature dependence of properties of LBE coolant such as specific heat capacity (C_p), density (ρ_m), viscosity (μ), and thermal conductivity k_c are being computed as per the equations given in OECD/NEA handbook, 2007 [5, 15]. The steady-state simulation with temperature feedbacks has been performed before initiating the simulation of ATWS, and steady-state temperature profile in various region are shown corresponding to graphs at $t = 0$ (Fig. 4c–h). In the test case of LORA in CHTR, about 2.5 mk positive reactivity is to be added within 2.8 s due to inadvertent withdrawal of a control rod in the hot critical core operating at full power during BOC. The transient is followed by failure of scram (ATWS), and power rise is being arrested by only fuel temperature feedback. The transient power with time has been predicted up to 600 s from initiation. The total power is rising during initial 22 s and then decreasing due to negative fuel temperature feedback and stabilizing about 2.5 times of initial value (Fig. 4a). The maximum fuel temperature is found to be rising up to 1300 °C and stabilizing around 1275 °C (Fig. 4b).

The axial temperature profile of fuel and coolant of hottest channel and its variation during ATWS transient are shown in Fig. 4c–f. Figure 4c shows the variation of axial profile of fuel temperature during heating of the core from 0 to 160 s of transient; here, it has been observed that maximum fuel temperature location is shifting toward the center of the fuel assembly, and it is because of the center-peaked power production profile in the channel. Whereas from approximately 160–600 s (during cooling of the core at almost constant power production, Fig. 4d), the maximum fuel temperature point shifts toward the top of the core. In case of the coolant temperature profile (Fig. 4e), during about 20–160 s of transient, maximum coolant temperature point is before the outlet point; this is because fuel temperature are lesser than the coolant inlet temperature near top of the channel, and coolant heat is being transferred to fuel. From Fig. 4f, it can be seen that as the core is approaching to the steady-state operation after transient (about after 500 s), the maximum coolant temperature is in the top mesh/outlet point of the channel.

The radial temperature profile in central axial mesh of the hottest fuel assembly has also been predicted and shown in Fig. 4g, h. The steady-state radial temperature profile (at 0 s of transient), the maximum temperature point is at outer surface of

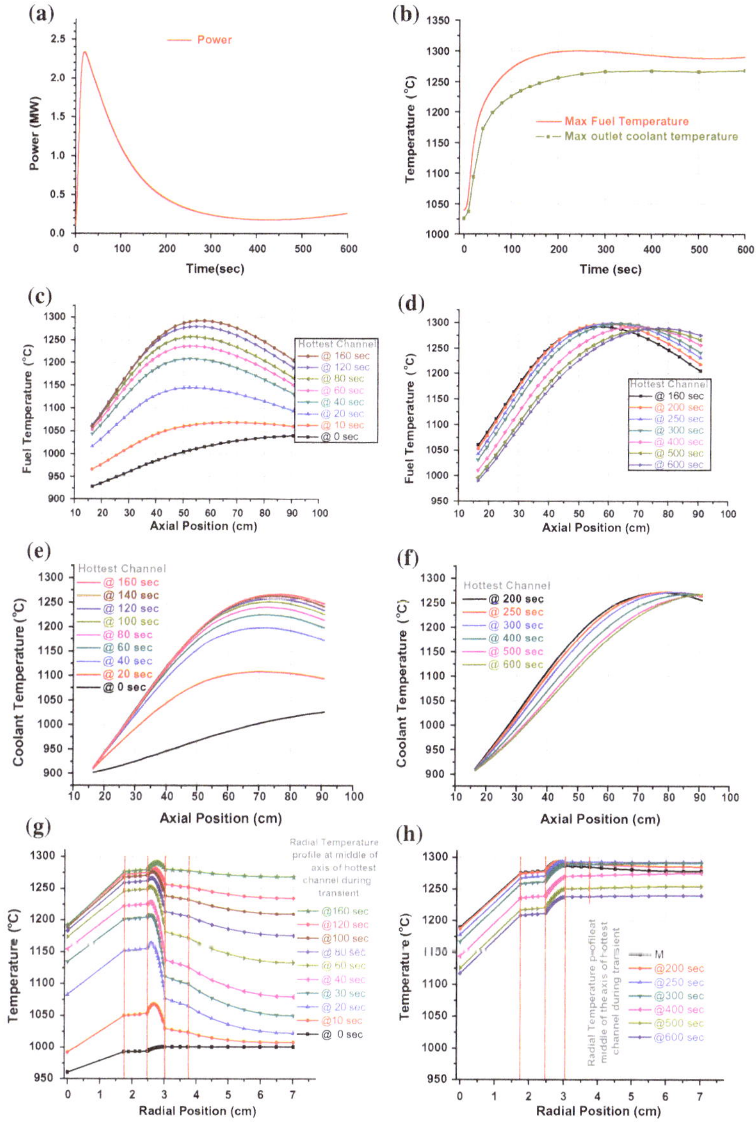

Fig. 4 **a** Variation of power with time in case of ATWS in CHTR. **b** Variation of maximum fuel temperature and maximum outlet coolant temperature versus time during transient. **c** Variation of axial profile of fuel temperature in the hottest channel from 0 to 160 s of transient. **d** Variation of axial profile of fuel temperature in the hottest channel from 160 to 600 s of transient. **e** Axial profile of coolant temperature in the hottest channel during 0–160 s of transient. **f** Axial profile of coolant temperature in the hottest channel during 200–600 s of transient. **g** Variation of temperature radial profile in the hottest channel from 0 to 160 s of transient. **h** Variation of temperature radial profile in the hottest channel from 180 to 600 s of transient

the annular fuel region and all outer point are at the same temperature, because it is assumed in the model that no heat transferring at the outer radial surface of the fuel assembly. As long as transient started and power started rising, the maximum temperature point is found to be in the annular fuel region which has asymmetric parabolic temperature profile (i.e., region of heat source) during time $t > 0$ to 160 s of transient (Fig. 4g). In Fig. 4h, it shows that the heat is finally transferred to the coolant of each channel, and BeO moderator region has maximum temperature from 250 to 600 s of transient.

4 Conclusions

The cases of ATWS in critical core of CHTR during first approach to criticality and in hot operating core have been carried out with Doppler feedback capability of 3D space-time analysis code ARCH-IQS. Results of benchmark analysis of ARCH-IQS have been found to be matching well with other benchmarked codes. The results of 3D space-time analysis of LORA followed by failure of scram (ATWS) case during approach to criticality of CHTR with adiabatic fuel heating feedback model in ARCH-IQS have been compared with point kinetics analysis code PATH. Transient analysis is also carried out with recently developed TH module with 1D radial heat conduction model for temperature feedback in IQS module. The 3D space-time analysis of ATWS case in full power operating core of CHTR during BOC has also been carried out, and in this case 1D radial heat conduction-based TH module of ARCH-IQS is utilized for Doppler feedback. The transient power, maximum fuel temperature, maximum coolant temperature, moderator temperature, and radial as well as axial temperature profile in each fuel assembly of hot operating CHTR core have been estimated and analyzed. It has been found that fuel and coolant temperatures are to be well within the safety limits during analyzed cases of CHTR transients.

References

1. R.K. Sinha, Advanced nuclear reactor systems—an Indian perspective. Energy Procedia **7**(2011), 34–50 (2011)
2. R.K. Sinha, I.V. Dulera, Carbon based materials—applications in high temperature nuclear reactors, in *IJE&MS-17, 2010*, pp. 321–326
3. D.K. Dwivedi, A. Gupta, P.D. Krishnani, Challenges in core reactivity management and control optimization in physics design of compact high temperature reactor, in *BARC Newsletter*, vol. 317, Nov–Dec 2010
4. D.K. Dwivedi, A. Gupta, P.D. Krishnani, Physics design of compact high temperature reactor, in *International Conference on Peaceful Use of Atomic Energy 2009, New Delhi*
5. D.K. Dwivedi, A. Gupta, K. Umasankari, 3D space-time analysis of ATWS in compact high temperature reactor with integrated thermal-hydraulic model in code ARCH-TH. Ann. Nucl. Energy **110**, 306–316 (2017)

6. A. Gupta, ARCH: A 3D space time analysis code in cartesian and hexagon geometries, in *NSRP-19, Mamallapuram, TN, India, 12–15, Dec 2012*
7. D.K. Dwivedi et al., Incorporation of improved quasi static (IQS) scheme module in 3-D space-time analysis code ARCH, in *ARP 2013: Advance in Reactor Physics: Simulation Techniques and Analysis Methodologies, Mumbai, India, October 23–25, 2013*
8. N.E. Todreas, M.S. Kazimi, *Nuclear Systems*, 2nd edn, vol. 1: *Thermal Hydraulic Fundamental*
9. D.K. Dwivedi, A. Gupta, P.D. Krishnani, *Technical Report: RPDD/HTR/65/2015*
10. V.K. Jain, Point-kinetics model with one-dimensional (radial) heat conduction formalism, in *BARC-1452, 1989*
11. A. Borgohain, N.K. Maheshwari, P.K. Vijayan, R.K. Sinha, CFD analysis on heat transfer in low Prandtl number fluid flows, in *NURETH14-457, 2011*
12. D.K. Dwivedi, A. Gupta, P.D. Krishnani, *Technical Report: RPDD/HTR/57 & 60 /2014*
13. http://aerbench.kfki.hu/aerbench/Dyn002.pdf
14. P.D. Krishnani, Ann. Nucl. Energy **9**, 255 (1982)
15. OECD/NEA, Nuclear Science Committee, Working Group on Lead-Bismuth Eutectic, *Handbook on Lead-Bismuth Eutectic Alloy and Lead Properties, Materials Compatibility, Thermal Hydraulics and Technologies*. ISBN 978-92-64-99002-9 (2007)

Part V
Molten Salt Reactors

Investigations of BaF_2–ThF_4 System

Sumanta Mukherjee, Smruti Dash, S. K. Mukerjee and K. L. Ramakumar

Abstract The standard molar Gibbs energy of formation of $BaThF_6$(s) has been experimentally determined using a solid electrolyte galvanic fluoride cell: (−)Pt, NiO(s) + NiF_2(s)|CaF_2(s)|$ThOF_2$(s) + $BaThF_6$(s) + BaF_2(s), Pt(+), with CaF_2(s) as solid electrolyte. From the measured emf values, the Gibbs energy of formation has been calculated as $\Delta_f G^o_m$ ($BaThF_6$, s, T) {±5.1} (kJ mol^{-1}) = −3379.9 + 0.5411 (T/K). Using the experimentally obtained data, the both axis potential diagram of Ba–Th–F–O system has been calculated and binary phase diagram of BaF_2–ThF_4 system has been computed.

Keywords MSR · $BaThF_6$ · Emf · Phase diagram · Potential diagram

S. Mukherjee (✉) · S. Dash (✉)
Homi Bhabha National Institute (HBNI), Bhabha Atomic Research Centre, Mumbai 400085, India
e-mail: sumantam@barc.gov.in

S. Dash
e-mail: smruti@barc.gov.in

S. Mukherjee · S. Dash
Fuel Chemistry Division, Bhabha Atomic Research Centre, Mumbai 400085, India

S. K. Mukerjee
Product Development Division, Bhabha Atomic Research Centre, Mumbai 400085, India
e-mail: skmukerjee@barc.gov.in

K. L. Ramakumar
Radiochemistry and Isotope Group, Bhabha Atomic Research Centre, Mumbai 400085, India
e-mail: karanam.ramakumar@gmail.com

A. K. Nayak and B. R. Sehgal (eds.), *Thorium—Energy for the Future*,
https://doi.org/10.1007/978-981-13-2658-5_32

1 Introduction

Molten salt breeder reactors (MSBRs) are generation IV-type reactors where the fissile and fertile isotopes are dissolved in molten salt system and circulated through the reactor core and the primary loop generating high energy steam with high thermal efficiency (~45%) [1]. In a molten salt reactor, a mixture of fissile component, UF_4(s) and fertile component, ThF_4(s) are dissolved in molten fluoride medium which is nothing but the fluoride salts of alkali and alkaline earth metals. Here the uranium is present as ^{233}U which on fission in the coolant salt produces a number of fission products and heated up the reactor core. These fission products could be grouped by the several classes as shown in the table.

Noble gases	Rare earths	Noble and semi-noble metals	Volatile fluorides	Stable fluorides
Kr, Xe	La, Ce, etc.	Mo, Nb, Ru, Pd, Te, Sb	Nb, Ru, Tc, Mo, I, Te	Rb, Sr, Y, Zr, Cs, Ba

The molten fluoride is an ionizing solvent medium. The formation of complex ion in this medium increases its conductivity, hence the heat transfer properties, but formation of compound decreases the electrical conductivity of the medium. So, the interaction of dissolved fission product with coolant or fuel salts is important for heat transport properties of the molten salts. The present study is focused on BaF_2–ThF_4 system which is characterized with single complex fluoride: $BaThF_6$. The thermodynamic information of $BaThF_6$(s) is not reported in the literature. The $\Delta_f G^o(T)$ of $BaThF_6$ has been experimentally measured in this study using solid electrolyte galvanic cell containing CaF_2 electrolyte. The pseudobinary phase diagram of BaF_2–ThF_4 system has been computed using measured $\Delta_f G^o(T)$ of $BaThF_6$(s) and required data estimated in this study. The both axis potential diagram or chemical potential diagram of Ba–Th–F–O system has been calculated to determine the stability domain of $BaThF_6$(s).

2 Experimental

2.1 *Preparation of the Compound ($BaThF_6$(s))*

The compound has been prepared by the co-precipitation route. The aqueous solutions of barium carbonate ($BaCO_3$) and thorium nitrate ($Th(NO_3)_4$) were prepared separately and were mixed in equivalent amount, and to the mixture, 40%

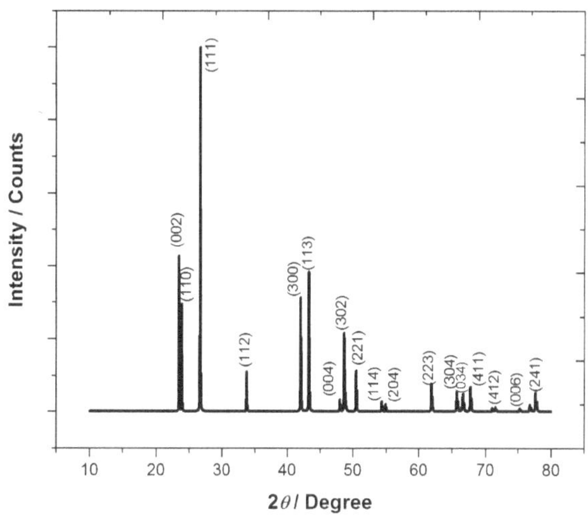

Fig. 1 XRD pattern of $BaThF_6$(s)

hydrofluoric acid solution was added in stoichiometric amount which resulted in the formation of a white precipitate. This precipitate was filtered and dried under IR lamp and characterized by X-ray diffraction technique. The experimentally obtained XRD pattern (Fig. 1) of $BaThF_6$(s) shows good agreement with the reported XRD data in JCPDS file no: 44-0244 [2].

2.2 *Establishment of Equilibrium Phase Mixture for Emf Study*

For the establishment of phase field, ternary phase mixtures of dried BaF_2(s), $ThOF_2$(s), and $BaThF_6$(s) contained in inconel crucibles were sealed in quartz ampoules under the dynamic vacuum of 10^{-6} mbar pressure. The sample in the ampoules were equilibrated 800 K for 200 h and slowly cooled to the room temperature. XRD analysis (Fig. 2) of the equilibrated mixture containing BaF_2(s), $ThOF_2$(s), and $BaThF_6$(s) did not show any extra XRD peak other than the starting material.

2.3 *Emf Cell Setup*

For the measurement of $\Delta_f G^o_m$ ($BaThF_6$, s, T), a single compartment cell has been constructed, and the cell has been represented as:

$$\text{Cell: I} \quad (-)\text{Pt, NiO(s)} + \text{NiF}_2\text{(s)}|\text{CaF}_2\text{(s)}|\text{ThOF}_2\text{(s)} + \text{BaThF}_6\text{(s)} + \text{BaF}_2\text{(s), Pt(+)} \quad (1)$$

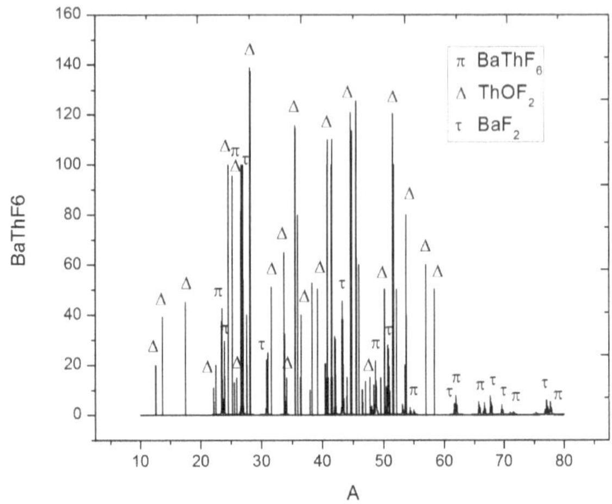

Fig. 2 XRD pattern of the phase mixture after heating

The ternary phase mixture {$ThOF_2(s) + BaThF_6(s) + BaF_2(s)$} and binary phase mixture {$NiO(s) + NiF_2(s)$} were separately pelletized to make cathode and anode of the emf cell (Eq. 1). Prior to the emf measurements, fluoride cell assembly has been calibrated using the following cell:

$$\text{Cell: II} \quad (-)\text{Pt, Fe(s)} + \text{FeF}_2\text{(s)}|\text{CaF}_2\text{(s)}|\text{Ni(s)} + \text{NiF}_2\text{(s), Pt(+)} \tag{2}$$

The cathode and anode of the cells were kept pressing against the two polished surface of the $CaF_2(s)$ electrolyte by an alumina tube and a quartz holder. The detail of the fluoride cell is given in our previous study [3]. The emf of the cell was measured using a high impedance Keithley model 614 electrometer.

3 Results and Discussion

3.1 Calibration

The emf values produced for the cell II were due to the overall cell reaction:

$$\text{NiF}_2\text{(s)} + \text{Fe(s)} = \text{FeF}_2\text{(s)} + \text{Ni(s)} \tag{3}$$

Emf values obtained from the cell II as a function of temperature were given in Table 1 and were expressed by a linear expression:

$$E/\text{V} \pm 0.001 = 0.3107 + 2.634 \times 10^{-5}(T/\text{K}) \tag{4}$$

Table 1 Emf at different temperature for cell: (−)Pt, Fe(s) + FeF_2(s)|CaF_2(s)|Ni(s) + NiF_2(s), Pt(+)

T/K	mE/V	T/K	mE/V	T/K	mE/V	T/K	mE/V
917.3	334.8	951.8	335.6	954	335.8	975.3	336.6
927.9	335.2	946.1	335.6	958.3	335.9	976.2	336.2
936.9	335.4	949.1	335.7	964.5	336.1		
942.9	335.5	953	335.9	968.9	336.2		

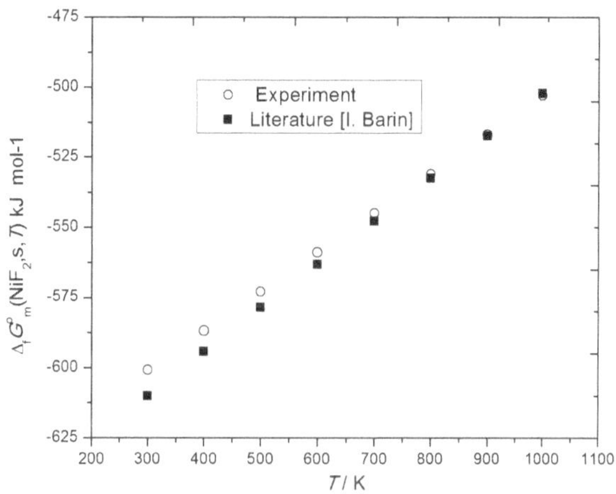

Fig. 3 Comparison of $\Delta_f G^o_m(NiF_2, s, T)$ obtained from experimental technique and that of from the literature [4]

The Gibbs energy of reaction (3) can be calculated from the relation:

$$\Delta_r G^o_m(T) = -n \cdot F \cdot E \quad (5)$$

where n is the number of electrons involved in the reaction, F is the Faraday constant, $F = 96{,}484$ °C/F. $\Delta_r G^o_m$ (T) can be expressed as:

$$\Delta_r G^o_m(T)\{\pm 0.2\}\text{kJ mol}^{-1} = -61.3 + 5.14 \times 10^{-3}(T/\text{K}) \quad (6)$$

The measured emf values (from Eq. 6) were used to calculate $\Delta_f G^o_m$ (NiF_2, s, T) values and it is given as

$$\Delta_f G^o_m(NiF_2, \text{s}, T)\,\text{kJ mol}^{-1} = (-643.6 \pm 0.2) + (0.1401 \pm 0.0003)T/\text{K} \quad (7)$$

The Gibbs energy value (−537.1 kJ mol^{-1}) calculated from Eq. 7 at 750 K has been found to match well (within 2.4 kJ mol^{-1}) with that (−539.5 kJ mol^{-1}) of reported by Barin et al. [4] which showed reasonable agreements. The comparison of experimental data and the literature has been shown in Fig. 3. It demonstrates that thermodynamic data generated by the cell is reliable.

3.2 $BaThF_6$

The emf of cell I has been used for the measurements of $\Delta_f G^o_m$ ($BaThF_6$, s, T). The emf values were generated due to following cell reactions at anode and cathode:

$$\text{At cathode,}\quad BaThF_6(s) + (1/2)O_2(g) + 2e^- = BaF_2(s) + ThOF_2(s) + 2F^- \quad (8)$$

$$\text{At anode,}\quad NiO(s) + 2\,F^- = NiF_2(s) + (1/2)O_2(g) + 2\,e^- \quad (9)$$

The net cell reaction is given as:

$$BaThF_6(s) + NiO(s) = BaF_2(s) + ThOF_2(s) + NiF_2(s) \quad (10)$$

The generated emf values as a function of temperature can be given as:

$$E/\text{V} \pm 0.007 = 0.4996 - 4.07 \times 10^{-4}(T/\text{K}) \quad (11)$$

The Gibbs energy of reaction for the net cell reaction (Eq. 10) can be calculated from the relation of Eq. 3 and can be expressed as:

$$\Delta_r G^o_m(T)\,\text{kJ mol}^{-1} \pm 0.8 = -96.4 + 0.0786(T/\text{K}) \quad (12)$$

$\Delta_f G^o_m$ ($BaThF_6$, s, T) has been calculated by incorporating $\Delta_f G^o_m$ (T) for NiF_2(s), NiO(s), BaF_2(s), and $ThOF_2$(s) from the literature [4]. The calculated Gibbs energy expression of $BaThF_6$(s) can be given as:

$$\Delta_f G^o_m(BaThF_6, s, T)\{\pm 5.1\}\left(\text{kJ mol}^{-1}\right) = -3379.9 + 0.5411(T/\text{K}). \quad (13)$$

4 Computation of Phase Diagram of BaF_2–ThF_4 System

The phase diagram is a representation of the thermodynamic properties of a system. The Gibbs energy functions of all phases involved in the system must be known for the calculation of the same. In some cases, where the Gibbs energy functions for a phase are not known, using the optimization procedure the thermodynamic description of the phases can be calculated. Such optimization method is known as CALPHAD method [5, 6]. The phase diagram can be constructed by the minimization of Gibbs energy of possible phases present in BaF_2–ThF_4 system where all the phases obey the laws of minimum Gibbs energy. The binary phase diagram described the stable phases which were formed due to interaction of different mole fraction of BaF_2 and ThF_4 over a range of temperatures. For the stoichiometric compounds, ThF_4, BaF_2, and $BaThF_6$, Gibbs energy of both solid liquid states and solid and liquid solutions between BaF_2 and ThF_4 must be known for the computation. The Gibbs energy of a solution phase is calculated from the Gibbs energy of the pure BaF_2 and ThF_4

weighted by their mole fractions and a mixing term and can be represented by the following equation:

$$G^{sol}(T) = x_{BaF_2} \cdot G^{o}(BaF_2, T) + x_{ThF_4} \cdot G^{o}(ThF_4, T) + R \cdot T \cdot \left(x_{BaF_2} \cdot \ln x_{BaF_2} + x_{ThF_4} \cdot \ln x_{ThF_4}\right) + G^{ex} \quad (14)$$

The excess contribution (G^{ex}) is the only unknown term (in Eq. 14) which is responsible for the deviation of the real systems from their ideal behavior. So it is highly important to determine the unknown excess Gibbs energy term. The excess Gibbs energy of the liquid solution can be described using the classical polynomial model as:

$$G^{ex} = \sum_{ij} x_{BaF_2} \cdot x_{ThF_4} \cdot L_{ij} \quad (15)$$

The term $L_{i,j}$ can be described with the general equation

$$L_{i,j} = \mathrm{A} + \mathrm{B} \cdot T + \mathrm{C} \cdot T \cdot \ln T + \mathrm{D} \cdot T^2 + \ldots \quad (16)$$

where the parameters (A, B, C, D …) are subject of the optimization during the thermodynamic assessment of experimental and estimated data from different sources.

The classical polynomial model has been used to describe the $(Ba,Th)F_x$ liquid solution. Two main routes have been followed to calculate the coefficients of G^{ex}, trial and error approach and mathematical iteration. In the first path, the coefficients are selected arbitrarily on experience basis and then refined. The later route is mathematical iteration approach. Both methods were used to calculate G^{ex} values for computation of the binary phase diagram of BaF_2–ThF_4 system which is shown in Fig. 4. It shows that there is no solid solubility found in the system, and $BaThF_6(s)$ compound is formed in the solid state. This system has two eutectic points. The enthalpy of mixing has been calculated from the phase diagram. The lines on the phase diagram represented various equilibrium phase boundaries.

5 Activity Versus Mole Fraction Plot

Activity of the ThF_4–BaF_2 mixture above eutectic temperature (here 1300 K) has been calculated. It has been plotted as a function of mole fraction, and a linear plot (Fig. 5) has been obtained.

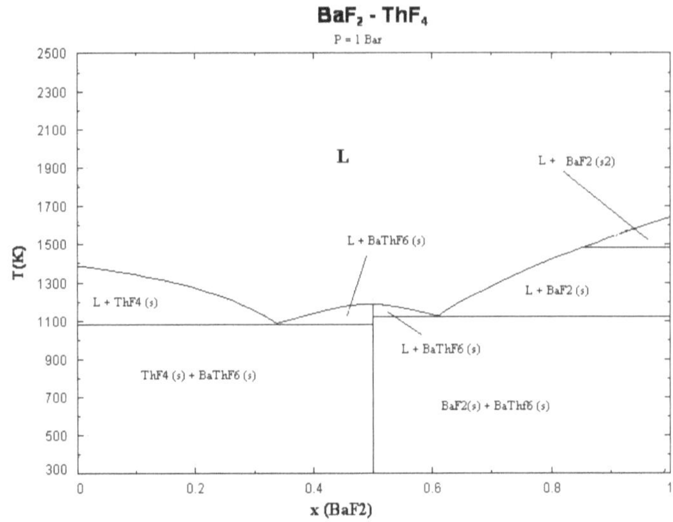

Fig. 4 The pseudobinary phase diagram of BaF_2–ThF_4 system

Fig. 5 Activity versus mole fraction

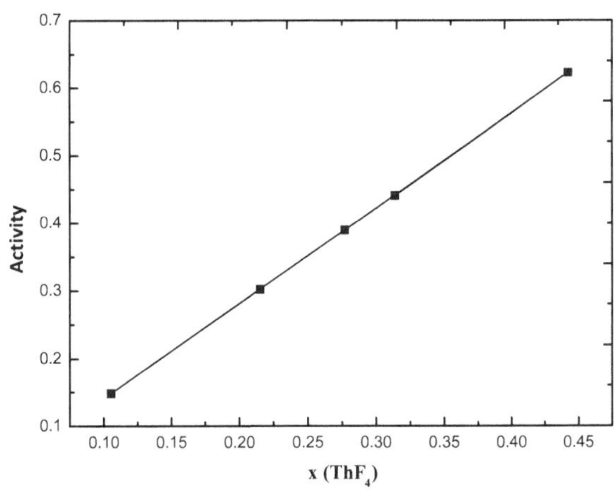

6 Enthalpy of Mixing Versus Mole Fraction Plot

Enthalpy of mixing of ThF_4 and BaF_2 system has been calculated. ΔH_{mix} has been plotted with mole fraction of ThF_4, and it shows a linear plot as given in Fig. 6.

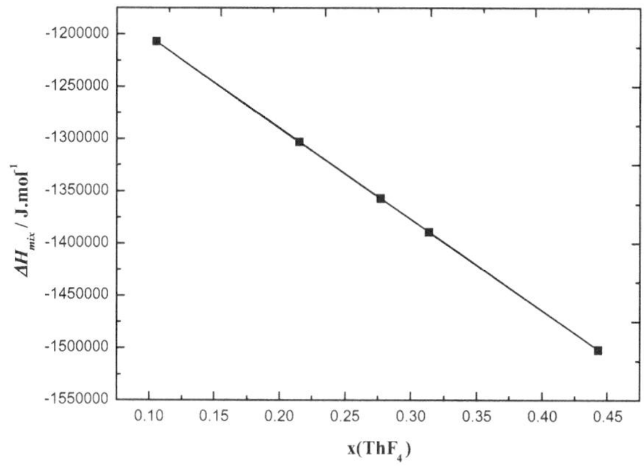

Fig. 6 Enthalpy of mixing versus mole fraction

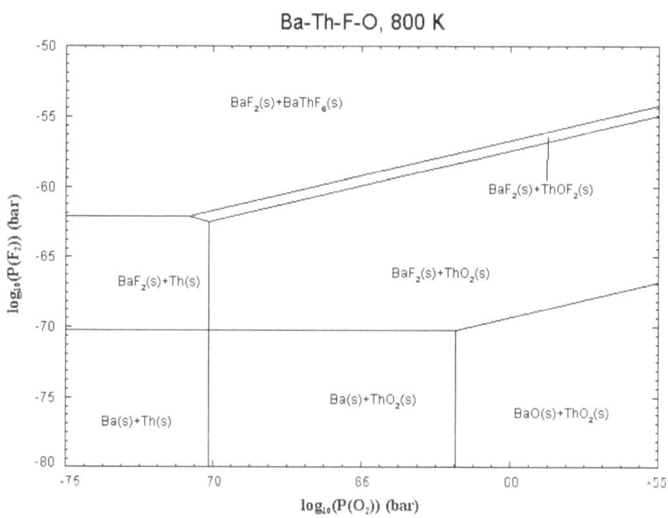

Fig. 7 Chemical potential diagram of Ba–Th–F–O system

7 Computation of Stability Domain of $BaThF_6$(s) in Ba–Th–F–O System

The chemical potential diagram of Ba–Th–F–O system gives the stability domain of the various phases as a function of the chemical potentials of oxygen and fluorine that have been computed based on Gibbs energy minimization technique. In both axis potential diagram, the phases of Ba–Th–F–O system has been plotted as a function of $\log(p(F_2))$ along y-axis and $\log(p(O_2))$ along x-axis. The required Gibbs energy data of unary and binary compounds in Ba–Th, Ba–F, Th–F, Ba–O, Th–O, and F–O systems have been taken from the literature. The ternary compounds in Ba–Th–F, Ba–Th–O, and Th–O–F systems have been estimated in this study using the method given in the literature. The $BaThF_6$ was found to be stable above $\Delta\mu(F_2) = -957.35$ kJ mol^{-1}. It shows that $BaThF_6$(s) coexists with BaF_2(s) (Fig. 7).

8 Conclusion

$BaThF_6$(s) was synthesized and characterized in this study. The Gibbs energy formation of $BaThF_6$(s) and binary phase diagram of BaF_2–ThF_4 system have been determined for the first time. The phase diagram shows $BaThF_6$(s) is stable and melts congruently. The enthalpy of mixing between BaF_2 and ThF_4 in the liquid state is found to be negative. There is no solid solubility in this system. The chemical potential of Ba–Th–F–O system shows that barium-thorium fluorides were found to be stable above $\Delta\mu(F_2) = -957.35$ kJ mol^{-1}.

References

1. R.K. Sinha, Advanced nuclear reactor systems—an Indian perspective. Energy Proc. **7**, 34–50 (2011)
2. PCPDFWIN version 2.2, Joint Committee on Powder Diffraction Standards, JCPDS-ICDD (2001)
3. R. Prasad, S. Dash, S.C. Parida, Z. Singh, V. Venugopal, Thermodynamic studies on $SrThO_3$(s). J. Nucl. Mater. **312**(1) (2003)
4. I. Barin, *Thermochemical Data of Pure Substance*, 3rd edn. (VCH, Weinheim, Federal Republic of Germany, 1995)
5. L. Kaufman, H. Bernstein, *Computer Calculation of Phase Diagrams* (Academic Press, New York, NY, 1970)
6. A.D. Pelton, P. Chartrand, G. Eriksson, The modified quasi-chemical model: part IV. Two-sublattice quadruplet approximation. Metall. Mater. Trans. A **32**(6), 1409–1416 (2001)

The iThEC Strategy

Jean-Pierre Revol

Abstract Fossil fuels are present in finite quantity in the Earth's crust and will therefore run out rather soon on a human timescale. In addition, burning fossil fuels contributes to global warming and has a tremendous health impact through atmospheric pollution. Fossils fuels have to be replaced as energy sources as soon as possible. However, achieving this requires a major and systematic R&D effort. This R&D effort must not be biased; it must include nuclear energy, which is abundant, is energy intensive, produces no greenhouse gases or air pollution, and still has a great potential for improvement. The international Thorium Energy Committee (iThEC) is at the origin of two new initiatives in this domain: a first fast-neutron ADS experiment of substantial power ($\geq$1 MW), at INR Troitsk in Russia, with the goal of demonstrating the possibility of destroying nuclear waste, using it as fuel in a thorium matrix, and a highly innovative high-power proton accelerator based on superconducting cyclotron technology. These two projects are well within the reach of present technologies and would constitute important milestones in the development of intensive, abundant, and sustainable energy sources for the future. The future energy supply of the world is such a crucial challenge that all options should be considered, with the goal of choosing what will turn out to be the best solution in the end.

Keywords Thorium · Accelerator-driven systems · ADS · iThEC · Nuclear waste

1 Introduction

The international Thorium Energy Committee (iThEC) [1] is a Geneva-based, non-profit association, founded under Swiss law in 2012, strongly associated with CERN. Its members, from 14 countries, include physicists, engineers, politicians concerned by energy issues. Several iThEC members were involved in pioneering CERN

J.-P. Revol (✉)
international Thorium Energy Committee (iThEC),
Rue François-Dussaud 17, 1227 Geneva, Switzerland
e-mail: jean-pierre.revol@cern.ch

A. K. Nayak and B. R. Sehgal (eds.), *Thorium—Energy for the Future*,
https://doi.org/10.1007/978-981-13-2658-5_33

accelerator-driven experiments: FEAT [2] and TARC [3], carried out by Nobel Prize Laureate Carlo Rubbia and his team, as well as in the design, construction, and exploitation of the CERN neutron test facility n_TOF [4]. iThEC achieved an important milestone by signing an MoU with IBeL SA, a company involved in nuclear physics, accelerator technology, nuclear medicine, biology, and information technologies, with significant expertise in marketing, business, and operations management, with the goal to conduct R&D activities in the framework of an international collaboration at INR Troitsk.

iThEC is acting to promote R&D in the use of thorium in order to transmute nuclear waste and produce safe, clean, and abundant energy, in particular with accelerator-driven systems (ADS). The first task for iThEC was the organization of the ThEC13 conference, in the CERN Globe of Science and Innovation, in October 2013. This conference, attended by the major world players in thorium R&D, provided a review of the status of the field, crucial to the iThEC for the definition of its strategy.

2 iThEC'S Motivations

Fossil fuels are present in finite quantities in the Earth's crust and inevitably they will run out sooner or later. At the rate they are used today, the timescale for exhausting these sources of energy is of the order of 50–100 years for oil and gas, and of 100–150 years for coal. If consumption continues to increase, this will occur sooner. These are short time periods considering the implications. However, there are strong reasons for replacing fossil fuels by better fuels even sooner. Indeed, it does not make sense to burn fossil fuels till supplies have run out:

- In the first place, there is an increasing consensus that global warming due to the release of greenhouse gases issuing from human activities is a serious problem. Even if this is doubted, precaution should prevail.
- The immediate major problem associated with burning fossil fuels is air pollution. This is already very costly to society. For instance, burning coal costs Europe alone more than 40 billion Euros in annual health care expenses. The European Environmental Agency declared that: "Air pollution poses the single largest environmental health risk in Europe today." More than half a million deaths per year in China are due to air pollution [5]. World Health Organization (WHO) estimated that one in eight of total deaths in the world are the result of exposure to air pollution. This means that the use of fossil fuels kills 10,000 people each day.
- In addition, fossil fuels could be put to much better use. For instance, instead of burning oil, it can be used in the manufacturing of plastics, rubber, tires, paints, glue, drugs, cosmetics, and detergents.

In conclusion, there is every reason to stop burning fossils fuels, as soon as possible. Today, fossil fuels represent 86% of the primary energy consumption [6]. The problem is that they are cheap and abundant, so the current tendency is still to increase their consumption (Fig. 1). It is expected that, given a rising standard of living and

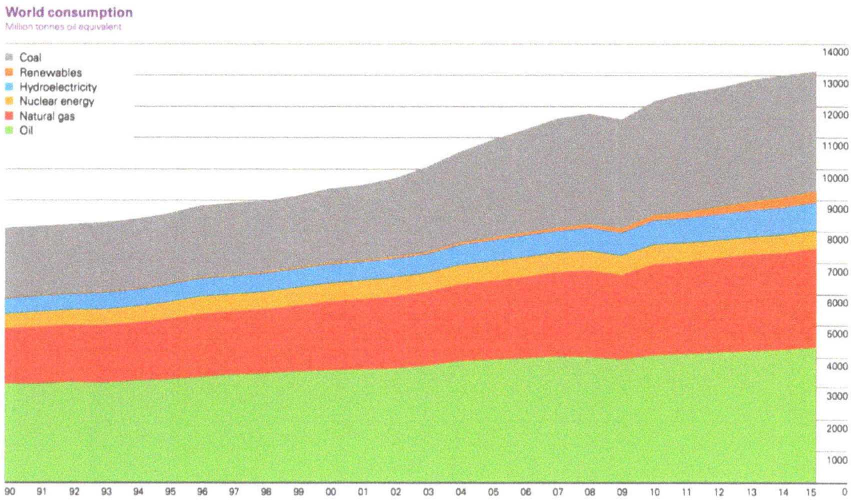

Fig. 1 World primary energy consumption from 1990 to 2015, showing that coal, gas, and oil represent 86% of the world consumption, hydro 6.5%, nuclear 4.4%, and renewable 2.8% [6]

a global population plateauing around 10–11 billion inhabitants, the world energy consumption would have to increase by a factor 3 or more in the next 100 years to keep pace with demand. Therefore, replacing fossil fuels is a formidable challenge to society that can only be met through vigorous and systematic research and development (R&D). As far as the energy domain is concerned, this R&D must be pursued without prejudice and must include nuclear energy. There is no reason to believe that nuclear energy could not be made acceptable to society, in terms of safety, waste management, and proliferation. Its incomparable advantages are that: it is abundant and energy intensive; it does not produce greenhouse gases or create air pollution.

This challenge may also turn out to be an opportunity, as the innovation required will drive economic growth. This is particularly true for developing countries that do not yet have a fully developed nuclear industry. These countries have the choice of copying what was done in countries with advanced nuclear industry or taking the innovative option. In this way, they would control the energy market in the future, which is clearly the intention of China.

3 Why Thorium?

Given the expected huge energy demand, any potentially important source of energy must be considered, in particular thorium, as the required technologies represent only a modest extrapolation of existing nuclear and particle accelerator technologies.

As stated by Carlo Rubbia [7] at the ThEC13 international conference, thorium constitutes "a sustainable energy resource on the human timescale."

Table 1 Estimated world thorium resources [8]: There are sufficient thorium reserves in India to power 846 ADS systems of 1 GW of electric power for 1000 years

	Country	Tonnes		Country	Tonnes
1	India	846,000	10	South Africa	148,000
2	Brazil	632,000	11	China	100,000
3	Australia	595,000	12	Norway	87,000
4	USA	595,000	13	Greenland	86,000
5	Egypt	380,000	14	Finland	60,000
6	Turkey	374,000	15	Sweden	50,000
7	Venezuela	300,000	16	Kazakhstan	50,000
8	Canada	172,000		Other countries	1,725,000
9	Russia	155,000		World total	6,355,000

Indeed, thorium exists in abundance. There are 1.2×10^{14} tonnes of thorium in the Earth's crust, which is about the same as lead and three to four times more than uranium. Thorium resources are broadly distributed on the Earth, and India is particularly well endowed, with the largest thorium resources (Table 1). Natural thorium is isotopically pure, found mostly in monazite ores (Fig. 2), and relatively cheap as often a by-product of rare earth mining. Thorium produces smaller amounts of long-lived nuclear waste (TRU)[1] than uranium. In addition, it has excellent physical properties such as higher melting point for both metallic and oxide forms and better thermal conductivity when compared to uranium. This means that there are higher safety margins for design and operation. When used in ADS, thorium offers the most efficient method to destroy TRU through fission [9] and allows the system to be subcritical, with obvious safety advantages. Most importantly, the thorium fuel cycle has the great advantage of being proliferation resistant.

Mainly because of neutron inventory reasons, thorium cannot simply be substituted for uranium in critical reactor fuel. Uranium-233 has to be bred from thorium in order to produce energy. Three basic approaches are envisaged [10]:

1. A chain of critical reactors to first breed plutonium (in CANDU reactors). Afterward, use the plutonium in fast sodium-cooled critical reactors around which a thorium blanket allows the breeding of uranium-233, so it can be used to manufacture fuel for advanced thermal rectors. This scheme has been adopted by India; it technically works but requires maintaining three different reactor technologies. Moreover, it does not solve the problems of nuclear waste management and sustainability.
2. Moving the fuel continuously so that there is always fresh fuel in the core. This can be done in pebble-bed reactors or in molten salt reactors, both of which have serious safety issues. In traveling wave reactors, which have yet to be developed, it is not the fuel that moves, but the neutron breeding and fission wave.

[1]Transuranium elements (TRU): chemical elements with atomic numbers greater than 92 (the atomic number of uranium).

Fig. 2 Thorium occurs in several minerals including thorite ($ThSiO_4$), thorianite (ThO_2+UO_2), and monazite ([Ce, La, Nd, Th]PO_4). The photograph shows a 230 g monazite sample, containing 2–3% thorium, from the Steenkampskraal mine, in South Africa. (*Courtesy* Trevor Blench, CEO of STL)

3. Using a particle accelerator to produce the extra neutrons needed to sustain a chain reaction, in so-called accelerator-driven systems. C. Rubbia demonstrated that this is the most efficient and elegant solution for using thorium [9]. This is what motivates iThEC in its efforts to promote accelerator-driven thorium systems (ADTS).

4 Accelerator-Driven Systems

Unfortunately, at the moment, developed countries do not seem to be ready to move in the direction of thorium fuel. So far, in most countries with established nuclear power generation, the development of thorium technologies has been opposed for political reasons by green parties and for economic reasons by nuclear industry, already vastly invested in the development of a new generation of fast critical uranium breeder reactors (Generation IV [11]) and which benefits from institutional funding. However, the development of ADS is not in competition with present nuclear industry programs, and it is now time to start preparing for the future.

The recent history of ADS may be characterized by four phases. The first phase, which demonstrated the basic concepts of ADS, in particular at CERN at the initiative of Carlo Rubbia, was completed in the 1990s. The second phase, which tested the basic elements of an ADS separately, was completed in the 2000s: a proton beam of power exceeding 1 MW at PSI; a neutron spallation source of 1 MW, MEGAPIE, also at PSI, and various developments of high-power spallation targets for SNS, ESS, and

EURISOL; the development of high-power superconducting accelerating cavities by the particle physics community; the EUROTRANS R&D program of the European Union on partitioning and transmutation for the uranium fuel cycle.

It would seem natural to follow these with a phase 3 consisting of the first coupling of a proton beam to a fast-neutron subcritical core at a significant power ($\geq$ 1MW). However, this step is still missing, more than 20 years after the first pioneering FEAT experiment. Phase 3 should also include the development of an accelerator optimized for industrial applications of ADS. Phase 3 is concentrating all the efforts of iThEC at the moment.

The good news is that there exist two important ADS projects, which are aiming directly at what could be defined as the industrial fourth phase: the MYRRHA project [12] in Europe, which should be the flagship of ADS systems. Unfortunately, at this time, it is not fully funded, and the intended date of completion, 2025, will most likely be exceeded. The other important project is ADANES [13] in China, which has the ambition of reaching 1000 MW of electrical power by 2032. Given their long timescales, and the large leap into industrialization, both projects would benefit from a phase 3. In addition, there are many other ADS-related activities in progress, in particular in Ukraine with an electron beam of 100 kW driving a subcritical core [14] ready to be commissioned. Given the importance of the energy issue, the lack of coordination and collaboration between these various efforts is regrettable.

5 iThEC's Initiatives

In this context, iThEC is at the origin of two new initiatives: a first ADS experiment of substantial power ($\geq$1 MW) in collaboration with INR Troitsk [15], near Moscow, and a high-power superconducting cyclotron [16].

Through a systematic review of the status of thorium R&D in the world at ThEC13, visits of the Institute of Nuclear Research RAS, Moscow (INR), and detailed discussions with INR management, iThEC became convinced of a rather unique opportunity in Russia, at the Troitsk INR laboratory. Here, the existing infrastructure (Fig. 3) would allow, for the first time at significant power (a few MW), the coupling of a proton accelerator to a fast-neutron subcritical core. The accelerator exists and only needs a relatively modest refurbishment. Troitsk is already operating a neutron spallation source. A beam line toward an available experimental pit could be implemented quickly. In addition, the infrastructure exists for the manipulation of radioactive materials. It is estimated that in 5-year time a landmark experiment could be carried out which would validate the properties of an ADTS, demonstrate safety, demonstrate the destruction of minor actinides, and test new possibilities for the production of radioisotopes for medicine. This would be at a cost of 5% of currents projects such as MYRRHA [12]. A road map of the project has been prepared jointly by INR management and iThEC. The Troitsk experiment would constitute in addition a versatile fast-neutron test facility, the subcritical character of which guarantees safety. Discussions with contacts in China, the Czech Republic, India,

Fig. 3 Photograph taken at INR Troitsk of the beam target area showing the proposed ADS pit (9) and the presently operating pulsed neutron source cell (10) and its beam line (8). Taken from [15]

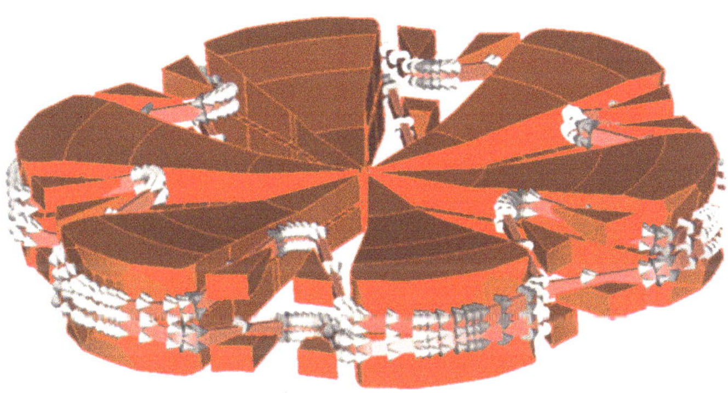

Fig. 4 3D view of the six-sector single-stage cyclotron with reversed valley B-field, considered for the Horizon 2020 FET proposal [16] (patented under the name S2CD by AIMA DEVELOPPEMENT)

South Korea, Japan, Switzerland, Turkey, Chile, as well as with IAEA indicate that such a project, if approved, would get international support, and a collaboration on the CERN experimental collaboration model would be possible. Global cooperation, which iThEC is actively stimulating, is highly desirable in this domain.

The second iThEC initiative is a project submitted to the European Union, in the framework of the Horizon 2020 FET program. This is a collaboration with CERN, PSI, and leading European industrial companies to design an innovative single-stage, high-power superconducting cyclotron [16] (Fig. 4). Reliability, minimal beam losses, and a much lower cost than other technologies are the main goals of the study.

6 Conclusion

In conclusion, thorium is an important energy resource that cannot be neglected. It is most important not to leave thorium ADS out of energy R&D programs. The development of thorium technologies is an opportunity for developing countries, such as India. The ADS experiment at Troitsk and the development in parallel of an innovative high-power superconducting cyclotron are both necessary steps toward an industrial ADS prototype and the exploitation of thorium resources. It is the priority of iThEC to obtain approval for these two projects. As Russian authorities would like to see international support for the Troitsk experiment, iThEC and INR are seeking partners from the international scientific community.

Acknowledgements I would like to thank all my iThEC and INR colleagues. Without their contributions and enthusiasm, none of the projects described here would have been possible.

References

1. iThEC's Web page: www.ithec.org
2. S. Andriamonje et al., Experimental determination of the energy generated in nuclear cascades by a high energy beam. Phys. Lett. B **348**, 697–709 (1995)
3. A. Abánades et al., Results from the TARC experiment: spallation neutron phenomenology in lead and neutron-driven nuclear transmutation by adiabatic resonance crossing. Nucl. Instrum. Methods Phys. Res. A **478**, 577–730 (2002)
4. F. Gunsing, Nuclear Data for the Thorium Fuel Cycle and Transmutation, in *Thorium Energy for the World, ThEC13 Proceedings of CERN Globe of Science and Innovation*, Geneva, Switzerland, 27–31 Oct 2013 (Springer, 2016)
5. GBD MAPS Working Group, Burden of Disease Attributable to Coal-Burning and Other Air Pollution Sources in China (Tsinghua University Report, 2016)
6. 65th edition of the BP Statistical Review of World Energy (June 8, 2016). https://www.bp.com/content/dam/bp/pdf/energy-economics/statistical-review-2016/bp-statistical-review-of-world-energy-2016-full-report.pdf; https://www.bp.com/en/global/corporate/media/press-releases/65th-editionof-bp-statistical-reviewof-world-energy.html
7. C. Rubbia, A Future for Thorium Power?, in *Thorium Energy for the World*, in *ThEC13 Proceedings of CERN Globe of Science and Innovation*, Geneva, Switzerland, 27–31 Oct 2013 (Springer, 2016)
8. Data taken from "Uranium 2014: Resources, Production and Demand", OECD Nuclear Energy Agency and the International Atomic Energy Agency
9. C. Rubbia; et al., Conceptual Design of a Fast Neutron Operated High Power Energy Amplifier, *CERN/AT/95-44 (ET)*, 29 Sept. 1995; see also Rubbia, C. A High Gain Energy Amplifier Operated with Fast Neutrons, *AIP Conference Proceeding* 346, International Conference on ADT Technologies and Applications, Las Vegas (1994)

10. J.-P. Revol, Accelerator-Driven Systems (ADS) Physics and Motivation, in *Thorium Energy for the World, ThEC13 Proceedings of CERN Globe of Science and Innovation*, Geneva, Switzerland, October 27–31 2013 (Springer, 2016)
11. Generation IV nuclear energy systems, see: www.ne.anl.gov/research/genIV
12. H. Aït Abderrahim, MYRRHA: A Flexible and Fast Spectrum Irradiation Facility, in *Thorium Energy for the World*, ThEC13 Proceedings of CERN Globe of Science and Innovation, Geneva, Switzerland, 27–31 Oct 2013 (Springer 2016)
13. P. Luo, S.C. Wang, Z.G. Hu, H.S. Xu, W.L. Zhan, Accelerator driven sub-critical systems—a promising solution for cycling nuclear fuel, Institute of Modern Physics, Chinese Academy of Sciences, Lanzhou, 730000, China. https://doi.org/10.7693/wl20160903
14. Y. Gohar, I. Bolshinsky, D. Naberezhnev et al., Accelerator driven sub-critical facility: conceptual design development. Nucl. Instrum. Meth. Phys. Res. A. **562**, 870–874 (2006)
15. S.F. Sidorkin, A.D. Rogov, L.I. Ponomarev, E.A. Koptelov, Proposal of the ADS Research Stand Based on the Linac of the Institute for Nuclear Research of the Russian Academy of Sciences, in *Thorium Energy for the World*, ThEC13 Proceedings of CERN Globe of Science and Innovation, Geneva, Switzerland, 27–31 Oct 2013 (Springer 2016)
16. M. Conjat, J. Mandrillon, P. Mandrillon, Cyclotron Drivers for Accelerator-Driven Systems, AIMA DEVELOPPEMENT Company, in *Thorium Energy for the World*, ThEC13 Proceedings of CERN Globe of Science and Innovation, Geneva, Switzerland, 27–31 Oct 2013 (Springer 2016)

Fission Product Removal by Vacuum Spraying in a Heavy Water Moderated Molten Salt Reactor

Thomas Jam Pedersen

Abstract In the MSRE at ORNL volatile, fission products were removed from the salt in the pump chamber via a helium flow gas and a spraying mechanism. Copenhagen Atomics (CA) proposes a new partial vacuum spraying mechanism for removal of volatile fission products from the salt and show that it is likely that up to 50% of fission products can be removed with this simple method. Combining partial vacuum spraying with a heavy water moderated molten salt reactor, CA suggests that it would be possible to develop a breeder reactor in thermal spectrum without online pyro or wet chemistry salt cleanup. Doing this by introducing ceramic materials and new production techniques includes aerographite insulation materials.

Keywords Thorium · MSR · Fission products · Xenon extraction
Helium bubbling · Nozzle spraying · Aerosol · Volatile fission products
Heavy water moderator · Breeder reactor · Aerographite · LIBS
Laser-induced breakdown spectroscopy · Copenhagen Atomics · ThEC15
Mumbai · BARC

1 Introduction

The world is challenged in many ways but production of energy seems to be at the core to many of the solutions.

The team behind Copenhagen Atomics (CA) believes Thorium MSR technology could play an essential role in solving many of these global problems. We root this in the strong correlation and there is between prosperity and access to energy as shown in Fig. 1. Therefore, CA urged the Thorium MSR groups around the world to work together. The main contributions from CA to such collaborations would be within the fields of Measurement Technologies, Chemistry, and Control Systems. The presentation at ThEC15 included two examples of subsystems where CA would be able to

T. J. Pedersen (✉)
Copenhagen Atomics, Copenhagen, Denmark
e-mail: thomas@copenhagenatomics.com

A. K. Nayak and B. R. Sehgal (eds.), *Thorium—Energy for the Future*,
https://doi.org/10.1007/978-981-13-2658-5_34

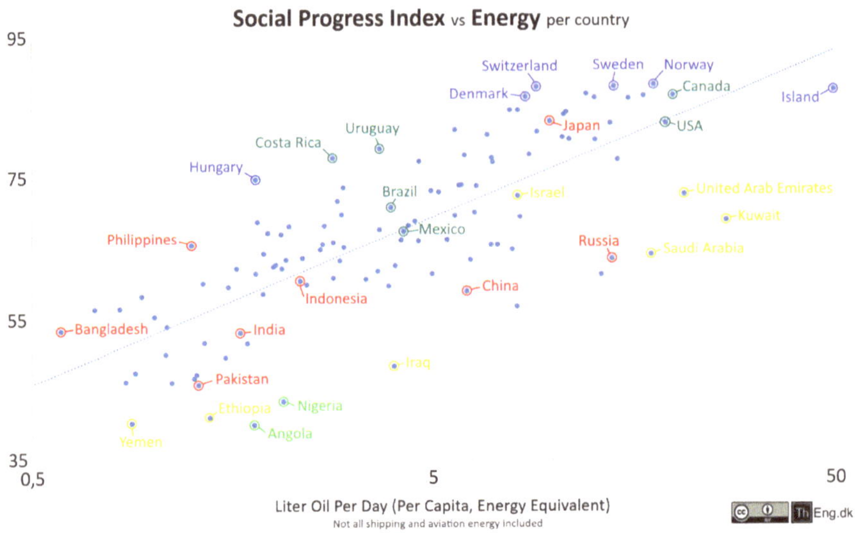

Fig. 1 Social Progress Index 2015 plotted against International Energy Agency global energy production numbers from 2013 for 115 countries. Several countries indicated with the name. Blue dotted line is best fit to dataset [1]

provide valuable knowledge and rapid progress to other teams. Laser-induced breakdown spectroscopy (LIBS) and fission product (FP) removal by vacuum spraying are two examples of areas where CA is leading the global effort.

2 Laser-Induced Breakdown Spectroscopy

LIBS is considered a non-destructive technique—less than one-billionth of a gram of material is consumed during a typical test. Laser pulses generate a high-temperature micro-plasma on the surface of the sample. After this excitation, light that is characteristic of the elemental/isotopic composition of the material is emitted and analyzed using a spectrometer. The advantages of this technique include no need for sample preparation and that the technique can be applied on both solids and liquids/melts. Furthermore, the method does not require direct physical contact with the sample making it a true remote sensing technique.

This enables chemical analysis without having to remove material from the, e.g., reactor or reprocessing system. This measurement technology has successfully been used for quantitative analysis of, e.g., molten salts, contaminants in water, aerosols, and solid materials (metals, glasses, and salts).

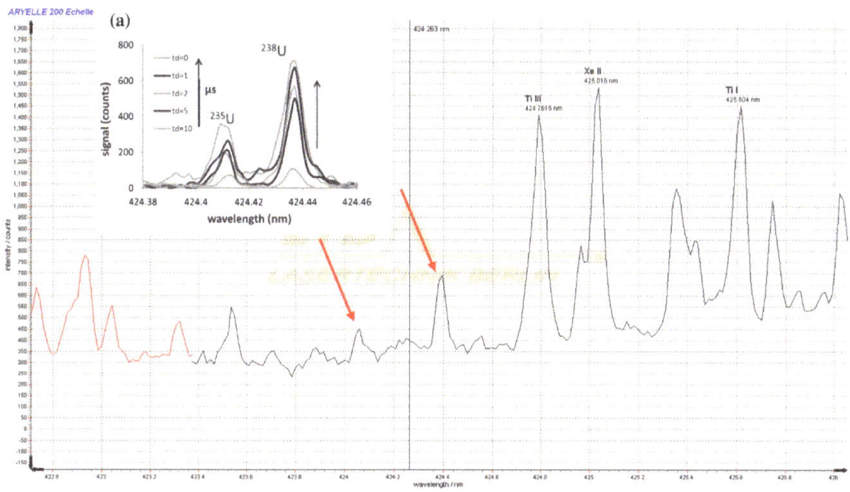

Fig. 2 LIBS measurement on metallic Th sample (measurement performed by CA); zoom in on the 424–425 nm and insert of characteristic peaks of 235U and 238U

CA Has Used LIBS to Successfully Determine

- Accurate quantitative chemical analysis of fluoride melts above 1000 °C;
- Isotope detection of traces of 235U and 238U in a metallic Th sample (Fig. 2);
- Isotope detection of 6Li and 7Li in solid LiF (Fig. 3).

CA is currently calibrating our equipment to analyze Thorium molten salts in a sealed environment with samples containing several hundred fission product isotopes.

LIBS Accuracy Potential

In the CA LIBS measurement system, we use a system of optical fibers, lenses, and mirrors to be able to make accurate measurements where the electronic equipment is separated from radiation and the hot salt.

The typical detection limit for relevant elements and isotopes is ~10 PPM and isotopes can typically be analyzed with a relative accuracy of 98–99% (demonstrated on fluoride and silicate melts by CA).

The CA LIBS system will be used in house to develop and validate the range of fission products, which can be removed from the salt in the vacuum spraying system described below. CA will also market the LIBS system as a component for other MSR manufactures and to be used for inspection and fissile inventory control by IAEA or other authorities.

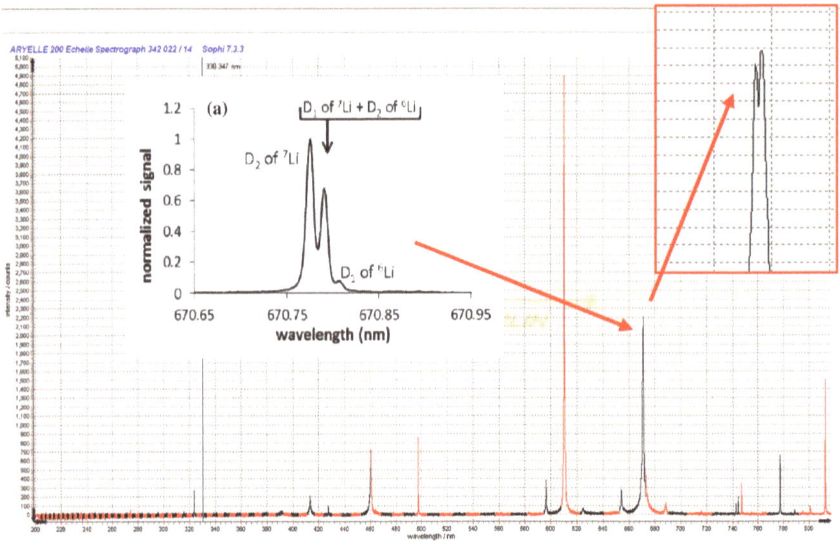

Fig. 3 LIBS measurement on solid LiF sample (measurement performed by CA), incl. a zoom in on the 671 nm range and the characteristic peaks of 6Li and 7Li

3 Fission Product Removal by Vacuum Spraying

Vacuum spraying is an advancement of the traditional helium bubbling method, which will theoretically make it possible to remove a large fraction of the fission products from the fuel salt online. By doing so, the need for online wet chemistry reprocessing of the reactor fuel salt can be avoided. This method relies on evaporation of noble gases and evaporation of volatile fission product compounds. The ability for a large extraction ratio of fission products stems from the fact that the fission products will decay through multiple elements that are either noble gases or volatile compounds. This is the same mechanism that helium bubbling relies on, but the key difference in the proposed method is to spray the salt through nozzles with a flow gas, rather than bubbling helium through the salt. This inversion of geometry has the advantage that salt droplets in a gas increase the surface area to volume ratio, thereby significantly improving the fission product extraction through increased evaporation.

Figures 4 and 5 show an example of some of the FP decay chains. All 3500 decay chains can be found online [2].

Many of these elements are in volatile form below 700 °C at low pressures; some isotopes only stay in the volatile state for seconds, before they decay further down the decay chain. In order to reach a high salt cleanup ratio, the system for removing these volatile isotopes is preferably located close to the reactor core with significant high fuel salt flow speed to reach this FP evaporation chamber in seconds after the fission has occurred.

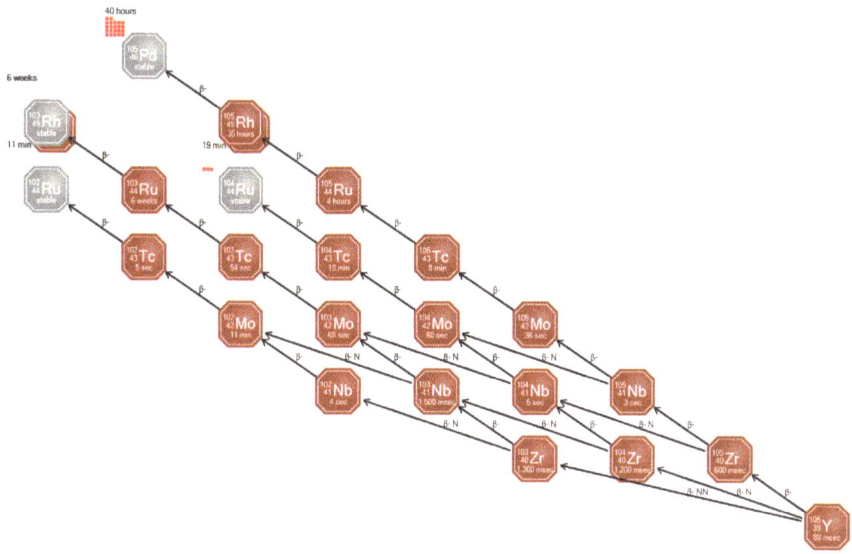

Fig. 4 Typical decay chain for fission product isotopes. Nb and Mo are volatile and can be removed

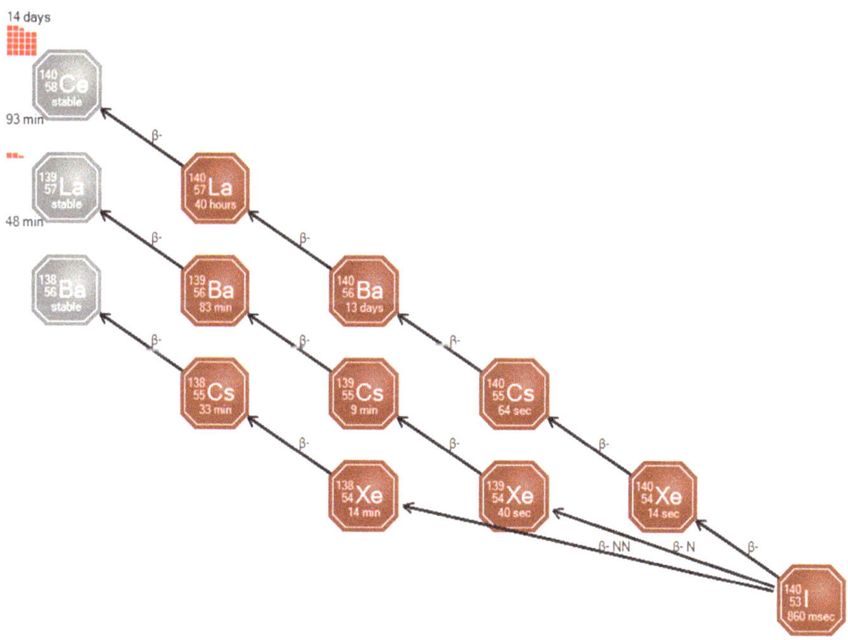

Fig. 5 Typical decay chain for fission product isotopes. Only Xe is volatile and most of it must be removed within seconds else it is not possible later

Table 1 Boiling point and reduction potential of different fluoride species typically found in Thorium fuel salt

Num	Symbol	Name	E0	Charge	Fluoride	Fluoride boiling Point C
2	He	Helium	0	0	He	-267
18	Ar	Argon	0	0	Ar	-185
36	Kr	Krypton	0	0	Kr	-153
54	Xe	Xenon	0	0	Xe	-108
33	As	Arsenic	-0.225	3	AsF3	60
34	Se	Selenium	-0.67	4	SeF4	101
35	Br	Bromine	1.087	-1	BrF3	125
32	Ge	Germanium	0.26	2	GeF2	130
42	Mo	Molybdenum	-0.2	5	MoF5	215
41	Nb	Niobium	-1.1	5	NbF5	234
40	Zr	Zirconium	-1.45	4	ZrBr4,ZrI4,ZrF4	357-431-912
51	Sb	Antimony	-0.51	3	SbF3	376
23	V	Vanadium	-1.13	3	VF3	483
53	I	Iodine	0.5355	-1	IF3	493
52	Te	Tellurium	0.57	4	TeF4	647
47	Ag	Silver	0.7996	1	AgF	700
50	Sn	Tin	-0.14	2	SnF2	850
31	Ga	Gallium	-0.56	3	GaF3	1000
26	Fe	Iron	-0.44	2	FeF2	1100
4	Be	Beryllium	-1.99	2	BeF2	1175
49	In	Indium	-0.34	3	InF3	>1200
55	Cs	Cesium	-2.92	1	CsF	1251
94	Pu	Plutonium	-2.031	3	PuF4	>1277
24	Cr	Chromium	-0.424	2	CrF2	1300
27	Co	Cobalt	-0.277	2	CoF2	1400
37	Rb	Rubidium	-2.98	1	RbF	1410
92	U	Uranium	-1.798	3	UF3,UF4	>1417
30	Zn	Zinc	-0.7618	2	ZnF2	1500
29	Cu	Copper	0.52	2	CuF2	1676
3	Li	Lithium	-3.04	1	LiF	1676
90	Th	Thorium	-1.899	4	ThF4	1680
48	Cd	Cadmium	-0.403	2	CdF2	1748
28	Ni	Nickel	-0.257	2	NiF2	1750
25	Mn	Manganese	-1.17	2	MnF2	1820

It is important to realize that these fission products are created as two individual atoms in a very large body of fuel salt. Independent of the techniques used to remove the fission products from the salt, these single atoms will have to reach the surface of the salt. Thus from the first principle, the chance of removing the fission products from the body of fuel salt is improved when the surface area of the salt can be increased.

The fundamental idea behind vacuum spraying is to spray the salt out through a special type of nozzle that create ultra-small droplets, thereby maximizing the surface area of the salt. This is sometimes referred to as atomizing nozzles, aerosol nozzles, or ultrasonic nozzles. Fission products that are in gas phase (or have a measurable vapor pressure) at the reactor salt temperature can then be separated from the salt simply by having a flow of gas that carry them away from the partial vacuum spraying chamber. Table 1 shows typical boiling points at atmospheric pressure of some of the fluoride species in the salt. Note that some tin and gallium is also likely to be removed at low pressure and 800 °C salt temperature.

4 Simulations

The author has developed new simulation software applications in order to be able to simulate which of the FP can be removed. These simulations take into account the information about half-life, decay chain, boiling points, vapor pressure, salt flow speed, droplet size, vacuum chamber pressure, and neutron captures cross section, to simulate how many of FP can be removed from the salt and how this will affect the neutron economy of a reactor. Results show that between 30 and 50% (molar) of the FP can be removed from the fuel salt using this vacuum aerosol cleanup method. This video shows some of the simulation results [3].

4.1 Validation of Results—Future Work

CA plans to validate the simulation results by adding many of the different elements from Table 1 to fuel salt samples, e.g., LiF-ThF4-UF4. These fuel salt samples will not use enriched lithium and uranium. The samples will be heated in a furnace to a range of different temperatures and with a range of different cover gas flow rates and pressures. In another setup, CA plans to pump FLiNaK salt containing stable elements from Table 1 through a loop with a vacuum spray chamber at different temperatures and pressures to measure the amount of the different elements that can be removed at different flow speeds, droplet sizes, and cover gas flow rate.

5 Heavy Water Moderated Molten Salt Reactor

Heavy Water Moderated Molten Salt Reactor (HWMSR) has been proposed many decades ago and several examples of documents on the subject exist [4]. New developments in material sciences make it possible to construct aerogel-like materials out of graphite called aerographite [5]. Research is currently progressing to show that these aerographites can be produced with very good thermal insulation properties while at the same time being able to withstand the neutron flux and temperature in a Thorium MSR core. Constructing a MSR core with high-temperature salt in pipes insulated from heavy water at 50 °C now becomes possible. Figure 6 suggests thermal Thorium MSR core with heavy water moderation and aerographite as insulation material between the water tank zirconium wall and the ceramic pipe made with a very high content of carbon material. Preliminary studies by CA show that such a HWMSR would become a thermal breeder reactor. By using a highly optimized vacuum aerosol salt cleanup technique described above, it would become possible to run the breeder only with batch cleanup of the salt. Assuming daily or weekly batch cleanup of the salt using pyro and wet chemistry is likely to result in a HWMSR with a better neutron economy than other MSRs, which claim doubling times less than

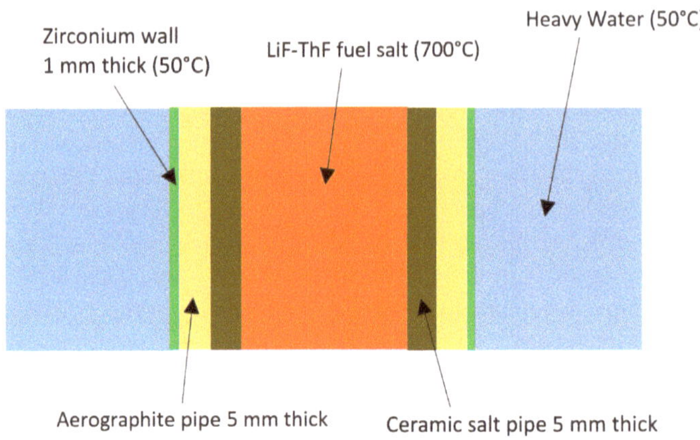

Fig. 6 Cross-sectional view of HWMSR core pipes made of ceramics surrounded by heavy water

20 years. Another positive attribute is that heavy water is less costly than graphite when reuse and decommissioning are taking into account. Note that startup fissile inventory in thermal MSRs is much smaller than fast MSRs, and thus, thermal MSRs are likely to outperform fast MSRs economically.

6 Conclusions

In order to solve global energy shortage and to move away from fossil fuels, engineers, and scientists around the world must team up and push technologies like Thorium MSR forward. CA has shown that significant progress has be made on salt purification and measurement systems, thus contributing to bringing the technology forward. The CA talks at ThEC15 in Mumbai gave an overview of the areas in which CA is working and this paper gives enough details for other MSR developers to evaluate if there is a possibility of a win-win collaboration between their team and CA.

Acknowledgements Thanks to Aslak Stubsgaard and Thomas Steenberg for providing valuable information which has been presented in this paper. The main sponsor for this work was Th Engineering (theng.dk)

References

1. Social Progress Index vs Energy Use Per Capita. https://en.wikipedia.org/wiki/List_of_countries_by_energy_consumption_per_capita (2016)
2. Display 3500 Different Decay Chains Graphically. http://theng.dk/Simulation/DecayChains (2016)

3. Molten Salt Reactor Off Gas Simulations. https://www.youtube.com/watch?v=sDMFONiFAaI (2015)
4. Design and Analysis of a Conceptual Thorium Fueled Heavy Water Moderated Molten Salt Reactor. http://nuclearpoweryesplease.org/pub/thunder%2013-12.pdf (2008)
5. M. Mecklenburg et al., Aerographite—a network of tetrapod-shaped and coherent interconnected graphite tubes, in *Annual World Conference on Carbon—Carbon 2013*

Radiation Dose Rate Mapping of Molten Active Fluoride Salt Loop at UED, BARC

Tanmay Sarkar, Pradeep Bhargava, B. Ananta, A. K. Srivastava and R. K. B. Yadav

Abstract Molten salt breeder reactor test facility is being designed and developed at BARC, with main objective of utilizing thorium abundance in India. Experimental facility, namely molten active fluoride salt loop (MAFL), is proposed to be set up in UED, BARC, to study the thermal-hydraulic behaviour of fuel salt as well as blanket salt and their compatibility with different structural materials. Radiation shielding requirements for this facility have been carried out using point-kernel code, QAD-CG. A validation exercise with the same active salt is carried out by measuring the dose at different detector location. The results show that no shielding is required for the dose criterion of 10 μSv/h. In validation exercise, the observed value is matching well with calculated value after a short distance from the source.

Keywords Thorium · Radiation shielding · Point kernel · Molten salt Radiation dose

1 Introduction

Molten salt breeder reactor (MSBR) test facility, designed and developed in BARC, is very important in context of India's future nuclear energy utilization scenario using thorium resources. It is having enhanced safety features like strong negative temperature coefficient of the reactivity, improved sustainability, increasing the

T. Sarkar (✉) · P. Bhargava
Health Physics Division, Bhabha Atomic Research Centre, Trombay, Mumbai 400085, India
e-mail: tanmays@barc.gov.in

B. Ananta · A. K. Srivastava
Reactor Engineering Division, Bhabha Atomic Research Centre, Trombay, Mumbai 400085, India

R. K. B. Yadav
Radiation Safety Systems Division, Bhabha Atomic Research Centre, Trombay, Mumbai 400085, India

A. K. Nayak and B. R. Sehgal (eds.), *Thorium—Energy for the Future*,
https://doi.org/10.1007/978-981-13-2658-5_35

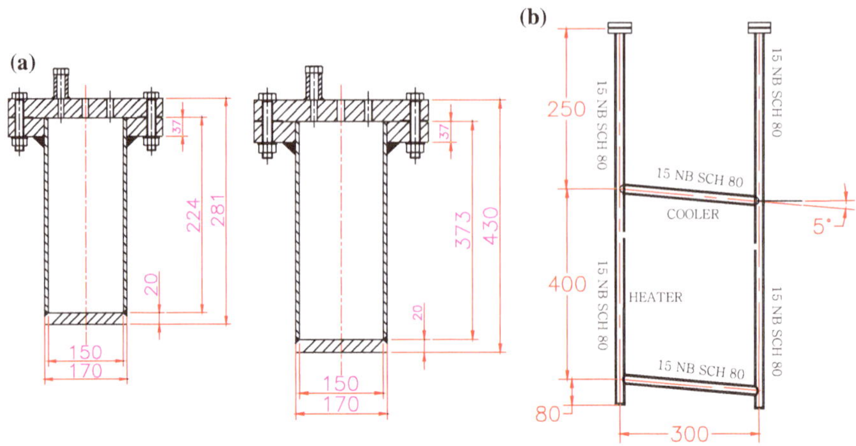

Fig. 1 **a** Molten active fluoride salt corrosion test facility (MAF-Corr). Dimensions are in cm, **b** molten active fluoride salt loop (MAFL). Dimensions are in cm

proliferation resistance of nuclear energy, stable coolant low-pressure operation that do not require expensive containment, easy to control passive heat cooling and unique characteristics of actinide burning and waste reduction.

MSBR is based on the molten salts, which act as fuel, blanket and coolant for the reactor. The composition of the fuel salt for the test facility was LiF–ThF_4–UF_4 (22.5–68.9–8.6 wt%). The blanket salt was LiF–ThF_4 (22.5–77.5 wt%). Two types of facility are there in which first one is to test the thermal-hydraulic behaviour of molten salt and blanket salt and to study their natural circulation flow pattern. This facility is called molten active fluoride salt loop (MAFL) [1]. Two loops are there for this test facility. These loops were operated at a pressure of 0.5 kg/cm^2 (g) and temperature of 800 °C. The loops were fabricated with Hastelloy N material. The second type of facility is molten active fluoride salt corrosion test facility (MAF-Corr) [2]. The purposes of this experiment were to get handling experience of molten active fluoride salt eutectic mixture, study the corrosion behaviour of structural materials, study the melting and solidification behaviour of liquid fuel salts. Two vessels fabricated with Incoloy 800H were used for doing this experiment. Upon the two vessels one was used for fuel salt corrosion studies and another for blanket salt studies. In Fig. 1, the schematic diagram of the vessels and loops of the facility with dimension is given.

2 Material and Methods

Dose rate at any detector location outside the vault is calculated by using the QAD-CGGP code [2]. This is a Fortran 77 code for calculation of dose rate due to a volume source in arbitrary complex geometry of shield configuration. It uses the point-kernel method, in which the original source volume is divided into a large number of small volumes of voxels. From each voxel, the optical path traversed in the source/shield material to the detector point is estimated and suitable corrections are made for the scattered contribution to obtain the dose rate based on the basic point source formula. The dose rates from all voxels are summed up to give the dose rate due to the entire volume of the source. Several build-up factor estimation techniques are optionally available, e.g. the GP method and Kapo's method. The advanced versions of the code can take a limited number of multiple identical sources. The dose conversion factor is taken from ANSI/ANS 6.1.1-1991. It has limited capability for visualization of 2-D slices through the 3-D geometry.

ORIGEN2 code [3] is run to obtain the decay and photon release rate from these materials. ORIGEN2 is a widely used computer code for calculating the build-up, decay and processing of radioactive materials. The code outputs a combined 18-group gamma spectrum in three groupings: fission products, actinides and activation products; this data is extremely useful in gamma shielding calculations. The data is available as photons/s or MeV/s against 18 energy groups (with effective energy ranging from 15 keV to 11 meV) at different cooling periods.

There are mainly two modelling tools for radiation shielding calculation. The first one is the point-kernel code QAD-CGGP; and the second category includes the Monte Carlo-based codes. The first group uses analytical techniques in the code for calculating radiation dose. QAD-CG is one of these types of code.

The QAD codes are a series of point-kernel codes developed by Los Alamos Scientific Laboratory to calculate the fast neutron and gamma-ray penetration through various shield configurations. When the version with the combinational geometry subroutine is used, the code is identified as QAD-CG. Point-kernel method is macroscopic approach used for gamma radiation exposure rate calculations. In this approach, gamma radiation propagation is assumed to be beam-like. Effects of radiation interaction in matter are described using macroscopic linear attenuation factors. This method is much less computationally intensive than Monte Carlo method. Series of calculations necessary for shielding optimization could be conducted at reliable time using point-kernel method. But due to macroscopic approach to radiation transport, this method lacks consistency. The main problem point-kernel method encounters is account for scattered radiation which is usually implemented through semi-empirical approximation. Additional 'build-up' factor must be introduced as a multiplier to the attenuated dose. Determination of the appropriate build-up factor can be rather complex as it depends upon the energy, the thickness and type of material. Uncertainties in determining build-up factor essentially limit the accuracy of point-kernel method. According to point-kernel method used in QAD codes, source volume is made of a number of point isotropic sources (point kernels). Each point kernel gives contribution to the dose rate at detecting point. In this method, the point kernel representing the transfer of energy by the un-collided flux along a line-of-sight path is combined with an appropriate build-up factor to account for the contribution from the scattered photons. With a distributed source, the point kernel is integrated over the source volume for each energy source. The gamma-ray dose rate at any point due to an isotropic source emitting S photons of energy E per second per unit volume is

$$D(\vec{r}) = K \int_V \frac{S\left(\vec{r'}\right) B\left(\mu' \left|\vec{r} - \vec{r'}\right|, E\right) \exp\left(-\mu \left|\vec{r} - \vec{r'}\right|\right)}{4\pi \left|\vec{r} - \vec{r'}\right|^2} \mathrm{d}V$$

where

$\vec{r}$	Point at which gamma dose to be calculated,
$\vec{r'}$	Location of source in volume V,
V	Volume of source region,
μ	Total attenuation coefficient at energy E
$\left\vert\vec{r} - \vec{r'}\right\vert$	Distance between source point and point at which gamma intensity is to be calculated,
$B\left(\mu \left\vert\vec{r} - \vec{r'}\right\vert, E\right)$	Dose build-up factor,
K	Conversion factor flux to dose rate.

The build-up factors used in this code are based on the Goldstein and Wilkins moment method calculations for gamma-ray transport in infinite homogeneous media. The code uses Capo's fit to the Goldstein-Wilkins data with bivariant polynomial expressions to calculate the approximate build-up factors as a function of gamma-ray energy and number of mean free paths from the source to the detector. For high atomic number materials, the polynomial fit has the form

$$B\left(\mu \left|\vec{r} - \vec{r'}\right|, E\right) = \sum_{j=0}^{j=4} \sum_{i=0}^{i=3} C_{ij} \left(\mu \left|\vec{r} - \vec{r'}\right|\right)^i E^j$$

And for low atomic number materials it has the form,

$$B\left(\mu \left|\vec{r} - \vec{r'}\right|, E\right) = \sum_{j=0}^{j=4} \sum_{i=0}^{i=3} C_{ij} \left(\mu \left|\vec{r} - \vec{r'}\right|\right)^i \left[\frac{1}{E}\right]^j$$

where C_{ij}'s are the coefficients of the expansion.

3 Result and Discussion

ORIGEN2 code [3] is used to obtain the decay and photon release rate from these materials. The photon spectrum of 50 and 5 years of decay is used for these calculations and tabulated in Table 1. This result is used in calculating the dose rate for the facility as well as for the validation exercise.

Table 1 Photon spectrum for 1 kg of thorium after 5 years and 50 years of decay

S. No.	Energy (MeV)	No. of photons/s 5 year	No. of photons/s 50 year
1	1.50E-02	2.45E+06	6.52E+06
2	2.50E-02	1.66E+05	5.09E+05
3	3.75E-02	1.19E+05	3.75E+05
4	**5.75E-02**	**1.65E+05**	**4.88E+05**
5	**8.50E-02**	**5.15E+05**	**1.90E+06**
6	**1.25E-01**	**6.80E+04**	**2.09E+05**
7	**2.25E-01**	**8.17E+05**	**2.90E+06**
8	**3.75E-01**	**3.58E+05**	**9.12E+05**
9	**5.75E-01**	**5.32E+05**	**1.90E+06**
10	**8.50E-01**	**1.41E+06**	**3.62E+06**
11	**1.25E+00**	**9.38E+04**	**2.38E+05**
12	**1.75E+00**	**1.80E+05**	**4.81E+05**
13	**2.25E+00**	**4.47E-03**	**1.20E-02**
14	**2.75E+00**	**3.46E+05**	**1.38E+06**
15	3.50E+00	1.61E-03	4.27E-03
16	5.00E+00	4.78E-04	1.27E-03
17	7.00E+00	3.10E-05	8.24E-05
18	1.10E+01	1.96E-06	5.21E-06

Table 2 Validation exercise

Distance from source (cm)	Measured value* (μSv/h)	Calculated result (μSv/h) (after 5 years of decay of thorium)	Ratio of calculated to measured dose rate
10	0.13	0.16	1.24
20	0.035	0.04	1.14
30	Bkg	…	Nil

*Excluding statistical variation

A validation exercise has been carried out to validate the results by considering the 5 year decay of the fresh Th. Measurement of dose rate at different locations has been performed with a source and compared with the calculated dose rates. This source used for the measurement is a glass vial of thickness 1 mm containing $LiF+ThF_4$ in the powder form. The source is in powder form and cylindrical shape having diameter of 2.54 cm and height of 0.8 cm. The mass of Thorium in the salt taken was 4.88 g from measurement. The comparison of dose rates from measurement and calculation is presented in Table 2.

The inventories of the fuel material handled are taken from the design report of the above-mentioned facility. The geometry of each component (loops and vessels) in the loop with the source (highlighted 11 energy groups) is modelled using point-kernel code, QAD-CG. The dose rate at desired locations is given in Table 3. The dose rate for achieving shielding requirements was taken to be 10 μSv/h.

Table 3 Dose rate at typical detector locations (decay period 50 year)

S. No.	Source details	Distance at which dose rate computed	Photon dose rate (μSv/h)
1	Loop 1 fuel salt, ID 15 mm, OD 22 mm, 0.83 kg Th	Surface	525.0
		2.0 cm	126.1
		3.0 cm	60.0
		4.0 cm	37.7
		30 cm	3.8
		100 cm	0.8
		200 cm	0.3
2	Loop 2 fuel salt, ID 15 mm, OD 22 mm, 0.94 kg Th	Surface	595.0
		2.0 cm	143.0
		3.0 cm	68.0
		4.0 cm	42.7
		30 cm	4.3
		100 cm	1.0
		200 cm	0.3
3	Vessel 1, 22.4 cm height, ID 15 cm, OD 17 cm, 4.4 kg Th	Surface	222.3
		6 cm	94.8
		30 cm	15.0
		100 cm	1.9
4	Vessel 2, 37.3 cm height, ID 15 cm, OD 17 cm, 8.33 kg Th	Surface	266.0
		6 cm	132.0
		30 cm	26.6
		100 cm	3.6

4 Conclusions

It is observed from Table 3 that 30 cm distance from the source is sufficient to achieve the design criteria of 10 μSv/h for loop 1 and loop 2. Shielding is not required for the distance greater than 30 cm from the source for loop 1 and loop 2. It has been observed that a distance of 100 cm is required for vessel 1 and vessel 2 for achieving criteria of 10 μSv/h.

Acknowledgements The authors thank Shri. Kapil Deo Singh, Head, Safety Studies Section, Health Physics Division, and Dr. R. M. Tripathi, Head, Health Physics Division, for their keen interest in this work.

References

1. Design Basis Report for Molten Active Fluoride salt Loop (MAFL), UED, BARC (2014)
2. Design Basis Report for Molten Active Fluoride Salt Corrosion Test facility (MAF-Corr), UED, BARC (2014)
3. QAD-CG QAD-CGGP, Fast Neutron and Gamma Penetration in Shields with Combinatorial Geometry, CCC-0493/01
4. ORIGEN2 A.G. Croff, ORIGEN2: A versatile computer code for calculating the nuclide compositions and characteristics of nuclear materials. Nucl. Tech. **62** (1983)

CFD Analysis of Molten Fluoride Salt Natural Circulation in a Rectangular Loop

Rakesh Chouhan, A. Borgohain, A. K. Shrivastava, N. K. Maheshwari and P. K. Vijayan

Abstract Indian Molten Salt Breeder Reactor (IMSBR) is being designed and developed in BARC with main objective of utilizing thorium abundance in India. In order to design the reactor, it is necessary to study the thermal-hydraulic behavior of various molten salts. Moreover, considering the importance of natural circulation in high-temperature systems, experimental studies on natural circulation of molten salts are required. This paper deals with the fabrication of loop with molten salt, $LiF + ThF_4$ (22.5–77.5 wt%), which is proposed as blanket salt for IMSBR program. Loop is fabricated with 15 NB Sch 40 pipe made of Hastelloy-N material having height of 400 mm and width of 300 mm. CFD analysis for natural circulation of the same loop is also carried out. CFD simulation is performed for steady state and transient cases at different heater powers. Various transients, such as step power change, heater trip, and start-up of natural circulation, are also carried out. Comparison of CFD results with standard correlation for steady state flow has been done.

Keywords MSBR · Molten fluoride salt · OpenFOAM · Computational domain

Nomenclature

c	Constant in steady state correlation
g	Acceleration due to gravity, m/s^2
Gr_m	Modified Grashof number
h	Heat transfer coefficient, W/m^2 K
L_t	Total length, m
N_G	Geometry parameter

R. Chouhan (✉) · A. Borgohain · A. K. Shrivastava · N. K. Maheshwari · P. K. Vijayan
Reactor Engineering Division, Bhabha Atomic Research Centre,
Trombay, Mumbai 400085, India
e-mail: rakeshc@barc.gov.in

A. K. Nayak and B. R. Sehgal (eds.), *Thorium—Energy for the Future*,
https://doi.org/10.1007/978-981-13-2658-5_36

Q	Total heat input, W
r	Constant in steady state correlation
Re_{ss}	Steady state Reynolds number, vD_h/υ
t	Time, s
T	Temperature, °C
U_z	Velocity along heater
U_y	Velocity along cooler

1 Introduction

Molten Salt Breeder Reactor (MSBR) is based on molten salts, which acts as fuel, blanket, and coolant for the reactor, as it has high boiling point, enables us to operate the system at low pressure and high temperature [1]. There are two types of molten salt-based reactor concepts, one is liquid fuel and other is based on solid fuel with molten salt as a coolant. The technology used for MSRs based on a liquid fuel is fundamentally different from the solid fuel technologies currently in use. Some of the advantages specific to MSRs (in terms of safety/reliability, for example) result directly from this characteristic. Furthermore, this type of reactor is particularly well adapted to the thorium fuel cycle (Th–^{233}U) which has the advantage of producing less minor actinides than the Uranium–Plutonium fuel cycle (^{238}U–^{239}Pu). The fuel of the MSR is based on the dissolution of the fissile material (^{235}U, ^{233}U, or ^{239}Pu) in an inorganic liquid that is pumped at a low pressure through the reactor vessel and the primary circuit, and thus also serves as the primary coolant. The heat generated in the fission process is transferred in a heat exchanger to a secondary coolant, which is also generally a molten salt. This intermediate loop is introduced for safety reasons: to avoid direct contact between the steam and the fuel. The operating temperature of the MSR is in between 550 and 750 °C, the lower limit being determined by the fusion temperature of the salt and the upper one by the corrosion rate of the structural material.

The present work deals with the CFD simulation of the rectangular loop with open-source OpenFOAM CFD tool. Vertical heater and horizontal cooler arrangement for natural circulation of salt in computational domain is considered. Hexahedral meshing in all four legs and tetrahedral meshing in corners are used in the loop, and mesh independency test was also carried out. Laminar model for steady state and transient cases was chosen to carry out simulations [2]. CFD simulation is performed for steady state and transient cases at different power levels. Results obtained in CFD simulation are analyzed and discussed in detail. The transient CFD studies performed are step power change, heater trip, and start-up of natural circulation.

2 Literature Review

Most of the previous investigations related to molten salt studies were based on the experimental works. These studies were carried out to measure the physical properties of molten salts [3] and heat transfer characteristics [4]. The heat transfer performance of a proposed MSBR coolant salt [NaBF–NaF (92–8 mol%)] and fuel salt [$LiF–BeF_2–ThF_4–UF_4$ (72–16–12–0.3 mol%)] was measured in forced convection loop by Silverman et al. [5]. The results of these experiments were similar for both salts and indicate that the proposed salts behave as normal heat transfer fluids with an extended transition region. Navier–Stokes and energy conservation equations to estimate thermal gradients in a molten salt were also used [6]. Khokhlov et al. [7] proposed analytical equations to evaluate the density, viscosity, heat capacity, and thermal conductivity of various molten salt mixtures of Li, Na, Be, and Zr fluorides. Recently Graydon et al. [8] performed experimental studies using fluoride salt (FLiNaK) operating at up to 700 °C; however, the heater provided temperature differentials of up to 30 °C. Experiments were performed to examine the use of Laser Doppler velocimetry, infrared videography, and visualization using standard videography through sapphire windows. Ferng et al. [9] also presented CFD methodology for investigating the thermal-hydraulic characteristics of FLiNaK salt. The proposed model was validated through comparison with existing correlations and experimental data of friction factor.

The significant novelties of this work are: first, to study thermal-hydraulic behavior of molten fluoride salt in passive mode inside a rectangular loop, which is not available in open literature; second, to use OpenFOAM CFD code to predict the behavior of natural circulation phenomena for high-temperature molten fluoride salt after some modifications; third, to compare CFD results with standard correlations available in the literature of rectangular loops for validation.

3 Description of Experimental Setup

As the natural circulation data for molten salts are scarce, a molten salt loop is designed, fabricated, and under commissioning. A loop is fabricated with 15 NB Sch 40 pipe made of Hastelloy-N material having height of 400 mm and width of 300 mm as shown in Fig. 1. Extra lengths are provided on each vertical leg for accommodation of molten salt expansion. A melt tank is positioned on top of loop for initial filling of salt inside the loop. Six surface heaters, four on each leg of rectangular loop and two on each expansion leg, are mounted. One tape heater on melt tank and connecting pipe is also wrapped. Proper arrangement for continuous argon gas purging to maintain inert atmosphere has been done. Adequate instrumentation is provided for molten salt temperature, heater control, pressure and oxygen monitoring in the argon gas.

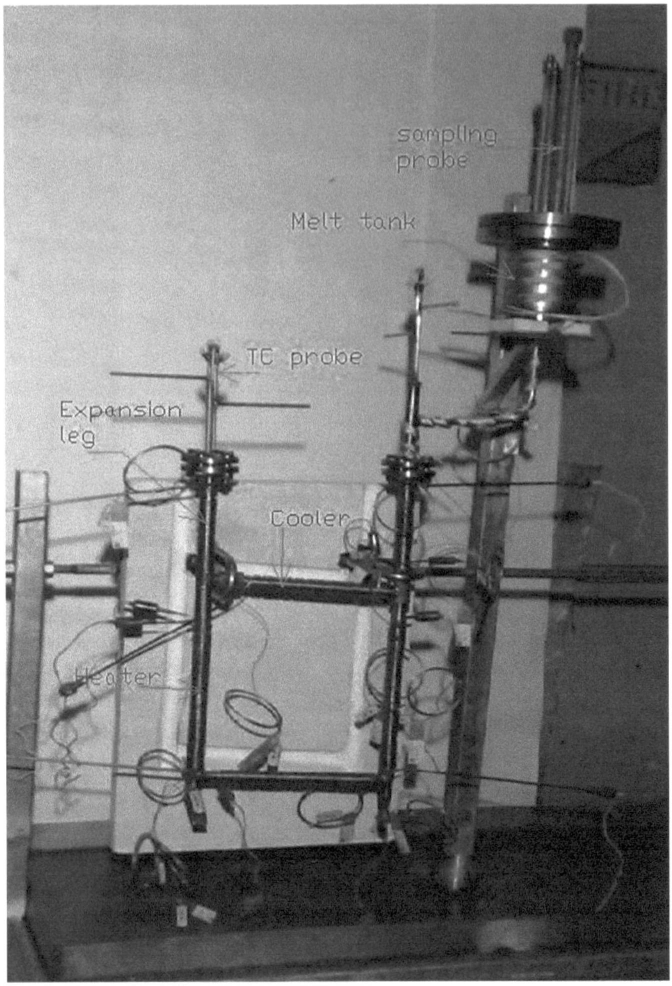

Fig. 1 LiF + ThF_4 experimental setup

4 Model Descriptions

CFD simulations for natural circulation of molten fluoride salt in a rectangular loop having height 400 mm, width 300, and 14 mm internal diameter as a computational domain have been performed in OpenFOAM CFD tool. Computational domain having both horizontal leg is tilted at 5° inclination with horizontal vertical heater (red) and horizontal cooler (blue) configuration to support natural circulation of the molten fluoride salt as shown in Fig. 2.

In rectangular loop section, different types of mesh were generated with different number of finite volume grids. Keeping in mind the CFD simulation time, tetrahedral

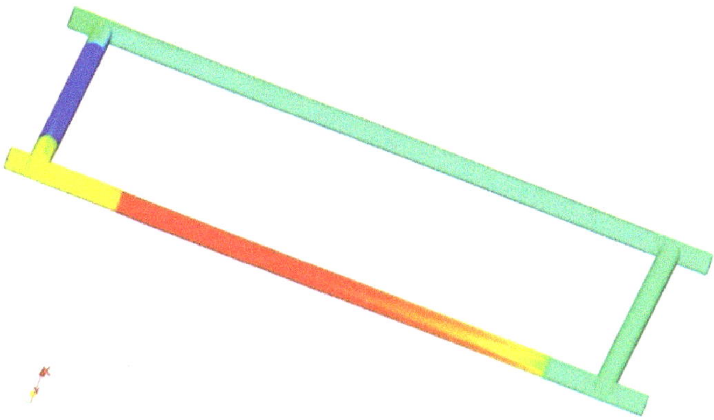

Fig. 2 Computational domain in OpenFOAM

Fig. 3 Hexahedral mesh and tetrahedral mesh for loop section in OpenFOAM

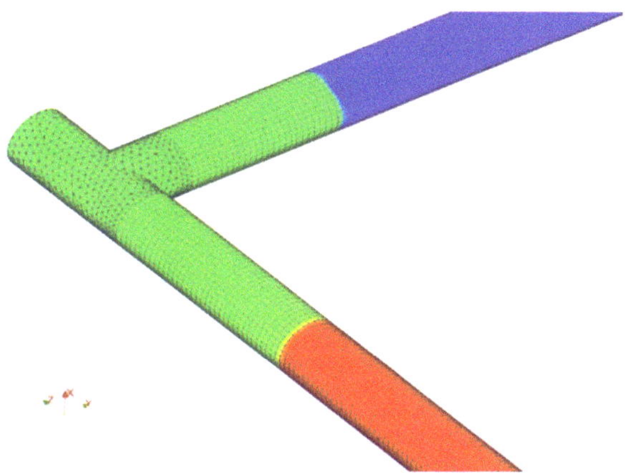

mesh is generated at all four corners, whereas hexahedral and pyramids mesh are generated in pipe section. Observations of flow in pipes are of much more significance than corners. Figure 3 represents the types of mesh generated in rectangular loop as computational domain.

The Reynolds-averaged model used to analyze the thermal-hydraulic characteristics for the incompressible molten salt of $LiF + ThF_4$ (22.5–77.5 wt%) includes equations for continuity, momentum, and energy. The initial temperature for all simulations is set at 600 °C, and the operating pressure is at 1 atm. A uniform heat flux is defined on the heater surface as per simulation requirements. Heat rejection on cooler surface was modeled by heat transfer coefficient and ambient temperature as input to model various cooler capacities. Table 1 shows thermophysical property values used in simulations.

Table 1 Thermophysical properties of eutectic mixture of LiF_4–ThF_4 at 700 °C

Thermophysical property	Values at 700 °C
Density (kg/m^3)	4213
Kinematic viscosity (m^2/s)	3.79E−6
Dynamic viscosity (kg/m s)	0.0159
Thermal conductivity (W/m K)	1.0013
Specific heat capacity (J/kg K)	1315

4.1 Model Assumptions

The following assumptions have been made:

- Fluid is single-phase Newtonian and incompressible.
- Boussinesq approximation is valid, i.e., density differences are only important in producing buoyancy.
- Fluid properties are constant except in the formulation of buoyancy term.
- An adiabatic boundary condition at the rectangular loop walls is applied.

4.2 Initial and Boundary Conditions

4.2.1 Initial Condition

- Temperature of whole domain is specified as 600 °C, which is above melting point temperature (~565 °C) of molten fluoride salt.

4.2.2 Boundary Conditions

- Different heat fluxes at heater are specified as per the practical applicability of loop as will be specified during experiment.
- Ambient temperature and appropriate heat transfer coefficients are specified at cooler.
- No-slip condition specifies that velocity of fluid particles is zero on the wall of pipe.
- Walls of loop except heater and cooler are considered adiabatic.

4.3 Methodology

CFD simulations have been performed using the open-source CFD code Open-FOAM 2.2.2 (2013). The standard available steady state buoyantBoussinesqSimple-Foam solver and transient buoyantBoussinesqPimpleFoam solver using finite volume method are used for steady state and transient simulations. The basic governing steady state and transient equations are used for simulations. Both SIMPLE and PIM-PLE algorithms [10] were used to solve coupled equations of pressure, velocity, and temperature. Upwinding method was used for velocity, temperature, and turbulence equations. All the solutions were considered to be fully converged when the sum of scaled residuals was below 10^{-6}. All the computations have been performed on an Intel i7 motherboard computer having a quad processor with 16 GB RAM and 2.8 GHz processor speed.

4.4 Mesh Independency

Three-dimensional transient simulations are highly expensive. Hench, mesh independence test was performed to estimate the coarsest grid which can give mesh-independent solutions. Such a grid imparts minimum computational expense and time without compromising on accuracy of the results. Three different grids, Mesh 1: 93857; Mesh 2: 119495, and Mesh 3: 145243, were considered to check the grid independence using temperature variation along the centerline of heater leg. It can be seen that the mesh size of Mesh 1 shows slightly different profiles than the results of Mesh 2 and Mesh 3 (Fig. 4). It can be seen that the results of Mesh 2 and Mesh 3 are matching and hence it was concluded that model with gird size Mesh 2 and above can be used for simulation.

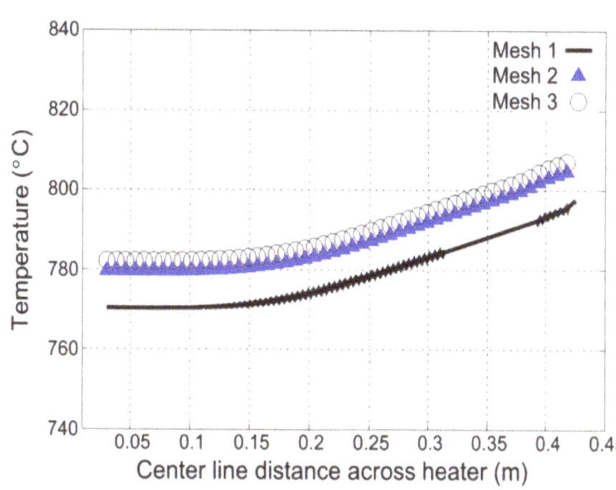

Fig. 4 Temperature values obtained at centerline of heater leg for the three different grids

5 Results and Discussions

5.1 Steady State Results

Variable magnitude should be constant with respect to spacial locations in the computational domain to achieved steady state condition. During CFD simulation in OpenFOAM, temperature and velocity data at different locations are generated in postprocessing. Figure 5 represents steady state temperature distribution along the center plane of the computational domain. Steady state velocity distribution can be seen in Fig. 6a, b along with heater and cooler central planes, respectively.

Figure 7 represents the plot of variation of temperature with iteration time for different locations. Simulation parameters like power 500 W at heater, heat transfer coefficient (h) 79 W/m^2 K at cooler, and ambient temperature 303 K are used as input. Initially, some fluctuations in the graph can be observed due to start-up of natural circulation. Further increase in temperature of loop is observed, and finally, temperature curve becomes flat which means amount of heat rejection at cooler surface becomes equivalent to heat addition at heater surface. Hence, steady state condition is reached in the computational domain.

Figure 8, velocity variation along the cross-sectional distance for same case as above, showing magnitude of velocity at heater outlet, is less at center as compared to near wall velocity. It is due to heating effect near surface of heater which causes the local density gradient near surface. Natural convection phenomena predominant

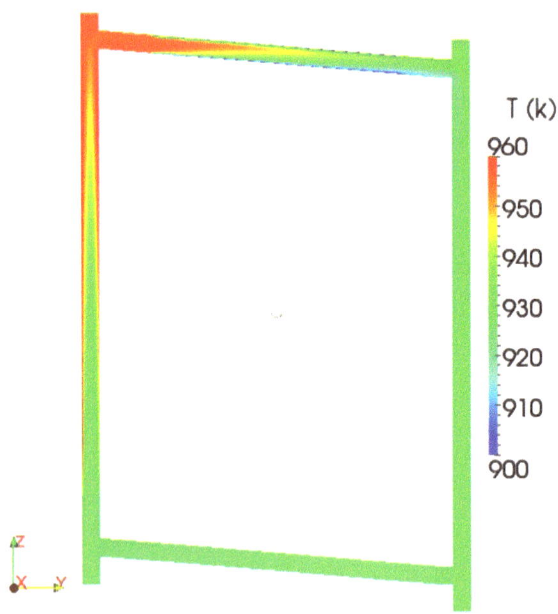

Fig. 5 Steady state temperature distribution in the loop

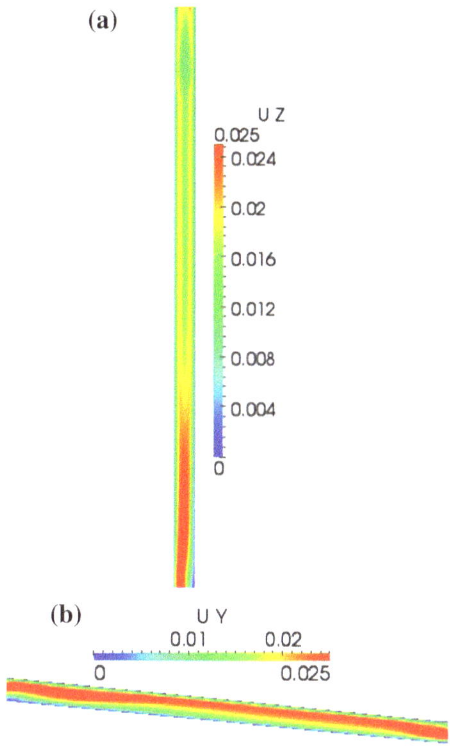

Fig. 6 **a** Steady state velocity (m/s) distribution along heater central plane and **b** steady state velocity (m/s) distribution along cooler central plane

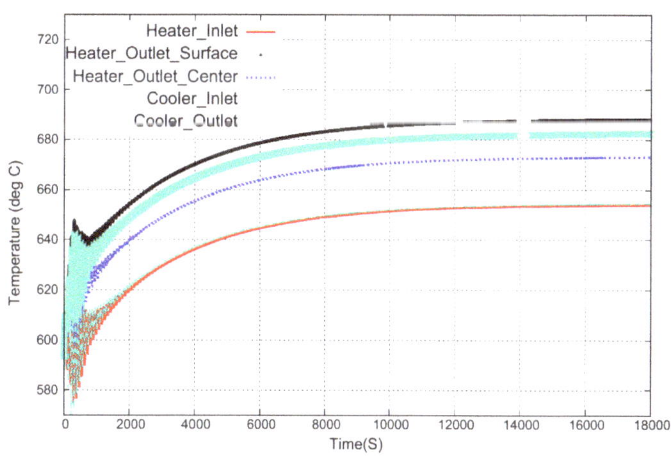

Fig. 7 Variation of temperature with iteration time for heat input $Q = 500$ W at heater for steady state case

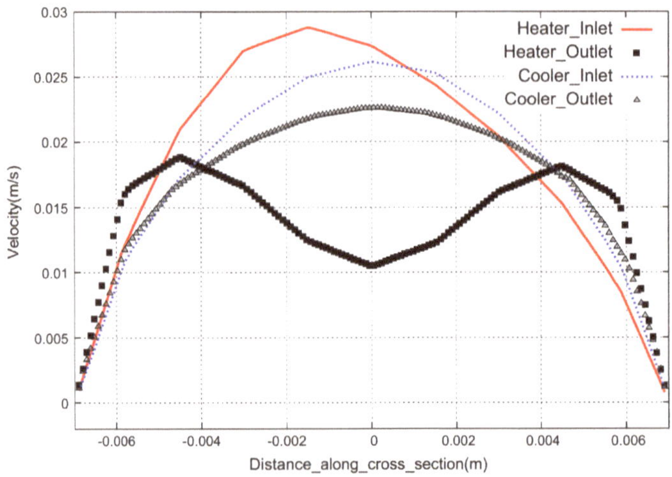

Fig. 8 Variation of velocity cross-sectional distance for heat input $Q = 500$ W at heater for steady state at time step 14,000 s

due to above reason lead to less velocity at center which is different from general pipe flow phenomena. At other locations, i.e., heater inlet, cooler inlet, and outlet velocity profile deviation is either negligible or very little as compared to general pipe flow velocity profile.

5.2 Transient Results

Transient studies are carried out to simulate some postulated accidental scenario in the molten salt-based systems. The main objective is to observe the behavior of molten salt during various transient conditions. Figure 9 shows temperature distributions along with radial distance at input and output locations of both heater and cooler.

5.2.1 Heater Trip Transient

In the power failure scenario, the consequences of sudden non-availability of heat source are simulated using heater trip transient. This transient was realized by setting the main heater power to zero value from steady state at some power. Initial conditions for this transient are as follows: Loop was in steady state at 500 W power. The start of the heater trip was simulated at 18,000 s. At this time, the main heater power was set back to zero value from 500 W.

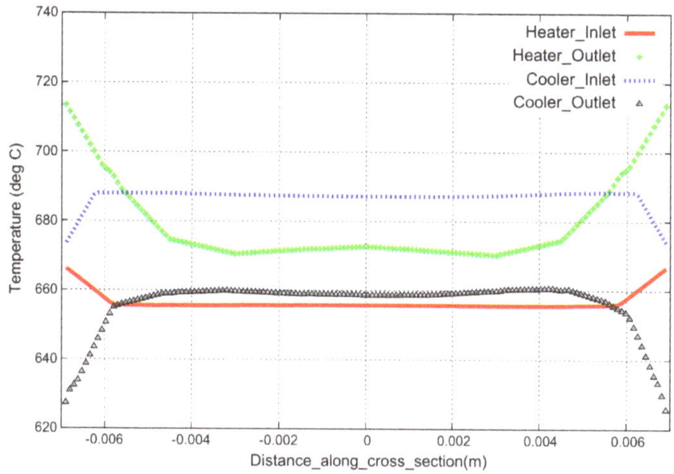

Fig. 9 Variation of temperature cross-sectional distance for heat input $Q = 500$ W at heater for transient case

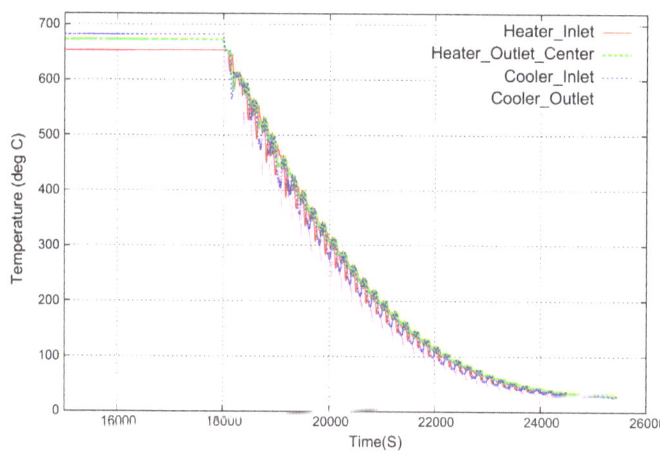

Fig. 10 Inlet–outlet temperature variations in heater and cooler after 18,000 s for heater trip transients

As shown in Fig. 10 at the start of the transient, heater outlet temperature decreases rapidly and came closer to the heater inlet temperature due to the loss of heat source. This temperature drop was accomplished quickly, and the inlet and outlet temperatures of the heater then decreased together.

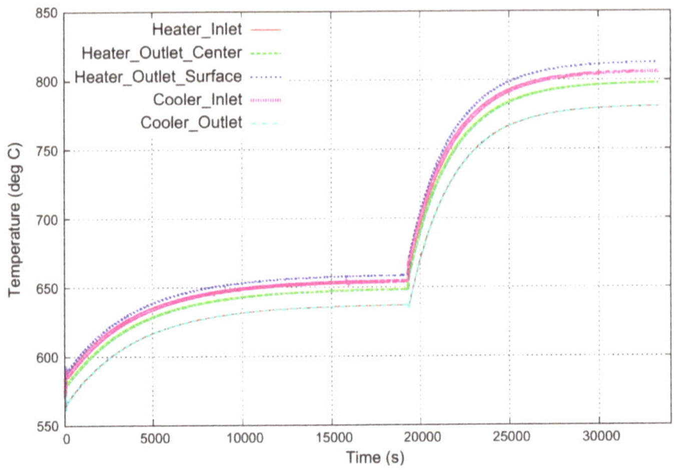

Fig. 11 Step power change transient from 250 to 500 W

5.2.2 Step Power Change Transient

Initially, simulation is started in the loop with heat flux 250 W at heater surface and heat transfer coefficient 48 W/m^2 K at cooler surface. Steady state was achieved with this condition in the loop. So to know the heater power change effects in the loop, transient condition was started with changing the main heater power to 500 W from 250 W. The effect of this power change is shown in Fig. 11 in terms of change in temperature with time. Heater outlet and cooler inlet temperatures increase more sharply compare to cooler outlet and heater inlet just after increasing the main heater power and again steady state condition attained in the loop.

6 Validation of CFD Results

An analytical approach is used for comparison of results in steady state condition for various heater power levels. Steady state flow prevails in a natural circulation loops when buoyancy force is balanced by the retarding frictional force. In this case, retarding frictional forces are pipe frictional losses due to flow and losses due to change in geometrical parameters, whereas generated buoyancy head as a driving force is considered. Based on this general approach, analytical calculations were carried out for steady state mass flow rates for a rectangular loop. The flow rates obtained in CFD analysis are compared with analytically calculated values as shown in Fig. 12.

Obtained mass flow rates were used in standard correlation and compared with CFD results as shown in Fig. 13. Vijayan et al. [11] showed that steady state flow

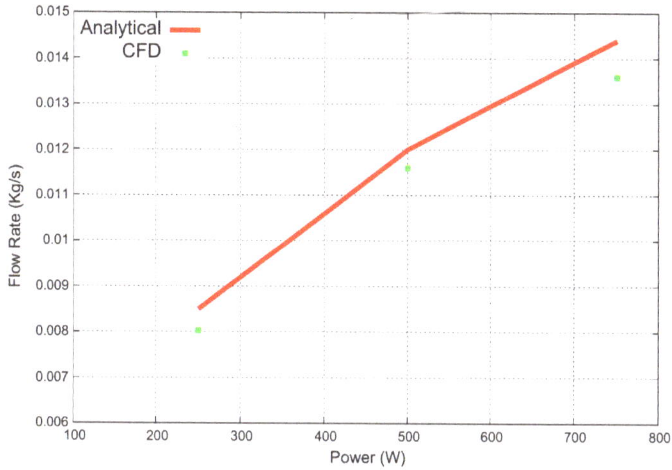

Fig. 12 Steady state natural circulation flow rates at different power levels

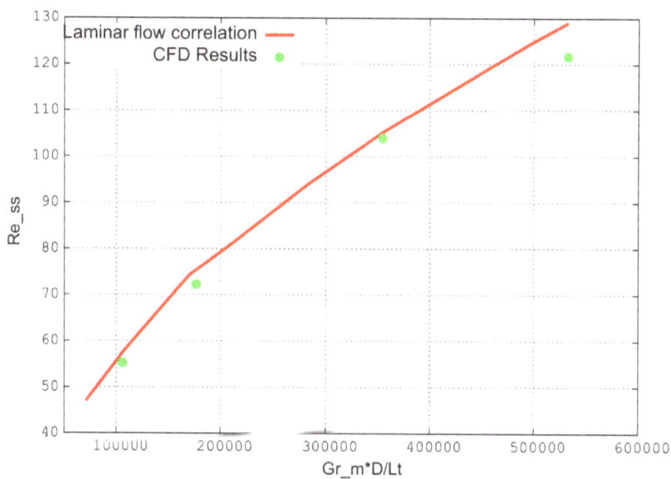

Fig. 13 Comparison of CFD data with steady state correlation given by Vijayan [11]

in single-phase uniform or non-uniform diameter natural circulation loops can be expressed as

$$Re_{ss} = c\left(\frac{Gr_{m}}{N_{G}}\right)^{r}$$

where constants 'c' and 'r' depend upon the nature of flow (i.e., laminar or turbulent). Parameter N_G depends upon geometry of the loop. It can be observed from above graph that CFD results are matching with the analytical results.

7 Conclusions

- A rectangular loop is made for natural circulation studies in molten salt LiF+ ThF_4. The loop is under commissioning.
- CFD studies of the loop are carried out to simulate various steady state and transient cases and found that proposed CFD model can capture the thermal-hydraulic characteristics of proposed blanket salt for IMSBR.
- The natural circulation steady state mass flow rates for different heater power levels are compared with standard available correlation and analytical calculations. Comparison results are in good agreement ($\pm 10\%$).
- Further validation of the CFD data will be carried out when the experimental results of the loop are available.

References

1. K. Furukawa, K. Arakawa, L.B. Erbay, Y. Ito, Y. Kato, H. Kiyavitskaya, A. Lecocq, K. Mitachi, R.W. Moir, H. Numata, J.P. Pleasant, Y. Sato, Y. Shimazu, V.A. Simonenco, D.D. Sood, C. Urban, R. Yoshioka, A road map for the realization of global scale thorium breeding fuel cycle by single molten fluoride flow. Energy Convers. Manag. **49**, 1832–1847 (2008)
2. H.K. Versteeg, W. Malasekra, An Introduction to Computation Fluid Dynamics. The finite volume method (1995)
3. J.W. Cooke, Development of the Variable Gap Technique for Measuring the Thermal Conductivity of Fluoride Salt Mixtures, ORNL-4831. Oak Ridge National Laboratory (1973)
4. L. Bin, W. Yu-ting, M. Chong-fang, Y. Meng, G. Hang, Turbulent convective heat transfer with molten salt in a circular pipe. Int. Commun. Heat Mass Transf. **36**, 912 (2009)
5. M.D. Silverman, W.R. Huntley H.E. Robertson, Heat transfer measurements in a forced convection loop with two molten fluoride salts: LiF–BeF_2–ThF_4–UF_4 and EUTECTIC $NaBF_4$–NaF, in Report No. ORNL/TM-5335, pp. 16–21 (1976)
6. P. Mandin, H. Belachgar, A. Nuttin, G. Picard, Hydrothermal Modeling for the Molten Salt Reactor Design Optimization, in *The 11th International Topical Meeting on Nuclear Reactor Thermal-Hydraulics (NURETH-11)*, France, 2–6 Oct 2005
7. V. Khokhlov, V. Ignatiev, V. Afonichkin, Evaluating physical properties of molten salt reactor fluoride mixtures. J. Fluor. Chem. **130**, 30 (2009)
8. G.L. Yoder, D. Heatherl, Jr., D. Williams, Liquid Fluoride Salt Experiment Using a Small Natural Circulation Cell. ORNL/TM-2014/56 (2014)
9. Y.M. Ferng, K.Y. Lin, C.W. Chi, CFD investigating thermal-hydraulic characteristics of FLiNaK salt as a heat exchange fluid. Appl. Therm. Eng. **37**, 235–240 (2012)
10. S.V. Patankar, *Numerical Heat Transfer and Fluid Flow* (Hemisphere Publishing Corp, New York, 1981)
11. P.K. Vijayan, Experimental observations on the general trends of the steady state and stability behavior of single phase natural circulation loops. Nucl. Eng. Des. **215**, 139–152 (2002)

The Stable Salt Reactor—A Radically Simpler Option for Use of Molten Salt Fuel

Ian R. Scott

Abstract All reactor designs over the past 50 years designed to use molten salt fuel have shared a common misconception that such fuel must be actively pumped between a reaction chamber and a heat exchanger. This design philosophy originated with the Aircraft Reactor experiment and the Molten Salt Reactor experiment and has not been challenged since. Application of modern tools of computational fluid dynamics to the question shows that in fact the use of molten salt fuel in fuel tubes very similar to those used in conventional solid-fueled reactors is entirely practical and makes possible a radically simpler design of molten salt-fueled reactor, the Stable Salt Reactor.

Keyword Stable salt fission fuel reactor

1 Introduction

The concept of molten halide salts as fuel for nuclear reactors first appeared in the years after the Manhattan Project in the context of developing a nuclear reactor to power a strategic bomber. This led to the development of the Aircraft Reactor Experiment and Molten Salt Reactor Experiment. In both the cases, the reactor design was to pump a molten salt-based nuclear fuel through a moderator block and then to a heat exchanger.

There was, however, an earlier concept whereby the molten salt-based fuel would be put into static tubes with coolant passed through the array of tubes, very much in line with current thinking for solid-fueled reactors. This concept was, however, abandoned undeveloped, because simplistic calculations based on heat transferring through the molten salt merely by conduction and not convection indicated the tubes could be no more than 2 mm in diameter [1].

I. R. Scott (✉)
Moltex Energy, London, UK
e-mail: ianscott@moltexenergy.com

A. K. Nayak and B. R. Sehgal (eds.), *Thorium—Energy for the Future*,
https://doi.org/10.1007/978-981-13-2658-5_37

Despite it being common knowledge that convection is the dominant heat transfer mechanism in liquids, no attempt was made to calculate the convective contribution to heat transfer in such static molten salts. In the context of developing an airborne nuclear reactor that was logical as gravity is an unreliable force in such conditions but the fact that computational fluid dynamics (CFD) as a practical science was still 30 years from development contributed to the decision at a purely practical level.

Application of modern CFD methods shows in fact that heat transfer from molten salt fuel is perfectly adequate and that pumping of the fuel to achieve turbulent flow and hence high heat transfer rates is not necessary.

This simple insight opens the door to a radically simpler and safer way to use molten salt fuel which is described in this paper.

2 The Stable Salt Reactor

2.1 General Reactor Design

The stable salt reactor is a fast spectrum reactor using as fuel UCl_3 (24%), $PuCl_3$ (16%), and NaCl (60%). The plutonium has an isotopic vector typical of spent PWR fuel. The choice of chloride salt rather than the commonly preferred fluoride salt is driven by two factors, lower melting point and the possibility to adjust the redox state of the chloride salt to a far more reducing state than is possible with fluoride salts. This allows the virtual elimination of any thermodynamic driving force for corrosion of steels by the salt (see below).

The tubes of fuel are grouped into fuel assemblies strongly resembling those used in sodium fast reactors. Those fuel assemblies are immersed in a pool of coolant which is composed of ZrF_4 (42%), NaF (10%), and KF (48%). The coolant is contained in a steel tank located in a subsurface concrete-lined pit.

The coolant is circulated by pumps through the reactor core and through heat exchangers that are also immersed in the coolant pool. The secondary coolant is a similar composition to the primary coolant but remains non-radioactive since it is screened from the neutron flux by the coolant by virtue of the hafnium contamination of the zirconium in the coolant coolant. This makes the coolant an effective neutron shield (10^4 screening by 1 m of coolant). The secondary coolant is utilized to generate electricity or not of transfer process heat depending on the reactor purpose.

The reactor core departs from the convention of fast reactor cores being cylindrical and is in fact rectangular as shown in Fig. 1. The core is composed of a set of identical modules, each module containing an array of 10×10 fuel assemblies. The modules contain all the key reactor components and are factory manufactured thereby radically simplifying on-site assembly. The rectangular core structure also permits a very simple and mechanically robust procedure for refueling whereby fuel assemblies are migrated one step at a time across the core's short dimension with spent fuel assemblies being removed from the core, allowed to cool for several months still

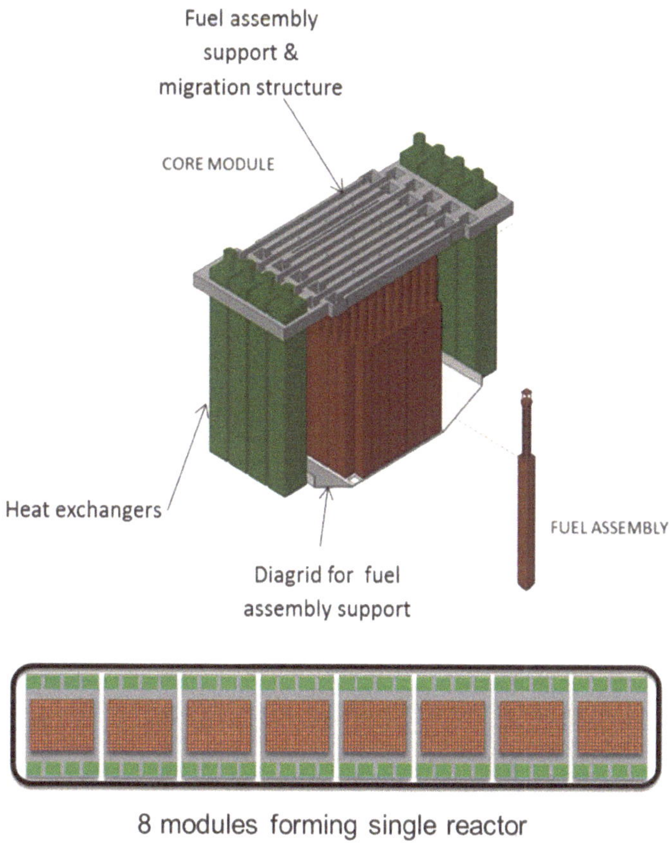

Fig. 1 Modular structure of the Stable Salt Reactor

immersed in the coolant and then removed from the reactor. Fresh fuel assemblies are added to the opposite end of the row of assemblies.

Reactivity control is achieved primarily via the high (−6 pcm/K) negative temperature reactivity coefficient, with total core reactivity controlled via the refueling frequency and shut down achieved using steel clad boron carbide shut down blades with are inserted under gravity between the modules. Standard passively safe magnetic release mechanisms are used for the shutdown blades which are not required to exert fine control over reactivity under any circumstances. Controlled reactor start up at commissioning and after infrequent shut downs is achieved by moving fuel assemblies out from and into the core rather than removing neutron absorber from the core.

This refueling process can be carried out at full reactor power making the reactor continuously refueled and therefore not requiring control rods or reactivity shims to control excess reactivity added in batch refueling.

2.2 *Thermal Hydraulics of the Fuel Salt*

The thermophysical properties of the fuel salt are given in Table 1.

CFD analysis using ANSYS/FLUENT software was carried out by Wilde Analysis Ltd on the basis of a vertical power distribution in the tube rising from 0.4 at the ends to 1.4 at the center. Figure 2 shows the peak temperatures for the fuel salt and fuel tube wall at varying mean power levels for a 10mm tube diameter. The power level selected for the Stable Salt Reactor was 150 kW/l which provides an almost 400 K temperature margin for the fuel salt and tube wall temperatures well within the capability of standard nuclear-grade steels known to have high tolerance for neutron damage such as PE16 and HT9.

2.3 *Gas Evolution from the Fuel Salt*

Fission produces relatively large amounts of gaseous fission products. A huge advantage of molten salt fuel is that the dangerous isotopes of cesium and iodine which form the major hazard from solid nuclear fuel in an accident scenario are non-volatile salts. However, the noble gasses xenon and krypton are still evolved as is cadmium which is gaseous at the temperatures involved. One option was to trap these gasses within the fuel tubes as is done with metallic nuclear fuel (from which the gasses also evolve freely). This, however, leads to pressurization of the tube and consequently a

Table 1 Parameters used in the CFD model

Composition of fuel salt	$NaCl/UCl_3/PuCl_3$ 60/20/20 mol%
Melting point of fuel salt	730 K [1]
Boiling point of fuel salt	1837 K
Density of fuel salt	$4.1690–9.014 \times 10^{-4}$ g ml^{-1} K^{-1} [2]
Viscosity of the fuel salt	$\log_{10}(\eta[cP]) = -1.2675 + \frac{1704}{T[K]}$
Thermal conductivity of fuel salt	0.5 W m^{-1} K^{-1}
Specific heat capacity of fuel salt	Rises from 520 J kg^{-1} K^{-1} at M.Pt. to 670 at B.Pt.
Composition of coolant salt	$KF/ZrF_4/NaF$ 48/42/10 mol%
Density of coolant salt	2770 kg m^{-3}
Specific heat capacity of coolant salt	1050 J kg^{-1} K^{-1}
Kinematic viscosity of coolant salt	1.7×10^{-6} m^2 s^{-1}
Thermal conductivity of coolant salt	0.7 W m^{-1} K^{-1}
Coolant velocity between fuel tubes	4 m s^{-1}
Coolant inlet temperature	450–460 °C

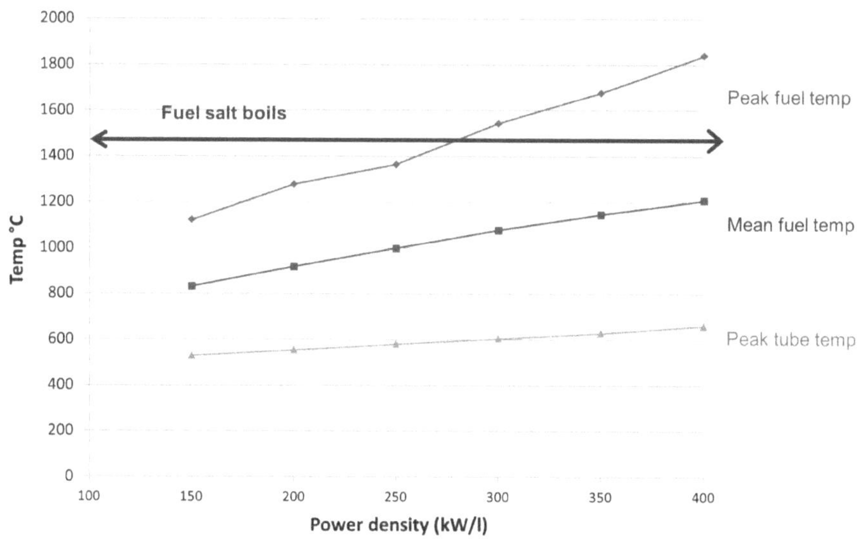

Fig. 2 Temperatures in a 10-mm-fuel tube calculated by CFD

hazard of tubes failing. The possibility that the tubes could be safely vented into the coolant was therefore investigated.

The calculations were carried out using HSC software from Outotek on the basis that the surface of the fuel salt in the tube would be at approximately 1000 °C and evolve vapor characteristic of that temperature but that the evolved vapour would condense on the wall of the fuel tube above the level of the fuel salt, the wall there having the same temperature as the coolant of 600 °C.

The vapor remaining then represents what is actually released from the tube.

Figure 3 summarizes the results of this analysis, note the logarithmic scale. Only the noble gasses, $ZrCl_4$ and Cd, are released in significant quantities into the coolant salt. A separate analysis of the extent of indirect release of Cs-137 via release and subsequent decay of Xe-137 showed that hold-up times in the fuel salt and gas spaces in the tube were sufficient to reduce Xe-137 release to less than 10^{-6} of the total. It was therefore concluded that the reduction of hazard represented by venting the tubes to avoid pressurization was substantially greater than any hazard created by venting the gasses.

2.4 *Redox Control of Fuel and Coolant Salts to Eliminate Corrosion*

Corrosion of steels by molten salts is a major problem for molten salt reactors, necessitating development of special alloys that are slower to corrode. The fundamental

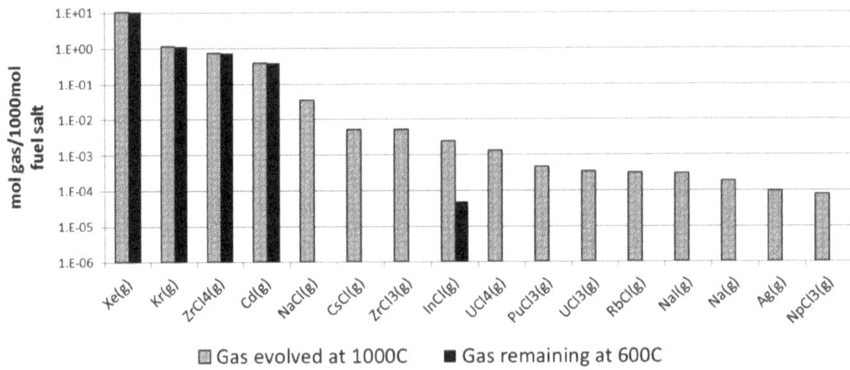

Fig. 3 Gas evolution from fuel salt and condensation on fuel tube wall

problem is that the normal mechanism whereby steel avoids corrosion in corrosive environments such as water is by formation of a passivating layer of chromium oxide which shields the steel from contact with the corrosive liquid. Molten salts dissolve such oxide layers leaving the steel exposed to the corrosive potential of the liquid.

A different approach to corrosion control is therefore needed for molten salts. One option is to make steels with much lower concentrations of relatively reactive alloy components such as chromium and high levels of less reactive components such as nickel. Hastelloy N is the archetypal alloy of this type. Unfortunately, this approach is only partially successful as chromium in such alloys still dissolves in the molten salt though at a slower rate.

A better approach is to ensure that the material of the alloy is intrinsically resistant to corrosion by ensuring that the thermodynamically stable form of the alloy components is actually the metal—so that there is no driving force for the metal to dissolve in the salt. Using metals such as molybdenum or tungsten is one approach—but data on the behavior of such metals in reactor radiation fields is very limited making the obstacles to their use very high.

Instead of changing the alloy, one can however change the salt. The relevant parameter is the redox potential; a strongly reducing salt will not dissolve even chromium from steel. This is where choice of salt becomes critical. For the fuel salt, use of fluoride salts of uranium is limited because redox potentials low enough to prevent solubilisation of chromium will reduce uranium trifluoride to uranium metal. That is a principal reason why the Stable Salt Reactor uses chloride salts for fuel. The redox potential of the fuel salt is controlled by adding metallic zirconium to the inside of the tube. This not only drops the redox potential to the level of the Zr/ZrCl couple but holds it there despite net chlorine release from the uranium trichloride due to fission. Under these conditions, the equilibrium concentration of $CrCl_2$ in the salt is below 10^{-8} M ensuring undetectable corrosion.

A similar approach to redox control was used for the coolant salt. Use of metallic zirconium for redox control is not possible in a pumped system as zirconium will migrate and deposit on vulnerable components. However, zirconium has both a divalent and tetravalent fluoride, and addition of 1% ZrF_2 to the coolant salt provides the ZrF_2/ZrF_4 redox couple that maintains a low redox potential such that equilibrium concentration of CrF_2 in the coolant salt is below 10^{-6} M which prevents detectable corrosion. Importantly, this redox buffer is resistant to contamination of the coolant salt with oxygen or water which will inevitably leak into the system under real conditions.

2.5 *Thorium and the Fuel Cycle for the Stable Salt Reactor*

The Stable Salt Reactor is a fast reactor designed to burn reactor-grade plutonium. In its initial form, it is not designed to be a net breeder but has been designed explicitly to make development of a net breeding version as simple as possible.

Moltex Energy believes that a reactor using a pure thorium/U-233 fuel cycle would be extremely challenging from a nuclear safeguards perspective. U-233 is eminently weapons usable and would probably be regarded by the IAEA as similar in risk to weapons-grade plutonium. It is also chemically relatively simple to isolate from molten salt systems—much more so than plutonium which is much harder to separate from the lanthanide fission products. The problems of contamination with U-232 are in truth minimal for an organization intending to use the weapon rather than store it unused. We therefore consider that U-233 should be used as fuel only if it can be denatured immediately it is formed with U-238.

This will be achieved in the Stable Salt Reactor by replacing the zirconium-based coolant salt with a ThF_4-based salt. The thorium is equally as effective as hafnium in making the coolant layer an effective neutron shield and will produce Pa-233 in a highly dilute form in the coolant, largely protecting it from further neutron capture. When the Pa-233 decays to U-233, the U-233 can be continually removed from the coolant salt by reductive extraction into a thorium/uranium–238/bismuth alloy. The bred U-233 is thus immediately denatured.

However, a substantial part of the breeding of new fissile material will occur in the fuel salt itself due to its uranium content and it is expected that axial uranium breeding elements will be added to the fuel assemblies to take advantage of the substantially superior neutron yield from plutonium in fast reactors. The net breeder version of the Stable Salt Reactor will therefore operate in a combined uranium/thorium fuel cycle.

2.6 *Economics of the Stable Salt Reactor*

The Stable Salt Reactor was designed from the outset to change the current economic paradigm of nuclear energy under which only subsidy of one sort or another allows deployment of reactors. That can only be done by reducing the huge capital costs associated with nuclear reactors.

Several factors are brought together making major cost reductions possible. These are

- The intrinsic safety of molten salt fuel essentially eliminating the cesium and iodine fission product risks to the population in the event of an accident
- Eliminating high pressure or high chemical reactivity coolants that can provide the driving force to disperse radioactivity in an accident
- Simplifying reactor control by avoiding excess reactivity after refueling and taking advantage of the intrinsic effect of negative temperature reactivity coefficient to exert primary reactor control
- Use of standard steels instead of new exotic alloys
- Use of modular factory-based construction methods to reduce on-site activity
- Achieving passive decay heat removal to atmosphere after reactor shut down instead of complex redundant active systems
- Use of standard superheated steam turbines as used in conventional power stations instead of the bespoke low-temperature turbines required for PWR's.

On the basis of this, Moltex Energy commissioned Atkins Ltd to independently estimates the capital cost of a 1 GW size Stable Salt Reactor-based on UK safety requirements and UK building costs. Their estimate for the complete nuclear island (including reactor building, emergency diesel generators and building, control room, fresh and spent fuel buildings) was £714/kW. Moltex Energy with input from Seimens has extended this figure to include the turbine and its building, ancillary buildings, site cost, etc., and come to a full plant cost of £1300 which compares extremely favorable to the most recent cost estimate for a PWR in the UK of £5600/kW (for a 3.2 GW plant).

Even with the uncertainties associated with any new reactor design, this figure provides a huge margin before the reactor becomes merely competitive with a PWR. The current estimate in fact means that electricity production would be cheaper than both gas and coal, and this is the precise target that nuclear energy must reach to make the impact on global power production that is desperately needed if we are to stop climate change.

3 Conclusions

For nuclear energy to become a real competitor to fossil fuels, its cost has to come down dramatically. Molten salt fuel has an intrinsic advantage over solid fuel forms due to the transformation of dangerous volatile fission products into stable non-volatile salts. The challenge has been to find a way to use molten salt fuel that does not create the very engineering complexity that has driven nuclear costs so high. Use of molten salt fuel in static fuel tubes opens up that possibility. Static fuel avoids the complexity of pumped molten salt system—the pumps, valves, drain tanks, protection against freezing, leak management systems, refueling systems, chemical treatment systems, degassing systems and so on. It allows the design of exceptionally simple reactors where intrinsic safety carries the burden of the safety case instead of engineered safety systems.

The Stable Salt Reactor has been designed from the very start to achieve the cost level to allow it to drive coal and gas out of the electricity generation market. That is the goal of Moltex Energy.

Acknowledgements The advances in reactor design set out in this paper would not have occurred without the generous support of Tim Abram, Olga Negri, Hugues Lambert, and Joel Turner from Manchester University. Prof. Derek Fray FRS from Cambridge University provided invaluable advice on chemical thermodynamics. Simon Leefe from Wilde Analysis Ltd. carried out the CFD studies.

Reference

1. ORNL Report 919. http://web.ornl.gov/info/reports/1951/3445605662896.pdf

Molten Salt Reactor Research in Switzerland

Jiri Krepel, Boris Hombourger, Carlo Fiorina, Sergii Nichenko, Jarmo Kalilainen, Peter Burgherr, Horst-Michael Prasser and Andreas Pautz

Abstract The enhancement of safety, sustainability, and social acceptance is the main requirements for a nuclear renaissance. Two Gen IV concepts, the molten salt reactor and the high-temperature reactor, have a potential for the paradigm shift in nuclear safety, and Switzerland is involved in the international research efforts dedicated to both of these reactors. Due to the liquid fuel state, especially molten salt reactor (MSR) can act as a safe, sustainable, and acceptable reactor for the future. Main aim of the MSR research program in Switzerland is the safety. Nonetheless, possible MSR designs and so the fuel cycles are also evaluated. At the same time, tools are being developed for MSR transient analysis and thermochemical simulations.

Keywords MSR · GEMS · GeN-Foam · EQL0D · EQL3D

1 Introduction

Switzerland is a member of the Generation Four International Forum (GIF) and joined the GIF Molten Salt Reactor (MSR) Project in November 2015. The participation in the Gas Fast Reactor project was simultaneously terminated; however, the participation in the very high-temperature reactor (HTR) project is ongoing. Since the beginning of the membership, Paul Scherrer Institut (PSI) acts as the Swiss-implementing agent in GIF.

J. Krepel (✉) · B. Hombourger · C. Fiorina · S. Nichenko · P. Burgherr · H.-M. Prasser · A. Pautz
Division of Nuclear Energy and Safety, Paul Scherrer Institute (PSI), 5232 Villigen, Switzerland
e-mail: jiri.krepel@psi.ch

B. Hombourger · C. Fiorina · A. Pautz
Laboratory for Reactor Physics and Systems Behaviour, Ecole Polytechnique Fédérale de Lausanne, 1015 Lausanne, Switzerland

J. Kalilainen · H.-M. Prasser
Laboratory of Nuclear Energy Systems, ETH Zürich, Sonneggstrasse 3, 8092 Zurich, Switzerland

A. K. Nayak and B. R. Sehgal (eds.), *Thorium—Energy for the Future*,
https://doi.org/10.1007/978-981-13-2658-5_38

HTR and MSR reactors have a potential for the paradigm shift in the nuclear safety. Compared to the other Gen IV reactors, HTR and MSR can avoid: criticality and re-criticality accident, cladding melting, wild exothermic reactions, and steam pressurization. HTR fuel is formed by TRISO particles in graphite matrix and is chemically and thermally stable up to 1600 °C and even above this temperature there is not a violent chemical reaction or sudden burst with radioactive material release. MSR relies on liquid fuel in the form of molten fluorides or chlorides salts. The liquid fuel state opens new technological horizons and obviously brings many challenges; yet, it has also very appealing safety and fuel cycle features. Both these reactors have excellent neutronics characteristics and can profit from the thorium fuel cycle. Nonetheless, based on the fuel form, HTR is rather dedicated to once-through cycle with high fuel utilization. The discharged fuel has a stable chemical form suitable for the final repository; however, it is not straightforward to recycle. The molten salt fuel of MSR enables a variety of fuel cycle, but cannot be stored in final repository "as is." On the other hand, its pyrochemical cleaning and recycling may be simpler. Hence, MSR has a potential to combine unparalleled safety with the long sustainability of the closed cycle.

In the past between 1973 and 1980, there was a project called [1] Salt reactor On site reprocessing Fast converter Task (SOFT) at Swiss Federal Institute for Reactor Research (EIR, the predecessor of PSI) looking on chlorides-based fast MSR. It proposed a reactor with 3 GW_{th} power, four-loop layout design, and is fueled by natural chlorides (see Fig. 1). A closed U-Pu fuel cycle with reprocessing and breeding ratio around 1.04 was foreseen. There was 75 m^3 of the salt in the core, 32 m^3 outside with a composition: 1 $PuCl_3$–8 UCl_3–10 NaCl. The core was reflected by 122 m^3 of $CaCl_2$–NaCl and steel.

The MSR research in Switzerland is nowadays carried out within the ETH board, predominantly at PSI. Nonetheless, the research is strongly student-oriented and the universities ETH Zurich and EPFL in Lausanne are thus also involved in the research. Within PSI, the Nuclear Energy and Safety Division (NES) is the competence center for nuclear reactors. Together with the two major topics, safety of current reactors and waste management, the advanced reactor research belongs to the NES missions. The motivation for advanced reactor research, especially for MSR, is driven by:

1. Unique combination of safety and sustainability features.
2. Variety of novel and multidisciplinary research topic.
3. Attractiveness for PhD students and postdocs projects.
4. Potential for alternative financial funding sources.

The MSR-based innovative research project represents an important asset to maintain excellence in nuclear engineering, to ensure successful recruiting of highly motivated and efficient workforce in the coming years and to provide an attractive working area for future generations of scientists and engineers.

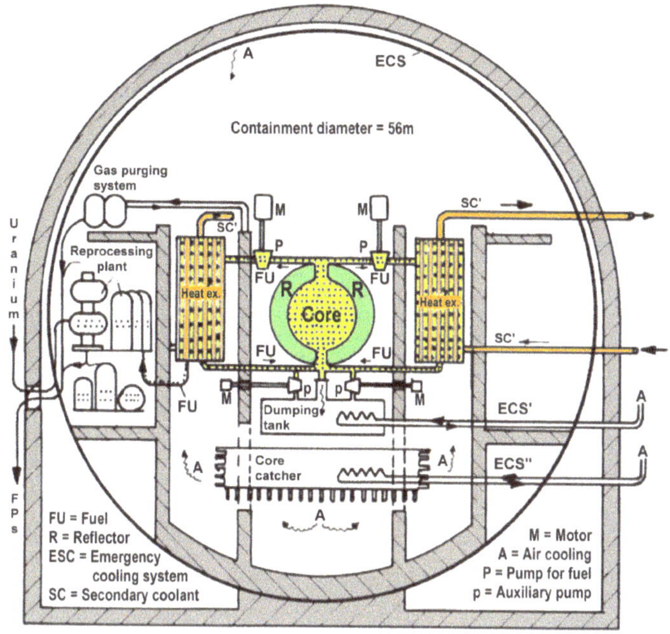

Fig. 1 Layout of the SOFT reactor containment (EIR report nr. 411, 1980)

2 NES Research Projects

Recently, a project on MSR was established at NES as an umbrella for several ongoing national and international projects. In this respect, the most important is the PSI participation in the European Horizon 2020 project SAMOFAR. There are also four smaller ongoing national projects. Swiss National Science Foundation (SNSF) financed in 2014 a PhD project dedicated to Modular MSR Designing for Low Waste Production and another PhD project aiming at Nuclear Data Assimilation in Reactor Physics, where the experimental results for Pu and Th containing PROTEUS reactor experiment should be used to estimate uncertainty of initial Pu concentration in Th-U cycle-based MSR. Another project entitled "Feasibility and plausibility of innovative reactor concepts in an European electricity supply environment" was financed in 2015 by ETH Zurich and the Fund for Projects and Studies of the Swiss Electric Utilities. Finally, the Swiss Nuclear, the nuclear energy section of *swiss*electric, recently financed a project focusing on "Chemical thermodynamic aspects of LWR Pu and MA burning in MSR."

The NES project is structured into four working packages (WPs) of similar research topics, which can be further divided into general or design dependent. The safety of MSR, as the key point and main interest of NES, cannot be evaluated without knowing the system design, fuel chemistry, salt thermal hydraulics features, safety and fuel cycle approach, and the relevant material and chemical limits. Not all of

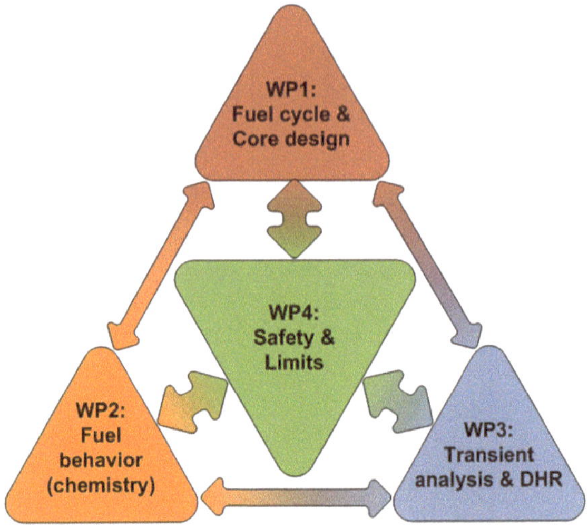

Fig. 2 Illustration of four research areas and their interdependency

these issues will be addressed at PSI; however, there are several NES competences related to thermal and fast solid fuel reactors which are foreseen to be extended on the MSR analysis. The four topical areas or actually WPs are:

1. Core design and fuel cycle.
2. Fuel behavior at nominal and accidental conditions.
3. Transient behavior and decay heat removal system.
4. Safety, fuel stream, and relevant limits.

These activities are interconnected, and safety represents the central point of interest (see Fig. 2). At the same time, the structure is proposed so that there are research topics which can be independently studied within each of the groups. In the WP1, performance of several MSR designs was evaluated. For other working packages, the necessary tools and codes are being prepared.

3 Fuel Cycle and Core Design

The fuel cycle and core design working packages are dedicated to the evaluation of various MSR concepts from the fuel cycle perspective, aiming at sustainable energy production and simultaneously limiting the waste production as well as granting a high level of safety. The ongoing work is based mainly on the SNF PhD study and PSEL project.

The main criterion for sustainability evaluation is the reactivity excess of different designs in closed cycle equilibrium. For that purpose, several equilibrium cycle procedures have been developed. Initially, the ERANOS-based [2] EQL3D routine [3, 4], developed originally for solid-fueled fast reactors, has been modified [5] to simulate the continuous fission products (FPs) removal in MSRs. The removal is simulated by virtual decay constants for both FPs and actinides. Recently, another procedure named EQL0D specifically dedicated to MSR simulation has been developed [6–8]. There are three versions of this routine:

1. EQL0D v1: MATLAB–ERANOS (ECCO) coupling through reaction rates, cell-level routine.
2. EQL0D v2: MATLAB–SERPENT coupling through reaction rates, cell or core-level routine.
3. EQL0D v3: MATLAB–SERPENT coupling through the SERPENT burn-up matrix, cell or core-level routine, unlike *v*1 and *v*2 it also includes fission products.

EQL0D v1 and v2 are MATLAB scripts coupled through the actinides reaction rates with the cell transport code ECCO [2] or Monte Carlo code SERPENT [9], respectively. It solves the Bateman equation by scaling and squaring method [10]. EQL0D v3 adopts the burn-up matrix directly from the Monte Carlo code SERPENT 2 and solves it by the same CRAM [11] as the code. Depending on the calculation scheme, reprocessing and fertile material feed can be periodic or continuous. After each MATLAB calculation, the newly computed fuel composition is provided to SERPENT for the next iteration. The code can be run to predict the fuel composition over user-provided burn-up steps or using an iterative scheme converging directly to the equilibrium composition.

Several fast-spectrum MSR options have been evaluated by EQL0D procedure. The most promising from the fluorides is the $Na^{37}Cl$ carrier salt. Its reactivity excess in equilibrium closed cycle may allow for a breed-and-burn fuel strategy [12]. Fluorides salts provide lower equilibrium reactivity excess. The best of them is the LiF carrier salts, the reference carrier salt of the SAMOFAR project. It was used for several fuel cycle studies. A parametric spectral study at equilibrium cycle and at full core level was accomplished with the EQL3D procedure [3]. A hybrid spectrum MSR core was evaluated [13], the simplified reprocessing scheme was assessed [14], and the continuous and batch-wise reprocessing schemes were compared in [15].

Due to the large excess reactivities of several fast-spectrum salts at equilibrium in the above-mentioned analyses, the feasibility of operating a MSR in a breed-and-burn (BNB) cycle was investigated. A BNB reactor is operated on an open fuel cycle without or with minor fuel reprocessing, typically without fissile elements in the feed. To operate continuously, the reactor must thus be capable to breed fuel excess which is equal or grated than the discharged fissile fuel. Therefore, an excellent neutron economy is needed, which is only achievable in a hard neutron spectrum. For a MSR, this means that chloride salts with chlorine enriched in ^{37}Cl are needed. However, such a fast neutron spectrum also implies a large migration area, i.e., a strong tendency for neutrons to leak out. To obtain a MSR of reasonable size, salts with a high actinide density should thus be used.

Salts containing both thorium and uranium were investigated. In the case of uranium, two chlorides have eutectics with low melting points: UCl_3 and UCl_4, but the latter is less stable at high temperature. Nonetheless, eutectics containing both

Table 1 Selected salts for BNB and their properties

Fuel	Total and ac. densities			Melt. T (°C)
68NaCl–32UCl$_3$	3.32	1.66	g/cm^3 at 900 K	520
60NaCl–40UCl$_3$	3.64	1.97		590
80UCl$_4$–20UCl$_3$	3.79	2.38		545
UCl$_4$	3.56	2.20		590
55NaCl–45ThCl$_4$	3.15	1.61		375
ThCl$_4$	3.82	2.33		770
55NaCl–40ThCl$_4$–5PuCl$_3$	~3.15	~1.76		~375

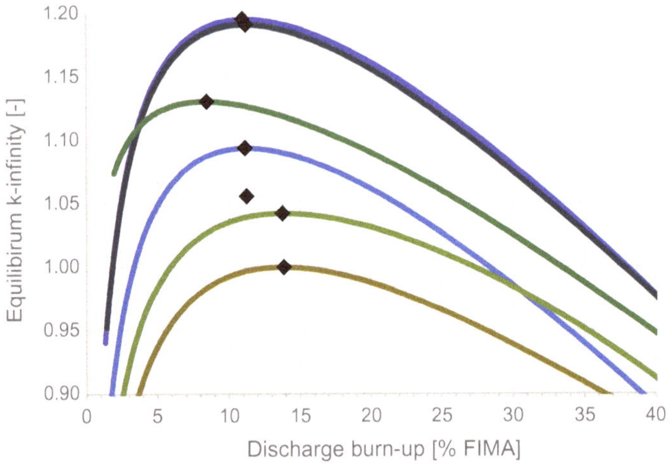

Fig. 3 Equilibrium k_{inf} of selected salts as a function of the discharge burn-up. The colors correspond to the salts in previous table, and maxima are depicted by the point

compounds were chosen as possible fuel compositions in an effort to obtain a higher actinide density. The uranium-bearing salts selected for investigation included the common eutectic-containing sodium chloride, NaCl–UCl$_3$ (68–32 mol%), and a composition with a higher uranium trichloride fraction (60–40 mol%) coming at the cost of a higher melting point, as well as pure UCl$_4$ and the eutectic of the pure uranium chlorides, UCl$_3$–UCl$_4$ (20–32 mol%). For the thorium-bearing salts, NaCl–ThCl$_4$ (55–45 mol%), pure ThCl$_4$, and a composition containing 5 mol% PuCl$_3$ were selected. Their properties are reported in Table 1. Of particular interest is the achievable value of k-infinity at equilibrium, particularly its maximum, which would give the smallest core. It is depicted as a function of the discharge burn-up in Fig. 3.

As expected, the least diluted salts containing more actinides perform better than the more diluted ones due to less parasitic captures on non-fertile nuclides.

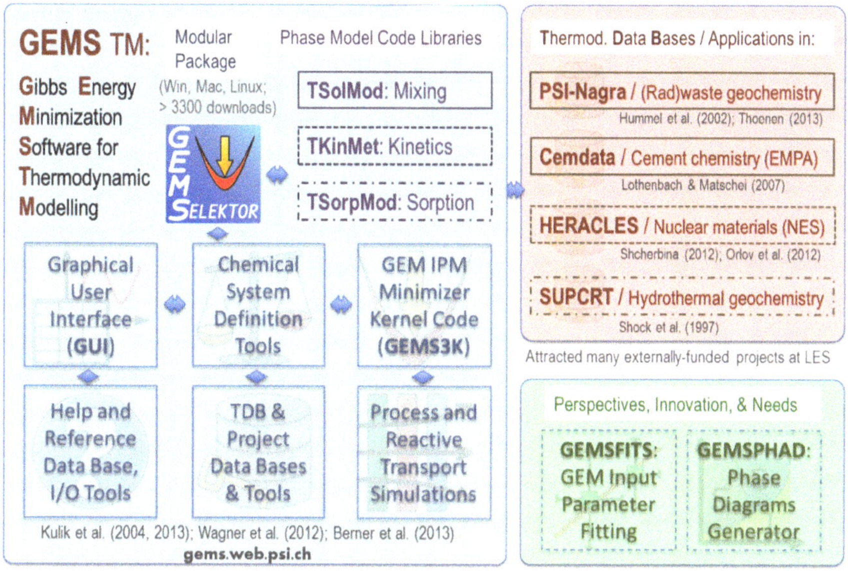

Fig. 4 Structure of the collaborative project GEMS

Thorium-bearing salts perform systematically worse than the uranium-bearing one, unless supported by a small amount of plutonium in the feed. Essentially, all maxima are reached at a discharge burn-up of 10–15% FIMA.

4 Fuel Behavior and Chemistry

The second working package of the NES project focuses on the MSR fuel behavior at nominal and accidental conditions, which can be analyzed by thermochemistry of molten salts. The variety of possible salt composition identifies the need for a tool that is able to predict physical and thermodynamic properties of a wide range of mixtures. The PSI in-house open-source Gibbs Energy Minimization Software package (GEMS) package [16] thus plays a vital role (see Fig. 4) in this question. The PSI activities grouped in this area are financed via the EU projects ACSEPT, SACSESS, and SAMOFAR and through MSR project financed by the Swiss Nuclear "Chemical thermodynamic aspects of LWR Pu and MA burning in MSR."

The utmost interest in safety evaluations are the physical properties related to accidental conditions such as phase changes (freezing or boiling). Moreover, the behavior of specific fission products such as ^{99}Mo can be of great interest for medical application, in particular for isotopes production. The GEMS code with the extended HERACLES database [17] is applied to model the above-mentioned MSR thermodynamic properties and behavior. GEMS has a modular structure and flexible

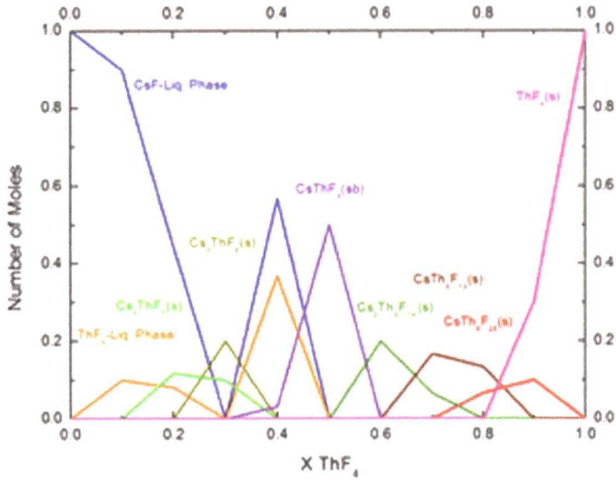

Fig. 5 Speciation in CsF-ThF_4 system at 1073 K

database with an interface built around an efficient open-source GEMS3K numerical kernel code [18] for solving phase equilibria in complex non-ideal systems. Moreover, GEMS can be coupled with other transport or multi-physics codes that significantly broaden extent of GEMS code applicability. It is developed at NES since the year 2000 under the lead of D.A. Kulik.

GEMS3K also contains a so-called TSolMod code library [19] with more than 25 state-of-the-art mixing models for aqueous, gaseous, and solid solutions and melts. This gives GEMS a broad applicability to various types of materials, e.g., fused salts, if extended with the necessary datasets for phases and their components and mixing models for binary interactions between the compounds.

Thermodynamic database HERACLES has been extended with the newly obtained information on binary interaction parameters for various systems. This information, together with the information on thermophysical properties of pure compounds, has been used to study complex systems behavior, speciation in selected molten salts, and Gibbs energies of formation of different species in selected melts. Figure 5 demonstrates the speciation plot in the CsF–ThF_4 system as a function of mole fraction of ThF_4. Further on, this approach been used to assess the standard potentials for different species in the selected melts, like Mo in LiCl–KCl and MgO in LiCl–KCl melts, and to build the E-pO^{2-} stability diagrams for the systems of interest [20]. The performed theoretical study provides a broad and thorough picture of Mo and MgO speciation and electrochemical behavior in LiCl–KCl and LiF–AlF_3 melts.

In cases when the information on binary interactions between different compounds in binary systems is missing, the so-called atomistic modeling is used to fill the information gap. Usually, two main approaches are considered: molecular dynamics (MD) and density functional theory (DFT). In particular, molecular dynamics is used to study mixing enthalpies, excess volumes for different binary systems, polar as well as nonpolar [21, 22], structure of solid and liquid phases. For example, the information on the radial distribution functions (RDFs) (see Fig. 6) for Li–Th–F

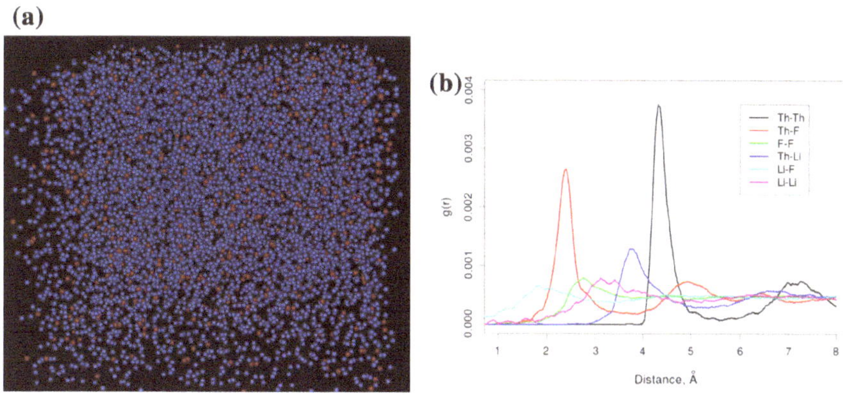

Fig. 6 Structure of molten LiF–ThF_4: **a** snapshot of the molten salt and **b** radial distribution function for the salt

molten salt at 1600 K allows for direct estimation of coordination numbers (Cns) and structure factors which can then be directly compared to experimentally determined data from X-ray or neutron diffraction.

MD serves a reliable tool in the prediction of mixing properties, as well as structural changes effects between different compounds. Moreover, MD was also successfully applied to study dynamic properties like thermal conductivity [23, 24], dynamic viscosity, and diffusion, providing the atomistic-level information and improving the understanding of underlying processes.

The work being performed within the "Chemical thermodynamic aspects of LWR Pu and MA burning in MSR" project is directed toward the acquisition of the necessary thermochemical data needed for the thermodynamic modeling of chemical equilibria in fused fluoride salts along the selected burn-up path. Selected approach allows for simulation of actinides and fission products behavior, solubility, and speciation in phases with ideal and non ideal mixing. In addition, thermodynamic modeling fosters quantitative understanding of systems at different conditions and serves as a predictive tool to complement experimental studies. Thus, obtained information is safety-relevant because some of the salt components may precipitate or evaporate at a certain temperature, altering in this way composition and stability of the fuel/salt mixture.

The ultimate aim of the second working package of the NES is the development of the thermodynamic database for actinides, fission products, and molten salts potentially used in MSR. The database will be used to study speciation, actinides solubility, and chemical stability of different species in fuel salt melt. The developed phase diagrams for the fused salt will be used to understand the options for LWR Pu and MA burning as an initial MSR fuel. In general, MSR core cannot be properly designed without the appropriate knowledge of the salt chemical behavior. The phase diagram and the respective salt behavior have influence not only on the fuel cycle but also on the reprocessing strategy and safety of the reactor. The developed database of force field models for molten salts, actinides, and fission products can be further used for MD modeling of complex systems to study the atomistic behavior (structure and stability of atomic structures, reaction processes, the effects produced by energetic neutrons and ions) and physical properties (like viscosity and thermal expansion). The activities within the work package promote collaboration between research institutes on an international level (TU Delft, ITU, and PSI), between laboratories within NES, as well as within relevant international projects (SAMOFAR, GIF MSR).

5 Transient Behavior and Decay Heat Removal System

In this working package, two different simulation tools for MSR transient analysis are developed and applied. These are OpenFOAM based on multi-physics solver GeN-Foam [25–27] and on the system code TRACE–PARCS [28–30]. MSR thermal hydraulics is peculiar from several perspectives, and there are only a few accessible codes which can address these peculiarities. The tasks accomplished or planned in this working package are:

- Initial GeN-Foam solver development [25–27].
- Pre-evaluation of the heat exchangers with TRACE and salt properties inclusion [28].
- Development of TRACE point-kinetics delay neutron precursors drift model [29].
- Development of TRACE–PARCS delay neutron precursors drift model and application to MSRE transients [30].
- Decay heat distribution in the fuel stream of MSR [31].
- Application of the GeN-Foam [32] and the TRACE–PARCS [33] to MSRE and MSFR, respectively.
- In 2017, PhD project was started for finalization, verification, and validation of the two codes.

The achieved results are illustrated here by a TRACE-PARCS results example in Fig. 7. It shows the steady-state distribution of three selected groups of delayed neutron precursors [33] in the MSFR benchmark geometry. Another example is the impact of anti-swirl vanes on salt flow in MSRE core [32] shown in Fig. 8. In the selected example, the porous media approach is used and the flow is not restricted in the radial direction. Both these tools are still in the development phase and will need thorough verification and validation in the future.

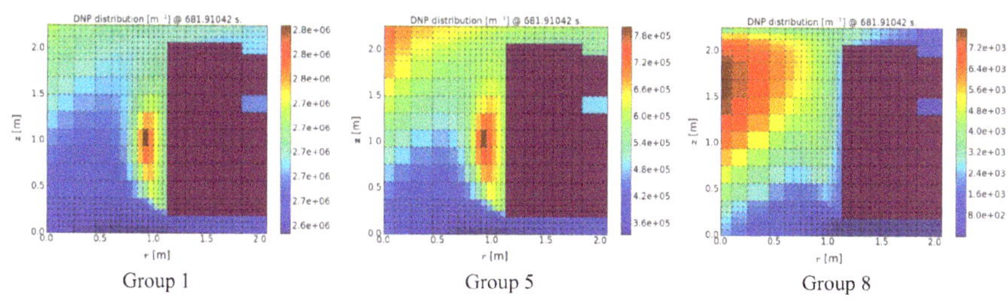

Fig. 7 Distribution of three selected groups of delayed neutron precursors in the MSFR benchmark geometry

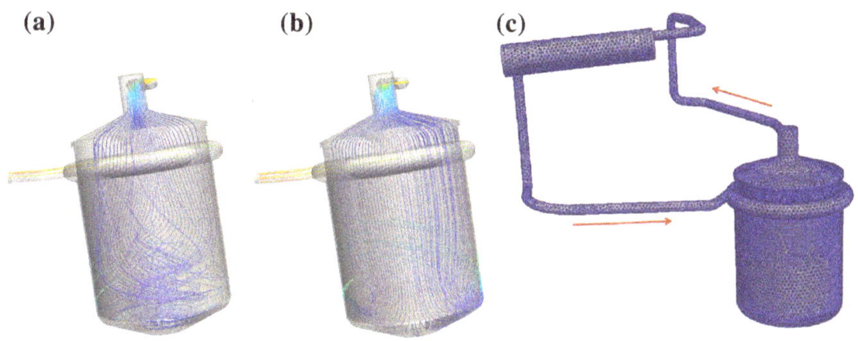

Fig. 8 GeN-Foam modeling of MSRE. Illustration of the porous media approach and impact of the anti-swirl vanes inclusion. Core with the free flow (**a**), included anti-swirl vanes (**b**), and the circuit model (**c**)

6 Safety, Fuel Stream, and the Relevant Limits

The ultimate objective of the three working packages is the MSR safety. It is linked to all other topics. Only the WP1 addresses also the sustainability, not only safety issues. Even though safety is the major aim of the project at PSI, the activities in WP4 are rather limited. Within the SAMOFAR project, the aerosols formation and migration in the MSR containment atmosphere should be addressed accounting for the formation by nucleation and growth by condensation, coagulation, and agglomeration [34]. Another example is the probabilistic safety assessment (PSA) of MSR. A simplified level-3 PSA should address the consequences of hypothetical MSR accident. The simulation will be based on PSI's Energy-related Severe Accident Database (ENSAD) and the experience from EU project NEEDS [35]. One MSc study related to this WP was already accomplished. The Mini Fuji MSR reactor [36] was analyzed as a reference design using the PSA level-1. The main events leading to vessel damage (see Fig. 9) were quantified. The salt draining system, often highlighted as a major safety advantage, is responsible for the high core damage frequency by all major events.

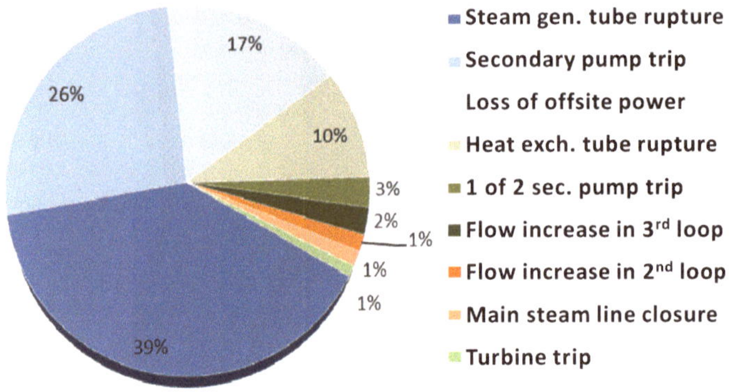

Fig. 9 Main events with vessel damage for Mini Fuji MSR design

7 Summary

MSR research in Switzerland focuses on selected MSR safety and sustainability aspects. PSI as the implementing agent for GIF MSR project participation is leading the Swiss research and cooperates with the universities through the students' works. PSI is also a member of SAMOFAR project consortium. The key tools in the MSR research portfolio at PSI are: MD and GEMS code with HERACLES database for the chemistry, TRACE and GeN-Foam for the transient analysis, EQL3D and EQ0D for the fuel cycle analysis, and ENSAD database-related tool for the PSA analysis.

References

1. M. Taube, W. Heer, *Reactor with Very Low Fission Product Inventory*. Report Nr. 411 (Eidgenössisches Institut für Reaktorforschung (EIR), Würenlingen, AG, Switzerland, 1980)
2. G. Rimpault, et al., The ERANOS code and data system for fast reactor neutronic analyses, in *PHYSOR 2002*, Seoul, Korea, 7–10 Oct 2002
3. J. Křepel et al., Fuel cycle advantages and dynamics features of liquid fueled MSR. Ann. Nucl. Energy **64**, 380–397 (2013)
4. J. Křepel et al., EQL3D: ERANOS based equilibrium fuel cycle procedure for fast reactors. Ann. Nucl. Energy **36**, 550–561 (2009)
5. C. Fiorina et al., Investigation of the MSFR core physics and fuel cycle characteristics. Prog. Nucl. Energy **68**, 153–168 (2013)
6. B. Hombourger, Parametric lattice study for conception of a molten salt reactor in closed thorium fuel cycle, Master's thesis, joint EPFL-ETHZ Master of Science in Nuclear Engineering, 2013
7. B. Hombourger, et al., Parametric lattice study of a graphite-moderated molten salt reactor. J. Nucl. Eng. Radiat. Scince. **1** (2015)
8. B. Hombourger, et al., The EQL0D procedure for fuel cycle studies in molten salt reactors, in *ICAPP 2016*, San Francisco, USA, 17–20 April 2016
9. J. Leppänen, *Serpent a continuous-energy monte carlo reactor physics burnup calculation code* (VTT Technical Research Centre of Finland, Finland, 2013)

10. A.H. Al-Mohy, N.J. Higham, A new scaling and squaring algorithm for the matrix exponential. SIAM. J. Matrix Anal. Appl. **31**(3), 970–989 (2009)
11. M. Pusa, J. Leppänen, Solving linear systems with sparse gaussian elimination in the chebyshev rational approximation method. Nucl. Sci. Eng. **175**(3), 250–258 (2013)
12. B. Hombourger, et al., Fuel cycle analysis of a molten salt reactor for breed-and-burn mode, in *ICAPP 2015*, Nice, France, 03–06 May 2015
13. J. Křepel, et al., Hybrid spectrum molten salt reactor, in *PHYSOR 2014*, Kyoto, Japan, 28 Sept–3 Oct 2014
14. J. Křepel, et al., Molten salt reactor with simplified fuel recycling and delayed carrier salt cleaning, in *Proceedings of ICONE 22*, Prague, Czechia, 7–11 July 2014
15. J. Křepel, et al., Comparison of several recycling strategies and relevant fuel cycles for molten salt reactor, in *ICAPP 2015*, Nice, France, 03–06 May 2015
16. D.A. Kulik, S.V. Dmytrieva, T. Wagner, G. Kosakowski, T. Thoenen, U. Berner, et al., Gibbs energy minimization software (GEMS) homepage (2004–2014), http://gems.web.psi.ch
17. N. Shcherbina, N. Kivel, I. Günther-Leopold, D.A. Kulik, J. Bertsch, HERACLES project and thermodynamic database homepage (2012), http://www.psi.ch/heracles/heracles
18. D.A. Kulik, T. Wagner, S.V. Dmytrieva, G. Kosakowski, F.F. Hingerl, K.V. Chudnenko, U. Berner, GEM-Selektor geochemical modeling package: revised algorithm and GEMS3K numerical kernel for coupled simulation codes. Comput. Geosci. **17**, 1–24 (2013)
19. T. Wagner, D.A. Kulik, F.F. Hingerl, S.V. Dmytrieva, GEM-selektor geochemical modeling package: TSolMod library and data interface for multicomponent phase models. Can. Mineral. **50**, 1173–1195 (2012)
20. S. Nichenko, M. Streit, Thermodynamic modelling of molybdenum behaviour in chloride molten salt, in *Proceedings of TopFuel-2015*, Zurich, Switzerland, 13–17 Sept 2015
21. J. Dai, X. Li, L. Zhao, H. Sun, Enthalpies of mixing predicted using molecular dynamics simulations and OPLS force field. Fluid Ph. Equilib. **289**, 156–165 (2010)
22. J. Dai, C. Wu, X. Bao, H. Sun, Prediction of the heat of mixing for binary fluids using molecular dynamics simulation. Fluid Ph. Equilib. **236**, 78–85 (2005)
23. S. Nichenko, D. Staicu, Molecular dynamics study of the mixed oxide fuel thermal conductivity. J. Nucl. Mater. **439**(1–3), 93–98 (2013)
24. S. Nichenko, D. Staicu, Thermal conductivity of porous UO_2: molecular dynamics study. J. Nucl. Mater. **454**(1–3), 315–322 (2014)
25. C. Fiorina et al., GeN-Foam: a novel OpenFOAM® based multi-physics solver for 2D/3D transient analysis of nuclear reactors. Nucl. Eng. Des. **294**(2015), 24–37 (2015)
26. C. Fiorina, K. Mikityuk, Application of the new GeN-Foam multi-physics solver to the European sodium fast reactor and verification against available codes, in *Proceedings of ICAPP 2015*, Nice, France, Paper 15226, 03–06 May 2015
27. C. Fiorina, et al., Development and verification of the neutron diffusion solver for the GeN-Foam multi-physics platform. Ann. Nucl. Energy **96**, 212–222 (2016)
28. V. Ariu, Heat exchanger analysis for innovative molten salt fast reactor. M.Sc. thesis, EPFL ETHZ joint M.Sc. in Nuclear Engineering, Switzerland, 2014
29. M. Zanetti, et al., Extension of the FAST code system for the modelling and simulation of MSR dynamics, in *Proceedings of ICAPP 2015*. Paper 15226, Nice, France, 03–06 May 2015
30. H. Kim, Static and transient analysis of molten salt reactor experiment using Serpent/TRACE/PARCS codes. M.Sc. thesis, EPFL ETHZ joint M.Sc. in Nuclear Engineering, Switzerland, 2015
31. J. Choe, Empirical decay heat correlation and fission product behavior in MSR. M.Sc. thesis, EPFL ETHZ joint M.Sc. in Nuclear Engineering, Switzerland, 2015
32. J. Bao, Development of the model for the multi-physics analysis of molten salt reactor Experiment using GeN-Foam code. M.Sc. thesis, EPFL ETHZ joint M.Sc. in Nuclear Engineering, Switzerland, 2016

33. E. Pettersen, Coupled multi-physics simulations of the MSFR using TRACE-PARCS. M.Sc. thesis, PSI defended at Universite Paris-Saclay, France, 2016
34. C. Bosshard, et al., Large eddy simulation of the differentially heated cubic cavity flow by the spectral element method. Comput. Fluids **86**, 210–227 (2013)
35. P. Burgherr, S. Hirschberg, E. Cazzoli, Final report on quantification of risk indicators for sustainability assessment of future electricity supply options, in *NEEDS Deliverable no D7.1—Research Stream 2b. NEEDS project "New Energy Externalities Developments for Sustainability"* (Brussels, Belgium, 2008)
36. D. Pyron, Safety analysis for the licensing of molten salt reactors. M.Sc. thesis, EPFL ETHZ joint M.Sc. in Nuclear Engineering, Switzerland, 2016

Part VI
Water Cooled Reactors with Thorium Fuel

Hydrodynamics of AHWR Gravity-Driven Water Pool Under Simulated LOCA Conditions

I. Thangamani, Seik Mansoor Ali and Vishnu Verma

Abstract The advanced heavy water reactor (AHWR) employs a double containment concept with a large inventory of water within the gravity-driven water pool (GDWP) located at a higher elevation within the primary containment building. GDWP performs several important safety functions in a passive manner, and hence, it is essential to understand the hydrodynamics within this pool during a LOCA. To achieve this objective, several LOCA scenarios involving RIH break sizes ranging from 2 to 50% are postulated, and the AHWR containment system response is obtained. An in-house containment thermal hydraulics code called "CONTRAN" is used for this purpose. The blowdown mass and energy discharge data for each break size, along with the geometrical details of the AHWR containment, form the main input for the analysis. Apart from obtaining the pressure and temperature transients within the containment building, the focus of this work is on simulating the hydrodynamic phenomena of vent clearing and pool swell occurring in the GDWP. The variation of several key parameters such as pressure and temperature within the primary containment volumes V1 and V2, differential pressure, BOP rupture time, vent clearing velocity, effect of pool swell on the V2 air space pressure, GDWP water level, etc., is presented and discussed in detail, and important findings are highlighted. Further, the effect of neglecting the modeling of pool swell phenomenon on the containment transients is also clearly brought out by a comparative study. The numerical studies presented here give insight into AHWR containment transients that would be useful to both the system designer as well as the regulator.

Keywords AHWR · Containment thermal hydraulics · GDWP · Pool swell
Numerical analysis

I. Thangamani (✉) · V. Verma
Reactor Safety Division, Bhabha Atomic Research Centre, Trombay,
Mumbai 400085, India
e-mail: tmi@barc.gov.in

S. M. Ali
Safety Research Institute, Atomic Energy Regulatory Board,
Kalpakkam 603102, India
e-mail: seik@igcar.gov.in

A. K. Nayak and B. R. Sehgal (eds.), *Thorium—Energy for the Future*,
https://doi.org/10.1007/978-981-13-2658-5_39

1 Introduction

AHWR employs several passive safety features to assure a high level of safety during accident conditions. One of these is the availability of a large inventory of water (~9000 m^3) in the GDWP located at a higher elevation within the primary containment building. It facilitates the removal of core decay heat, long-term core cooling by direct injection, limiting of containment peak pressure and temperature, passive cooling of the containment system, moderator system, concrete structures as well as provides radionuclide management by dissolving, trapping, entraining or scrubbing away part of the fission products, etc., for prolonged period. Further, it also supports safety-related systems like the isolation condenser for decay heat removal and the passive containment isolation system for quick containment isolation. Since a number of important safety functions are required to be performed by the GDWP, it is essential that the availability of GDWP should be ensured during normal and accident states.

AHWR uses a double containment envelope, i.e., a primary and a secondary containment as shown in Fig. 1. The primary containment is completely surrounded by the secondary containment. The primary containment is divided into two volumes, i.e., V1 and V2, for control of accessibility and efficient accident management. The arrangement is such that volume V2 completely surrounds the volume V1. These two volumes are interconnected by a vent system via GDWP [1]. The volume V1 houses all the high enthalpy systems like the reactor core, steam generator, and other associated systems, and it is inaccessible during normal operation. The volume V2 contains low enthalpy systems like GDWP and is generally accessible during reactor operation. V1 is maintained at a pressure slightly lower than V2 for ventilation purpose due to which fixed quantity of fresh air from V2 is continuously drawn into V1 through available leak area between these volumes. During a large break loss of coolant accident (LOCA) involving double-ended rupture of the reactor inlet header, steam is released into V1 volume of the primary containment, and subsequently steam-air mixture is transported to the V2 volume due to pressure difference. When the steam-air mixture bubbles through the pool of water in GDWP, the ensuing heat and mass transfer causes steam to condense, and the cooled air finally reaches the V2 volume air space. The GDWP internal structures are subjected to various hydrodynamic loads during this process. These loads are induced in a particular sequence; first due to jet impingement on the floor during the vent clearing phase, followed by pool swell and fall back, steam-air flow, and finally due to steady and oscillatory condensation (steam chugging). These loads are in addition to the general thermodynamic loading of the containment due to mass and energy released during LOCA [2]. A blow-out panel (BOP) is provided as a protection measure between V1 and V2 for pressure equalization during a LOCA, when the differential pressure exceeds a threshold value. Numerical studies conducted earlier [3, 4] revealed that the BOP ruptures for pipe break sizes greater than 50%, for which the V1–V2 differential pressure exceeds 50 kPa. Pressure equilibrium is then quickly established between V1 and V2, and flow of steam-air mixture through GDWP ceases. The conclusion was that BOP would remain intact for pipe break sizes less than 50%, and the blow-

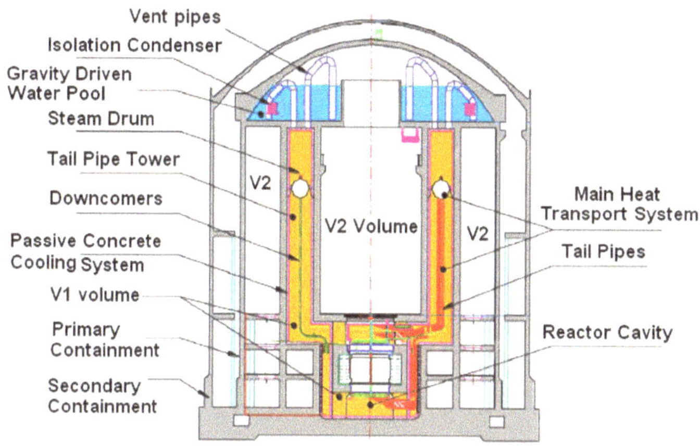

Fig. 1 **a** Schematic diagram of AHWR containment system

down steam-air mixture passes through GDWP vent pipes continuously, thereby maintaining the GDWP effective for a longer duration. These studies, however, did not account for the variation in GDWP water level and the phenomena of pool swell.

Recently, the need was felt to reevaluate the containment transients considering the vent clearing and pool swell events. The aim was to investigate the effect of modeling these events on the overall containment pressure and temperature transients, and to obtain an estimate of the efficacy of GDWP in absorbing the energy of steam-water mixture discharged during LOCA. To achieve these objectives, the capability of in-house containment thermal hydraulics code "CONTRAN" (CONtainment TRansient ANalysis) was first enhanced by incorporating a pool swell model [5]. LOCA scenarios involving RIH break sizes ranging from 2 to 50% were postulated for investigation, and some of the earlier analyses were revisited. The transient variations of several key parameters of interest were obtained and are presented and discussed here in detail. Important findings and some safety issues are also highlighted. Further, the effect of neglecting the pool swell phenomenon on the containment transients is clearly brought out by a comparative study.

2 Mathematical Model

The in-house computer code CONTRAN is a multicompartment, transient containment safety analysis code. The entire containment geometry can be divided into a network of volumes (i.e., compartments) which are interconnected by junctions (representing the inter-compartmental openings). Each compartment consists of a liquid and vapor region. The conservation equations of mass and energy together with equation of state are iteratively solved for all the prevailing gas/liquid species in all the volumes of the network. Appropriate initial conditions are provided at the start of the transients. The blowdown mass, energy discharge as well as activity release rates for any break size can be given as input. The sequence of events following a simulated LOCA is as follows; first, the V1 volume gets pressurized, thereby establishing a flow path for steam-air mixture from V1 to V2 through the vent pipes submerged in GDWP. The pressures exerted on the free surface in the vent pipe pushes the water level downward, until the vent is fully cleared of water (vent clearing phase). Once the vents are cleared, the expelled air causes a swell in the pool level. The air trapped in the bubble is ultimately transported to the V2 air space. If BOP ruptures, during the transient, due to high V1–V2 differential pressure, then an alternate flow path is established between V1 and V2, in addition to the normal path described above.

The vent clearing model consists of solution to the one-dimensional momentum equation for incompressible flow utilizing the V1 and V2 pressure transients as boundary conditions [6]. The calculations are continued till the vent pipes get cleared of water. Vent flow velocity, change in water level in vent pipe as well as in GDWP can be obtained as output. Once the vents are cleared, subsequent calculations are performed by the pool swell model. An air bubble is assumed to form at the end of the vent pipe and detaches if its diameter is twice the vent pipe diameter [7, 8]. The bubble growth, detachment, translation in the GDWP pool, water level increase, etc., can be predicted. The bubble growth (i.e., bubble radius) is calculated by solving Rayleigh–Plesset equation [9] by considering a spherical bubble formation. Isothermal conditions are assumed to prevail within the air bubble. The bubble radius calculated at every instant is used for calculating the rise in pool level. The detached bubble is assumed to have initial pressure same as V1 volume pressure. After detachment, the vertical migration velocity of free bubble due to buoyancy is calculated by solving one-dimensional transient momentum equation for the bubble transport in an infinite medium.

3 Numerical Analysis

The entire numerical analysis is carried out using the upgraded version of CONTRAN described earlier. The primary containment is divided into two volumes (V1 and V2), which are interconnected by four different types of junctions, i.e., break junction, vent junction, BOP, and a flow junction representing a ventilation unit as shown in Fig. 2. Although a single vent pipe is modeled, the net effect of several vent pipes is obtained by multiplying the single vent results with total number of vent pipes. The fluid mass and energy blowdown data for a given break size of LOCA are the main inputs to this code, apart from the geometrical details of AHWR containment building, GDWP and vent system, etc. [10, 11]. Within each time step, the mass and energy conservation equations in each of the two volumes are solved to obtain the pressure and temperature. The momentum equation for vent clearance, Rayleigh–Plesset equation for bubble formation and growth, and bubble migration velocity in the vertical direction are solved using fourth-order Runge–Kutta method. The pressure in the upstream volume (V1) and downstream volume (V2) form the

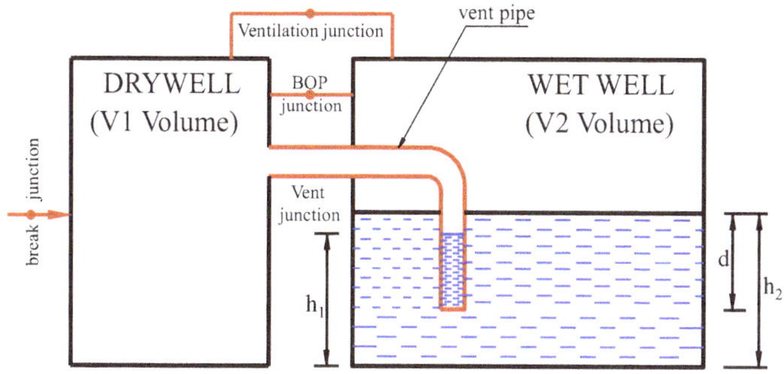

Fig. 2 Volume connectivity and nodalization of AHWR containment

boundary conditions for the solution of the momentum equations (for flow in vent pipe) in each time step. For the vent clearing phase, the level transient in the vent pipe as well as the interface velocity is obtained. For the pool swell phase, transient variation of bubble radius/volume before its detachment, free bubble volume and velocity during translation, GDWP water level rise, and V2 volume air space pressurization due to water level swell are all obtained. The total change in GDWP level during the pool swell phase is obtained by accounting for the contribution due to bubble formation/growth at vent pipe exit, its growth and translation as a free bubble. The time step sizes chosen for capturing the vent clearing and pool swell transients are 10^{-5} and 10^{-6} s, respectively.

4 Results and Discussion

Numerical studies are conducted for LOCA break sizes ranging from 5 to 50%. To highlight the effect of pool swell phenomena on the containment pressure transients, the analyses have been carried out with and without modeling of the phenomena.

Figure 3a–d shows the comparison of pressure transients with and without the modeling of pool swell phenomena for LOCA involving RIH break sizes of 50, 30, 10, and 5%, respectively. When pool swell phenomenon is not considered, it was found that the differential pressure across BOP (pressure in V1-pressure in V2) exceeds 50 kPa for the 50% LOCA case and causes BOP to rupture. This is evident from Fig. 3a. In all the remaining cases, where the break size is <50%, the BOP does not rupture, and the steam-air mixture continues to pass through GDWP as long as the driving force exists. When the pool swell phenomena is appropriately modeled, back pressure in V1 volume increases due to increase in GDWP water level created by vent clearing and formation and transport of bubbles. Consequently, the BOP ruptures and establishes a flow path, bypassing the GDWP. Rupture of BOP is observed for all the break sizes, and the predicted containment peak pressures are higher. Further, for break sizes of 50 and 30%, flow from V1 volume takes place

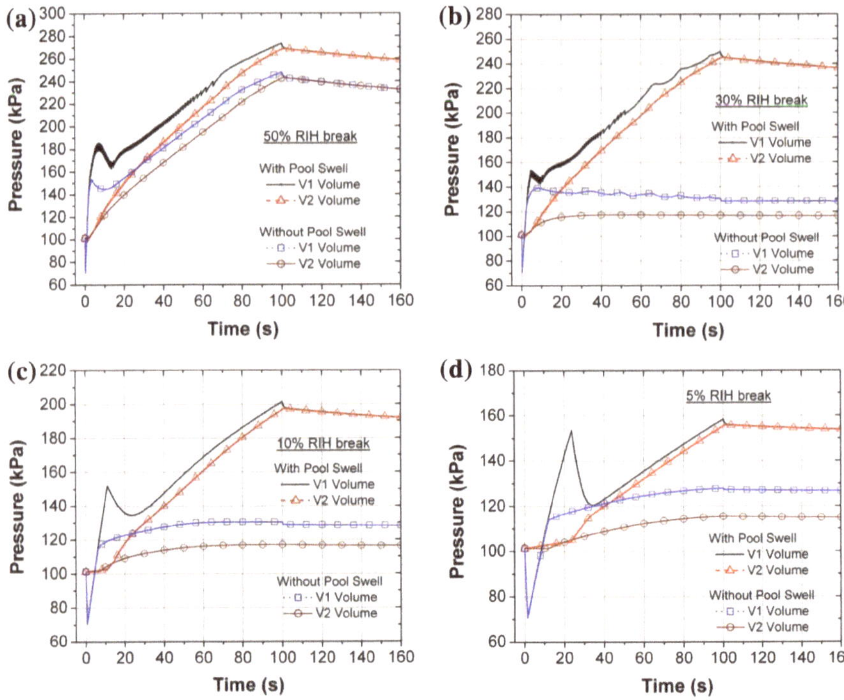

Fig. 3 Containment pressure transients with and without pool swell model for various LOCA cases; **a** 50% RIH break, **b** 30% RIH break, **c** 10% RIH break, and **d** 5% RIH break

through GDWP as well as BOP, whereas for the 10 and 5% break sizes, flow takes place through BOP only.

Figure 4a, b shows the variation of differential pressure between V1 and V2 for different break sizes with and without accounting for pool swell phenomena. For all break sizes, during initial phase of transient, it is observed that the V1 volume pressure reduces from 100 to 70 kPa for a short duration. This is due to decrease in V1 volume temperature from initial values of 285 °C down to ~70 °C. This reduction in V1 pressure has the potential to cause siphoning of GDWP water into V1 volume. If pool swell phenomenon is not modeled, then the flow through GDWP is continuous. However, when pool swell is modeled, the flow becomes oscillatory due to formation and detachment of bubble from vent pipe exit. Figure 4a depicts this behavior for the 50 and 30% break cases. For the 10 and 5% break sizes, differential pressure oscillations are not observed since no flow occurs through the vent pipes. Pressure equalization between V1 and V2 takes place after 100 s in all the cases, as the blowdown inputs is provided till that time only. As high enthalpy coolant flashes into V1 volume at atmospheric pressure, it is assumed to mix uniformly with V1 volume air, and thereafter the temperature is calculated based on energy balance in each time step. The temperature decreases during initial stages of transient because

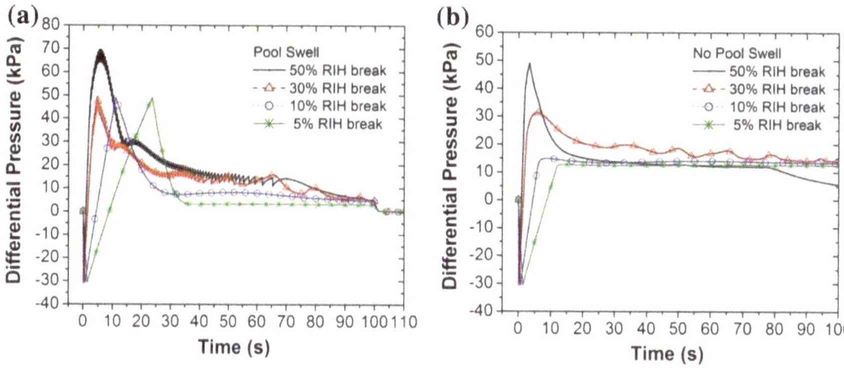

Fig. 4 Containment differential pressure transients for various LOCA cases; **a** considering pool swell, **b** without considering pool swell

steam absorbs energy from air (which is initially at 285 °C), due to latent heat and high specific heat of steam.

Figure 5a, b shows the containment temperature variation in V1 and V2 volume for different break sizes accounting for pool swell phenomenon. Figure 5c, d shows the corresponding plots without considering the pool swell. These figures show that temperatures are higher in the former case. This is due to low flow of steam-air mixture through GDWP (i.e., less energy absorption in GDWP) and more flow through BOP. It is also evident from Fig. 5b that for 50% break size, V2 temperature is much higher. This can be attributed to the early opening of BOP.

Figure 6a shows the variation of interface velocity in vent pipes for different break sizes. As per the modeling convention, negative velocity indicates the flow of water in downward direction. It is clear that the interface velocity at vent pipe exit is quite low for the all break sizes. The hydrodynamic load induced by emerging jet may not be significant during this phase of the transient. After the vent clearing phase, bubble formation takes place at the exit of vent pipes, and it has been assumed that the air bubble, after reaching two times the vent pipe diameter, detaches from vent pipe and a new bubble will be formed thereafter. The detached bubble moves freely in the pool water with initial pressure being equal to V1 pressure. The bubble then starts to grow if bubble pressure is higher than the combined pressure of V2 volume and hydrostatic pressure due to GDWP water level. However, if the bubble pressure is lesser than sum of V2 and hydrostatic pressure, it undergoes contraction. Therefore, after release the bubble size undergoes oscillation till it breaks into V2 volume air space. The variation in bubble size is shown in Fig. 6b for the 50% break size case. It can be observed that the bubble radius reaches up to 1.4 m before it breaks into V2 volume. The frequency of this bubble formation and detachment process is approximately 3 Hz (i.e., three times per second). This process of bubble formation, growth, and subsequent movement of free bubble causes the water level to oscillate in the pool. This is depicted in Fig. 6c for the 50% break case. As the bubble breaks into V2 volume, the water level falls back to original level, and it then

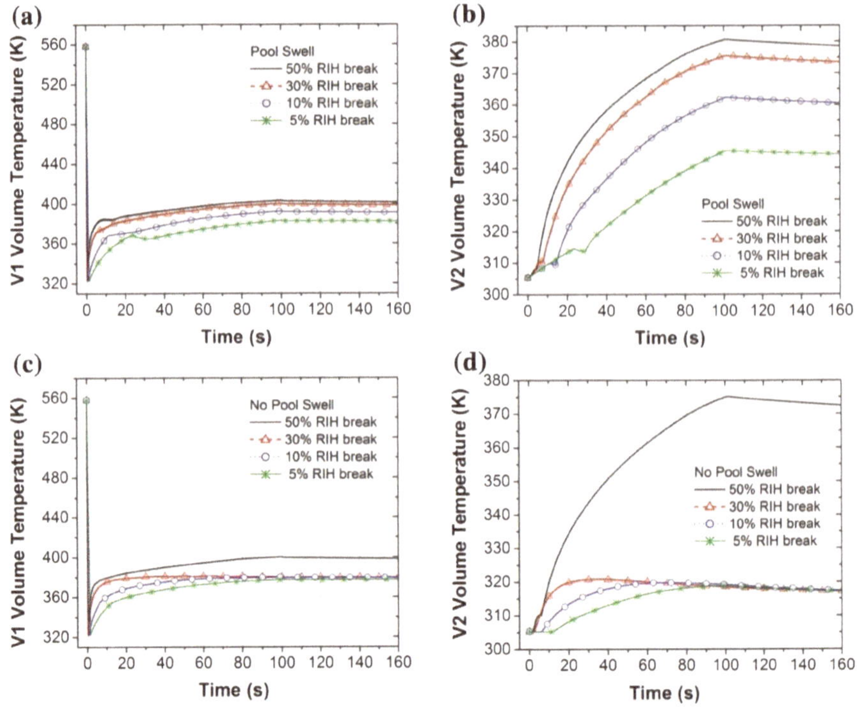

Fig. 5 Containment temperature transients for various LOCA cases; considering pool swell **a** V1-volume, **b** V2-volume; without considering pool swell **c** V1-volume, **d** V2-volume

rises due to the growth of subsequent bubble. The rise and drop in pool level causes hydrodynamic load over pool structures. Figure 6d shows the variation of pressure within the bubble, V2 volume air space, and the GDWP hydrostatic pressure, for the 50% break LOCA case considering pool swell. The bubble formation, detachment, and migration to V2 volume air space including the bubble break constitutes one cycle. The gradual overall rise in V2 volume air space pressure is due to added mass of air. The minor pressure fluctuation is due to GDWP level variation. The GDWP level change influences the hydrostatic pressure as shown in Fig. 6d. The transients for bubble radius, GDWP level, pressure, etc., for the 30% case are similar to 50% break LOCA. Since BOP panel gets ruptured for all the break sizes, the flow through GDWP gets reduced. Since large quantities of noncondensable (air) passes from V1 to V2 along with steam, the possibility of occurrence of steam chugging phenomena and associated loads are minimal. The calculated average rise in pool level is around 1.4 m which is less than the available clearance height of 1.9 m.

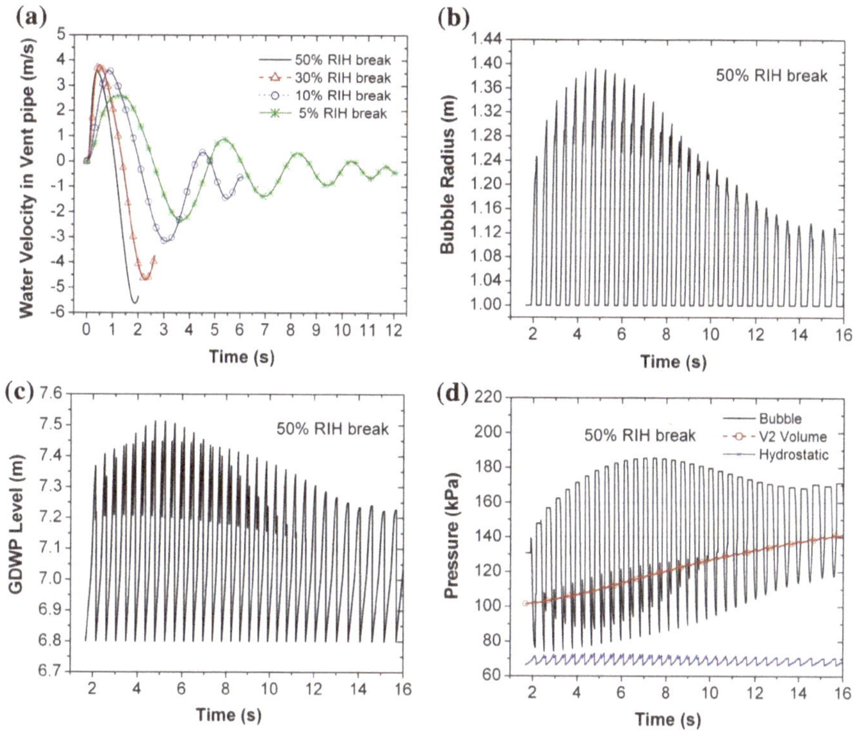

Fig. 6 **a** Vent clearing transients in GDWP; for 50% LOCA case, **b** variation of bubble radius in GDWP; **c** water level in GDWP; **d** pressure transients in GDWP

5 Conclusions

This paper presents a detailed thermal hydraulic analysis for AHWR containment transients under simulated LOCA scenarios involving RIH breaks of various sizes (2–50%). At the outset, a brief description of the AHWR containment system is given. Then a description of the salient features of the in-house computer code CONTRAN that is used for the analysis is given, followed by the analysis methodology. The variation of several key parameters such as primary containment pressure, temperature and V1–V2 differential pressure with time, vent clearing velocity and time, effect of pool swell on the V2 air space pressure, GDWP water level, etc., are presented and discussed. The effect of neglecting the pool swell phenomenon on the containment transients is clearly brought out by a comparative study. It was found that by accounting for the pool swell phenomenon in the modeling, the BOP ruptures for all LOCA cases, whereas it is not so if that effect is neglected. Therefore, the GDWP efficacy in condensing the steam discharged during LOCA is perhaps over predicted if the pool swell phenomenon is neglected in the modeling. Another issue that came to light is that the pressure in volume V1 is very low in the initial phase of the containment transients, which indicates the possibility of siphoning of GDWP water into V1. This needs to be taken into account during the design of the vent system. The numerical studies presented this paper give insight into containment transients that would be useful to both the system designer as well as the regulator.

References

1. R.K. Sinha, A. Kakodkar, Design and development of the AHWR—the Indian thorium fuelled innovative nuclear reactor. Nucl. Eng. Des. **236**, 683–700 (2006)
2. Atomic Energy Regulatory Board (AERB), *Vapor Suppression System (pool type) for PHWR*. Safety Guide No AERB/SG/D-22, June 2000
3. S. Mansoor Ali, V. Verma, R.K. Singh, A.K. Ghosh, Analysis of AHWR containment to assess the performance of passive containment isolation system, in *Proceedings of the 19th National & 8th ISHMT-ASME Heat and Mass Transfer Conference* (JNTU Hyderabad, India, 2008)
4. I. Thangamani, V. Verma, R.K. Singh, A.K. Ghosh, Performance analysis of passive containment isolation system, in *37th International & 4th National Conference on Fluid Mechanics and Fluid power FMFP2010*, IIT Madras, Chennai, India, December 16–18, Paper No. 254 (2010)
5. I. Thangamani, V. Verma, R.K. Singh, A.K. Ghosh, Effect of suppression pool swell on containment pressure transients during LOCA, in *Proceedings of SMiRT-21*, New Delhi, India, November 6–11, Paper No. 254 (2011)
6. A.K. Ghosh, R.B. Grover, *Analysis of Vent Clearing Transient in a Pressure Suppression Type Containment System*, BARC Report No. BARC/I-814, 1984
7. G.B. Wallis, *One Dimensional Two Phase Flow* (McGraw Hill, New York, 1969)
8. A.K. Ghosh, A simplified analytical model for pool swells, in *Proceedings of 9th SMiRT*, Lausanne, August 17–21 (1987)
9. C.L. Chen, V.K. Dhir, Hydrodynamics of a bubble formed at vent pipe exit. Int. J. Multiphase Flow **8**(2), 147–163 (1982)
10. S.K. Haware, A.K. Ghosh, H.S. Kushwaha, Analytical studies on optimization of containment design pressure, in *Proceedings of NURETH-11*, Avignon, France, pp. 1–10, Paper No. 482 (2005)
11. S.K. Haware, A.K. Ghosh, V. Venkat Raj, A. Kakodkar, Analysis of CSNI benchmark test on containment using the code CONTRAN, in *Proceedings of Third International Conference on Containment Design and Operation*, vol. 2, Toronto, Canada (1994)

Experimental Measurement and Analysis of Natural Circulation Flow Oscillations in a Vertically Heated Channel

J. Aparna, A. K. Nayak and A. Rama Rao

Abstract Fuel utilization and passive safety are some of the important design aspects of advanced reactors. Thorium due to its abundance and importance in providing sustainable energy has been considered as future fuel for advanced reactors in India such as advanced heavy water reactor (AHWR) and compact high-temperature reactor. AHWR uses thorium-based fuel and has passive natural circulation of water in parallel channels in its primary heat transport and decay heat transport systems. The two-phase boiling heat transport system exhibits oscillations and flow instabilities in varying scales in operating and accident conditions and they influence the thermal and reactivity margins in design. Experimental studies on natural circulation at low pressures assume importance in understanding the heat transport mechanisms and system behavior in case of a loss of coolant accident or a pressure boundary failure. The present research work highlights the multimodal oscillations, measured in a boiling natural circulation loop and experimental studies on the physics behind the components contributing to the flow oscillations. The boiling parallel channels of the facility are vertical with central heater simulating a typical vertical coolant channel with fuel in a nuclear reactor. The bubbles and slugs in the flow channel have been observed to form and coalesce around the length of the heater and flashing has been observed at a higher elevation. The sequence of events has been nearly similar, but appeared faster at increased heating rates. The overlapping temporal and spatial events in the channel lead to new nonlinearities and multimodal chaotic oscillations. Advanced signal analysis methods have been applied to the signals to study the time-evolving characteristics of flow oscillations at different channel heating rates.

Keywords Natural circulation · Multimodal oscillations
Empirical mode decomposition · Fast Fourier transform

J. Aparna (✉) · A. K. Nayak · A. R. Rao
Reactor Engineering Division, Bhabha Atomic Research Centre, Mumbai, India
e-mail: aparnaj@barc.gov.in

A. K. Nayak and B. R. Sehgal (eds.), *Thorium—Energy for the Future*,
https://doi.org/10.1007/978-981-13-2658-5_40

1 Introduction

Multimodal oscillations are observed in natural circulation flows, due to various concurrent temporal and spatial phenomena. The instabilities such as flow excursion or flow reversal, geysering instability, flashing induced instability, and type I low steam quality density wave oscillation (DWO) have been reported in the literature [1]. Flow reversal instability has been reported when the channel pressure drop decreases with increasing flow. Geysering occurs due to condensation effects and flashing due to reduced static pressures with elevation. Density wave oscillations appear in a vertical channel due to the buoyancy effects. A small change in quality results in significant change in void fractions and buoyancy head, resulting in flow oscillations identified as low steam quality DWOs. Flashing oscillations have been observed in vertical channels with tall risers in natural circulation loops [2]. Geysering, density wave and pressure drop oscillations have also been reported in vertical channels.

Two-phase flow instabilities and stability margins in advanced nuclear reactors have also been extensively reported in the literature; however, a detailed measurement and flow visualization of dynamic physical phenomena responsible for the multimodal flow oscillations have not been reported widely. In this present study, flow visualization of physical phenomena in an annular flow channel and the measured flow oscillations has been discussed. The vertical heated channel of the facility is one meter in length and the fluid around the central heater underwent a series of events like nucleation of bubbles, coalescing of bubbles and condensation, coalescing of bubbles to slugs, formation of slugs, flashing, steam separation and repetition of the sequence, after a gestation period. In the literature, geysering is described to occur due to condensing of vapor bubbles and flashing is described to be a sequence of vapor formation at an elevation where the fluid is at saturated conditions. The channel based on the observation of oscillatory patterns, appeared to undergo mostly geysering and flashing instabilities at low heating rates up to 800 W. At higher channel heating rates, the event appeared to occur in rapid succession.

2 Description of the Experiments

Flow in natural circulation system is induced by the differences in fluid density due to heating. The flow oscillation in a ten channel natural circulation experimental setup has been taken up to study the individual heated channel flow oscillations. The parallel channels in the facility are connected individually to a common steam drum and a header [3]. The individual vertical channels simulate coolant channels housing fuel, being simulated by a centrally located heater, as illustrated in Fig. 1. The heated channels are made of transparent borosilicate glass for flow visualization. The heater power in individual channels can be varied and the steam generated in the steam drum is cooled and the condensate is returned to the steam drum. The cooling section has a radiator with a constant speed fan and a cooler. The variables in the

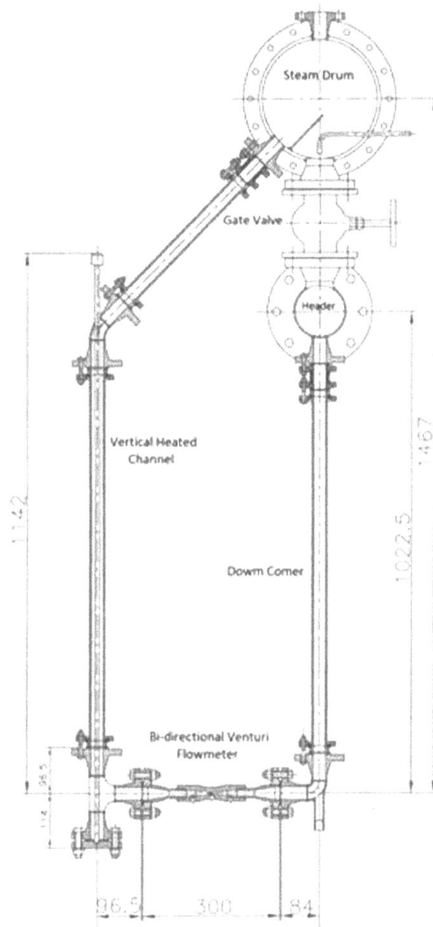

Fig. 1 Natural circulation loop schematic of the parallel channel experimental setup

loop are the steam drum level and heater power. The loop design is largely to aid flow visualization and hence has not been insulated except for the steam line to the cooling section. The loop is designed for atmospheric pressure and boiling conditions. Flow oscillations in the heated vertical channels have been measured at different heating rates and the corresponding high-speed photography of the phenomena in the annular flow channel around the heater simulating fuel has been described in the following sections.

The parallel channels are instrumented with flow, temperature, and power measurements. The inlet flow of individual parallel channels is measured by bidirectional venturi flow meters in the horizontal leg where there is no boiling and the flow is in a single phase. The differential pressure across the flow meter has been measured using fast sampling pressure transducers and data recorders for capturing the transients at different channel powers. Flow oscillations in the channels have also been photographed at a rate of 1000 frames per second [4].

2.1 Time-Evolving Nature of Flow Oscillations

The heaters in individual flow channels have been maintained at 400 W. Initially convection from the central heater to the fluid in the annulus could be seen and as the inlet temperature reached nearly 98 °C, intermittent flow oscillations appeared in the flow annulus. The sequence of events in a single oscillation cycle started with the formation of bubbles. Initially few bubbles are formed and are seen to move downstream and condense. Subsequently, the bubbles coalesce into small slug, followed by condensation and formation of longer slugs in succession. The slugs move downstream due to the buoyancy head, followed by successive steam packets appearing at a higher elevation in the heated channel. Slugs move downstream and boiling appears initially at the trailing end of the slugs. This is followed by formation of steam packets flashing in quick succession at a higher elevation and restoration of single-phase condition in the vertical channel. The cycle repeated at regular intervals depending on the gestation time required for heating up the channel fluid to saturation.

The flashing boundary appeared to vacillate continuously. The local vacillation during flashing, for visual observation, looks to be due to instantaneous rollover of colder fluid from above, as steam separates and moves to the steam drum. High-speed photographs have also been investigated for substantiating the observations. The frequency of oscillations in the channel due to slugs and flashing events was distinguishable by signal analysis methods. As the power in the channel is increased from 400 to 2000 W, the successive events appeared rapidly and the boiling boundary shifted to lower elevations. The vacillations were intense with a wider length and are of higher frequency.

The formation of vapor bubbles, slug formation events in adjacent channels were not simultaneous and had time delay between them. This observation has been measured as phase shift in flow measurements in adjacent channels. The boundary conditions in the steam drum and header have been the same for the channels and the phase shift among the channels was random with time.

The reported literature in flow oscillations has been mostly described using the channel temperatures and a qualitative description of the flow oscillations as bulk phenomena based on discrete spatial observation using void fraction measurements. The measurement of channel temperature and void fraction is discrete in nature and does not carry spatial information of a loop, as in the case of pressure measurement. Flow or pressure measurement in one location in a closed loop carries the space-shifted and time-shifted information due to mass/flow conservation within the loop. In this research work, an attempt has been made to explain the events individually to understand the multimodal oscillations observed in the channel flow in a quantitative manner.

The sequence of events resulting in flow oscillations at 800 W channel power has been described in the first section with high-speed photographs. The next section deals with the empirical mode decomposition of the flow signals into individual events.

2.1.1 Parallel Channel Flow Oscillations

Parallel channel oscillations have been widely reported in the literature for up to four channels. The flow oscillations in adjacent channels have been reported to occur in different points in time generating phase shift between them. In this experiment, the first six consecutive parallel channels had been maintained at individual channel powers of 400 W and flow oscillations have been recorded.

The middle channels have recorded higher peak-to-peak amplitude compared to the outer ones. In the duration of recorded flow, channel 1 oscillates first followed by channel three and two, as illustrated in Fig. 2. The time delay observed in the first set of oscillations between the three channels tends to reduce in the next set of oscillations. After two cycles, channel two and three are in phase. After six cycles, channels one and three are in phase. Subsequently, the flow oscillation amplitudes have been observed to decrease. Similarly, channels four and five and four and six have also been observed to be in phase, in between the oscillations, as illustrated in Fig. 3. As the power is increased, the randomness in the oscillations also increases with varying degrees of peak-to-peak amplitudes, oscillation periods, and a time delay between the oscillations in parallel channels. These oscillations in individual channels are communicated to the steam drum and bulk oscillations have been visualized through the borosilicate glass windows of the steam drum.

Visually, the flow oscillations had been observed to constitute a systematic set of events like formation of bubbles, slugs, and beat type to and fro oscillations in the channel at low power and as the channel power is increased the interactions between the events appeared to increase in intensity. At low powers, the channel had boiling at higher elevations, and at higher power, nearly 75% of the channel had slugs and void. The parallel channels displaying spatial and temporal oscillatory events at a single instant of time have been illustrated in Fig. 4.

The flow oscillations in a single channel at 800 W power have been taken up and the individual events constituting one oscillation have been discussed. The sequence of events appears to be nearly the same with increased channel power, except for the intensity and timescale.

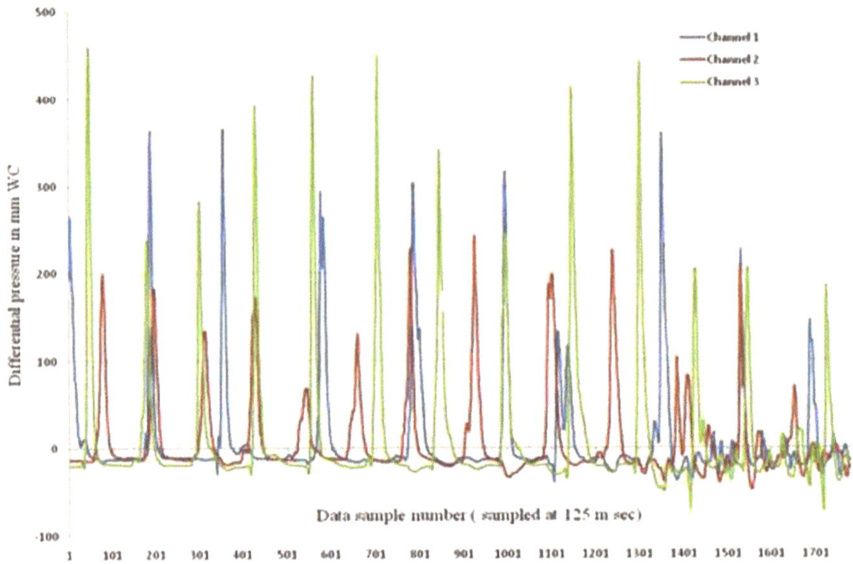

Fig. 2 Flow oscillations in channels 1, 2, and 3 at a channel power of 400 W

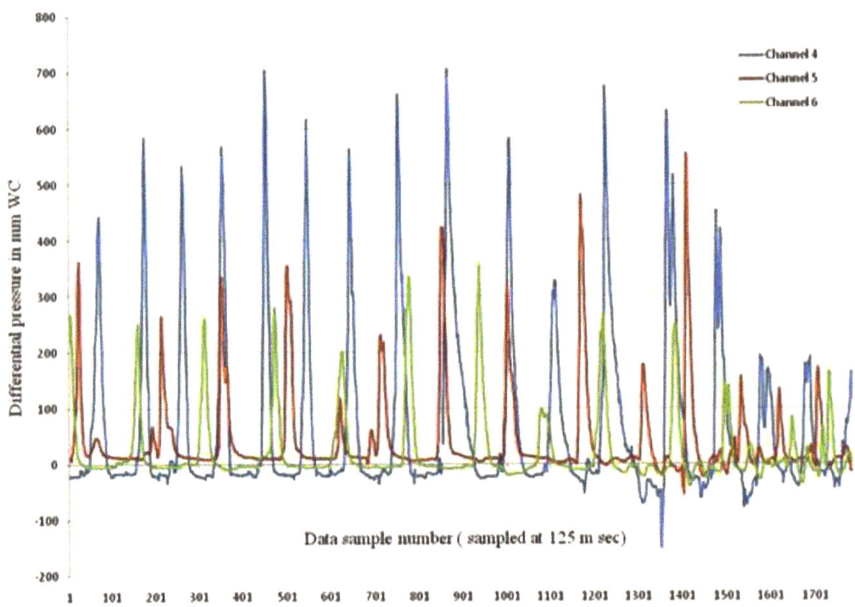

Fig. 3 Flow oscillations in channels 4, 5, and 6 at a channel power of 400 W

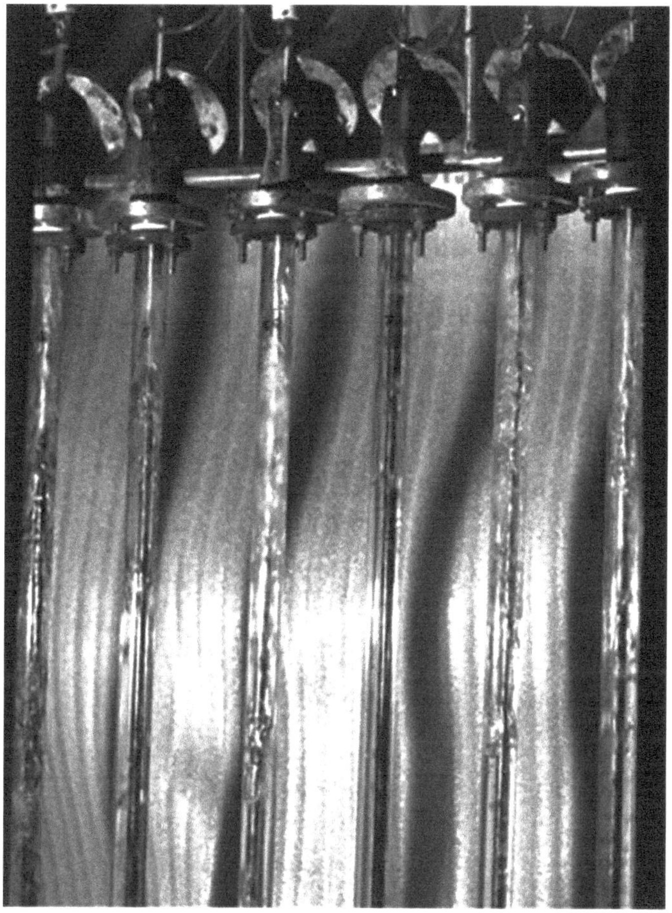

Fig. 4 Natural circulation in adjacent parallel channels with different local phenomena at the same instant

3 Analysis of Flow Oscillations Using Time-Frequency Techniques

The flow dynamics measured in the channel at any time has the instantaneous signals smudged by space-shifted signals due to flow. These flow oscillations observed in a natural circulation system are structurally different from the flow turbulence, which are conventionally low-pass filtered in industrial measurements for obtaining a stable reading. Hence to capture the constituent high-frequency components of the oscillations, one of the channel flows was measured using a Keller make differential pressure transducer across the Venturi flow meter, at a high frequency with a sampling rate of 40 Hz.

Since the signal had both temporal and spatial components, the FFT analysis methods were not suitable as the time information is lost in integration and would only give the frequency content. Hence to preserve the time and space information, time-frequency signal analysis methods such as wavelet transform and empirical mode decomposition methods have been applied to the signal. Empirical mode decomposition derives the scaling function from the signal itself and would give more meaningful insights into the oscillations, when compared to using standard basis wavelets in the wavelet transforms.

3.1 Empirical Mode Decomposition of Flow Oscillations

Empirical mode decomposition (EMD) is a signal analysis technique to decompose a time-evolving signal into its simple intrinsic modes of oscillations [5]. EMD is an adaptive technique emphasizing the local embedded characteristics of the signal and each segregated oscillatory mode is called an intrinsic mode function (IMF). The sifting of the signal for intrinsic modes is terminated, when the residue becomes a monotonic function or when the standard deviation criterion is achieved between the consecutive intrinsic mode functions. The measured flow signals corresponding to a single flow oscillation and its constituent five intrinsic mode functions obtained by EMD technique at a channel power of 800 W have been illustrated in Fig. 5. The measured flow signal at 800 W is a periodic signal with embedded temporal oscillations. The physical phenomena corresponding to each of them have been identified by a comparative analysis of empirical mode decomposed signals and the high-speed photographed events in the same time frame.

The first and second peaks in the signal and the corresponding inset figures illustrate the formation of slugs. The third inset image illustrates the flow reversal and the formation of vapor bubbles at the tail end of the reversing slug. The fourth inset indicates the growth and movement of the slug in the upward direction and formation of more vapor bubbles at its tail end. The fifth in sequence indicates the formation of another slug at the trailing end of the first slug and all these events had been discussed in text along with Fig. 14. The fifth and sixth inset figures illustrate the upward movement of the slug and the formation of vapor bubbles in the channel. The seventh inset illustrates the flashing and formation of steam which have been discussed in text along with illustrations in Fig. 15. As the slugs and steam move to the steam drum, buoyancy head decreases and the channel flow also decreases resulting in a decrease in the amplitude of the periodic oscillation. Due to flashing, quick succession of vapor bubbles is formed in the channel, and they appear as oscillatory modes in the falling edge of the measured signal. Further, the appearance of successive steam packets at different elevation creates a visually vacillating boiling boundary. The flashing and the upward movement of the boiling boundary are illustrated by the inset images 8–11, embossed on the falling edge of the measured signal in Fig. 5.

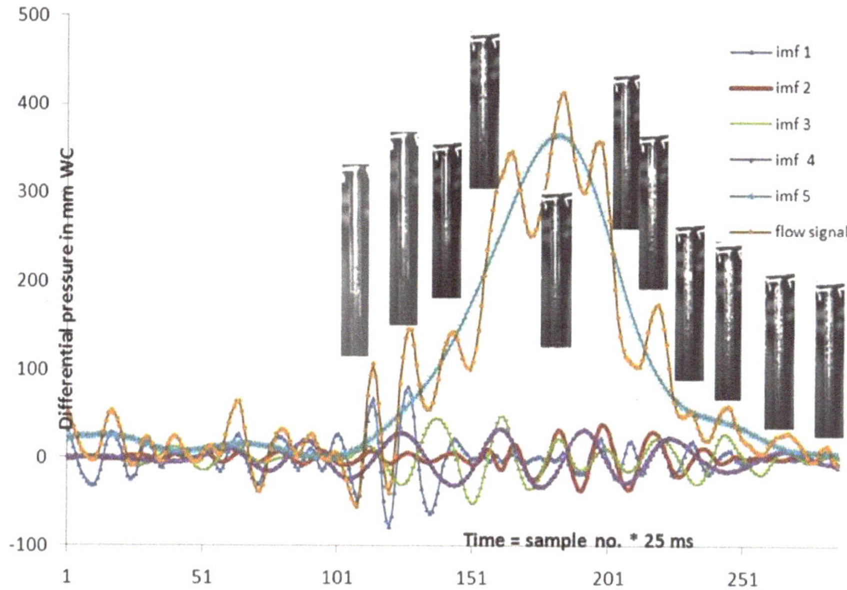

Fig. 5 Empirical mode decomposition of flow signals at a channel power of 800 W with corresponding inset photographs of local phenomena (inset images 1–11 viewed from left)

3.2 *Description of the Physical Phenomena Based on Intrinsic Mode Functions*

The intrinsic mode functions derived from the measured flow signal when analyzed with the high-speed video images appear to represent the physical phenomena in the channel. The first four intrinsic mode functions of the signal have the temporal and spatial oscillation-related information and the fifth intrinsic mode function indicates the resulting cumulative channel flow (Fig. 5). The initial oscillations in the first intrinsic mode function occur first in time trace and they correspond to the first slug in the channel and its growth. The next temporal event appears in the second intrinsic mode function, which corresponds to the formation and growth of the second slug, flow reversal, and formation of vapor bubbles in the trailing edge of the slugs. These two events are closely followed by a rate of change of flow in the fifth intrinsic mode function, which gives the cumulative flow in the channel. The substantiation of the phenomena is based on the observation that the change in intrinsic mode function IMF 5 trails the events in the other intrinsic mode functions.

The next event in time axis is the successive formation of vapor in flashing. There are five distinguishable successive vapor formations, indicated by five oscillations in the trailing end of the first intrinsic mode function. This observation has been supported by the high-speed photographs as indicated as insets in Fig. 6. The shifting

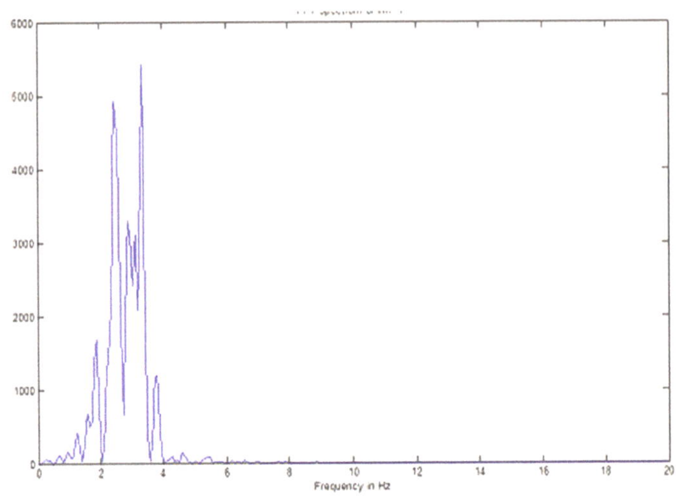

Fig. 6 Energy versus frequency (Hz) spectrum of intrinsic mode function 1

Fig. 7 Energy versus frequency (Hz) spectrum of intrinsic mode function 2

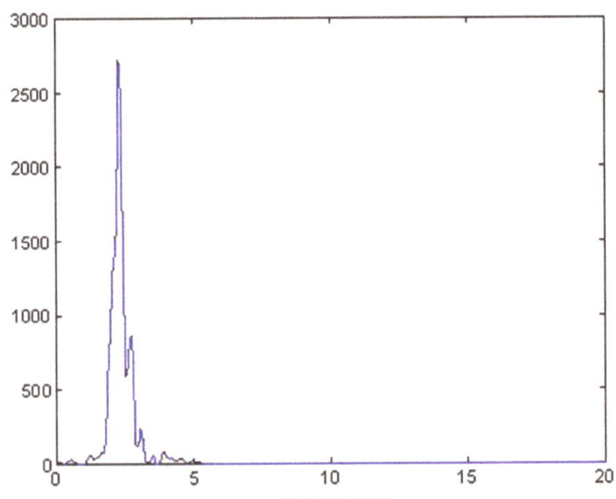

in successive flashing location appears to be due to the rollover of hot flashed fluid from below the cooler fluid layer (Fig. 7).

The third intrinsic mode function is a sub-harmonic of the first IMF and the fourth IMF is a sub-harmonic of the second intrinsic mode function. This observation is substantiated by their dilated replication in time domain and by their frequency spectrum. The frequency spectrum of the first intrinsic mode function as given in Fig. 6 has its components between 1.8 and 4 Hz with very close frequency peaks, indicating beat patterns in the flow signals. The beat patterns generated in the flow signals are due to the superposition of oscillations with very close frequencies. When the superposition is constructive, the amplitudes get added, and when in out of phase,

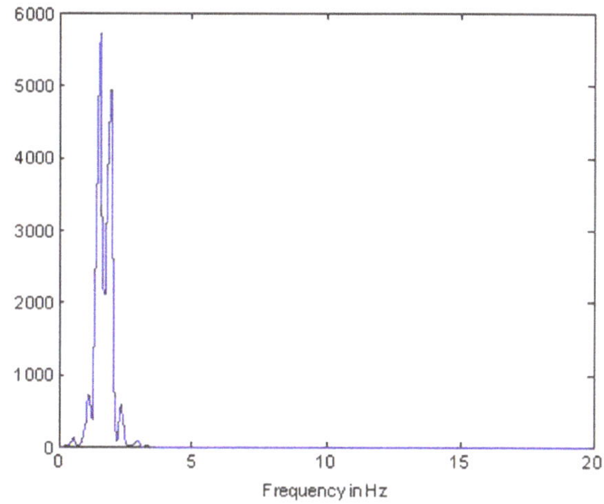

Fig. 8 Energy versus frequency spectrum of intrinsic mode function 3

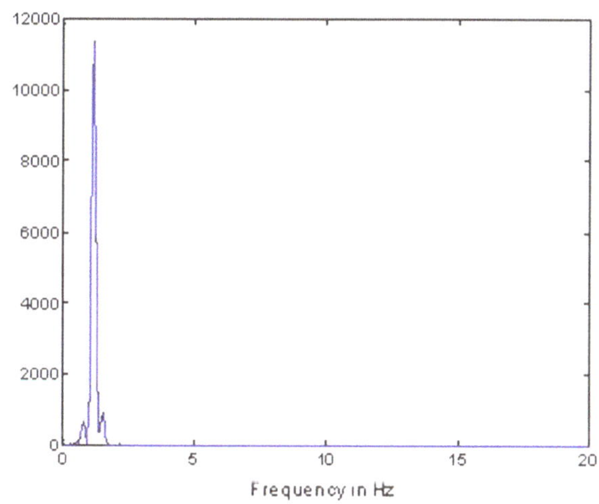

Fig. 9 Energy versus frequency (Hz) spectrum of intrinsic mode function 4

the amplitudes cancel each other, and the envelope of the beat cycle is at a differential frequency of the two oscillations. The frequency spectrum of the third intrinsic mode function as indicated in Fig. 8 has its components lay between 0.9 and 2 Hz, indicating the sub-harmonic content of IMF 1 in the spectrum. The frequency spectrum of IMF 3 indicates close peaks in frequency content similar to IMF 1, indicating beat cycles in the flow.

The frequency spectrum of IMF 2 as indicated in Fig. 7 has its maximum energy peak at 2.5 Hz, and intrinsic mode function 4, as indicated in Fig. 9, has its maximum energy peak at 1.25 Hz, indicating the sub-harmonic contents of IMF 2. The beat frequencies in the signals and the sub-harmonics in the intrinsic mode functions appear to be due to the transit time of the to and fro flow oscillations.

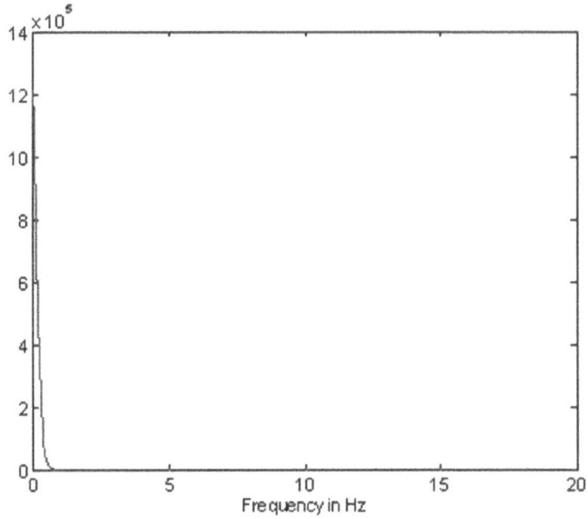

Fig. 10 Energy versus frequency spectrum of intrinsic mode function 5

The intrinsic mode function 5 is indicated in Fig. 10 and represents the cumulative flow in the channel due to the local physical events. The observation of beat cycle phenomena in the signals is substantiated by the high-speed photograph of the slugs, their growth, forward motion which is followed by flow reversal, as illustrated in Fig. 5. The natural circulation flow in the channel is due to the cumulative effects of all the intrinsic mode functions, which represent the spatial and temporal physical phenomena in the channel.

4 Description of the Sequence of Events in a Single Flow Oscillation Using High-Speed Photography

High-speed video imaging of the flow oscillations with 1000 frames per second has been carried out for generating insights into the visually observed and physically measured events in the flow annulus around the heater. Images representing observed phenomena that can be compared to the signal have only been illustrated for brevity. The inset images in Fig. 5, giving the intrinsic mode functions corresponding to slugs, flow reversal, and flashing, have been discussed with more images for giving a complete perception of events constituting a single flow oscillation.

Figure 11 illustrates the adjacent channels at 800 W channel power with different stages of observed flow sequences. The first inset image indicates the formation of bubbles and the second channel indicates the bubbles coalesced to a small slug. The phase difference in the measured flow oscillations of adjacent parallel channels could be explained by this observation. Figure 11 also illustrates the measured elevations in displacement of slugs and vacillations in the boiling boundary. The actual elevation

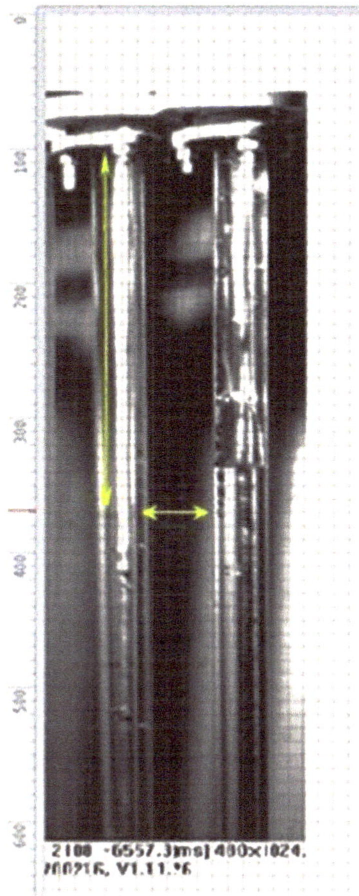

Fig. 11 Different stages of boiling phenomena observed in a single time instant in parallel channels

from the center of the Venturi flow meter (Fig. 1) is obtained by subtracting the indicated scale value in the photograph, from 1100 mm.

4.1 Channel in Single-Phase Condition

Before the inception of oscillations and during the gestation period between oscillations, the channel has convective currents, which could be seen visually in the channels.

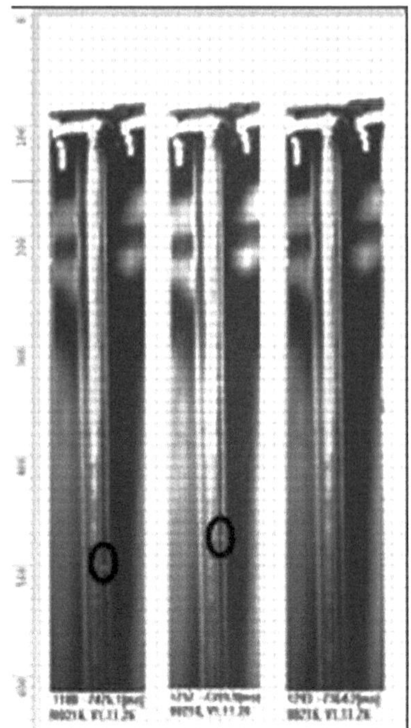

Fig. 12 Formation of bubbles from the heater surface and their upward motion

4.2 *Formation of Bubbles from the Heater Surface and Their Upward Motion*

In this phase, convection currents become prominent adjacent to the heater and bubbles start forming closer to the heater, move up and condense at a cooler region above the heater. Figure 12 comprises photographs of three consecutive events as insets. The first inset indicates the formation of vapor bubble at an elevation of 600 mm, the second inset figure indicates its vertical movement up to 625 mm, and the third figure indicates condensation and disappearance of the bubble. The formation and condensation of a bubble in the channel have taken 113 ms.

4.3 *Coalescing of Bubbles*

The condensation results in a lower channel pressure and accelerates the vapor formation with an increased enthalpy of the fluid due to heating. The bubbles formed in quick succession, move up, or condense depending on the fluid temperature [6]. Subsequently, the bubbles coalesce into slugs and from this event, the measured channel flow appears to increase, which could be observed by the upward movement of the slug as illustrated in the inset photographs of Fig. 13.

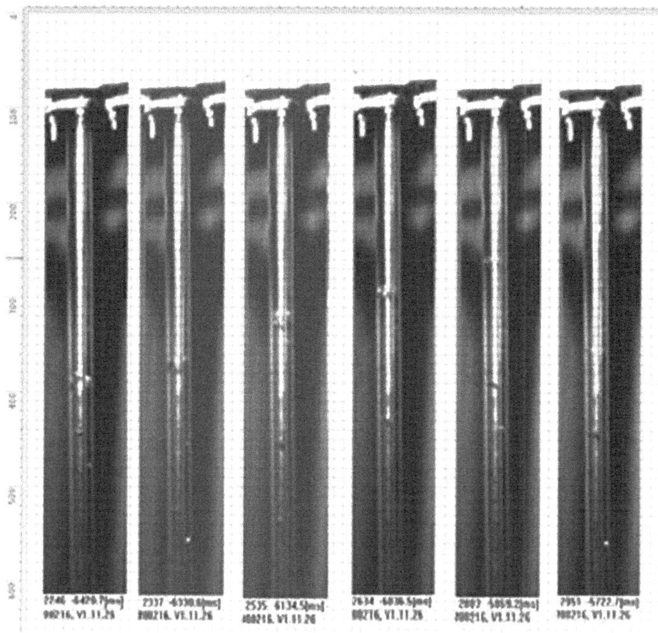

Fig. 13 Coalescing of bubbles into slug, their upward movement and condensation

4.4 Formation of Slug and Its Growth

The first inset in Fig. 13 indicates three bubbles before coalescing at an elevation of 700 mm, the second inset picture illustrates the coalesced slug at 720 mm, the third to fifth inset pictures illustrate the movement of slug from 770 to 840 mm elevation in the channel, and the sixth inset photograph indicates the condensation and disappearance of the slug. The coalescing and condensation of slug and a vertical travel of 140 mm occurred in a duration of 705 ms. Longer slugs form as the fluid enthalpy increases.

4.5 Flow Reversal and Formation of Bubbles at the Reversed Trailing Tail of the Slug

Figure 14 illustrates events associated with slugs. The slug formed in the channel grows in the upward direction, initially resulting in increased buoyancy head, leading to increased channel flow. The increased channel flow results in simultaneous forward motion of the growing slug. This forward surge of the slug is followed by a flow reversal which appears to be due to the inertial effects of the channel flow. The flow

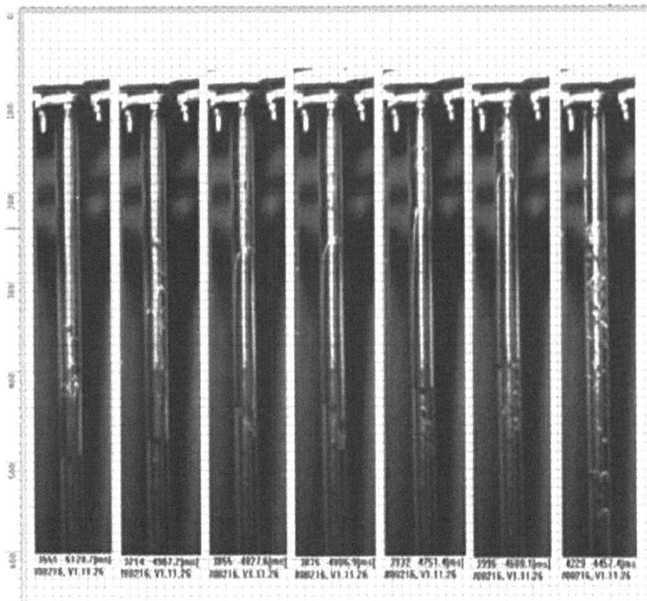

Fig. 14 Formation of slug, growth, flow reversal followed by formation of bubbles at its trailing tail

reversal results in a falling fluid film which is at a higher temperature, resulting in the formation of vapor bubbles, i.e., bulk boiling of low frequency at the tail end of the slug. Due to the increased buoyancy head by the formation of bubbles and the continuous heat input in the channel, the inertial flow reversal is followed by continued slug growth, due to reduced elevation head, accompanied by an upward motion due to the dynamics in the channel. The inset images 1–5 are illustrated in Fig. 5 besides the corresponding signal peaks.

4.6 Formation of Bubbles at Quick Succession at an Elevation of 900 mm During Flashing and Continuous Oscillation of Boiling Boundary

The first inset picture in Fig. 15 illustrates the flashing at an elevation of 900 mm where the single-phase fluid flashes to vapor in quick succession. The second inset picture illustrates the flashing at an elevation of 850 mm, and the third inset picture illustrates the flashing at an elevation of 950 mm. In the fourth inset picture, the flashing occurs at an elevation of 920 mm. Flashing at different but adjacent elevations appears as oscillating boiling boundary in the channel. Flashing at different elevations appears to be due to convection rolls [7]. During flashing, the hot vapor–water phase lies below

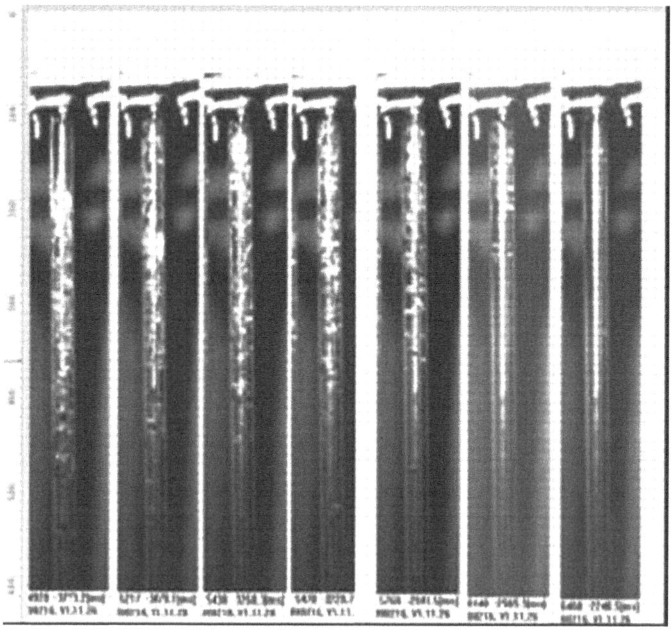

Fig. 15 Flashing and oscillating boiling boundary

the relatively cooler fluid, resulting in rollover of the fluid pocket followed by steam separation. This results in local temperature gradients and shifting the successive boiling elevations. This continuous chugging and oscillating boiling boundary has been visually observed and can also be seen in Fig. 15. The inset images 1–5 are illustrated as inset images 7–11 in Fig. 5, besides the corresponding signal peaks.

4.7 Upward Movement of the Boiling Boundary and Channel Returning to Single-Phase Condition

The lower boiling boundary moves up due to the flow in the channel from an elevation of 610 to 900 mm, i.e., by 290 mm in 1208 ms, as illustrated in Fig. 15. The seventh inset picture indicates the channel returning to single-phase condition.

5 Summary

Natural circulation flow oscillations in a vertical heated channel have been observed to be due to the agglomeration of distinct and interacting spatial and temporal physical phenomena. Experimental work has been carried out in vertical heated channels at 400 W and at 800 W power to substantiate the observation of multimodal oscillations and to further describe the constituent temporal and spatial physical events in the channels. Empirical mode decomposition technique has been used to split the flow signals into intrinsic mode functions and their spectral content has been identified by fast Fourier transforms. Visual observation, measured flow oscillations, intrinsic mode functions, and the corresponding high-speed photographs have been studied concurrently for identifying significant physical events in the channel such as the slug formation, slug growth, flow reversal, flashing, and oscillating boiling boundaries in the channel. The empirical mode decomposition of the flow signal, their frequency spectral analysis, and high-speed photography has given valuable insights into the time-evolving nature of the flow oscillations, contributing physical phenomena, sub-harmonics oscillations, and beat cycles in their flow pattern.

Acknowledgements The authors thank Dr. P. K. Vijayan and Shri I. V. Dulera, for their valuable support in carrying out the experimental work. The authors also acknowledge with thanks the support extended by the operations and maintenance personnel of the Reactor Engineering Division for conducting the experiments.

References

1. J.A. Boure, A.E. Bergles, L.S. Tong, Review of two-phase flow instability. Nucl. Eng. Des. **25**, 165 (1973)
2. A. Manera, Experimental and analytical investigations on flashing-induced instabilities in natural circulation two-phase systems-applications to the startup of boiling water reactors. Research thesis, 2003, ISBN: 90-73861-44-6
3. J. Aparna, B.V.S.G. Sharma, Design & development of experimental measurement and analysis methods for the natural circulation flow distribution studies in transparent parallel channels, in *Proceedings of NCFMFP 2006-33rd National and 3rd International Conference on Fluid Mechanics and Fluid Power*, 2006
4. J. Aparna, A.K. Nayak, I.V. Dulera, P.K. Vijayan, Experimental measurement and interpretation of natural circulation two-phase flow signals in a vertically heated channel during flashing induced oscillations, in *New Horizons in Nuclear Reactor Thermal Hydraulics and Safety*, Mumbai, India, 13–15 Jan 2014
5. H.L. Hu, J. Zhang, J. Dong, Z.Y. Luo, T.M. Xu, Identification of gas-solid two-phase flow regimes using Hilbert-Huang transform and neural network techniques. Instrum Sci. Technol. **39**(2), 198–210 (2011)
6. M. Furuya, F. Inada, T.H.J.J. van der Hagen, Flashing-induced density wave oscillations in a natural circulation BWR—mechanism of instability and stability map. Nucl. Eng. Des. **235**, 1557–1569 (2005)
7. A. Libchaber, J. Maurer, A Rayleigh Benard experiment: helium in a small box, in *Nonlinear Phenomena at Phase Transitions and Instabilities* (1983), pp. 259–286

Estimation of Large Early Release Frequency for Advanced Heavy Water Reactor

T. V. Santhosh, I. Thangamani, A. Srivastava, Gopika Vinod, Vishnu Verma and P. K. Vijayan

Abstract A framework for estimation of large early release frequency of a nuclear power plant is presented in this paper. The study aims at evaluating the reliability of containment safety features and analysing the physical processes such as timing and magnitude of radioactivity release. The release categories were formulated based on the amount of release of fission products (FPs) and their time of release. Reliability analysis of engineered safety features (ESFs) of containment was carried out using fault tree technique. The end states of containment event trees were formulated qualitatively based on the expert judgement, and the release categories were assigned accordingly. For interfacing with Level 1 probabilistic safety assessment (PSA), the plant damage categories were modified to include aspects such as time and extent of core damage and amount of metal–water reaction. Containment event trees were developed for these plant damage states, and the large early release frequency (LERF) was estimated for the advanced heavy water reactor (AHWR).

Keywords Probabilistic safety assessment · Large early release frequency
Loss of coolant accident · Emergency preparedness · Nuclear power plant

1 Introduction

Safety is of paramount importance in nuclear industry. Safety analysis is conducted, both deterministically and probabilistically, to establish and confirm the design basis of equipment important to safety. Probabilistic safety assessment presents a structured framework for quantitative evaluation of safety of nuclear power plants (NPPs). The assessment is conducted based on internationally accepted guidelines such as IAEA SSG-4 [1]. Level 2 PSA examines severe accidents through a combination of probabilistic and deterministic approaches to determine the release of radionuclides from containment including physical processes involved in the loss of structural

T. V. Santhosh (✉) · I. Thangamani · A. Srivastava · G. Vinod · V. Verma · P. K. Vijayan
Reactor Safety Division, Bhabha Atomic Research Centre, Trombay, Mumbai 400085, India
e-mail: santutv@barc.gov.in

A. K. Nayak and B. R. Sehgal (eds.), *Thorium—Energy for the Future*,
https://doi.org/10.1007/978-981-13-2658-5_41

integrity of the core. The probabilistic part of the study addresses the reliability evaluation of containment safety features, whereas deterministic part concentrates on the analysis of physical processes of an accident (i.e. timing and magnitude of radioactivity release), and evaluation of containment response. The important tasks involved are: (i) categorization and grouping of accident sequences into a number of plant damage states (PDSs), (ii) construction of containment event trees (CETs), (iii) development of CET top event and its quantification, (iv) assignment of release categories and (v) estimation of LERF for emergency preparedness.

2 Advanced Heavy Water Reactor

Advanced heavy water reactor is a 920 MWth vertical pressure tube type boiling light water cooled and heavy water moderated reactor [2]. The reactor core is housed in a low-pressure reactor vessel called Calandria which contains heavy water that acts as both moderator and reflector. This advanced reactor incorporates various passive features for decay heat removal, containment cooling and isolation. One of the salient design features of AHWR is the way the heat removal is achieved through natural circulation of primary coolant with no any primary coolant pumps. Natural circulation is achieved in systems such as main heat transport (MHT) system, containment, emergency core cooling system (ECCS) and isolation condenser (IC) circuit.

3 Interface with Level 1 PSA

Level 1 PSA culminates with the failure sequences that may lead to a situation of type core damage. Level 2 PSA starts with these failure sequences and models their accident progression. Such failure sequences owing to their large number with similar behaviour are typically grouped and defined as PDSs. From Level 1 PSA of AHWR, PDS categories are as shown in Table 1.

4 Development of Containment Event Trees

The sequence of accident progression from core damage to the containment, and subsequent release of radioactivity under each PDS needs to be modelled in Level 2 PSA. Typically, event tree approach is employed and the resulting event trees are referred as containment event trees. These event tree models take into the account of all physical and chemical processes that may occur upon severe accident.

Table 1 Plant damage states derived from Level 1 PSA of AHWR

Type	Fuel damage category (FDC)	Description	Event sequence
Severe accident beyond design basis	FDC3A	Temperature more than 1200 °C	Large break loss of coolant accident (LBLOCA) along with failure of shutdown system (SDS) 1&2, loss of regulation accident (LORA) (all control rod withdrawal failure) with failure of SDS 1&2
	FDC3B	Temperature more than 1200 °C	Transients with failure of SDS 1&2 and failure of passive poison injection system (PPIS)
	FDC3C	Temperature more than 1200 °C	LBLOCA with failure of ECCS and recirculation and failure of moderator cooling
	FDC3D	Temperature more than 1200 °C	Small break loss of coolant accident (SBLOCA) with liquid poison injection system (LPIS) not available and recirculation and failure of moderator cooling
Severe accidents within design basis	FDC2A	Temperature within 800–1200 °C	LBLOCA with failure of ECCS and recirculation
	FDC2B	Temperature within 800–1200 °C	Transients with failure of decay heat removal system
	FDC1A	Temperature less than 800 °C	SBLOCA with failure of ECCS and recirculation LBLOCA with failure of recirculation
Single-channel event	FDC2C	Temperature more than 1200 °C	Channel flow blockage with failure of SDS 1&2, Pressure tube-calandria tube (PT-CT) rupture with failure of SDS 1&2, Stagnation channel break with failure of SDS 1&2
Structural integrity failure	FDC4		End shield cooling failure with failure of SDS 1&2
Containment bypass	FDC5A	Containment bypass with mitigation system not successful	Main steam line break (MSLB) with main steam isolation valve (MSIV) failure

Table 2 Reliability analysis of containment ESFs

S. No.	System	Unavailability
1	Primary containment filtration and pump back system (PCFPB)	1.86e-2
2	Secondary containment filtration recirculation and purge system (SCFRP)	3.1e-3
3	Primary containment controlled discharge (PCCD)	7.87e-3
4	Calandria vault cooling system	1.6e-3

Table 3 Minimal cutset of PCFPB

Number	Order	Cutset	Unavailability
1	1	DA-FAIL-V1-1	2.00E-03
2	1	DA-FAIL-V2-1	2.00E-03
3	1	DA-FAIL-V1-2	2.00E-03
4	1	DA-FAIL-V2-2	2.00E-03
5	1	CCF-DAMPERS	2.00E-04
6	1	PCFB-CCF-FAN	3.30E-05
7	1	DEMISTER-1	2.50E-05
8	1	DEMISTER-2	2.50E-05

5 Reliability Analysis of Containment Engineered Safety Features

Reliability analysis of engineered safety features of containment has been carried out using fault tree approach for systems such as secondary containment filtration recirculation and purge system (SCFRP), primary containment filtration and pump back system (PCFPB) and primary containment controlled discharge (PCCD). The summary of reliability analysis of containment systems is shown in Table 2. A typical fault tree for PCFPB system is shown in Fig. 1, and associated minimal cutsets are shown in Table 3.

6 Estimation of Large Early Release Frequency

Large early release frequency as defined in [3] is the frequency of those accidents involving rapid and unmitigated release of airborne fission products from containment into the environment before the implementation of off-site emergency response such that there is potential for early health effects. LERF is essentially the expected number of large early releases per unit of time [3]. Several other definitions are available in the international scenario for LERF depending upon the type of core,

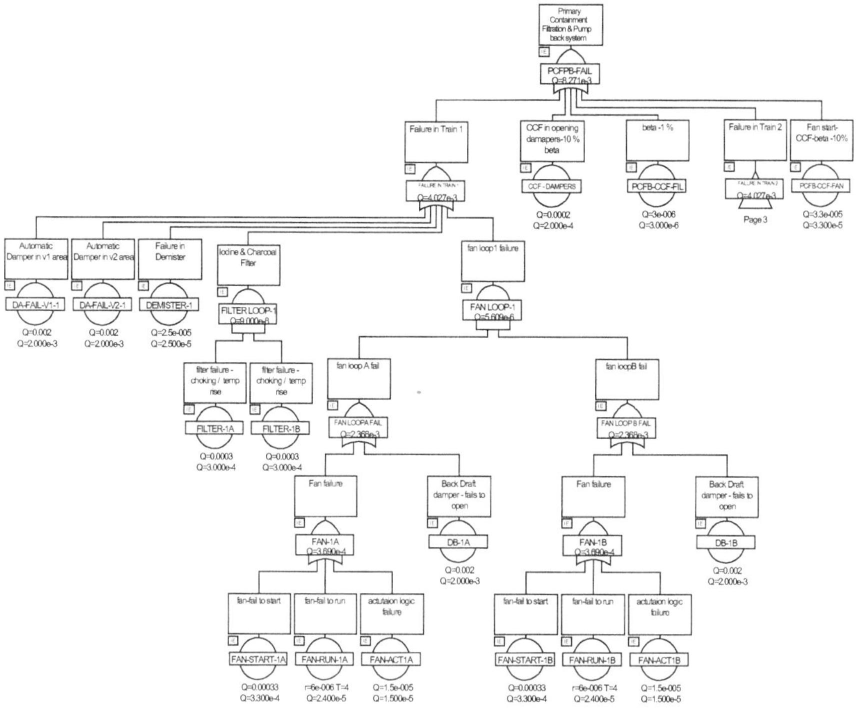

Fig. 1 Fault tree of PCFPB

containment and emergency response system. A few widely followed definitions are as below:

- The sum of frequencies of all event sequences that can lead to a release to the environment of more than 10^{14} Bq of caesium-137 is less than 1e-6 per reactor year. A greater release may require long-term relocation of the local population [4].
- LERF consists of the frequency of release classes associated with the early failure and bypass containment failure modes which have release fractions of the volatile/semi-volatile fission products (e.g. iodine, caesium, tellurium) equal to or greater than about 2.5–3% [5].
- A large release may be defined as a release corresponding in terms of activity to 5–10% of the core inventory of radioiodines. The activity associated with iodine corresponding to 10% of the core inventory of a large power reactor is an assumed threshold for prompt fatalities. This is especially the case when dealing with PDSs that could potentially involve large and early releases of radionuclides to the environment (for a large power reactor, releases of volatile species of iodine, and of caesium in excess of 5% of the core inventory, are typically considered significant for off-site consequences) [6].

Table 4 Release categories based on inventory of activity release and time duration

Sl. No.	Release category	Description	Quantity	Time
1	LER	Large early release	More than 10^{14} Bq of caesium-137 or 10^{15} Bq of iodine-131	Within 10 h
2	LIR	Large intermediate release		Between 10–24 h
3	LLR	Large late release		Beyond 24 h
4	SER	Small early release	From 10^{12} Bq of caesium-137 or 10^{13} Bq of iodine-131 to 10^{14} Bq of caesium-137 or 10^{15} Bq of iodine-131	Within 10 h
5	SIR	Small intermediate release		Between 10–24 h
6	SLR	Small late release		Beyond 24 h
7	VSER	Very small early release	Less than 10^{12} Bq of caesium-137 or 10^{13} Bq of iodine-131	Within 10 h
8	VSIR	Very small intermediate release		Between 10–24 h
9	VSLR	Very small late release		Beyond 24 h

In order for the quantification of LERF, the release categories were formulated based on the amount of release of fission products (e.g. large, small and very small) and time of release (e.g. early, intermediate and late) [1, 3, 4]. Table 4 summarizes the release categories adopted in this study. The term **Large** in LERF defines an amount of release more than 10^{14} Bq of caesium-137 or 10^{15} Bq of iodine-131, **Small** defines an amount of release within 10^{12} Bq of caesium-137 or 10^{13} Bq of iodine-131 to 10^{14} Bq of caesium-137 or 10^{15} Bq of iodine-131, and **Very small** defines an amount of release less than 10^{12} Bq of caesium-137 or 10^{13} Bq of iodine-131. The terms **Early** in LERF defines release within 10 h, **Intermediate** defines the release between 10 and 24 h and L**ate** defines the release beyond 24 h. The end states of containment event trees were decided qualitatively based on the expert judgement, and the release categories were assigned accordingly. For interfacing with Level 1 PSA, the plant damage categories were modified to include aspects such as time and extent of core damage and amount of metal–water reaction to analyse the containment behaviour with respect to energy management, radiological management, hydrogen management and corium management. Considering these aspects, various plant damage categories were introduced such as large break loss of coolant accident

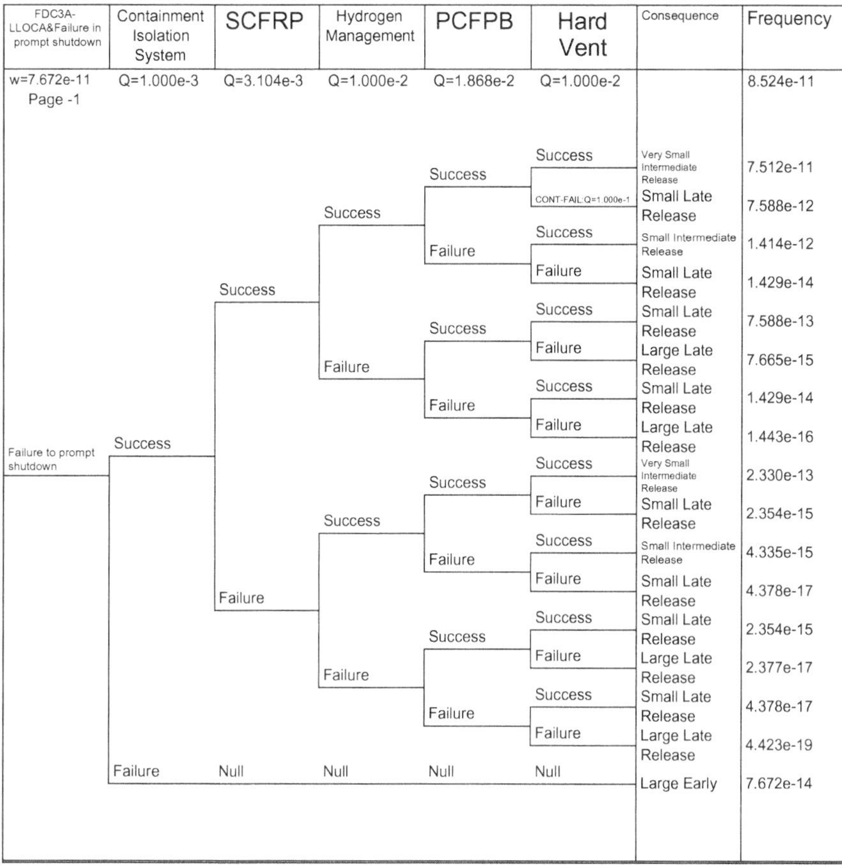

Fig. 2 CET for LB LOCA with the failure of prompt shutdown

(LB LOCA) with failure of prompt shutdown, transients with failure of prompt shutdown, LB LOCA with unavailability of all heat sinks, station blackout with loss of heat sinks, single-channel events and containment bypass events. Containment event trees were developed for these plant damage states. The containment event tree for the LB LOCA with the failure of prompt shutdown is shown in Fig. 2.

The consequence of various accident sequences has been identified through thermal hydraulic analysis and expert opinion. As a case study, containment event tree is discussed for plant damage state resulting LB LOCA with failure of prompt shutdown. For this scenario, containment isolation is demanded. Subsequently, containment engineered safety features such as PCFPB, SCFRP, hard vent and hydrogen mitigation measures will be called for. Since PCFPB and hard vent are assumed to come within 24 h, a very small intermediate release is assumed for successful operation of all systems. If there is a hard vent failure, it can lead to containment over-pressurization resulting in leakage. PCFPB reduces FP concentration and also provides circulation leading to uniform mixing. Since hydrogen management is available, no credit to hydrogen deflagration/detonation is given. Release of FPs to environment is by way of leakage. In case of unavailability of passive autocatalytic recombiner system (PARS), PCFPB and Hard vent, containment integrity may be threatened. Therefore, large but late release is expected. Role of SCFRP is limited; hence, the consequence is similar even if it fails.

Preliminary assessment of LERF, based on release categorization from qualitative expert judgement, has been carried out, and the estimated LERF is found to be 1e-13/year. The dominant contributors are: (a) LB LOCA with the failure of prompt shutdown coupled with containment isolation failure, (b) containment bypass event from main steam line break outside containment coupled with failure of main steam isolation valves and (c) LB LOCA with complete failure of emergency core cooling system (ECCS) and loss of moderator cooling.

7 Conclusions

Level 2 PSA is an integral part of safety analysis carried out for all types of NPPs worldwide. The results of Level 2 PSA help plant operators and regulatory authorities in decision making with regard to severe accident management. It also brings focus towards establishing the guidelines for prevention of core damage and early and late containment failure.

References

1. IAEA Safety Standards Series No. SSG-4, Development and Application of Level 2 Probabilistic Safety Assessment for Nuclear Power Plants, (International Atomic Energy Agency, Vienna 2010)
2. AHWR-LEU-Preliminary Safety Analysis Report, AHWR-LEU/PSAR/USI/01591, Section–4, Reactor Systems, Mar 2013
3. ASME/ANS RA-Sa–2009, Addenda to ASME/ANS RA-S–2008, Standard for Level 1/Large Early Release Frequency Probabilistic Risk Assessment for Nuclear Power Plant Applications (2009)
4. CNSI RD–337, Design of New Nuclear Power Plants (2008)
5. NUREG/CR-6595, Rev. 1, BNL-NUREG-52539, Rev. 1, An Approach for Estimating the Frequencies of Various Containment Failure Modes and Bypass Events (2004)
6. IAEA Safety Standards Series No. 50-P-8, Procedures for Conducting Probabilities Safety Assessment of Nuclear Power Plants (Level 2) (1995)

Transient Following Partial Loss of Feed Water for Thorium-based Natural Circulation Reactor

A. D. Contractor, A. Srivastava, Rajesh Kumar and B. Chatterjee

Abstract A new reactor system under design study is a vertical pressure tube-type boiling light water cooled and heavy water moderated. Main heat transport system consists of reactor core, core inlet and outlet extensions, inlet feeders, tailpipes, steam drums, downcomer, and inlet header. It incorporates several advanced passive safety features, e.g., heat removal through natural circulation both during normal operation and reactor shutdown without the use of any coolant pump. Safety remains the paramount consideration in the design, while plant availability is the next. Several reactor trip set points are provided in design to tackle the disturbances in operation, and many safety systems are provided to mitigate different accidental conditions. A major objective of the reactor design is to provide a capability to withstand a wide range of postulated events (PIEs) without exceeding the acceptance limits, thereby maintaining fuel integrity. Partial loss of feed water is one of such PIEs. Following article highlights the modeling and overall transient behavior and its safety implication.

Keywords Partial loss of feed · BFP trip · Natural circulation reactor · RELAP5

Abbreviations

BFP	Boiler feed pump
BWR	Boiling water reactor
CIES	Combined isolation and emergency safety
ECC	Emergency core cooling
GDWP	Gravity-driven water pool
IC	Isolation condenser
MHT	Main heat transport

A. D. Contractor (✉) · A. Srivastava · R. Kumar · B. Chatterjee
Reactor Safety Division, Bhabha Atomic Research Centre,
Trombay 400085, Mumbai, India
e-mail: ankurdc@barc.gov.in

A. K. Nayak and B. R. Sehgal (eds.), *Thorium—Energy for the Future*,
https://doi.org/10.1007/978-981-13-2658-5_42

MSIV Main steam isolating valve
PIEs Postulated initiating events

1 Introduction

India is developing a 300 MW(e) pressure tube-type BWR as described by Sinha and Kakodkar [1] which in its equilibrium condition will derive two-thirds of its energy from thorium-based fuel with plutonium as an external fissile feed and achieve self-sustaining characteristics in ^{233}U. This reactor incorporates various passive design features that would meet all objectives of a fourth-generation system. One of the important passive design features of this reactor is that the heat removal is achieved through natural circulation of primary coolant at all allowed power levels with no primary coolant pumps. Apart from this passive design feature, various passive safety systems include isolation condenser (IC) system for decay heat removal in case of unavailability of main steam condenser, emergency core cooling (includes both high pressure and low pressure ECC) system, passive containment cooling system, passive containment isolation, passive poison injection system, and passive automatic depressurization system. An inherent safe design feature of this reactor is negative void reactivity coefficient that helps in lowering the power generation with the increase in core void. A very high value of negative void reactivity feedback is not desired since in the case of decrease in void, it may result in power surge. A major objective of this reactor design has been to provide a capability to withstand a wide range of postulated events (PIEs) without exceeding the acceptance limits, thereby maintaining fuel integrity. One such PIE analyzed is partial loss of feed water.

Partial loss of feed may occur when one of the two working boiler feed pumps (BFPs) trips. However, standby pump should come within 14 s following the BFP trip and the reactor comes back to nominal condition. The course of the transient may differ if the standby pump is not initiated. This leads to gradual loss of inventory and reactor trip. The modeling and overall transient behavior and its safety implication are as follows.

2 Reactor Description and Modeling

As illustrated in Fig. 1, the MHT system of the proposed reactor consists of reactor core, tailpipes, steam drums, downcomer, inlet header, and inlet feeders. The reactor core consists of 452 pressure tubes. In each pressure tube, a 54-element single fuel bundle is being housed. A displacer rod is placed at the center of the bundle, and the fuel elements are arranged in concentric circles. In the first circle, second circle, and third circle, 12, 18, and 24 fuel elements are arranged, respectively. The coolant leaves the pressure tubes with an average exit steam quality of 18.7% and reaches four steam drums through 452 core top extensions and 452 tailpipes. The steam drum

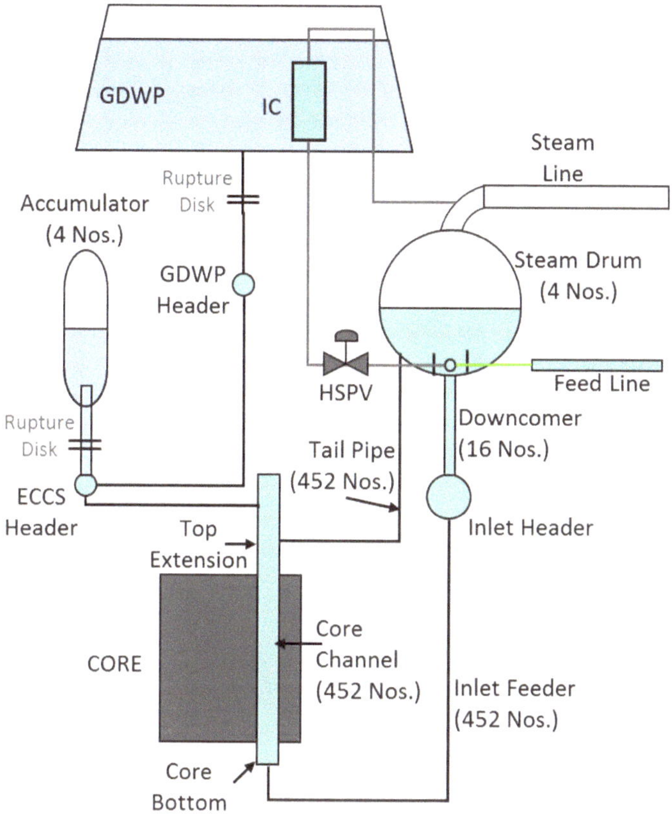

Fig. 1 Schematic of MHT system

provides a large interface area, which provides the natural separation without using mechanical separators. At the central location (between two plates) of the steam drum bottom, feed water enters which leads to a subcooled condition after mixing with the saturated water. This subcooled water leaves the steam drums through 16 downcomers to a single ring inlet header. From the inlet header, the coolant enters the 452 pressure tubes through 452 inlet feeders and 452 core bottom extensions.

RELAP5/MOD3.2 code, as described by Fletcher and Schultz [2], has been used worldwide for analyzing thermal-hydraulic behavior of nuclear reactor systems. It is based on two-fluid simulation using a non-equilibrium, non-homogeneous, six-equation representation. Constitutive models represent interphase drag, the various flow regimes in vertical and horizontal flow, wall friction, and interphase mass transfer. The code has also the capability to simulate the presence of slabs of material adjacent to the fluid. Thus, energy transfer to and from stationary slabs of material can be simulated. Control systems and component models permit simulation of equipment controller, balance of plant equipment (e.g., turbines, pumps, and condensers), and lumped-node representation of various processes. Further, code has

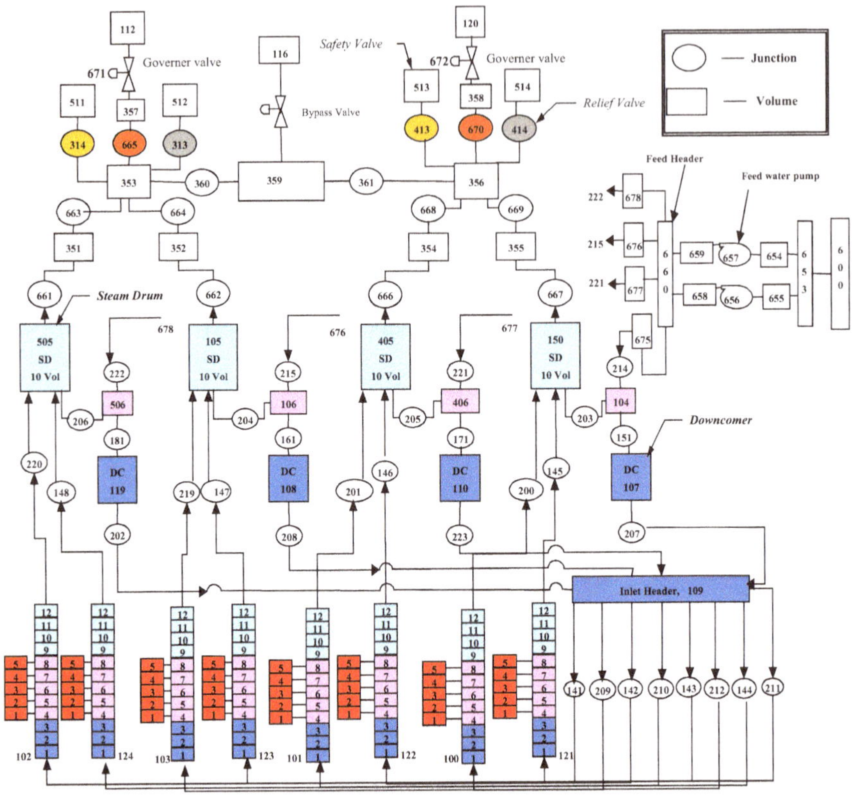

Fig. 2 MHT system nodalization

point reactor kinetics model. This model is good enough for wide range of scenarios where asymmetric situation does not happen.

The nodalization scheme of MHT system for natural circulation reactor is shown in Fig. 2. The core channels have been modeled by eight equivalent paths, representing four average (100, 101, 102, and 103) and four hot channels (121, 122, 123, and 124). Volumes 4–8 of each component represent core channels. Volumes 1 and 2 of these parallel paths represent vertical and horizontal portion of inlet feeders, while core bottom extension is represented by volume 3 of these pipe components. Volume 9 of these components simulates core top extension, while volumes 10, 11, and 12 represent tailpipes. There are four steam drums. Each steam drum is connected to 113 channels. These channels connected to each drum have been modeled by one hot channel and 112 average channels. Steam drum (3.75 m dia, 11 m long) has been simulated by saturated (150, 405, 105, 505) and subcooled volumes (104, 406, 106, 506). Saturated volumes of steam drums (components 505, 105, 405, and 150) are modeled separately using ten control volumes in each drum. The bottom part of the drum (subcooled portion in between two baffles is represented by subcooled

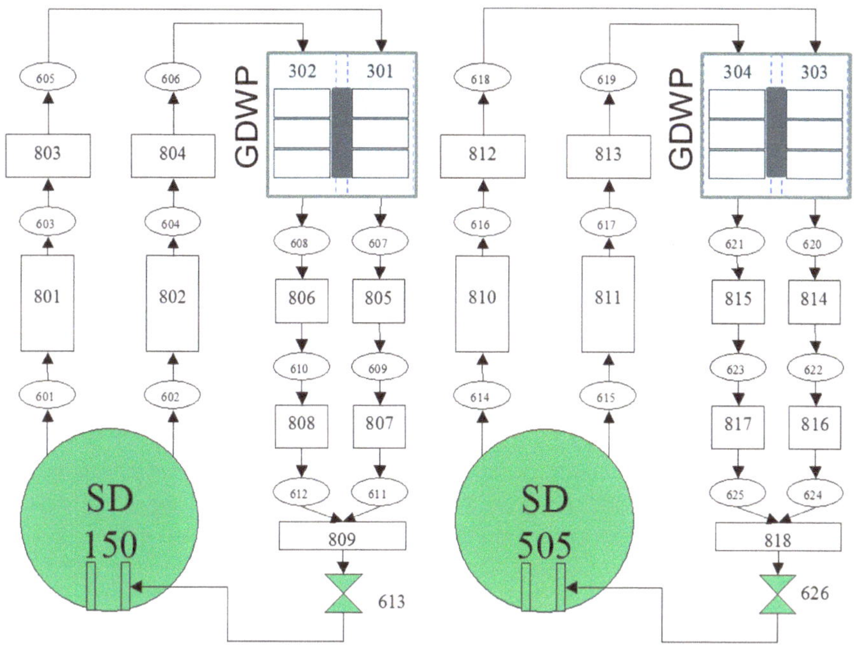

Fig. 3 Nodalization scheme for IC

volume components 406, 106, 506, and 104. The feed water along with recirculation flow enters into the subcooled part of the drums. Flow from this portion of steam drums enters the downcomers. Feed water conditions are simulated using component 600. Suction header of feed pumps is modeled by component 653. Two feed pumps are modeled by components 656 and 657. Component 660 models discharge header. Individual feed lines are modeled by components 675, 676, 677, and 678. Four downcomer lines from each steam drum are modeled by a single equivalent downcomer. There are four equivalent downcomer lines (107, 108, 110, and 119). Inlet header is modeled by component 109.

Steam lines from each steam drum have been modeled by volumes 351, 352, 354, and 355. Bypass valve is modeled by component 673. Bypass valve has been put in the pressure equalizing line (component 359) joining each banks' steam lines (353 and 356). CIES valves have been simulated by component 665 and 670. Governor valves have been modeled by component 671 and 672. Appropriate pressure controller scheme has been implemented on governor and bypass valve to maintain system pressure. Relief valves and safety valves mounted on steam lines are modeled by components 313, 314, 413, and 414. Reactivity feedbacks from different sources like fuel Doppler, coolant temperature, and void have been modeled with the reactor point kinetics of RELAP5. Three-element level controller scheme to maintain the drum water level has been simulated.

Nodalization scheme of IC system connected to two drums volume 150 and 505 is shown in Fig. 3. Each drum is connected to two ICs (301 and 302 with drum 150; 303 and 304 with drum 505), and outlet from IC is connected (613 and 626) to the corresponding subcooled volume of steam drums. Each IC representing 70 tubes is divided into three axial volumes. Sink of ICs is simulated by GDWP (709). Inlet cylindrical header of each IC is simulated with one component (803, 804, 812, and 813). Outlet cylindrical header of each IC is simulated with one component (805, 806, 814, and 815).

3 Results and Discussion

Transient has been initiated by tripping off one of the two BFPs after achieving steady state at full power condition. Immediately following one BFP trip, feed header pressure drops but this does not generate any reactor trip signal as may be the case if both the BFPs are tripped. Due to decrease in the feed header pressure, feed flow as shown in Fig. 4 through other working BFP is increased after initial drop; however, feed flow going to each steam drum is reduced by about 6.3 kg/s.

To analyze the reactor system behavior, it has been considered that the standby pump fails to start and the situation is prevailed for the rest of the transient. Net reduction in the feed flow causes heating up of the MHT system and thus negative reactivity feedback of void. Reactor power as shown in Fig. 5 reduces slightly. Though reactor power reduces, due to heating up of the system, steam flow rate does not reduce.

Net result is reduction in the MHT inventory. Steam drum level reduces gradually as shown in Fig. 6. Moment comes when Main Steam Isolation Valve (MSIV) closure is initiated on account of low steam drum level of 1.9 m. Subsequently, reactor also

Fig. 4 Feed water flow

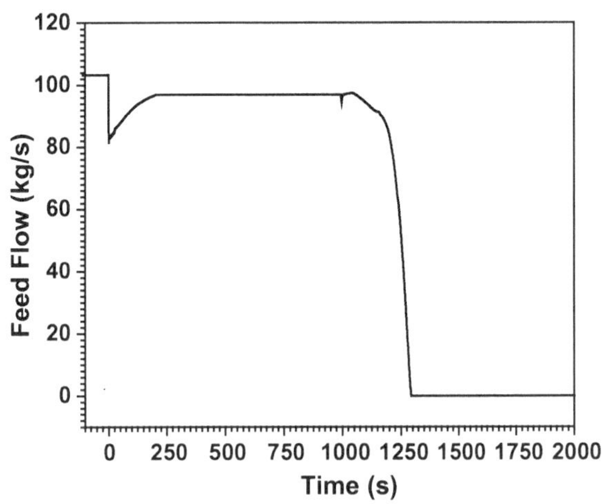

Fig. 5 Reactor power

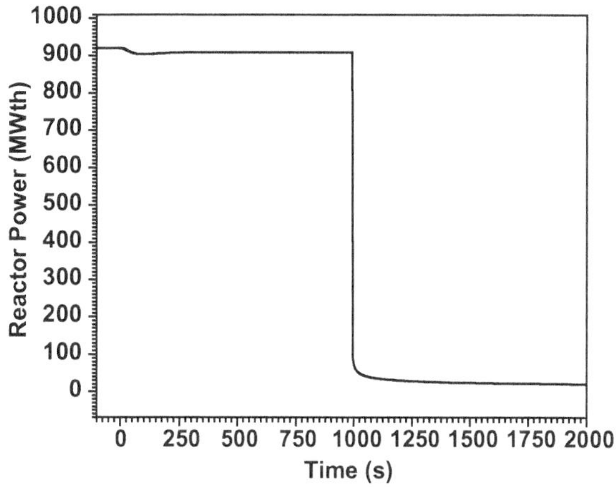

Fig. 6 Steam drum level

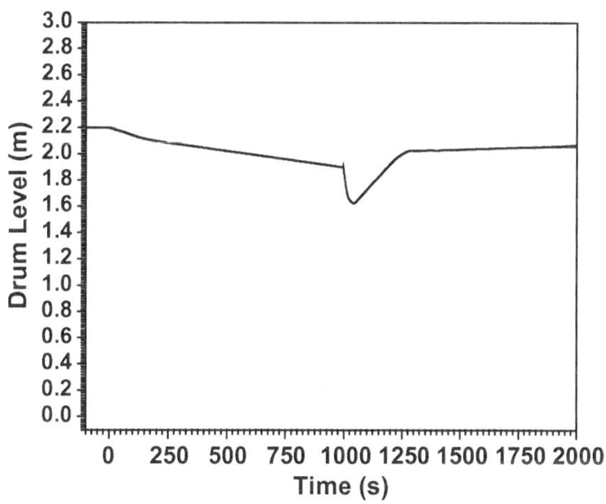

trips on 15% closure of MSIV and power comes down to decay power level. Closure of MSIV arrests the further loss of inventory.

Following MSIV closure, steam drum pressure as shown in Fig. 7 keeps on increasing and leads to opening of IC. Decay heat is then removed through IC, and system pressure is stabilized around 76.5 bar. Throughout the transient, fuel clad temperature as shown in Fig. 8 did not change and remained within safety limit.

In the course of the transient, other reactor trip signals observed are low steam drum level and high system pressure.

Fig. 7 Steam drum pressure

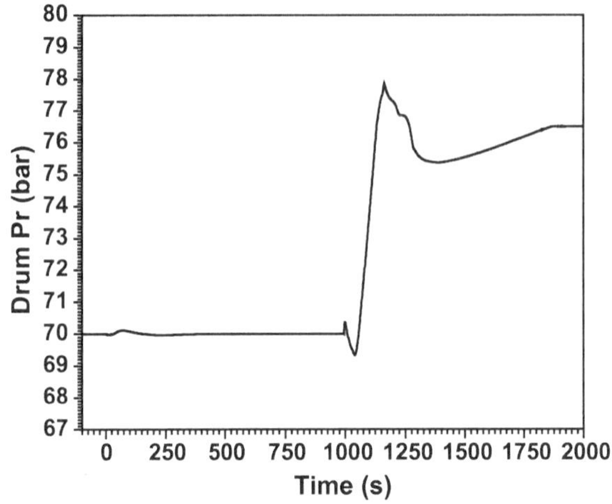

Fig. 8 Fuel clad temperature

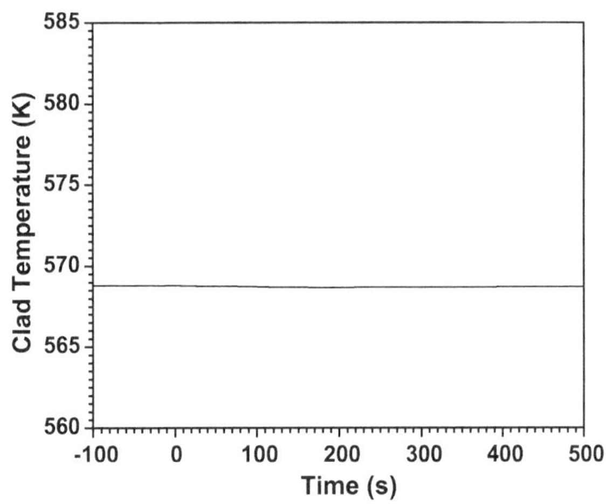

4 Conclusions

One BFP trip transient without standby pump initiation has been analyzed. This partial loss of feed flow causes MHT system heating up and gradual loss of inventory. Later, this loss of inventory is arrested by closure of MSIV. Closure of MSIV not only stops the loss of inventory but also trips the reactor and facilitates the decay heat removal through isolation condenser path. All other thermal-hydraulic parameters remain within safety limits.

References

1. R.K. Sinha, A. Kakodkar, Design and development of the AHWR-the Indian Thorium fuelled innovative nuclear reactor. Nucl. Eng. Des. **236**, 683–700 (2006)
2. G.D. Fletcher, R.R. Schultz, *RELAP5/MOD3.2 Code Manual* (Idaho National Engineering Laboratory, Idaho, 1995)

A Study on Premature Occurrence of Critical Heat Flux

A. K. Vishnoi, A. Dasgupta, D. K. Chandraker, A. K. Nayak, Nandan D. Hegde and A. Rama Rao

Abstract Two-phase natural circulation loops have extensive applications in nuclear and process industries. One of the major concerns with natural circulation is the occurrence of the various types of flow instabilities, which can cause premature boiling crisis due to flow and power oscillations. In this work, a transient computer code COPCOS (Code for Prediction of CHF under Oscillating flow and power condition) has been developed to predict the premature occurrence of critical heat flux (CHF) under oscillating flow conditions. The code incorporates conduction equation of the fuel and coolant energy equation. For CHF prediction, CHF lookup table developed by Groeneveld is used. Experimental investigation on CHF under oscillatory flow was also carried out in a facility named CHF and Instability Loop (CHIL). CHIL is a simple rectangular loop having a 10.5 mm ID and 1.1-m-long test section. The flow through the test section is controlled by a canned motor pump using a variable-frequency drive (VFD). The effect of frequency and amplitude of flow oscillation on premature occurrence of CHF has been investigated experimentally in CHIL loop. Further experimental data on degradation in CHF due to flow oscillations were compared with COPCOS's simulated results. Full paper covers description of the facility, experimental procedure, experimental results and data analysis using COPCOS.

Keywords Critical heat flux · Premature critical heat flux · Oscillatory flow

---

A. K. Vishnoi (✉) · A. Dasgupta · D. K. Chandraker · A. K. Nayak · A. Rama Rao
Bhabha Atomic Research Centre, Trombay, Mumbai 400085, India
e-mail: alokv@barc.gov.in

N. D. Hegde
VJTI, Matunga, Mumbai 400019, India
e-mail: nandanhegde71@gmail.com

A. K. Nayak and B. R. Sehgal (eds.), *Thorium—Energy for the Future*,
https://doi.org/10.1007/978-981-13-2658-5_43

1 Introduction

The natural circulation flow characteristics are different from that of a forced circulation system. The flow in the former is dependent on the driving head caused by the density difference between hot and cold legs of the system. Many times, the natural circulation flow is not fully developed and secondary flow may exist which may cause flow oscillations. So, under oscillatory conditions of such a system, the mechanisms of occurrence of the boiling crisis are more complicated and the flow oscillations can cause an early occurrence of the dry out as compared to that of steady circulation even for the same geometric and operating conditions. Experimental studies carried out by Rao et al. [1] show the premature occurrence of CHF under oscillatory flows. There was substantial degradation in the CHF value measured in the experiment during oscillatory flow conditions compared to the steady-state CHF value prediction by the lookup table. Kim et al. [2] conducted CHF experiments under flow oscillations of water at low pressure and low flow conditions. Mishima et al. [3] also conducted experiments in a natural circulation system under atmospheric pressure and concluded that the system in which flow rate changes due to oscillations has suffered from significant decrease in CHF. Vishnoi et al. [4] investigated premature occurrence of critical heat flux in 54-rod bundle under oscillatory flow and power conditions and studied the effect of frequency and amplitude on premature occurrence of CHF.

In the current study, a transient computer code COPCOS (Code for Prediction of CHF under Oscillating flow and power condition) has been developed to predict the premature occurrence of critical heat flux (CHF) under oscillating flow conditions, and an analysis for CHF under oscillatory flow has been carried out for tubular T/S having ID of 10.5 mm and heated length of 1.1 m using COPCOS. Further CHF experiments under oscillatory flow conditions were conducted in a facility named CHF and Instability Loop (CHIL). CHIL is a simple rectangular loop having a 10.5 mm ID and 1.1-m-long test section.

2 Description of the Loop

CHF and Instability Loop (CHIL) is an experimental loop set up with the objectives of studying natural circulation stability characteristics and critical heat flux (CHF) under oscillatory flow conditions. The schematic of the loop is shown in Fig. 1. The main loop consists of a vertical heater, steam drum, condenser and pump. The CHF (steady and oscillatory flow) experiments were performed through an electrically heated test section which has an inner diameter of 10.5 mm and a wall thickness of 1.1 mm.

A canned motor pump (3.7 kW) is used for forced circulation of the facility. The power to the pump is fed through a variable-frequency drive (VFD). The input frequency can be varied using this VFD to control impeller speed and thus the flow and

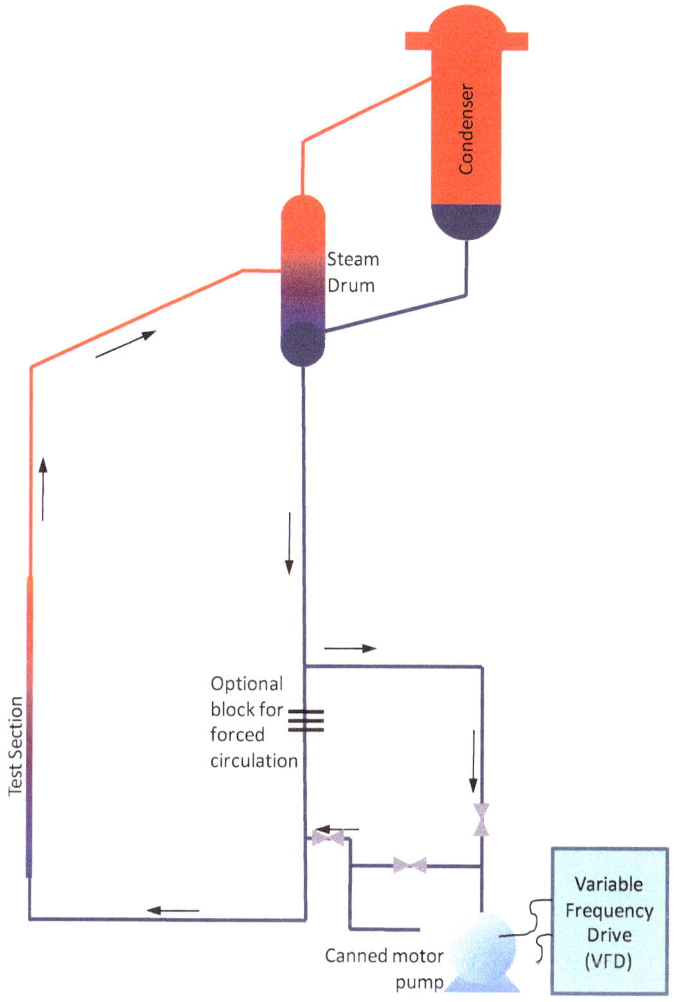

Fig. 1 Schematic of CHIL

developed head. The loop flow is modified by a combination of valve manipulation and pump input frequency adjustment. The VFD is also used to provide programmed oscillatory frequency to the pump, resulting in oscillatory flow variation in the loop. Surface-mounted T/Cs methodology is adopted for critical heat flux detection. Ten T/Cs have been used in the T/S for CHF detection.

3 Mathematical Formulation

COPCOS (Code for Prediction of CHF under Oscillating flow and power condition) has been developed to predict the premature occurrence of critical heat flux (CHF) under oscillating flow conditions. The code incorporates conduction equation of the fuel and coolant energy equation. For CHF prediction, CHF lookup table developed by Groeneveld et al. [5] is used.

3.1 *Radial Conduction Equation for Heater Tube*

For radial heat transfer inside the heater pin, one-dimensional radial conduction equation is solved by FDM.

$$\frac{\partial^2 T}{\partial r^2} + \frac{\partial T}{r \partial r} + \frac{1}{K} q = \frac{1}{\alpha} \frac{\partial T}{\partial t} \quad \text{for } r \neq 0 \tag{1a}$$

Inside boundary conditions at wall–coolant interface and outside boundary conditions at thermal insulation and outside air interface are given in Eqs. 1a and 1c.

$$K_{he} \frac{\partial T}{\partial r} + h_l T_1 = h_l T_l \quad \text{for } i = 0 \tag{1b}$$

$$K_{ins} \frac{\partial T}{\partial r} + h_a T_m = h_a T_\alpha \quad \text{for } i = m \tag{1c}$$

The simple explicit method is applied to discretise the one-dimensional diffusion equation with source term in cylinder geometry. On discretisation of the above equation, we get.

$$T_i^{n+1} = F_0\left(1 - \frac{1}{2i}\right) T_{i-1}^n + (1 - 2F_0) T_i^n + F_0\left(1 + \frac{1}{2i}\right) T_{i+1}^n + F_0 \frac{\Delta r^2 q_i^n}{k} \quad \text{for } i = 1, 2, \ldots m - 1 \tag{2a}$$

$$F_0 = \frac{\alpha \Delta t}{\Delta r^2} \tag{2b}$$

3.2 *Solution of Energy Equation for the Interface*

For heater and insulation interface, energy balance equation is solved as given below. Figure 2 shows the interface having two different materials. Energy balance is given below

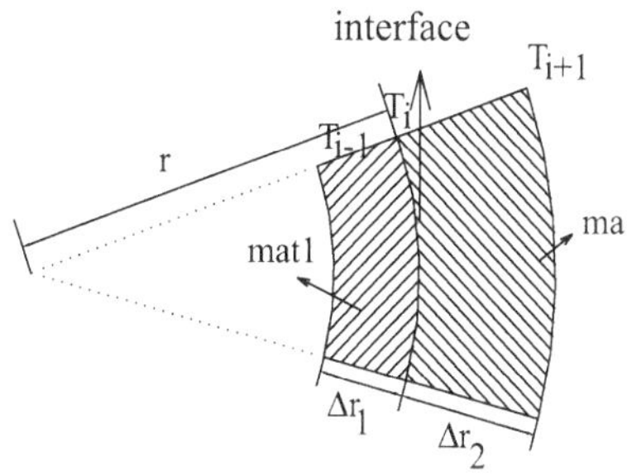

Fig. 2 Interface geometry

$$\frac{(\rho_1 c_1 \Delta r_1 + \rho_2 c_2 \Delta r_2) r}{2} \frac{(T_i^{n+1} - T_i^n)}{\Delta t} = k_1 (r - 0.5\Delta r_1) \frac{(T_{i-1}^n - T_i^n)}{\Delta r_1} - k_2 (r + 0.5\Delta r_2) \frac{(T_i^n - T_{i+1}^n)}{\Delta r_2} + \frac{(q_1 \Delta r_1 + q_2 \Delta r_2) r}{2} \quad (3)$$

Rearranging the above equation, we get

$$T_i^{n+1} = a T_{i+1}^n + b T_{i-1}^n + c T_i^n + d \quad (4)$$

where

$$\begin{bmatrix} a = \frac{k_2(r+0.5\Delta r_2)}{A\Delta r_2} \\ b = \frac{k_1(r-0.5\Delta r_1)}{A\Delta r1} \\ c = \left(1 - \frac{k_2(r+0.5\Delta r_2)}{A\Delta r_2} - \frac{k_1(r-0.5\Delta r_1)}{A\Delta r1}\right) \\ d = \frac{(q_1\Delta r_1 + q_2\Delta r_2)r}{2A} \\ A = \frac{(\rho_1 c_1 \Delta r_1 + \rho_2 c_2 \Delta r_2)r}{2\Delta t} \end{bmatrix} \quad (5)$$

For heater and insulation interface, properties are taken as given below

$$\begin{bmatrix} k_1 = k_f \text{ and } k_2 = k_{ins} \\ \rho_1 c_1 = \rho_f c_f \text{ and } \rho_2 c_2 = \rho_{ins} c_{ins} \\ q_1 = q \text{ and } q_2 = 0 \end{bmatrix} \quad (6)$$

3.3 Solution of Conservation of Energy Equation of Coolant

Conservation of energy equation for the coolant flow can be written as in Eq. (8). Figure 3 shows the energy balance for the coolant flow.

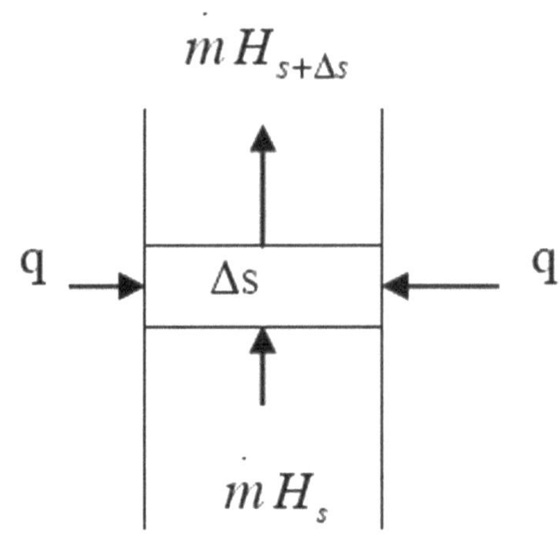

Fig. 3 Coolant energy balance

$$\dot{m}H_s - \dot{m}H_{s+\Delta s} + qP\Delta s = \rho A\Delta s\frac{\Delta H}{\Delta t} \tag{7}$$

Rearranging the above equation, we get

$$H_i^{n+1} = \left(1 - \frac{\dot{m}\Delta t}{\rho A\Delta s}\right)H_j^n + \frac{\dot{m}\Delta t}{\rho A\Delta s}H_{j-1}^n + \frac{Ph\Delta t}{\rho A}\left(T_w^n - T_j^n\right) \quad \text{where} \quad P = 2\pi R \tag{8}$$

3.4 Enthalpy–Temperature Hybrid Method

In Eq. (8), there are two variables, temperature T and enthalpy H, to be determined. These two variables are related by the thermodynamic enthalpy–temperature relations. The traditional approach chooses either enthalpy or temperature as an independent variable and incorporates the constitutive equations to eliminate the other. In order to do so, either enthalpy or temperature must be consistently and explicitly expressed as a function of the other over all possible phases. With phase change occurring at a single temperature, the value of enthalpy is undetermined at this point and cannot be expressed as a function of temperature. Thus, enthalpy is not an appropriate choice for the independent variable. Furthermore, if the specific heat varies with temperature such that the temperature cannot be expressed as an explicit function of enthalpy in the particular phase, then neither enthalpy nor temperature can be chosen as an independent variable to effectively eliminate the other from the energy equation. Therefore, the traditional approach of the enthalpy method fails in solving this problem. To overcome the difficulty, encountered in the traditional approach, enthalpy–temperature hybrid method is used. In this approach, both temperature and enthalpy are treated as independent variables. Thermodynamic temperature–enthalpy equations are incorporated and solved simultaneously in conjunction with the energy conservation equation. For the liquid whose phase change occurs at T_s, the enthalpy–temperature relation may be written as

$$H = \int_0^T C_{pl} \mathrm{d}T \quad H < H_l = \int_0^{T_s} C_{pl} \mathrm{d}T \tag{9a}$$

$$T = T_s \quad H_l \leq H \leq H_l + H_{fg} \tag{9b}$$

$$H = H_l + Hfg + \int_{T_s}^{T} C_{pg} \mathrm{d}T \quad H > H_l + H_{fg} \tag{9c}$$

3.5 *CHF Prediction Methodology*

In the current analysis, CHF is predicted by the recently published CHF lookup table by Groeneveld et al. [5]. However, the tube CHF lookup provides correction factors for different tube diameter. For this tube, correction factors have been used.

3.6 *Solution Procedure*

Both conduction equation of the fuel and coolant energy equations are solved together iteratively, using the procedure mentioned below.

1. The minimum of time steps calculated from conduction equation of solid and coolant energy equation is taken for calculation.
2. Discretised conduction equation is solved explicitly using the above time step to obtain T_0^{n+1}.
3. Equations (9a)–(9c) are augmented with the coolant energy Eq. (8) at every node point. At each iteration, the phase of every node point is checked and the constitutive equations are selected accordingly. Using T_m^{n+1} in discretised energy balance equation of the coolant, i.e. Equation (9a), fluid temperature T_1^{n+1} at new time level is obtained.
4. Again, conduction equation is solved using T_1^{n+1} fluid temperature and steps (2) to (4) are repeated until steady state is achieved.

Different inputs for the code are given below.

1) Power;
2) Inlet flow;
3) Inlet subcooling;
4) Pressure;
5) Axial flux distribution.

4 Results and Discussion

Critical heat flux experiments under oscillatory flow conditions were carried out in CHF and Instability Loop (CHIL). Initially, steady-state conditions were achieved in the loop at desired pressure, inlet subcooling and mean flow rate. Once steady-state conditions were achieved, oscillatory sinusoidal flow variation was given to inlet flow at desired frequency and amplitude and power was increased slowly in the steps of 0.5 kW till CHF occurs. Occurrence of CHF is identified with sudden temperature rise in the thermocouples provided at the top of the T/S. All T/Cs are provided with power supply trip at 325 °C. Figure 4 shows typical transient of oscillatory flow CHF experiments. This figure shows important parameters only for last 200 s so that flow oscillations' (frequency and amplitude) sudden rise in T/S temperature (to identify CHF) can be visualised. However, each set of experiments took around 2–3 h in which power was increased slowly (before raising the power, it was ensured that system should achieve steady state) in the steps of 0.5 kW till CHF occurs. It can also be observed that at the point of CHF there is a sudden rise in the surface temperature while test section power was constant before CHF. Oscillatory CHF experiments at various operating conditions (flow, pressure and inlet subcooling) were performed. Table 1 shows experiments' operating conditions and comparison of experimental CHF data with calculated CHF using COPCOS with LUT-1995. Figure 5 shows comparison of experimental and calculated CHF values. It can be observed that experimental CHF value is within 60% than the calculated CHF value. Experimental CHF value is coming lower than the calculated value; it shows degradation in CHF due to oscillation is much more than predicted by COPCOS using LUT-95. Large

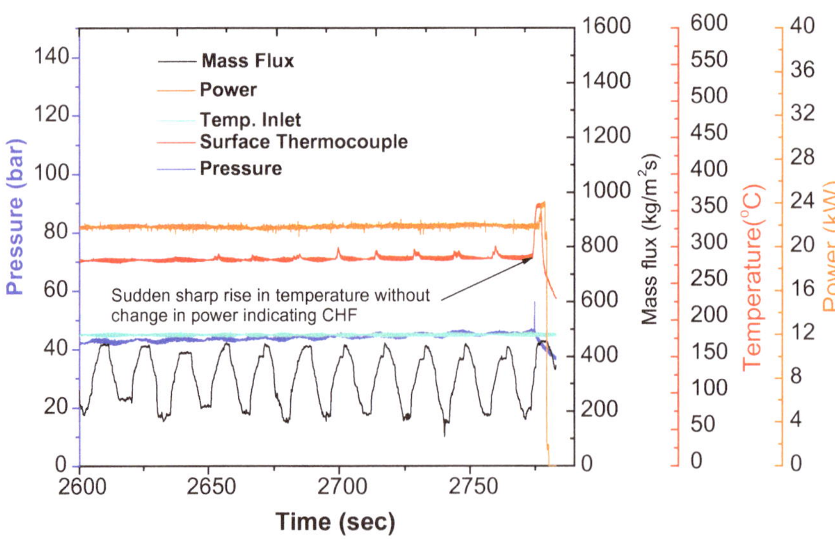

Fig. 4 Typical transient during CHF experiments

Table 1 Experimental operating conditions and comparison with COPCOS

Experiment S. No.	Pressure (bar)	Inlet Temp. (°C)	Mean mass flux of oscillation (kg/m^2 s)	Experimental CHF (kW/m^2)	Theoretical CHF by COPCOS using LUT-95
1	46	182	320	599	730
2	37	180	350	578	909
3	34	178	351	565	923
4	18	154	323	367	593
5	14	151	317	299	827
6	30	201	377	451	1103

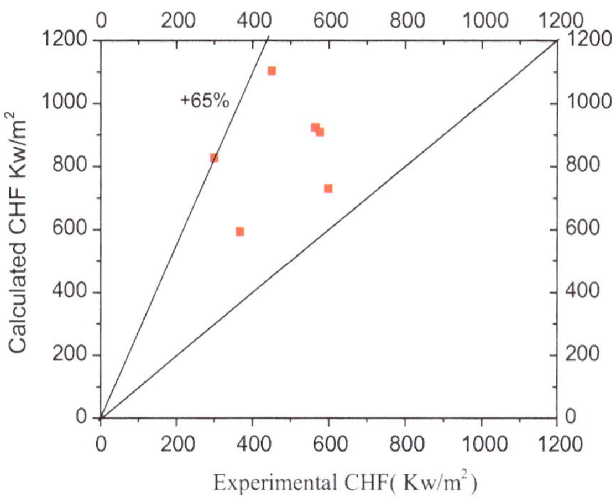

Fig. 5 Comparison of experimental oscillatory CHF with calculated CHF

deviation between experimental and predicted values could be due to the use of steady-state LUT-95 data in COPCOS code. CHF is a highly localised phenomenon, so prediction of oscillatory CHF with steady-state CHF data gives large deviation. However, Table 2 gives CHF values at oscillatory flow, mean flow of oscillation and minimum flow of oscillation. In all cases, it can be seen that CHF at oscillatory flow is lower than the CHF at minimum flow of oscillation. It indicates that COPCOS is able to predict degradation of CHF due to flow oscillations.

Table 2 Degradation in CHF due to flow oscillation

Experiment S. No.	Mean mass flux of oscillation (kg/m^2 s)	Min. mass flux of oscillation (kg/m^2 s)	Calculated CHF under flow oscillation using COPCOS (kg/m^2 s)	CHF at minimum flow of oscillation (by LUT-95) (kW/m^2)	CHF at mean flow of oscillation (by LUT-95) (kg/m^2 s)
1	320	190	730	846	1329
2	350	221	909	969	1400
3	351	219	923	965	1395
4	323	151	593	707	1265
5	317	190	827	850	1214
6	377	299	1103	1142	1387

5 Conclusions

CHF experiments have been carried out under oscillatory flow conditions in CHF and Instability Loop (CHIL) at various operating conditions. Following are the important conclusions of the study:

1. Sinusoidal flow oscillation was achieved successfully using VFD in pump, and oscillatory flow experiments were carried out at various pressure, flow and inlet subcooling.
2. CHF was successfully detected with surface-mounted thermocouple technique.
3. Experimental CHF value under flow oscillation is within 60% than the calculated CHF value using COPCOS. It shows degradation in oscillatory CHF is more than predicted by COPCOS.
4. COPCOS is able to predict the degradation in CHF as CHF under flow oscillation is coming less than CHF value at minimum value of flow oscillations.

References

1. G.S.S.P. Rao, D.K. Chandraker, P.K. Vijayan, D. Saha, An experimental investigation on critical heat flux in a natural circulation system under low pressure and low flow conditions, in *ISME-2003, XIII National Conference of Indian Society of Mechanical Engineers*, IIT Roorkee (2003)
2. Y.I. Kim, W.P. Baek, S.H. Chang, Critical heat flux under flow oscillation of water at low-pressure, low-flow conditions. Nucl. Eng. Des. **193**, 131–143 (1999)

3. K. Mishima, H. Nishihara, I. Michiyoshi, Boiling burnout and flow instabilities for water flowing in the round tube under atmospheric pressure. Int. J. Heat Mass Transf. **28**(6), 1115–1129 (1985)
4. A.K. Vishnoi, D.K. Chandraker, P. K. Vijayan, D. Saha, Investigation on premature occurrence of critical heat flux in 54-rod bundle under oscillatory flow and power conditions, in *19th National & 8th ISHMT-ASME, Heat and Mass Transfer Conference, Hyderabad, India* (2008)
5. D.C. Groeneveld, L.K.H. Leung, P.L. Kirillov, V.P. Bobkov, I.P. Smogalev, V.N. Vinogradov, X.C. Huang, E. Royer, The 1995 look-up table for critical heat flux in tubes. Nucl. Eng. Des. **163**, 1–23 (1995)

Intercode Comparison of SBO Scenario for AHWR

A. Srivastava, Rajesh Kumar, B. Chatterjee and P. K. Vijayan

Abstract The strong earthquake followed by tsunami, a low-frequency and high-consequence natural disaster, at Fukushima, has shaken the world's faith in nuclear energy. The new reactor design around the world has started looking ways to mitigate this scenario with innovative design features, whereas every nuclear country has started reviewing their existing reactor response to this scenario minimizing the consequences or limiting the damage progression. India has designed an advanced reactor concept, i.e., advanced heavy-water reactor (AHWR). This reactor has many passive design features to mitigate the consequences not only for the postulated design basis events but also for beyond design basis events. During station blackout (SBO) scenario leading to hot shutdown, the isolation condensers (ICs) are intended to remove decay heat with the help of overhead large pool of water, i.e., gravity-driven water pool (GDWP). This paper outlines the assessment of station blackout scenario for AHWR using the last version of the French best estimate computer code CATHARE2/V2.5_2 and its comparison with RELAP5/Mod3.2 findings. First, it explains the modeling of main heat transport system of AHWR and isolation condenser loop along with GDWP in CATHARE2 followed by thermal-hydraulic safety assessment of station blackout scenario and comparison of predictions with RELAP5 findings.

Keywords AHWR · SBO · RELAP5 · CATHARE · Boiling water reactor

1 Introduction

The proposed advanced heavy-water reactor is a 920 MW_{th} thorium-based vertical pressure tube-type boiling light-water-cooled and heavy-water-moderated reactor

A. Srivastava (✉) · R. Kumar · B. Chatterjee · P. K. Vijayan
Reactor Design and Development Group, Bhabha Atomic Research Centre,
Trombay, Mumbai 400085,
Maharashtra, India
e-mail: abhisri@barc.gov.in

A. K. Nayak and B. R. Sehgal (eds.), *Thorium—Energy for the Future*,
https://doi.org/10.1007/978-981-13-2658-5_44

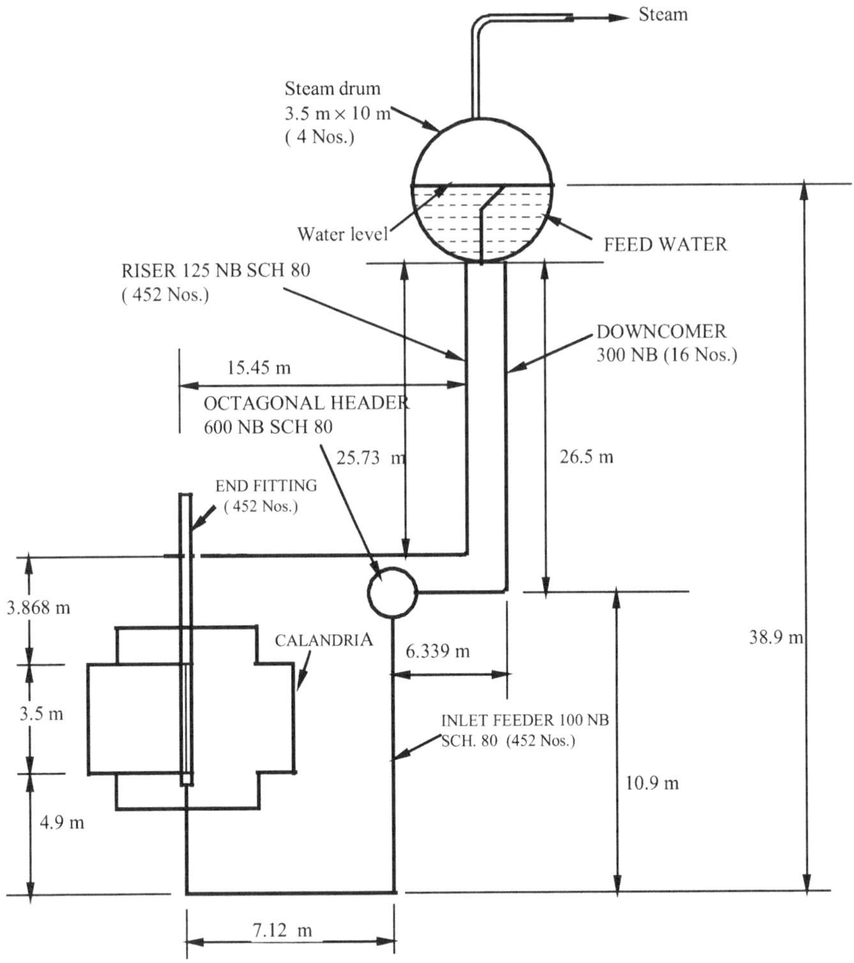

Fig. 1 Schematic of AHWR MHT

[1]. One of the important passive design features of this reactor is that the heat removal is achieved through natural circulation of primary coolant at all allowed power levels with no primary coolant pumps (Fig. 1).

Station blackout leads to feed water pump trip and loss of condenser vacuum which in turn results in turbine trip on low condenser vacuum signal. The bypass flow is not available due to loss of condenser vacuum. This leads to bottle up of the system, leading to pressure rise in main heat transport (MHT) system. Reactor trips on high MHT pressure signal. As the pressure continues to rise, IC valve starts opening at a pressure of 7.65 MPa to limit the pressure rise. The relief valve may open in the case of IC's capacity which is not adequate. This analysis highlights the thermal-hydraulic conditions following station blackout for long duration and highlights the design mitigating provisions to limit the pressure and clad temperature rise and intercode comparison of CATHARE2 predictions with RELAP5 [2] findings.

2 Modeling of AHWR

2.1 Modeling of AHWR in CATHARE

The analysis has been done using French best estimate code CATHARE. CATHARE-2/V2.5_2 [3] is a thermal-hydraulic system code that solves the conservation laws for water and steam for a wide variety of single- and two-phase flow conditions. The AHWR model includes reactor core, 452 feeders and 452 tailpipes, 4 steam drums (SDs), 16 downcomers, inlet header, and steam pipes up to turbine in CATHARE code. The ECC system comprising of four segmental ECC headers is modeled with axial volume. The accumulators are modeled with CATHARE-specific gadget ACCU along with discharge line. The IC inlet pipe, IC inlet header, further 1440 tubes of H-X, outlet header, and IC outlet pipe are modeled. At the steady-state condition, core flow, drum pressure, core inlet temperature, steam drum level, and total steam flow are found to be 2143 kg/s, 6.999 MPa, 259 °C, 2.2 m, and 414.8 kg/s, respectively. These values of different important parameters are matching well with nominal design values.

Isolation condenser (IC) loops have been modeled in CATHARE as shown in Fig. 2. The IC inlet pipe is modeled with axial element ICINLPIP having five control volumes. The IC inlet header is modeled with axial element ICINLHD. Further, 1440 tubes of H-X are modeled with ICHEX having seven control volumes. The five control volumes are passing heat to GDWP. The outlet header and IC outlet pipe are modeled with axial element ICOUTHD and ICOUTPIP, respectively. The exchanger model is used to model the heat transfer to the GDWP. The HSPV valve is simulated using CATHARE-specific control valve. This modeling of IC path is combined with an already developed model of AHWR MHT and ECC system, and proper steady state is obtained.

2.2 Modeling of AHWR in RELAP5

The nodalization scheme for the main heat transport (MHT) system for AHWR in RELAP5 is shown in Fig. 3. Out of 452 channels in the core, 20 high-power channels are clubbed as hot channel and remaining 432 channels are clubbed as average channels. The average channel path is represented by component 100, and hot channel path is represented by component 101. Volume 4 to volume 8 of each component represents core channels. Volume 1 and volume 2 of these parallel paths

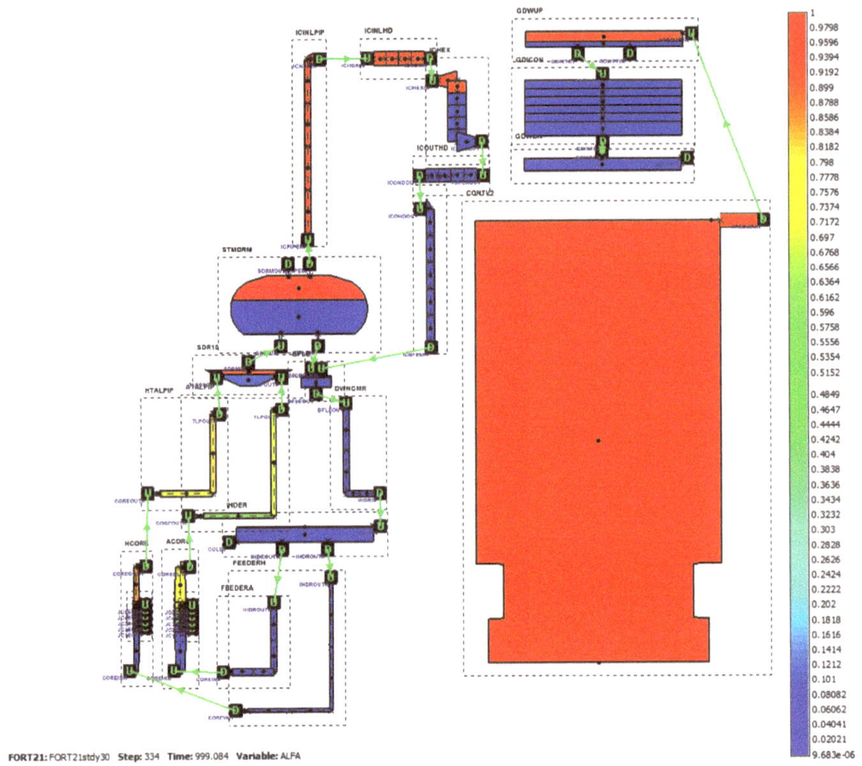

Fig. 2 AHWR MHT and IC nodalization in CATHARE

represent vertical and horizontal portions of inlet feeders, while core bottom end is represented by volume 3 of these pipe components. Volume 9 of these components simulates core top end, while volumes, 10, 11, and 12 represent tailpipes. The effect of nodalization was evaluated in a separate study, and it has been insured that the present nodalization scheme yields accurate and converged results.

Four steam drums (SDs) are modeled by a single equivalent SD, i.e., component 102. The saturated volume of steam drum 102 is modeled separately using ten control volumes. The bottom part of the drum (subcooled portion) in between two baffles is represented by subcooled volume component 103. The feed water along with recirculation flow enters into the subcooled part of the drums. Flow from this portion of steam drums enters the downcomers. Feed water conditions are simulated using component 600. The suction header of feed pumps is modeled by component 653. Two feed pumps are modeled by components 656 and 657. Component 660 models discharge header. Individual feed lines are modeled by component 675.

Four downcomers lines, from each steam drum, are modeled by a single equivalent downcomer, i.e., 104. Inlet header is modeled by component 105. Steam lines emerging from the steam drums are represented by volumes 351. Two steam lines

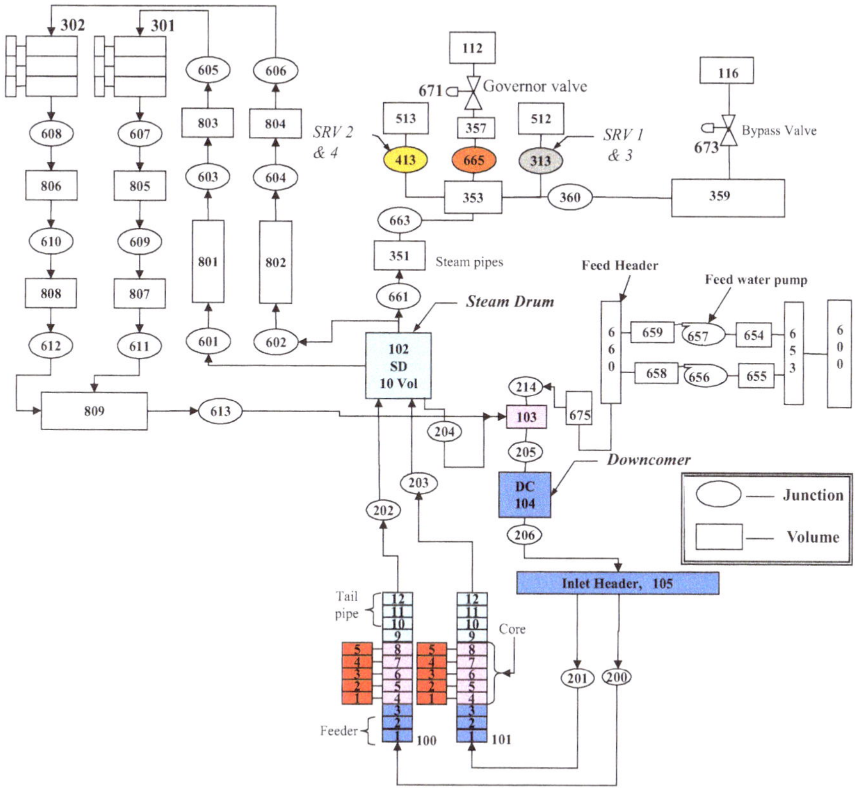

Fig. 3 AHWR MHT and IC nodalization in RELAP5

from each bank are joined together by volumes 353. Bypass valve is modeled by component 673. Combined isolation emergency stop (CIES) valves have been simulated by component 665. Governor valves have been modeled by component 671. Safety relief valves SRV 1 and SRV 3, and SRV 2 and SRV 4 mounted on steam lines are modeled by components 313 and 413, respectively. Appropriate core power modeling has been done considering power distributions (radial and axial), and point kinetics model has been used to simulate the reactor power. The thermal-hydraulic feedbacks are also modeled. Decay heat has been modeled using ANS 79 [2].

Each drum is connected to two isolation condensers (ICs), modeled by component 301 and 302, and outlet from IC is connected to subcooled volume 103 of steam drum. Each IC representing 70 tubes is divided into three axial volumes. Inlet cylindrical headers of each IC path are simulated with a single component (803, 804). Outlet cylindrical headers of each IC path are simulated with a single component (805, 806). The pipes in IC circuit are simulated by component 801, 802, 807, 808 as shown in Fig. 3.

Table 1 Steady-state comparison

Parameter	Design	CATHARE	RELAP5
Reactor power (MW_{th})	920.00	920.0	920.0
Total core flow (kg/s)	2145.00	2145.2	2145.2
Steam drum pressure (MPa)	7.00	6.9999	6.9999
Collapsed level in SD (m)	2.20	2.2027	2.1988
Hot core void		86.5	86.4
Average channel flow	4.7	4.74	4.745
Hot channel flow	4.9	4.89	4.9
Feed flow per SD (kg/s)	103.0	103.54	103.27
Total steam flow (kg/s)	412.0	414.16	413.08
Coolant inlet temperature (°C)	130.0	130.0	130.0
Feed water temperature (°C)	259.5	259.455	259.25

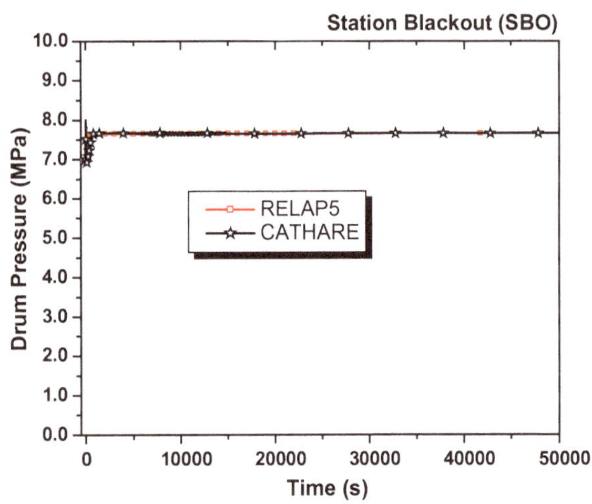

Fig. 4 SD pressure transient during SBO

The steady-state condition results of RELAP5 and CATHARE are compared with design nominal conditions. The comparison is given in Table 1.

3 Results and Discussions

The station blackout event begins at time $t = 0$ s. System pressure rises due to the bottling of MHT as shown in Fig. 4. The reactor trip is credited on high system pressure signal once system pressure reaches 7.6 MPa.

Further passive valve in the IC line opens at MHT pressure of 7.65 MPa. The heat removal from the ICs is initially small compare to reactor power. This results in initial

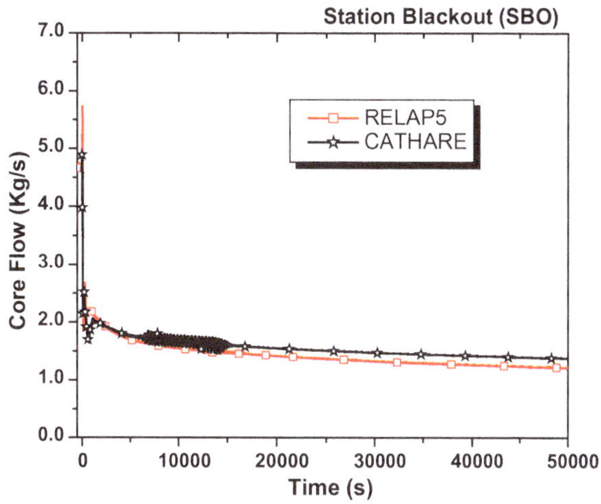

Fig. 5 Core flow transient during SBO

pressure rise as shown in Fig. 4. Once the reactor power comes to decay heat level and comparable to IC heat removal, the system pressure rise is arrested. The system pressure remains around 7.65 MPa due to the controlled opening of the IC valve based on system pressure. The initial pressure rise trend following initiation of the transient is well captured by both codes, and peak pressure value is comparable. Overall, the pressure response predicted by both the codes is having a good comparison.

The heat removed from the isolation condensers is transferred to GDWP large pool water having inventory of 6000 tonnes. Heat is removed passively by GDWP initially by sensible heating and later by boil off once the temperature reaches to saturation temperature. The water inventory in GDWP is sufficient to effectively remove the decay heat for more than grace period of 7 days. Figure 5 shows the core flow behavior during station blackout scenario. As the reactor trips on the high system pressure signal, the core flow reduces as shown in Fig. 5. The natural circulation flow depends on the heat coming from the core. As the core power keeps on decreasing over the time due to reduction in decay heat, the core flow also keeps on reducing over the transient as shown in Fig. 5. The core flow predictions by RELAP5 and CATHARE analyses match very well in the initial part of transient; however, CATHARE code predicts slightly higher flow than RELAP5 in the later part of transient. The clad temperature initially rises marginally corresponding to saturation temperature. In later part of transient, as heat removal is same as decay heat coming to the system, the clad temperature remains stable as shown in Fig. 6. The effect of slight difference in core flow does not reflect in clad temperature prediction as shown in Fig. 6.

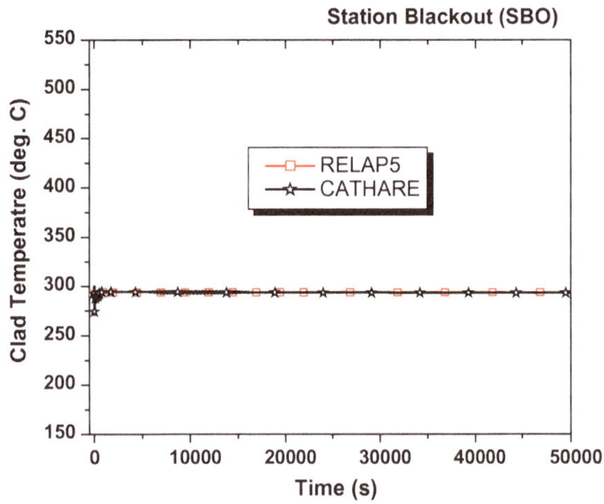

Fig. 6 Clad temperature transient during SBO

4 Conclusions

The prolonged SBO has been analyzed for AHWR using best estimate code CATHARE, and results are compared with already available results of RELAP5 code. The ICs are able to remove decay heat with the help of GDWP and maintain core temperatures well within the limit. Decay heat is removed passively by GDWP initially by sensible heating and later by boil off as seen in long-term calculation done with RELAP5. It has been found in analysis that IC system is capable to remove decay heat for more than 7 days. The decay heat removal through IC path along with passive moderator and end shield cooling keeps the integrity of different systems and maintains the core temperature well below the acceptance limit. Both the codes predict similar pressure response following SBO. The core flow prediction shows good match in the initial part of the transient; however, the core flow predicted by CATHARE code is little higher than the RELAP5 which results in later part of transient. The difference in core flow does not result in difference in clad temperature. Overall predictions made by CATHARE are in good agreement with RELAP5 results.

References

1. R.K. Sinha, A. Kakodkar, Design and development of the AHWR—the Indian thorium fuelled innovative nuclear reactor. Nucl. Eng. Des. **236**, 683–700 (2006)
2. C.D. Fletcher, R.R. Schultz, RELAP5/MOD3.3 Code Manual Volume I: Code Structure, System Models, and Solution Methods, NUREG/CR-5535/ Rev-I, Idaho National Engineering Laboratory, Idaho Falls, Idaho, Dec 2001
3. G. Geffraye et al., CATHARE 2 V2.5 2: a single version for various applications. Nucl. Eng. Des. **241**, 4456–4463 (2011)

Design and Development of Process-Powered Sensors on Wireless Sensor Network for Stand-alone Plant Critical Data Management During SBO and Beyond Design Basis Events

J. Aparna, Sebin Philip, Anita Topkar, I. V. Dulera, D. Das and A. Rama Rao

Abstract Considering the need to access crucial parameters in a nuclear reactor during station blackout condition (SBO), we have proposed self-powered or process-powered sensors with wireless data transmission. Low-power electronics comprising a piezoresistive pressure sensor, amplifier, and wireless transmitter was developed for demonstration of process-powered sensor operation. The required electrical power was derived using a thermoelectric generator as an energy harvester which converts heat energy from the process to electrical energy and drives the complete electronics along with the wireless transmitter. A power management circuitry has been used for management of power fed to the electronics. The complete instrumentation has been tested in the laboratory to demonstrate the operation of process-powered sensor and wireless instrumentation. Such approach would be suitable for wireless network of self-powered/process-powered sensors and also for wearable devices by operators. An overview of this development is presented in this paper.

Keywords Microelectromechanical system sensors · MEMS · SBO · BDBE
Self-powered sensors · Wireless networks · Process powered/self powered
Wearable devices

1 Introduction

Advanced reactors are designed with an aim of maximum safety, optimized fuel utilization, and effective system design. Safety aspects in reactor designs are being viewed for all possible vulnerabilities, and as a result, robust self-regulating passive

J. Aparna (✉) · I. V. Dulera · A. Rama Rao
Reactor Engineering Division, Bhabha Atomic Research Centre, Trombay, Mumbai 400085, India
e-mail: anita@barc.gov.in

S. Philip · A. Topkar · D. Das
Electronics Division, Bhabha Atomic Research Centre, Trombay, Mumbai 400085, India

A. K. Nayak and B. R. Sehgal (eds.), *Thorium—Energy for the Future*,
https://doi.org/10.1007/978-981-13-2658-5_45

safety features have been favored in Gen IV and advanced reactor designs. In addition to passive systems, the accident scenarios at Fukushima indicate the dire need of reliable and stand-alone self-powered/process-powered sensors, for monitoring plant critical parameters for effective damage control actions. There is a strong need for plant critical data management and situation awareness during the unavailability of all conventional power sources in a nuclear power plant, during extended station blackout (SBO) conditions [1]. These self-powered/process-powered sensors would assist the operators in managing events like SBO and help in containing any beyond design basis events (BDBEs) conditions, well away from the public domain.

In addition to the common mode failure of power source in Fukushima, during the post-accident analysis, it had been observed that, though the sensors were functional, the fire- and flooding related events had damaged the communication from sensors to the control room, i.e., the cables and terminations severely. Operators had used process calibrators for manually energizing the instruments to measure the process parameters. The series of events had also left many of the crucial reactor parameters, inaccessible for manual measurement. Hence, in addition to being passive, self-powered/process-powered sensors need a diverse communication channel, i.e., a rugged, stand-alone, independent channel of communication with a small footprint such as a proprietary wireless sensor network [2]. The difficulty in accessing instrument rooms during an SBO and accident also necessitates the design of low-power wearable devices for ease of measurement and plugs into the sensor networks wirelessly. Self-powered or process-powered sensors derive their energy passively from the process itself and do not depend on any external power supply systems. These sensors in effect derive their energy from the environment, in which they are deployed to measure. The energy harvested by the sensors is stored for their measurement and transmission functionality. In the case of nuclear reactor applications, these sensors need to perform under radiation environment, as applicable to the domain. The self-powered/process-powered sensors would be configured into a self-organizing fault-tolerant wireless sensor network for effective communication to a handheld mobile device for data monitoring and operator guidance in case of an extended SBO accidents. MEMS technology-based sensors which are compact, low powers, and give direct electronics readout would be most suitable for building such wireless sensor network or wearable devices. The review of measurement scenario during SBO and severe accidental conditions, requirements of self-powered/process-powered sensors, and their critical parameters are discussed in the subsequent sections. Finally, test results demonstrating the operation of a process-powered MEMS piezoresistive pressure sensor for wireless transmission of pressure data are presented.

2 Measurement Scenario During SBO and Severe Accidents

In case of a station blackout, the critical instrumentation is supported for 30 min by battery backup and few hours by DG sets. The current advanced reactor systems rely on passive natural circulation systems for decay heat removal, in case of a prolonged SBO like that of Fukushima and Three Mile Island accidents. However, the systems being passively safe, for ensuring safety and situation awareness of the nuclear core and sub-systems, need suitable instrumentation. Operators need to assess the status of the reactor continuously, for maintaining safe shutdown and to avert further progress by weighed operating decisions. Post Fukushima, reactor operators have relied on solar backed up power, enhanced battery backup, portable instruments, chargeable batteries, etc., for enhancing the operator's efficiency in case of an eventuality. However, in many situations in the past, all these external sources have been muted by a credible combination of events.

A nuclear reactor is the only credible man-made resource, which can give energy for years, even when it is under shutdown, by means of decay heat. Hence, power for the instruments that are crucial for reactor safety could be derived from the process itself, i.e., process heat, in process-powered mode. These instruments are indicated as self-powered/process-powered sensors, and the design of such sensors would focus on their availability during normal operation, extended SBOs, and severe accidents.

3 Requirement Evolution for Self-powered/Process-Powered Sensors for Nuclear Installations

Based on the post-accident assessment of the instrumentation in Three Mile Island and Fukushima, many of the sensors have maintained their integrity and functionality for their intended design and have given reasonable performance even under severe environmental conditions. In the case of Fukushima, even though the sensors were functional, the hydrogen fire and water inundation had left most of the cables non-functional. Operation personnel had been put in action to reach the actual instruments through debris, radiation exposure without electricity, and poor visibility. These activities by the personnel had resulted in heavy radiation exposure and delay in getting crucial information and loss of clues in situation awareness for the field personnel.

Hence, in addition to supporting the sensors with power, a diverse communication mechanism independent of the conventional cable route is essential considering the indomitable constraints posed during the Fukushima post-accident scenario. Self-powered /process-powered sensors in wireless sensor networks would aid operators in addressing the challenges posed during the post-accident handling of nuclear accidents.

In addition to wireless sensors, the stressed reactor needs operator scrutiny for many functional decisions, recovery of systems, pre-requisites such as water hammer, LOCA situations, breach in containments. In case of a mild/severe damage situations, the operators need navigational aid and easy access to data of the system and related parameters they are scrutinizing, a wholesome interface to the process, in situ for effective decision making. Hence, self-powered /process-powered wearable devices and battery-backed mobile devices could be evolved, with the capability to communicate to local instruments in WSN as well as to a centralized data device empowered with navigational aids and visual devices to transmit information back and forth to an authenticated station.

The challenges in the design and development of these sensors are compactness, compatibility for installation with least interference to the existing framework of sensors, low-power consumption, wireless communication among massive plant metal works and concrete structures in the reactor building, digital security, data integrity, and qualification for radiation.

4 Critical Parameters for Self-powered /Process-Powered Sensors

The design of the sensors needs to be guided by the choice of parameters which are crucial for reactor safety and auxiliary systems ensuring the containment integrity and effective emergency response. Some of the important parameters are the reactor coolant pressure, reactor coolant temperature, SG water level, main steam pressure, containment pressure, and spent fuel pool water level. The general guidelines in handling the accidents are to rely on installed plant systems [2] as the first phase of coping capabilities. In the second phase, the coping capabilities are extended to on-site portable systems and off-site capacity addition is considered as the final phase. Self-powered /process-powered sensors would be designed to fit in the installed coping capability available as standby for the operators during normal operation and emergency support in first-phase coping procedures to handle SBO and beyond design basis events. The design would also need to accommodate the dynamic in situ developments in personnel accessibility and weighed wireless communication routes.

5 Current Status of Wireless Sensor Networks and Applications

Wireless sensor networks have been widely employed in environmental monitoring, animal monitoring, and surveillance applications. The sensors are either with battery backup or use solar and allied energy harvesters which derive their energy from the environment. The concept of energy harvesters deriving energy from the environment to source the sensors in nuclear reactors combined with wireless sensor networks for deployment in nuclear reactors has been reported in the literature [3]. In nuclear reactor applications, deployment of wireless sensors has been suggested as an economical alternative to cable installations for varied applications such as real-time process variables, structural components, movable equipment, portable devices, and warehouse/inventory areas. Wireless sensors and their deployment challenges inside the containment have also been reported in the literature. The actual deployment of such sensors in the reactors would have to address various deployment challenges and regulatory clearances, and they do seem feasible with the current maturity level in sensor technology [4].

6 Compact, Low-Power Devices in Wireless Sensor Networks Using Wearable Devices and Mobile Technology

The wireless devices in addition to consuming low power need to be compact for easy installation and portability. In the present commercial mobile technology, wearable devices such as wristbands for physical fitness to health monitoring devices built with mobile phone compatibility and smart watches for professional tools have been developed. The lifestyle devices and the technology are in mature stage, and the same are proposed to be adapted in these sensors with proprietary security and data management. As an additional inherent security measure, hardware signatures and limiting the communication distance in design have been conceptualized as built-in features in the sensors. The commercial and open-source software provides multiple options for developing devices which can be deployment ready without much technological hassles and proprietary constraints in case of an emergency. A middle path needs to be worked out for deploying custom-designed platforms and commercial-scale devices for secure and safe applications. Figure 1 illustrates a self-powered /process-powered process sensor, powered to measure and wirelessly transmit the data from the process, using thermoelectric modules. The thermoelectric modules harvest energy from the temperature differential between the process piping and the ambient, to power the sensor.

7 Design and Development of Process-Powered Sensors

Figure 2 illustrates the design schematic of self-powered /process-powered sensors on a wireless sensor network for applications in nuclear reactors. These sensors would act as an operator empowering tool, for situation awareness, irrespective of the operating domain of the nuclear reactors. The self-powered /process-powered sensors are provided with a chargeable Li coin cell for efficient power management

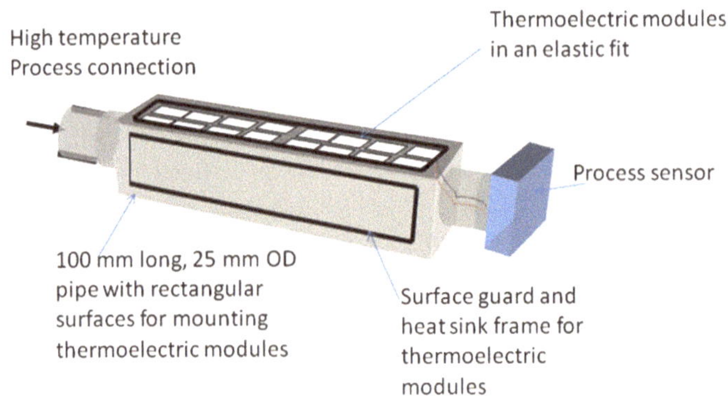

Fig. 1 Schematic of an add-on thermoelectric module to process-powered sensors

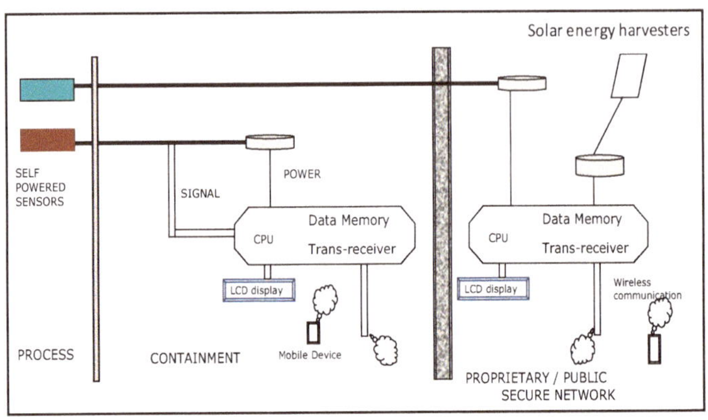

Fig. 2 Schematic of process-powered sensors on a wireless sensor network

and to support a local display device. The security issues inherent to the wireless communication would be addressed by device signature encryption for authentication and by limiting the transmission energy, thereby the communicable distance.

7.1 Design of Process-Powered Sensors

We have demonstrated the feasibility of using process-powered sensors for measurement of pressure as a parameter and wireless transmission of the data. The schematic diagram of the sensor instrumentation is presented in Fig. 3.

A low-power MEMS piezoresistive pressure sensor is used for measurement of pressure in the system. The sensor incorporates piezoresistors located over a thin silicon diaphragm in a Wheatstone bridge configuration which provides output voltage when the bridge is unbalanced. The resistance of piezoresistor changes due to

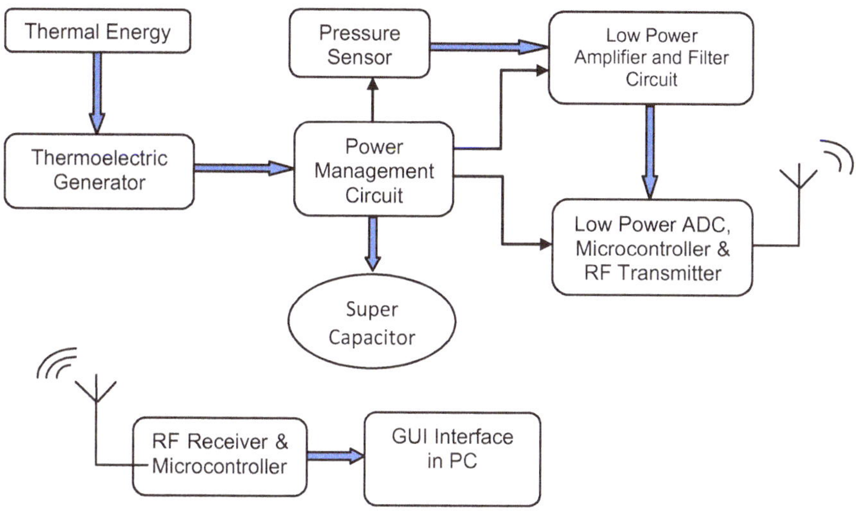

Fig. 3 Schematic of pressure sensor instrumentation

generation of stresses in the silicon diaphragm when pressure is applied. The pressure sensor was rated for 10 bar full scale, and output-voltage sensitivity of the sensor was ~20 mV/bar. During measurements, the sensor was operated up to 10 bar pressure. The output voltage was amplified using an amplifier of gain of ~13. The amplifier output voltage was digitized and transmitted wirelessly at a frequency of 2.4 GHz. The receiver data were received at a computer and displayed in graphical and numerical form. For process-powered operation, the power to the instrumentation including sensor was provided through a TEG mounted on an electric heater. Considering the power consumption of the complete electronics, the heart temperature was optimized to 130 °C for transmission at a rate of one sample/second. The complete process-powered sensor and related instrumentation were tested in the laboratory by providing pressure input to the sensor using a hand pump. The results of the tests conducted are presented in the subsequent section of this paper.

7.2 *Test Results of Process-Powered Pressure Sensor*

Initially, the measurement was done by simulating a Wheatstone bridge circuit of the piezoresistive MEMS pressure sensor. Calibration values of pressure sensor were used for setting the input voltage. The bridge was unbalanced using a variable resistor in the bridge circuit. The output voltage of the amplifier was transmitted wirelessly at 2.4 GHZ, and received values by the receiver were recorded in a computer. The power was provided using TEG at 130 °C. In Fig. 4, the output voltage received at the receiver is plotted against output voltage of the amplifier as transmitted by the

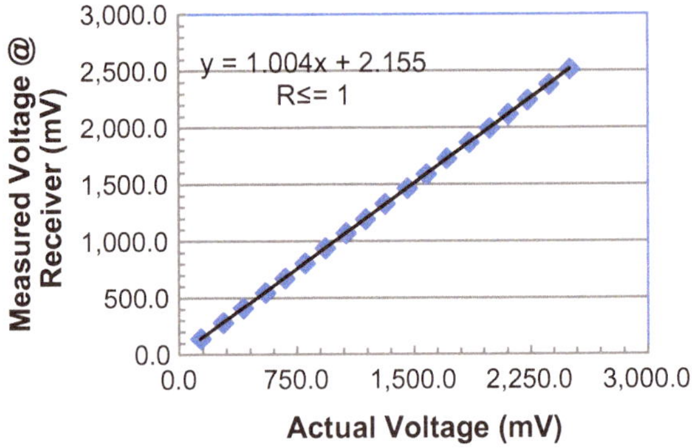

Fig. 4 Plot of data received at the receiver and data transmitted by the transmitter

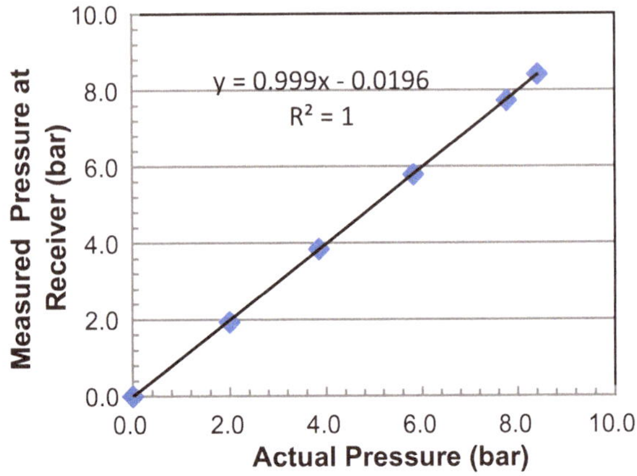

Fig. 5 Plot of pressure measured at the receiver and input pressure to the sensor

transmitter. As can be seen, the received and transmitted values are in very good agreement with $R^2 = 1$.

After verifying the operation of electronics and wireless transmission of the data, the measurements were carried out using a pressure sensor. During these measurements, the pressure of different magnitudes was applied using a hand pump. The plot of the actual pressure and pressure as recorded by the receiver is given in Fig. 5. As can be seen, the pressure recorded by the receiver is in good agreement with the input pressure applied to the pressure sensor using the hand pump.

Results presented in Figs. 4 and 5 show that the sensors and related instrumentation can be built for low-power wireless operation using an energy harvester such as TEG. In the tests presented in this paper, the heat was used as a form of energy for energy harvesting. However, a similar configuration can be built using piezoelectric energy harvesters optimized for the vibration spectrum of the environment.

8 Wireless Sensor Network

Self-powered /process-powered sensors derive their energy passively from the process itself and do not depend on any external power supply systems. These sensors in effect derive their energy from the environment, in which they are deployed to measure. The self-powered sensors would be configured into a self-organizing fault-tolerant sensor network for effectively communicating to a handheld mobile device for data monitoring and operator guidance. The self-powered /process-powered sensors on a wireless sensor network can act as an operator empowering tool, for situation awareness, irrespective of the operating domain. The self-powered /process-powered sensors are provided with a chargeable Li coin cell for efficient power management and to support a local display device. The security issues inherent to the wireless communication would be addressed by device signature encryption for authentication and by limiting the transmission energy, thereby the communicable distance.

9 Design Challenges and Approaches

Self-powered /process-powered sensors for nuclear environment would be able to address the measurement requirements during an extended SBO and should aid the operator remotely in providing the process information. Since the sensors would be configured in sensor networks [3], dynamic path for transmitting the signals exists and decision on the optimum weighed path is chosen by the local embedded intelligence based on local conditions. This configuration gives multiple paths and enhances the availability of information and adaptability.

Some of the formidable challenges in sensitive installations are the security vulnerabilities in wireless communication which needs to be addressed innovatively in an extensive manner integrating the hardware signatures and unique encryption methodologies. The design of communication devices with limited range and intelligence on the deployed environment would also need to be explored for ensuring security in communication.

Low-power embedded processors and wireless communication devices [4] are available in the commercial domain and are not qualified for radiation environment. MEMS sensors, which are well established for their meager power consumption, are suitable for low-power applications. However, the existing CMOS technologies and silicon devices have limited deployment potential in nuclear environment. Polysilicon MEMS devices and appropriately engineered shielding would render the available devices compact and compatible for nuclear applications.

10 Conclusions

We have designed a process-powered pressure sensor instrumentation and demonstrated its operation by conducting various tests. The new age instrumentation for advanced reactors needs to ensure its credible performance and availability in SBO and BDBE situations, in addition to their conventional capabilities, for safe maneuvering of the reactor in all its active domains. The other challenges posed by nuclear reactors are high-temperature performance, radiation tolerance, adherence to the ALARA principles, and economy. Self-powered /process-powered sensor designs would also need to address many new avenues of functionality and would also be throwing new challenges in the design of self-/process-harvesting in situ power sources, encryption, and security issues in wireless communication and simple wearable operator interface designs. The sensor packaging design suitable to withstand impacts, flooding, fire, and other hazards using advanced materials and lightweight designs also assumes importance in self-powered /process-powered sensors.

References

1. Severe accident management guidance technical basis report, vol. 2. (EPRI, 2012)
2. J.L. Rempe, D.L. Knudson, TMI-2—a case study for PWR instrumentation performance during a severe accident (2013)
3. D. Clayton, R. Williams, Communication requirements and concept of operation for sensors network. ORNL/TM-2013/180 (2013)
4. A. Bari, J. Jiang, Deployment strategies for wireless sensor networks in nuclear power plants. Nucl. Technol. **187** (2014)

FSC
www.fsc.org
MIX
Papier | Fördert gute Waldnutzung
FSC® C083411

Kolibri 360 GmbH
Ferdinand-Jühlke-Straße 7
99095 Erfurt, Deutschland
produktsicherheit@kolibri360.de